# the world of biology

# P. WILLIAM DAVIS
Professor of Life Science
Hillsborough Community College

# ELDRA PEARL SOLOMON
Instructor of Life Science
Hillsborough Community College

## McGRAW-HILL BOOK COMPANY

New York   St. Louis   San Francisco   Düsseldorf
Johannesburg   Kuala Lumpur   London   Mexico   Montreal   New Delhi   Panama
Paris   São Paulo   Singapore   Sydney   Tokyo   Toronto

# the world of biology

## LIFE, SOCIETY, ECOSPHERE

**Library of Congress Cataloging in Publication Data**

Davis, P   William.
    The world of biology.

    Includes bibliographies.
        1.  Biology.  I.  Solomon, Eldra, joint author.
II.  Title.  [DNLM:  1.  Biology.  QH307 D262w  1974]
QH308.2.D38      574        73-19724
ISBN 0-07-015548-8
ISBN 0-07-015547-X (pbk.)

1 2 3 4 5 6 7 8 9 0 V H V H 7 9 8 7 6 5 4

This book was set in Helvetica by Progressive Typographers.
The editors were Thomas Adams and David Damstra;
the designer was Anne Canevari Green;
the production supervisor was Joe Campanella.
The drawings were done by Vantage Art, Inc.
Von Hoffmann Press, Inc., was printer and binder.

*Cover photograph by hand-held camera aboard Gemini 8 shows thin veil
of earth's atmosphere with towering cumulus clouds in silhouette. (NASA)*

*For Freda, Nathan, and Seth,*
*Mical, and Amy*
*. . . and for their generation*

# CONTENTS

Preface xi
To the Student xv

**1**
LIFE AND SOCIETY:
AN INTRODUCTION
2
Man's dependence upon the environment 4
What is life? 8

**2**
HOW SCIENCE
WORKS
18
There *is* a scientific method 19
Why study scientific method? 19
Systematic thought processes 20
Scientific method 24
The ethics of science 36

**3**
THE ORGANIZATION
OF THE ORGANISM
40
The cell—building block of the body 41
Within the cell—structure and function 45
Differences in some major cell types 54
Cellular reproduction—mitosis 55
From single cell to complex organism 60

**4**
NUTRITION
66
The basic nutrients 68
Transfer of energy through the world of life 80
The problem of world food supply 89

**5**
PHYSIOLOGY OF
NUTRITION
100
Processing food 101
The problem of obesity 110
Food additives 113
Health foods 115

**6**
INTERNAL
TRANSPORT
118
The blood 119
The organs of circulation 124
The pattern of circulation 129
Cardiovascular disease 131
The lymph system 135

**7**
**DISPOSAL OF METABOLIC WASTES** — 146
The urinary system — 148
Other excretory structures — 156
Maintaining a steady state — 157

**8**
**RESPIRATION** — 160
Organismic respiration — 161
Cellular respiration — 169

**9**
**WHERE HAS ALL THE CLEAN AIR GONE?** — 180
Air pollution and human health — 184
Major pollutants — 188
Ecologic effects of air pollution — 194
The meteorology of air pollution — 196
Dealing with the problem — 198
Smoking: a serious form of air pollution — 200

**10**
**RESPONSIVENESS: NEURAL CONTROL** — 208
Organization of the nervous system — 210
How the nervous system works — 214
The drug scene — 232

**11**
**RESPONSIVENESS: ENDOCRINE CONTROL** — 240
How hormones work — 242
Hypothalamus and pituitary gland — 244
Thyroid gland — 249
Parathyroid glands — 250
Islets of the pancreas — 252
Adrenal glands — 254
Other hormones — 260

**12**
**HUMAN REPRODUCTION** — 262
Asexual reproduction — 263
Sexual reproduction — 264
The male — 268
The female — 274
Sexual intercourse — 281
Conception — 282
Sex of the offspring — 283
Contraception — 285
Population control — 294
Venereal disease — 296

**13**
**THE ORIGIN OF THE ORGANISM** — 300
The pattern of development — 301
Regulation of development — 320
Environmental influences upon the embryo — 323

## 14
## THE BLUEPRINTS OF LIFE

330

Giant cells 331
Miescher discovers nucleic acids 335
Mutations 340
Small wonders: the ribosomes 342
Genetic heretics 343
The control of genes 347

## 15
## INHERITANCE

352

The nature of heredity 353
The mendelian laws 358
Polygenic inheritance 361
Linkage 362
Multiple alleles 368
Pleiotropism 371

## 16
## HUMAN GENETICS

374

Inborn errors of metabolism 375
The Rh factors 377
Chromosomal defects 379
Consanguinity, or marriage between relatives 384
What can be done about human genetic illness? 385

## 17
## THE ORIGIN OF LIVING SYSTEMS

394

Two concepts of origins 395
The nature of evolution 395
Development of evolutionary thinking 396
Genetics and microevolution 397
Other aspects of microevolution 403
The origin of life 405
Evolutionary theory and the concepts of creationism 409

## 18
## BEHAVIOR

418

Tropisms and taxes 419
What is behavior? 424
Acquired and inherited behavior 424
Social behavior 440
Human implications of animal behavior 452

## 19
## OUR SPACESHIP EARTH

458

Our life-support system 459
The passengers and their quarters 478

## 20
## AQUATIC ECOLOGY

486

Properties of water 487
Aquatic communities 489
Succession in aquatic communities 502
Conservation of aquatic habitats 503

## 21
## WATER POLLUTION
510

What is pollution? 511

Kinds of water pollution 511

## 22
## CONTROLLING LAND
## AND WATER
## POLLUTION
530

Control of water pollution 531

Solid wastes and land pollution 536

Political and economic considerations 542

The solution to pollution 544

## 23
## OUR BIOLOGIC
## RESOURCES
556

Ecosystems and conservation 557

The impact of man upon ecosystems 559

The defense of the land 572

The conservation of natural communities 577

Endangered species 584

## 24
## ROOTS OF OUR
## ECOLOGIC CRISIS
592

Some possible causes of our ecologic crisis 593

How do we know we are overpopulated? 594

What to do about it 601

Affluence and economics 605

The role of technology 610

The upshot 612

The last word 614

## APPENDIX A
## BASIC PRINCIPLES
## OF CHEMISTRY
618

The atom 619

Molecules and their bonds 621

Chemical reactions 624

Acids, bases, and pH 626

The carbon atom 627

Solutions and colloids 629

Movement of molecules 629

The metric system 631

## APPENDIX B
## DIVERSITY OF
## LIVING SYSTEMS
632

Classification of the house cat and of man 633

Three kingdoms: Monera, Protista, and Fungi 635

Plant kingdom 638

Animal kingdom 642

Characteristics of vertebrate classes 650

Index 655

# PREFACE

Il est plus facil de dire des choses
nouvelles que de concilier celles qui sont
déjà été écrites.
— *Vauvenargues*

This book is the outgrowth of a two-term biology course intended to be taken as a "sampling" of natural science by students majoring in other fields. Since such students do not plan to become professional biologists, medical workers or biology teachers, the conventional textbook intended for the preparation of biologists is to them all too often an educational disaster.

In our teaching experience we have found that a practical, anthropocentric approach to biology motivates students to a greater extent than the more conventional theoretical approach. To this end we have designed the first half of the book to focus primarily upon the structure and function of the human body, not as a medical school cadaver, but as a living organism functioning in society and in the ecosphere. The unit on nutrition, for example, deals not only with the structure of the human digestive system, but also with problems of obesity, food additives, health foods, and world food supply. Similarly, the chapter on internal transport discusses heart disease, immune mechanisms, and organ transplants. The chapters on reproduction and development include venereal disease, abortion, contraception, pregnancy, childbirth, breast feeding, and environmental influences upon the fetus.

The second half of the book deals primarily with genetics, evolution, and environmental biology. A larger than customary amount of material on genetics is included for two reasons. First, this area of biology is likely to increase in importance radically during the lifetimes of most students now taking biology courses. Second, the material is seldom learned at all by students if it is not covered in sufficient depth to produce genuine understanding.

The environmental chapters cover water pollution, solid-waste disposal, energy problems, land use, conservation practices, and population problems. This is a far more detailed treatment than is usually included even in an "environmentally oriented" text. Such information is an indispensable part of the intellectual armament of the biologically literate citizen of today. The mention of a few ecologic "principles" scarcely equips him to deal with the current issues in any but a naive manner.

If some areas have received greater than usual emphasis, some perforce have received less detailed treatment than is customary. We present biology neither as a procession of phyla, nor as a commentary upon the tricarboxylic acid cycle. Such matters are covered,[1] for one can scarcely teach biology without reference to them, but their proportional representation has been reduced to a level commensurate with the needs of the student population to which the book is addressed. We have not, however, removed all intellectual rigor just to increase the appearance of "relevance." We have tried to take into account not only the wants which students express, but the needs (including abstract intellectual needs) which they do not always appreciate, but which the true educator is nevertheless responsible to meet.

Basic concepts and principles receive the greatest attention, and every attempt is made to avoid distracting the student and interfering with his learn-

---

[1] The basic principles of chemistry and of biologic classification are summarized in two appendices, which the instructor can either assign explicitly, or to which the student can refer when necessary. We believe that a student who is completely unacquainted with chemistry can gain the chemical knowledge he needs to understand basic biology from the chemistry appendix.

ing by providing a host of unimportant details. Thus in the chapter on scientific method we omitted elaborate and complicated examples which the student lacks the background to appreciate, and which so often succeed in clouding an appreciation for the principles themselves so that a basic and critical *understanding* of the scientific method eludes the student. Instead, our approach is conceptual so that hopefully, in the end, the student comes to view the scientific method not as part of biology alone, but as the cornerstone of our scientific and technologic society and a signal achievement of human thought.

We have tried to bear the educational background of the student in mind as we have written each chapter. By giving careful attention to the reading level, for example, we believe we have produced a text appropriate for the average entering college student. Biologic and technical terminology has been kept as simple as possible. Terms are introduced naturally in context, and can be readily located through the italicized page references in the index, making a separate glossary unnecessary.

A particularly helpful feature, we believe, is constituted by the learning objectives at the beginning of each chapter. These indicate to the student just what is expected of him as a result of reading the chapter. The chapter learning objective gives the learning requirement of the chapter as a whole; the interim learning objectives (ILOs) break down the learning process so that concepts can be individually mastered. By using them, the student can gain an idea of what he should get out of the chapter even before he commences it, and by referring back to them as he reads he can estimate his mastery of the material as he goes along. If the instructor wishes, of course, he may provide specific conditions and performance criteria for the mastery of each objective.

We have designed not only the textbook but also the accompanying study guide around these objectives. The student should get in the habit of using them as a guide. The study guide gives the student the opportunity to summarize and organize his knowledge in terms of each learning objective, and then helps him to assay his actual mastery of the objective. Our own class tests have assured us of the value of such an approach.

We have tried to reduce the unused academic "fat" in this book to a minimum. It is intended to be a vital or indispensable aid to the student learning biology, not merely an encyclopedia or "status symbol text" that is used only occasionally, gathering dust on the student's shelf while he studies the important material—the lecture notes. This should lighten the instructor's task, for if we have succeeded, he will be able to devote himself to truly creative interaction with the students, leaving much of the drudgery of routine pedagogy to us. Encyclopedic references may be consulted in the library; our book is intended to *teach*.

In a project of this sort many persons besides the authors make contributions without which the work would not have been possible. We wish to take this opportunity to express our thanks to those who have rendered particular aid to us in the conception, writing, revision, and illustrations of this book. We are especially indebted to Mrs. Karen Davis, whose critical pen has spared the reader much literary barbarism, and whose extensive work almost entitles her to listing as coauthor. Thanks also are due to Dr. E. Marshall Johnson of Jefferson Medical College, Dr. Wayne Frair of The King's College, and Dr. Charles Walker of University of Tampa, for their unstinting professional help and suggestions for revision. We are grateful to Miss Phala Fonte, who prepared many of our original illustration sketches. Most of the electron micrographs appearing herein are the work of Dr. Lyle Dearden of the University of California Medical School. We also wish to acknowledge the help of Dr. George Howe, Dr. E. H. Ketchledge, Dr. Harold Coffin, Dr. Robert Martin, Dr. Burton S. Guttman, Roy Lewis III, Joanne Cochran, Albert Brod, Mrs. Freda Brod, Edwin Solomon, Don Moores, Bea Kunda, Andrea Wells, Diane Morris, Marilyn Brown, Maribeth Alspach, Elaine Borquin, and Claudia Williams.

*P. William Davis*
*Eldra Pearl Solomon*

# TO THE STUDENT

At the beginning of almost any course, some attempt is usually made to justify it. The following points are apt to be covered to some degree:

1   This course is noble. The subject was studied in ancient Greece (or in the Middle Ages).
2   This course is broadening. It is a Liberal Art.
3   (Optional) This course has certain practical applications.
4   (Optional) This course is good for the soul.
5   (The clincher) One needs to take this course in order to graduate.

Convinced or not, the students are dismissed with an assignment and the vague expectation that the serious business will begin at the next meeting. No one expects the justification of the course to be brought up again; certainly it will not appear on the examination.

But it seems silly—to use no stronger word—to spend perhaps a year of concentrated work on a task when you really haven't the foggiest notion of why you are doing so other than reason (5) given above. So bear with us briefly as we try to show you some reasons for studying biology that we feel make sense.

Truthfully, we don't suppose that you are likely to be interested in biology just because the science goes back to King Solomon and Aristotle, or, for that matter, because you plan to become a physician or an ecologist. Some of you will, but most of you will not.

The question is not so much what you will become as what you are already. You are—right now—a living organism, a citizen, and a member of society, which demands your biologic participation (whether that contribution is positive or negative). You have a heart, a stomach, a brain, a family, a vote. If one should know something about the mechanism of an automobile, should he not know something about himself, under the hood, as it were?

Then too, biology is now in the limelight of public affairs as never before. Ecology, pollution, genetic engineering are concepts that every well-informed and responsible citizen should understand. Are you acquainted with the principal arguments about pesticides? Smoking? Water, land, and atmospheric pollution? Do you hold an informed opinion about drug use and abuse? Fluoridation? The use of chemical additives in food? Do you feel that you understand the relation that population, overpopulation, affluence, and the consumption of goods bear to conservational and environmental problems?

To use an overworked buzzword of our times, biology is *relevant*. Genuine relevance bears upon more than the satisfaction of immediately felt wants. It provides as much as possible for both present *and* future needs in a broad range of social and humane dimensions. Right now, for example, you may not see the importance of a discussion of breast feeding, menopause, or emphysema, but the time will come—probably sooner than you think—when these topics will be important to you.

You may not immediately see how relevant the citric acid cycle, DNA, and nitrogen fixation are, but if you do not understand the scientific basis for social and environmental biology, you will retain only preachments. A non-factual approach would be propagandistic and would insult your intelligence. Moreover it would produce opinions not solidly rooted in facts. And opinions not rooted in facts produce people easily swayed by anyone desiring to manipulate or mold them for his own purposes.

To be sure, you may feel that a few areas are *not* especially relevant to your own needs. In such cases we hope you will cultivate some sheer animal curiosity, and we think you will find the subjects we will be considering interesting as part of the world of living things around you.

You may even find that biology *is* broadening and *can* be good for the soul. Take a moment to think of how it may relate to your field of study. Are you a business major? Pollution control is a business expense which must be weighed. Are you a psychology major? Thought is a product of the brain. Are you an economics major? How would population stabilization affect our economy? One dare not risk becoming too narrow a specialist in any field of study, for no area of thought exists by itself. The countries of the individual disciplines form the continents of scientific knowledge and human thought.

The soul, we hope, may be improved not only by considering the explicit relation of philosophy and theology to our subject, but also through consideration of some of the mightiest moral and ethical dilemmas that confront our age—questions of genetic manipulation, overpopulation, mass starvation, and environmental deterioration. No one should remain unmoved by these issues.

This book will be, perhaps, somewhat controversial. At any rate we hope so. Otherwise we shall have said rather little that is worth saying.

Facts are facts, but they must be interpreted and their significance assessed. The resulting structures must be carefully appraised. We encourage, therefore, the development of your critical thinking. Though "science" has traditionally evoked the image of authority and certainty, ideas are rapidly changing, and today there is little in biology that may be considered exempt from controversy. As the great French mathematician Poincaré put it,

Science is built up with facts, as a house is with stones. But a collection of facts is no more a science than a heap of stones is a house.

Bear in mind that biology cannot explain everything. We have tried to be frank about this. Biologists even today have many opportunities to echo the sentiments of King Solomon, who observed (Proverbs 30:18–19):

There be three things which are too wonderful for me, yea, four which I know not: the way of an eagle in the air; the way of a serpent upon a rock; the way of a ship in the midst of the sea; and the way of a man with a maid.

P. William Davis
Eldra Pearl Solomon

# the world of biology

# 1 LIFE AND SOCIETY: AN INTRODUCTION

***Chapter learning objectives.*** The student must be able to (1) describe man's dependence upon physical and biotic conditions in the environment, and his interference with these conditions, and (2) describe the characteristics of living systems as outlined in this chapter.

*ILO 1* Identify physical conditions and biotic factors essential to life on our planet; describe the role and interdependence of producers, consumers, and decomposers.

*ILO 2* Identify three ways in which man is interfering with the stability of the ecosphere.

*ILO 3* Briefly describe the physical organization of a living system, including its cellular and molecular structure.

*ILO 4* Define metabolism, and describe the relationship between nutrition, synthesis, and cellular respiration.

*ILO 5* Describe the dependence of metabolic processes upon nutrients and oxygen from the environment.

*ILO 6* Describe the need for self-regulation in a biologic system, and relate responsiveness and movement to the process of maintaining a steady state.

*ILO 7* Give the function of reproduction, and describe the growth of an organism as a function of increase in mass and number of component cells.

*ILO 8* Define adaptation, and describe its function in promoting the perpetuation of a species.

*ILO 9* Define a living system.

*ILO 10* Define each new term introduced in this chapter.

Early man lived largely at the mercy of his environment. Most of his time and effort was devoted to seeking food, protection from wild animals, and shelter from the cold and rain. His world must have seemed a vast and hostile place. The situation has changed. Now man is apt to feel that at last he is master of his own destiny, that his tools and technology have gained him mastery over many of the forces of nature and enslaved them to his service. The discovery of fire, the invention of the wheel and of agricultural techniques, the development of printing and of the scientific method mark significant advances from our simple beginnings. Now we walk upon the face of the moon. We transplant hearts, kidneys, and corneas. We view viruses under electron microscopes, and generate electricity with nuclear power. An ever-growing knowledge of the workings of our own life processes has brought us face to face with the final needed mastery—that of ourselves.

The activities of early man had little influence upon the world around him. As human society has become increasingly technologic, however, man's activities have exerted a significant and damaging effect upon the environment. Coupled with explosive expansion of human population, our life style has transformed the earth and threatens to disrupt the delicate network of life upon its surface. Now the earth seems small—too small for us.

Mankind faces a critical choice. Either he will continue to abuse his great power, and finally make a shipwreck of his planet, or he will steer it back onto a survival course.

**FIGURE 1-1**
(a) Man walks on the moon (NASA); (b) performs open-heart surgery (Dr. Dennis Pupello); (c) yet makes a shipwreck of his planet. (U.S.E.P.A.)

(a)

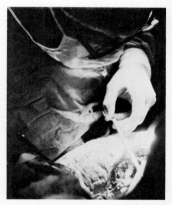

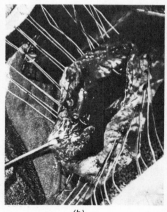

(b)

(c)

## MAN'S DEPENDENCE UPON THE ENVIRONMENT

We are not alone on our planet. Despite our technology, we depend upon our environment for the food we eat, the oxygen we breathe, the raw materials of our industry, and all that sustains our life and our society. Nor can we ignore our fellow organisms that inhabit the planet with us. Their well-being and ours are intricately interrelated.

### Conditions essential to life

Some of our surroundings are alive, and some are not. The physical and living (better called *biotic*) components of our environment interact with each other to produce the delicately balanced equilibrium that is essential to the continuance of life on earth. Some of the physical conditions upon which man's survival depends need only be named to be appreciated: a source of energy (such as the sun), an appropriate mixture of gases in the atmosphere, an appropriate temperature range, water and mineral nutrients, and protection from harmful ultraviolet radiation from the sun.

Scientists divide the biotic components of the environment into three broad categories, separated according to their mode of nutrition. For practical purposes, the only organisms able to nourish themselves by making their own food are green plants. Using the energy of sunlight, they combine carbon dioxide and water to form oxygen and food materials. Green plants are therefore termed *producers*. Most animals, including man, consume plants or other animals, and are thus *consumers*. Bacteria, fungi, and some

(a)

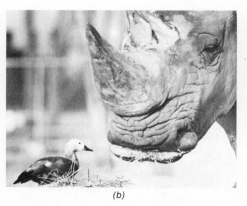
(b)

(c)

**FIGURE 1-2**
(a) Coconut palm tree—a producer. (b) Consumers (Busch Gardens). (c) Decomposers. Bacteria breaking down fragments of dead plant material in the soil. (T. Gibson, *World Crops*)

other organisms break down the decaying bodies of producers and consumers, and make their constitutents available for reuse. They are *decomposers*. Producers, consumers, and decomposers (Figure 1-2) are all essential parts of any self-sufficient community of organisms. In their interaction with one another and with their physical environment they constitute an ecologic system, an *ecosystem*. A pond, a forest, an ocean, a prairie, or any relatively self-sustaining group of living things may be considered an ecosystem. The largest ecosystem of all, and ultimately the only really self-sustaining one known, is the planet earth with all its inhabitants—the *ecosphere* (also called the *biosphere*).

During *photosynthesis* green plants capture light energy and use it to make complex food molecules from carbon dioxide and water. What they are doing is transforming light energy into chemical energy stored in the chemical bonds of the complicated compounds they produce. When men and other consumers eat plants, a portion of this energy becomes available to them. But green plants are also responsible for producing oxygen, which is required not only by consumers but also by the plants themselves.

All living things obtain energy by breaking down food molecules originally produced by green plants. The biologic process of breaking down these fuel molecules is termed *cellular respiration*. When chemical bonds are broken during cellular respiration, their stored energy is made available for life processes. Perhaps some word equations will make this clearer.

*Photosynthesis:*
Carbon dioxide + water + energy ⟶ food + oxygen

*Cellular respiration:*
Food + oxygen ⟶ carbon dioxide + water + energy

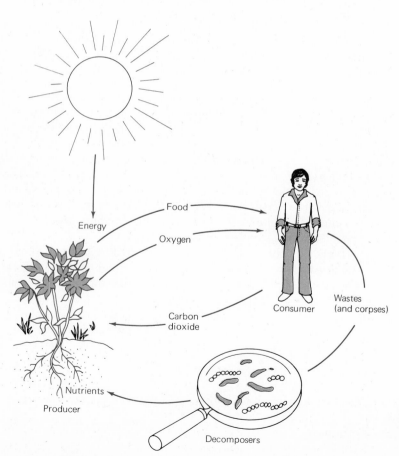

**FIGURE 1-3**
Interdependence of producers, consumers, and decomposers.

(a)

(b)

**FIGURE 1-4**
(a) Wastes from industry and motor vehicles are making the air unfit to breathe (OMIKRON). (b) Breathing fresh air from a machine in Tokyo. (OMIKRON)

The life-sustaining mixture of gases in the atmosphere is a consequence of gas exchange between plants and animals via the nonliving environment (Figure 1-3).

Although plants also carry on cellular respiration, the amount of carbon dioxide which they produce is insufficient to sustain photosynthesis. If it were not for animal respiration, all the carbon dioxide in the atmosphere would be removed by photosynthesis within about one year.[1]

Yet producers and consumers cannot by themselves comprise a complete partnership. If decomposers did not exist, nutrients would become progressively locked up in the dead bodies of plants and animals, and the supply of elements required by living systems would soon be exhausted. Even the maggot has its place.

**Man and the ecosphere**
The trouble is that man does not seem to know his place. His activities threaten the stability of the ecosphere. The third annual report of the Council on Environmental Quality states:

> Our power to build and destroy has become almost limitless, and the complexity of our technology and institutions has generated decisions with consequences often not apparent for years. For example, it may take many generations to rid the environment of chemicals which are discovered now to be a threat to man or the environment. The rapid development of new chemicals has strained our ability even to determine their effects.[2]

Exaggerated? Not really. Consider some instances. Life depends, after all, upon an appropriate mixture of gases in the atmosphere. When deprived of oxygen, cells die within a few minutes; that, obviously, is why one must breathe. How, then, can we escape if the air becomes unfit to breathe? Wastes from industries and automobiles have dirtied *most* of the air in the United States, and much of the rest of the world, and some scientists seriously propose that within 15 years residents of some urban areas may need to wear filter masks when they leave their homes, just as today we need to wear coats in the winter. Even now, coin-operated oxygen machines have appeared on the streets of Tokyo (Figure 1-4).

Unpleasant a thought as this is, perhaps we could adjust to it in time. But the ecosphere has no gas mask. Already forests inland from Los Angeles are dying from the adverse effects of smog, and decreases in photosynthesis, apparently related to air pollution, have been reported in other areas.

Consider another example: water. Essential as a component of all living things, water accounts for about two-thirds of the weight of the human body. In one sense, our supply of water seems limitless, for water is recycled con-

---

[1] LaMont C. Cole, "The Ecosphere," in *Man and the Ecosphere,* Freeman, San Francisco, 1971, p. II.
[2] *Environmental Quality—1972,* the 3rd annual report of the Council on Environmental Quality, August 1972, p. 344.

stantly from the ocean to the atmosphere and, via streams, to the ocean again. But the water we drink must be fresh and uncontaminated. It must not be salt water; yet less than 1 percent of the earth's water is sweet. As yet, no economically feasible way has been developed to create fresh water, so our supply is limited to that 1 percent. Yet the thirst of our society knows no bounds. Even now we are using twice as much water from underground reserves as nature returns to them. Sooner or later, we will miss the water. The well will have run dry.

On top of that, we waste large amounts of the water we have. Consider the spoilage of water produced by pollution, both from industrial and municipal sources. According to the Environmental Protection Agency, 29 percent of American stream and shoreline mileage is polluted. Yet the rate of pollution is accelerating. It is disconcerting to reflect seriously upon the fact that frequently the water in which we dump our wastes and poisons is also our drinking water. Much of it simply flows into the sea, hopelessly befouled.

**FIGURE 1-5**
Water is becoming unfit to drink . . . and to enjoy. (J. Paul Kirouac)

*(a)*

*(b)*

**FIGURE 1-6**
Though they may be very different in appearance, all living systems share certain basic characteristics. (*a*) An ameba. (*b*) Earthworm. (Roy Lewis)

Even the sea is not immune to pollution. A large part of the oxygen and food available to the inhabitants of the earth is produced by tiny floating aquatic plants called *phytoplankton*. Phytoplankton form the base of most aquatic nutritional chains, and they are adversely affected by even small concentrations of chlorinated hydrocarbon pesticides, which are now found in large amounts in many of our natural water resources.

Though the way was prepared by Osborne's book, *Our Plundered Planet*, published in the 1940s, and later by Rachel Carson's *Silent Spring,* public awareness has only recently focused on environmental problems.[3] Nations have only recently begun to discuss pollution as the global threat that it is. Organizations, politicians, media, and citizens that had never seriously concerned themselves with such issues have begun to take a welcome, if belated, interest in them. Some have counted the cost (the 1972 report of the Council of Environmental Quality estimates that in the United States alone the cost of meeting current none-too-high environmental standards over the decade 1970–1980 will be $287.1 million, or 2.2 percent of the gross national product); but others point to the needs. Each of us carries strontium 90 and lead in his bones, mercury in his blood and other tissues, asbestos and other particulate matter in his lungs, DDT in his body tissues, and, all too often, carbon monoxide in his blood. Hardly a proper legacy to leave to our children!

Whatever chance there is of dealing with pollution depends heavily upon our degree of commitment, and enlightened commitment demands sound knowledge. However overpopulated our world may be, energetic, enlightened, and committed persons are in short supply. Though biology is not all environmental, a large part of it *is* tied up with the environmental crisis in which we find ourselves. Many fields of biology impinge upon it.

We shall lay the foundations for our later consideration of environmental issues and the challenging and far-reaching implications of biology with an introduction to the basic characteristics of living systems.

## WHAT IS LIFE?

Anything that is alive may be referred to as a living system, but what does it mean to be "alive"? Biology is the science of life, yet even biologists have difficulty in defining such terms as "life," "living," or "alive." Several million kinds of living organisms share our planet and their diversity precludes lumping them together with a simple definition. There are, however, certain characteristics and activities that man has in common with an earthworm, a tree, or even a single-celled ameba (Figure 1-6). Taken together these features constitute life. They include specific types of molecules which form and react within a cellular structure, metabolism, self-regulation, reproduction and growth, and the ability to adapt to changes in the environment.

---

[3] For some that awareness dawned on the first Earth Day, Apr. 22, 1970.

Every living thing carries on these activities as a coordinated unit, or living system. It is hoped that the brief discussion of these characteristics that follows will provide perspective for a more detailed study of individual topics taken up in subsequent chapters.

### Physical organization

A man does not look like an earthworm or an ameba, much less a tree, yet he is made of the same basic materials.

*Cellular structure.* All living systems except viruses are composed of *cells,* the basic structural and functional units of life (Figure 1-7). Some living things, such as amebas, are made of only one cell, but the complex organisms with which we are most familiar are composed of billions of cells of many distinct types. Man, for example, is made up of an estimated 60,000 billion cells.

Most cells are microscopic in size, but a few, such as the yolks of birds' eggs, are quite large. Each cell consists of a discrete body of jelly-like *cytoplasm* surrounded by a limiting membrane. Cytoplasm is composed of water and many other chemical substances, some of which are organized into tiny structures (organelles) which perform specific cellular functions. Most cells contain a large body called the *nucleus* which holds the hereditary material containing information governing the structure and function of the organism.

*Chemical structure.* What are the chemical substances which make up cytoplasm? Although more than 100 chemical elements are known, only about 30 are commonly found in living systems. About 98 percent of the weight of cytoplasm is composed of only four elements—carbon, hydrogen, oxygen, and nitrogen.[4]

The smallest unit of a chemical element that retains its characteristic properties is called an *atom.* Atoms combine with one another by means of chemical bonds, a form of energy that holds them together, to form *molecules.* Thus when two atoms of oxygen combine chemically a molecule of oxygen is formed. Different types of atoms can combine with one another to form chemical *compounds,* which are molecules composed of at least two different types of atoms. Water is a chemical compound consisting of one atom of oxygen chemically combined with two atoms of hydrogen.

Compounds may be divided into two broad categories: *inorganic compounds,* which are relatively small, structurally simple substances, and *organic compounds,* which are generally quite complex molecules and which always contain carbon. Inorganic compounds include water and a variety of acids, bases, and salts. Organic compounds of major biologic importance may be classified into four principal groups: *proteins, nucleic*

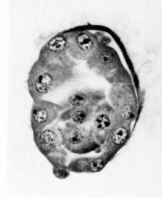

**FIGURE 1-7**
Cells are the basic structural and functional units of life: early human embryo (Carnegie Institution of Washington).

---

[4] If you are confused, consult Appendix A, Basic Principles of Chemistry.

*acids, carbohydrates,* and *lipids.* No single chemical is responsible for imparting life to a system. Rather it is the arrangement and interaction among many different types of atoms and molecules that constitute the dynamic equilibrium that we call life.

*Proteins.* Proteins are made of carbon, hydrogen, oxygen, nitrogen, and sometimes sulfur and phosphorus. They are large molecules that vary characteristically in different types of cells. The types of protein in a particular cell determine the structure, function, and activities of the cell. Many proteins function as *enzymes,* catalysts which regulate biochemical reactions in cells.

*Nucleic acids.* DNA (deoxyribonucleic acid) and RNA (ribonucleic acid) are the types of nucleic acids found in living systems. DNA is the basis of all *genes,* the hereditary material of the cell. DNA contains the "recipes" for making all the types of protein required by the organism. RNA functions in the actual process of protein synthesis. Nucleic acids are composed of carbon, hydrogen, oxygen, nitrogen, and phosphorus.

*Carbohydrates.* Familiar to us as sugars and starches, carbohydrates are composed of the elements carbon, hydrogen, and oxygen, and always contain twice as much hydrogen as oxygen. The simple sugar glucose is the most common fuel molecule found in living systems.

*Lipids.* Lipids, or fats, are richer in energy than carbohydrates and are the form in which excess food is stored in living systems. Like carbohydrates, lipids are made of carbon, hydrogen, and oxygen, but they contain relatively smaller amounts of oxygen per molecule.

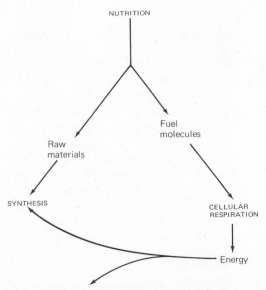

**FIGURE 1-8**
Some metabolic relationships.

## Metabolism

The production, arrangement, interaction, and breakdown of specific proteins, nucleic acids, carbohydrates, and lipids constitute *metabolism*. More simply, metabolism is the sum total of all chemical events taking place within an organism. It includes the chemical reactions essential to nutrition, growth and repair of the system, and conversion of energy into forms useful to the cells. Metabolic reactions occur all the time in every living system. When they cease, the organism dies. Metabolic reactions involve three related and interdependent processes: nutrition, cellular respiration, and synthesis (Figure 1-8).

*Nutrition.* Nutrition is the process of nourishing the organism, which means getting nutrients into its cells. Most living systems require the same types of nutrients—water, proteins, carbohydrates, lipids, and specific vitamins and minerals. Some of these nutrients are used by cells as components for making their own specific kinds of organic molecules, while others are utilized as fuel and broken down to provide energy for maintaining the system.

*Cellular respiration.* Broadly defined, respiration is the exchange of gases between organism and environment. At the organismic (whole-organism) level in complex animals, it includes breathing air into the lungs, transporting oxygen to the cells, and returning carbon dioxide from the cells to the lungs for exhalation.

At the cellular level the term respiration refers to the process by which each cell uses oxygen to "burn" fuel molecules. By a complex series of chemical reactions the cell breaks down nutrient molecules such as glucose, and, as chemical bonds are broken, energy is released. The cell captures this energy and packages it temporarily in high-energy bonds of a special energy-holding molecule called ATP (adenosine triphosphate). As a nutrient molecule is broken apart, its carbon-oxygen segments are released as carbon dioxide, and its hydrogen atoms eventually combine with the oxygen in the cell to produce water. Here is a summary equation for cellular respiration:

Carbohydrate fuel (e.g., glucose) + oxygen $\longrightarrow$
carbon dioxide + water + energy

Note that this equation is essentially the reverse of the equation for photosynthesis. While photosynthesis serves to store energy by producing carbohydrate, cellular respiration functions to break down carbohydrate, so that its energy may be released and utilized by the cell.

Almost all living cells require oxygen in order to carry on cellular respiration. When they are deprived of oxygen, energy cannot be packaged efficiently and the cellular machinery quickly grinds to a halt. Biologic energy is needed to carry out the activities of life such as movement, making molecules and cellular components, transporting materials, and thinking. Even reading this textbook requires an expenditure of energy.

*Synthesis.* Synthesis is the process of making molecules and may also involve arranging them to make new cellular components. As cell parts break down from normal wear and tear they must be replaced. Then, too, growing organisms require a steady supply of new molecules and components for new cells. Specialized cells require characteristic types of molecules for performing specific functions, and these must be chemically synthesized. For example, muscle cells produce the muscle proteins *actin* and *myosin* which function in contraction, while cells of the thyroid gland synthesize the hormone *thyroxine* which regulates the metabolic rate of the organism.

Building molecules requires energy, so synthesis depends on the availability of ATP molecules produced during cellular respiration. Some of the energy packaged within the ATP molecules is used to join atoms together, and is thereby transferred into the chemical bonds of the molecules fashioned by the cell.

Photosynthesis is a special type of synthesis carried on by green plants. Organic nutrients produced during photosynthesis are used both as materials for synthesizing other types of organic molecules required by the plant cells and as fuel molecules to be broken down during cellular respiration. Light energy from the sun, converted into chemical energy by the plant, is the ultimate energy source for all of life's activities. After being transferred from fuel molecules into more usable ATP molecules, energy is used to perform various types of work and finally dissipates in the form of heat into the earth's atmosphere, and ultimately to outer space.

### Self-regulation

Living systems must have the appropriate machinery for carrying on metabolic activities, and this machinery must function. But this alone is insufficient! Metabolic activities also must be regulated so as to maintain a balanced, or *steady state*. The organism must "know" when to synthesize what, or when more nutrients or extra energy is required. On the other hand it must not produce too much of any specific substance. When enough of a product has been made, the synthesizing mechanisms must be turned off. An organism does not remain on a table like a machine. It must be able to interact advantageously with diverse conditions in a constantly changing environment. Changes in the internal or external environment which threaten the stability of the organism must be dealt with on a continuous basis. Though living systems are not equipped with control panels covered with shiny buttons and levers, they do possess sophisticated and precise control mechanisms. And no one need push the buttons! The controls are self-regulating.

Even the simplest living thing is able to preserve its integrity by detecting stimuli and reacting to them. Cytoplasm itself is irritable—that is, sensitive to changes. When in need of nutrients (hungry) an ameba reacts positively to food in its aqueous surroundings by flowing toward and engulfing it. If poked, on the other hand, an ameba moves away from the threatening stimulus.

In complex organisms, cells regulate their own activities as well as respond to the needs of the organism as a whole. Certain structures work together, feeding information back and forth to each other regarding the chemical and energy requirements of the organism. Nerves, muscles, and glands function together to keep a complex animal in constant adjustment to changes that threaten to upset the steady state. Information is fed into the organism by specialized receptors, and various chemical and neural messages are sent from one part of the body to another, informing organs and their component cells of the needs of the entire organism.

Self-regulation is enhanced by an organism's ability to move. An ameba moves in a coordinated manner, thrusting out its cytoplasm to flow around food particles, or to retreat from hazardous conditions which threaten its well-being. Complex animals such as man possess groups of muscles which make complicated, purposeful movement possible.

A tree cannot pull up its roots and walk away, but it moves as it grows, opens its buds, transports food through its system, and responds to stimuli in its environment. Cytoplasm itself exhibits constant motion. Movement, though not necessarily locomotion, is characteristic of life.

Although movement and even instances of responsiveness do occur in the nonliving world, no inanimate object has the range of sensitivities or the coordinated reactions to stimuli that are characteristic of even the simplest living system. Response in biologic systems is functional, serving generally to enhance the well-being of the organism by maintaining its steady state.

**FIGURE 1-9**
A one-celled paramecium divides to form two new individuals (Carolina Biological Supply Company).

### Reproduction and growth

As an organism ages, self-regulating mechanisms begin to fail, resulting in an unsteady state. If unable to repair itself, the organism suffers disease and may, if failure is serious enough, die. Death of an individual, or even a generation of individuals, does not mark the end of a species,[5] however. Perpetuation of a species is provided for by the process of reproduction, for a biologically successful individual reproduces before it dies.

In simple organisms such as ameba, reproduction appears to be less a function of aging than of size. When an ameba has grown to a certain size it reproduces by dividing in half to form two new amebas. Before it divides, an ameba makes a duplicate copy of its hereditary material (genes), and one complete set of genes is distributed into each new cell. Except for size, each new ameba is identical to the parent cell. Unless eaten or destroyed by fatal environmental conditions such as pollution, an ameba does not die, but blends into each new generation. (See division of a paramecium in Figure 1-9.)

In man and other complex organisms the reproductive function is carried out by certain specialized cells. Usually two different types of reproductive cells fuse to form a fertilized egg. Generally, though not always, the sexual process involves two individuals, male and female.

Sex is biologically important because it introduces variety into a species.

---

[5] The term *species* refers to an exclusive group of potentially interbreeding organisms. For example, dogs belong to one species, horses to another.

Each offspring is not a duplicate of a single parent, but the product of various genes contributed by both the mother and father. As we shall consider later, variety is the raw material for the vital process of adaptation.

Following reproduction, new individuals undergo a process of growth and development. Living systems grow by taking in raw materials from the environment and refashioning them into their own specific types of molecules. In this way the amount of cytoplasm is increased until the cell grows to a maximum size characteristic of its own type. At that time the cell may divide into two. This process of cell division involves duplication of the genes, precise distribution of the genes into each new cell, and division of the cytoplasm. All cells come from preexisting cells by this process of cell division.

In a single-celled organism, such as an ameba, growth is restricted to an increase in the size of the cell; in more complex organisms growth is an increase in both the size and number of component cells. Man begins life as a single cell, the fertilized egg, which divides to form two cells, each of which divides again, and then again, until billions of cells have been produced.

In a multicellular organism growth is only one aspect of development. Cells must also arrange themselves to form characteristic structures, and must specialize, or become different from others, in order to perform specific functions.

A snowball rolling down a hill becomes larger as snow gathers around it. Crystals in solution aggregate to form an enlarging mass of rock, and cities expand as surrounding land is developed. None of these examples reflects growth in the biologic sense, however. These inanimate objects increase in size by adding on preexisting materials externally. In contrast, living systems take nutrients in and refashion them to increase *internal* mass. Biologic growth is always from the inside out.

## Adaptation

Adaptations are traits which enhance an organism's ability to survive in a particular environment. Adaptations may be structural, physiologic, or behavioral, or more often a combination of these. The long necks of giraffes are adaptations for reaching leaves of trees. Woodpeckers have the structural adaptations—powerful neck muscles, beaks fitted for chiseling, and long chisel-like tongues—which enable them to secure insects from the crevices they cut in tree trunks. Cactus plants have the adaptations needed for living in dry areas. Every biologically successful organism is actually a large collection of coordinated adaptations.

The process of adaptation involves changes in species, rather than in individual organisms. Many adaptations occur over long periods of time and involve many generations. They are the result of processes such as mutation (chemical change in a gene) and natural selection. Through time the ability of genes to mutate spontaneously has often been the key to survival. If every organism were exactly like every other individual of its species, any change in the environment might be disastrous to all, and the species would become extinct. Differences among individuals initiated by random mutation and enhanced by sexual reproduction provide for a differential in the ability of these individuals to cope with changes in their surroundings. Those which are best suited to cope with any specific change live to pass on their genetic recipe for survival.

One of the most interesting cases of adaptation has been documented in England during the past 100 years. In 1850 the tree trunks in a certain region of England were white because of a type of plant, a lichen, which grew on them. The common peppered moths were beautifully adapted for lighting upon these white tree trunks, for their light color blended with the trunks and protected them from predacious birds (Figure 1-10). At that time a captured dark mutant moth was regarded as an oddity. Then man changed the environment. He built industries that polluted the air with soot, killing the lichens and coloring the tree trunks black. The light-colored moths became easy prey to the birds, but now the dark-colored mutants blended with the dark trunks and escaped the sharp eyes of the predators. In these new surroundings, the dark moths were more suited ror survival. Today more than 90 percent of the peppered moths in the industrial areas of England are dark.

This example of adaptation provides a dramatic instance of man's inadvertent influence upon the environment. If a century ago man-made air pollution was significant enough to affect the survival of certain species, it is frightening to imagine the changes now in process as a result of the vastly greater amount of pollution presently being distributed throughout the environment.

**FIGURE 1-10**
Peppered moths of both light and dark varieties. On a sooty trunk with no lichens the white moth is at a disadvantage.
(Sir Gavin de Beer and *Endeavour*)

### A definition of life

Having discussed the characteristics of living systems, we are in a better position to attempt a definition of life. We might say that any system capable of self-regulated metabolism and able to perpetuate itself is alive. All such systems that we know of are composed of specific types of molecules, usually arranged to form cells. We should consider the possibility, however, that technologic man may some day construct a machine made of plastic and electrical circuits, capable of carrying on activities that constitute life. If such a system is self-regulating and self-perpetuating we will have to refer to it as living. Neither should we rule out the chance that man may eventually discover in some distant galaxy living things which do not share attributes common to life on earth. For the time being, however, the characteristics discussed above describe life as we know it.

**CHAPTER SUMMARY**

1 Man is part of a delicately balanced environmental system composed of physical and biotic components which interact and which are interdependent.

2 Nutritionally, man is a consumer and must rely upon producers, for organic nutrients, chemical energy, and oxygen. Decomposers break down essential nutrients so that they can be recycled through living organisms.

3 Man's activities are threatening the stability of the ecosphere; increased commitment to the task of solving environmental problems is needed.

4 Living systems are composed of molecules arranged into basic biologic units called cells.

5 Nutrition, synthesis, and cellular respiration, three metabolic processes, depend upon nutrients and oxygen from the environment.

6 Self-regulation of metabolic activities depends upon an elaborate system of internal controls. An organism maintains a steady state by responding appropriately to changes in its environment.

7 Perpetuation of a species depends upon reproduction.

8 Through the process of adaptation individuals acquire traits which enable them to survive in their specific environment.

9 Living systems are characterized by a controlled metabolism and the ability to perpetuate themselves.

**FOR FURTHER READING**

Asimov, Isaac: *Intelligent Man's Guide to the Biological Sciences*, Simon and Schuster, New York, 1968. A popularly written and fascinating introduction to biology.

Clement, Hal: *Star Light*, Ballantine, New York, 1971. (See also *Iceworld, Under Pressure*, and other titles now out of print, if you can obtain them.) A science fiction novel about creatures that breathe hydrogen instead of oxygen and whose metabolisms operate by reduction rather than by oxidation.

Ehrlich, Paul R., and Anne H. Ehrlich: *Population, Resources, Environment, Issues in Human Ecology*, 2d ed., Freeman, San Francisco, 1972. A readable in-depth discussion of man's relationship to the ecosphere and the environmental problems which modern society has caused.

Grobstein, Clifford: *The Strategy of Life*, Freeman, San Francisco, 1965. A stimulating introduction to biology.

Taylor, Gordon Rattray: *The Biological Time Bomb*, Signet Books, New American Library, New York, 1969. A fascinating account of recent and imminent discoveries in the biologic sciences.

Toffler, Alvin: *Future Shock,* Random House, New York, 1970. A discussion of the social changes which science and technology are helping to bring about. A disquieting view of the road ahead.

# 2 HOW SCIENCE WORKS

**Chapter learning objective.** The student must be able to analyze specific experimental situations, applying the concepts embodied in the scientific method. He must also be able to summarize critically the basic principles of the philosophy of science. Additionally, he must be able to draw conclusions from given experiments within the realm of his general knowledge and academic capability, and to decide whether the conclusions of others necessarily follow from the data they present.

*ILO 1* Describe and distinguish between the inductive and deductive modes of thought, contrasting the two with respect to basis, accuracy, and the generation of information.

*ILO 2* Analyze an experiment in terms of the scientific method (i.e., identify the hypothesis; decide whether the experiment really tests that hypothesis; decide whether the evidence confirms the hypothesis or whether considerable doubt still remains; identify the control; etc.).

*ILO 3* Define "hypothesis" and "experiment" precisely. Recognize each from examples.

*ILO 4* Distinguish between well- and poorly posed hypotheses (i.e., whether their predictions are testable or observable; whether experiments or observations designed to test them could be repeated; etc.).

*ILO 5* Be able to detect and identify the thought processes in an experimental design, and to assign each process to one or more of the following categories: inductive thought, deductive thought, and creative insight.

*ILO 6* Compare the concepts of hypothesis, theory, and law.

*ILO 7* Give reasons why science cannot achieve absolute truth.

*ILO 8* Outline the ethical dimensions of the scientific method. Give examples of possible ethical problems which may occur in the course of scientific investigation.

Almost every science course contains at least a bare summary of something called "The Scientific Method." Notwithstanding, some say that there *is* no such thing as *the* scientific method which so many of us learned in school. According to their view, there are as many scientific methods as there are scientists. In any case, one will seldom find an active researcher who consciously puts his ideas and experiments into the four, six, or eight steps that might be found in some freshman textbook.

## THERE *IS* A SCIENTIFIC METHOD

Just the same, there *is* a scientific method in the sense that we can reduce all good experimental work to a general formula or pattern. Granted, the practicing scientist does not always label his every thought "hypothesis," "control," or "conclusion," but if you had a record of his thought, you could put these and other labels in the appropriate places. The scientific method is a tool of thought, or rather, a whole tool kit. How does a master carpenter know which chisel to use or in which direction to plane the wood? He probably doesn't think back to the rules he learned as an apprentice, but he uses them automatically and unconsciously, along with the physical tools of his trade.

Similarly, scientific thought follows a pattern, even though the practicing scientist may be unaware of it. An understanding of any science is superficial without some grounding in the scientific method.

## WHY STUDY SCIENTIFIC METHOD?

As a citizen you should have a working knowledge of the methods of science for at least two reasons. First, though the scientific method will not make you a scientist (any more than knowing how to mix colors will make you an artist), it can sharpen your logical abilities and help you to think and to solve everyday practical problems.

Second, it can help you discriminate between genuine scientific advancement and unscientific nonsense which should be rejected or regarded with suspicion. Unfortunately, we are all deluged with a great mass of pseudoscientific rubbish. It originates from various quacks, advertisers, industry, government agencies, and even the respected scientific establishment. It is difficult for the average person to decide which items are valid and which are false or misleading. Increasingly, you will need to *question and challenge* what you hear and read. You must develop the ability to decide upon crucial issues *for yourself* on the basis of available evidence, not mere authority.

Consider, for example, the controversies over such diverse issues as fluoridation, leaded gasoline, the nutritional value of breakfast food, food additives, and the safety of the birth control pill. More bizarre but still relevant are questions concerning the accuracy of astrology, modern "prophets," tarot cards, and the like. In all these cases a working knowledge of the scientific method can help you to take an intelligent position. Without knowledge of the scientific method you are apt to have no opinion, or one based only on emotional values. Many issues today require informed citizens for their resolution. Pollution, urban problems, food production, overpopulation, and the rest of the vast tangle we call "public affairs" can be dealt with democratically only by an informed and critically thinking electorate.

The time has permanently passed when citizens can afford to have ill-informed opinions about public issues. Convictions should arise from informed opinions, and convictions are the basis of responsible social and political action. If we permit ourselves to remain apathetic, uncommitted, and indecisive, others more resolute than we will make decisions for us.


## SYSTEMATIC THOUGHT PROCESSES

It must be admitted that human existence is broader than the scope of the scientific method. There is much that is meaningful or even indispensable in our lives which is inherently outside the realm of scientific investigation. Nevertheless, underlying all our logical thought processes is what we may call *the law of rationality*. Simply stated, it is that conclusions should be based on adequate evidence. Some think that logical thinking is instinctive. As a matter of fact, to a large extent logical thought must be learned and then regularly practiced. The scientific method helps the scientist to use his thought processes in a rational, systematic, and error-minimizing way in order to discover, define, or delineate truth.


### Truth and objectivity

We may define *truth* as agreement with the facts, or reality. Underlying the search for truth or the "laws of nature" is the inherent belief that the world is rational and orderly, not chaotic. Without such an assumption, science is impossible. Such an idea may be upsetting to those who tend to think (erroneously) of science as being without assumptions, based only on pure reason, and infallible.

A difference of opinion about what constitutes truth does not mean that truth is relative. It simply means that in some instances we do not correctly perceive truth. This may be caused, among other reasons, by insufficient evidence, or because our ideas are untestable in the current state of scientific knowledge.

The scientist seeks for demonstrable truth which can be presented in a logically compelling way. He tries to distinguish between truth as it exists and belief or opinion about truth which cannot be demonstrated, or which may change as knowledge increases. Truth as it exists is *objective;* opinion about truth is *subjective*. In actuality pure objectivity is impossible, but the scientist strives to be as objective as he can.

Systematic thought processes can usually be broken down into two categories: *induction* and *deduction*. Both are used in the scientific method.

## Deductive reasoning

Deductive reasoning begins with supplied information called **premises** and draws conclusions on the basis of that information. Deduction tends to proceed from general principles to specific conclusions.

| | |
|---|---|
| All birds have wings. | *first premise* |
| Sparrows are birds. | *second premise* |
| Therefore, sparrows have wings. | *conclusion* |

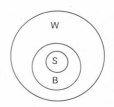

FIGURE 2-1

The above example is a **valid** argument. That is, the conclusion that sparrows have wings follows *inevitably* from the information given. No other valid conclusion is possible. If we were to diagram the argument with a series of circles to indicate the categories we are talking about, it would look like the circles in Figure 2-1.

All birds are within the category of winged animals. Sparrows are within the category of birds. Therefore, sparrows are also *necessarily* within the category of winged animals. There is no way that sparrows can be in the category of birds without also being in the category of winged animals. But consider the following example:

All birds have wings.
All sparrows have wings.
Therefore, all birds are sparrows.

It is immediately obvious that something is wrong. We have violated the rules of logic,[1] and so we come up with an *invalid* or unreliable conclusion, even though the two premises may be true. Such an argument could be diagramed in more than one way (see Figure 2-2).

A deductive argument is valid only when the premises *necessitate* (or inevitably lead to) the conclusion. In this case you can see by looking at the different sets of circles that the premises can be diagramed or interpreted in different ways. All birds are not necessarily within the sparrow category.

Notice that we called these logical arguments **valid** or **invalid, NOT true** or **false.** Truth refers to statements or propositions; validity refers *only* to the correctness of the logic we employ. Scientists may well question, for example, the *truth* of the statement that "all birds" (past and present) have wings.

Consider another example:

All Americans are rich.
John is an American.
Therefore, John is rich.

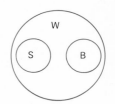

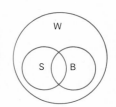

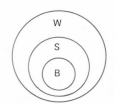

FIGURE 2-2

---

[1] In logic this error is technically known as an "undistributed middle term." A far more rigorous consideration of logic is needed before the nature of all logical errors can be precisely defined and understood.

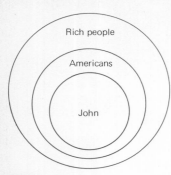

**FIGURE 2-3**

$$\begin{array}{r} 2 \\ +\ 2 \\ \hline 4 \end{array}$$ } Information

4    Conclusion

**FIGURE 2-4**

You might question the *truth* of these premises. (Can we indeed say that *all* Americans are rich?) Nevertheless, the above argument is a *valid* argument. That is to say, if the above premises are true, the conclusion follows of necessity.

This leads us to conclude that the method of deduction (logically applied) is only as strong as the premises with which we begin. If our premises are true, we may safely predict a conclusion. In the above example, given both premises, we may predict that John will be rich.

Deductive reasoning has some obvious limitations. First and most obvious, it is only as strong as the truth of its premises. Second, we cannot use the deductive method to discover new information, since the conclusion really contains no more information than that given in the premises. The conclusion is simply a *deduction* made on the basis of evidence. Deduction can be compared to simple arithmetic, in that if the information given is true, the conclusion also must be true.

In the scientific method, deductive reasoning is most valuable in helping the scientist to test his hypothesis, or tentative explanation. If his hypothesis is true, then a result or conclusion can be predicted. If the result or conclusion is not true, then the scientist must go back, question the truth of his hypothesis, develop another or a modified version, and test again.

### Inductive reasoning

Inductive reasoning, on the other hand, begins with specific examples and seeks to draw a conclusion or discover a unifying rule or general principle on the basis of those examples, such as in the following:

When released from support, apples, oranges, rocks, trees, and airplanes fall to the ground.

*Therefore:* a force acts upon all objects which attracts them to the ground (i.e., the force of gravity).

Even if many specific examples are used, and the results are very carefully reported, an inductive conclusion is still not necessarily (or infallibly) true.

When released from support, apples, oranges, rocks, trees, and airplanes fall to the ground.

*Therefore:* microscopic green men (smaller than atoms) occupy space; they hurl all objects to the earth whenever they are able to do so.

In the above illustration another *possible* explanation is given. Obviously, however, it is an unlikely proposition, and a poor choice for a scientific hypothesis, as you will discover in the section on what makes a good hypothesis.

Not only is it possible to choose a poor and logically unsatisfactory hypothesis, but inductive results can also be upset by new evidence which differs from the evidence upon which we based our conclusion. Furthermore, conclusions in the above examples contain more information than the reported facts on which they are based. We go from "many" examples to "all" when we draw up a general principle. This is known as the *inductive leap;* it is an unavoidable weakness of the inductive method.

Even if we base our conclusion on 100,000 observations, it is still possible that new observations will upset our conclusion. However, the greater number of cases we employ, the *more likely* we are to draw accurate scientific conclusions. The scientist is thus able to state that his conclusions have greater or lesser *statistical probability*.

Induction is subject to several characteristic errors. Among them are *sampling error* and *bias*.

**Sampling error.**  Sampling error arises when observations are inadequate in number or uncharacteristic of the thing we are attempting to study. One form of sampling error is often found in public polls and similar studies. In them, conclusions are sometimes drawn on the basis of a limited set of observations, or a small segment of the population which is not characteristic of the group as a whole.

For example, in an Associated Press release of August 5, 1970, Mr. Jim Adams reported the findings of the President's Commission on Pornography. This commission decided tentatively that exposure to pornography does not lead to sexual misbehavior among young people. The study, however, did not involve any young people!

> The finding that pornography does not corrupt youngsters' morals, the report says, is not based on actual studies of youngsters under 18 because of the (moral) sensitivity against such studies.

The scientist might well question the scientific value of such a study. Adults were polled, yet the results were applied to children! This introduces many possible sources of error. Basically, it assumes that adults are influenced in the same way and by the same things as children. Also, exposure over a long period of time (say from infancy to adolescence) could change the results.

**Bias.**  Since inductive conclusions are not inevitable in the sense that deductive conclusions are, they are susceptible to misinterpretation arising from subjective bias. Bias is a result of preconceptions, and it is unavoidable to some extent. Bias involves interpreting all data in terms of the hypothesis we wish to prove, and either ignoring or explaining away contrary evidence. No one is free from bias, but the scientist should try to dismiss as many preconceived ideas as possible before he begins his work. Such ideas can blind him completely to what is, one would think, obvious and self-evident.

Bias often occurs because a scientist has uncritically accepted recognized authorities. The early discoverers of sunspots seem to have delayed the publication of their findings because, according to accepted view, the sun's surface was pure and spotless! At first they could not believe their eyes. Copernicus and Galileo, also, had to reject the authorities accepted by almost everyone else in order to form the idea that the earth and other planets move around the sun.

The modern scientist walks a tightrope. He is obviously dependent on other authorities to a large extent. For example, a recent edition of the *Merck Index*, a scientific handbook, contains 1,121 pages describing the chemical and physical properties of thousands of substances. Lifetimes would not suffice to repeat all the experiments that led to these results. One must pretty well accept them.

On the other hand, from time to time a good scientist will reexamine even the most widely held theories. He will repeat experiments, consider other explanations for the data, search for new factors, and explain more fully the negative results that bear on the problem. Watson and Crick, the discoverers of the modern concept of DNA, had to reject several generally accepted chemical structural formulas before they could get their ideas to make sense.

One authority that is never fully rejected, however, is oneself! Of all prejudices, our own are hardest to shake. Philosophers call this the *egocentric problem* and claim that there is no escape from it. The scientist, however, seeks to minimize this source of bias.

That bias affects the results and interpretation of experiments is so well established that scientists often deliberately conceal some information from themselves. Often when drugs are tested, the physician administering them does not know whether he is giving a new medication or a harmless starch pill! Recently an experiment was designed to test the effect of bias in an experimental situation. A number of graduate students were divided into two groups. All were given rats to train, but one group was told that *their* rats were specially bred for high intelligence. The others were told that their rats were of subnormal intelligence. In actuality, all rats had been randomly selected from the same colony.

The students with the allegedly smart rats treated them very well. They gave them pet names, stroked them, and fed them snacks. Those with "stupid" rats, however, treated theirs roughly and tended to give them only minimum care. Not surprisingly, perhaps, the better-cared-for rats performed better and learned tasks more readily. Since the two groups of rats were the same, the expectations of the two groups of students in some way must have influenced or produced differences in the rats' behavior.

Both inductive and deductive logic are indispensable for the scientist. Neither type of reasoning, however, is reliable unless the scientist makes every attempt to obtain accurate information, and proceeds to use and interpret that information as carefully and objectively as possible.


## SCIENTIFIC METHOD

Now that we have looked at some characteristics of inductive and deductive reasoning, let us begin to look at how they function in the scientific method.

Let us take an elementary problem. John Jones enters the room. He staggers and gasps for breath. His coloring is gray, and he is clutching his left arm. He falls to the ground, unable to speak. A doctor is summoned. The doctor thinks that John has had a heart attack. John is rushed to the hospital. An electrocardiogram and blood-enzyme tests confirm that John has had a heart attack.

## Scientific procedures

Now let us restate the above problem in terms of the scientific method.

| Steps in the scientific method | Illustration |
|---|---|
| **1** *Recognize and state the problem.* | John is ill. We must discover what the illness is. |
| **2** *Collect information or data bearing on the problem.* | John's symptoms are gasping, staggering, gray color, apparent pain in left arm. |
| **3** *Formulate a creative hypothesis.* | John seems to have had a heart attack. |
| **4** *Make a deductive prediction on the basis of the hypothesis.* | If John has had a heart attack, a characteristic pattern will be revealed on an electrocardiogram, and certain blood-enzyme tests will be positive. |
| **5** *Make observations or design and perform experiments to test the hypothesis.* | Enzyme tests and the electrocardiogram are given. The results are in accordance with the prediction. |
| **6** *Formulate a conclusion.* | John has had a heart attack. |

Although the scientific method is primarily inductive, deductive reasoning has been used at least twice–after step 2 and after step 3. After step 2 the reasoning would have been:

**2a** People who have these symptoms are heart attack victims.
John has these symptoms.
Therefore, John is a heart attack victim.

The reasoning which led to the deductive prediction in step 4 would have been:

**3a** Heart attack victims have test results as in step 4.
(If) John has test results as in step 4.
(Then) John is a heart attack victim.

However, suppose that the blood-enzyme tests and the electrocardiogram seemed to indicate that John had *not* had a heart attack. Then the doctor would have to go back to step 3 and formulate a new hypothesis. His new hypothesis might be, "John has had an acute gallbladder attack," or "John is suffering from food poisoning." He would then use new tests to confirm or disprove his new hypothesis.

Creative thinking is needed at every step. If, for example, John were twelve years old instead of forty-five, that piece of information by itself might be sufficient to make a doctor try another hypothesis first. That fact would have been more significant evidence than the apparent pain in John's left arm.

The above illustration is greatly simplified. In most instances each step of the scientific method is far more complicated.

### Recognizing the problem

According to legend, Isaac Newton was first led to think about gravity while idly watching the fall of ripe apples. Doubtless millions before him had observed the same phenomenon, as well as thousands of other problems like it. But, as Pasteur said, "Chance favors the prepared mind." It would seem that Newton's mind was the first that was prepared to see anything more in that apple's fall than the simple fact that it fell.

Discoveries are usually made by those who are in the habit of looking sharply at nature, and whose minds have been prepared—usually by some preliminary idea of what they are looking for. It is also helpful to possess background knowledge of what others have discovered. This is why Nobel prizes in science are more often won by scholars than by laymen. Not that great scientific discoveries are hard to understand. Quite the contrary. Some of the most fundamental concepts are really quite simple. It is amazing how obvious the truly great hypotheses appear in retrospect, but how difficult they were to think of in the first place! Nothing is harder, it seems, than thinking a truly original, self-evident thought.

The discovery of penicillin is such an example. In 1929 the British bacteriologist Alexander Fleming noticed that one of his bacterial cultures had become invaded by a blue mold. He almost discarded it, but before he did, he noticed that the area contaminated by the mold was surrounded by a zone where bacterial colonies did not grow well. His culture looked something like the one shown in Figure 2-5.

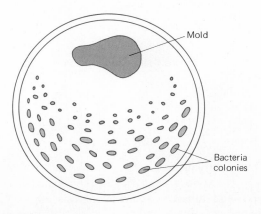

**FIGURE 2-5**
The discovery of penicillin.

The bacteria were disease bacteria of the genus *Staphylococcus*. Anything that could kill them was interesting! Fleming saved the mold, a variety of *Penicillium,* of which bread mold is a more familiar example. It was subsequently discovered that the mold produced a substance which slowed bacterial growth—penicillin, the first antibiotic.

We may well wonder how often the same thing happened to other bacteriologists who failed to notice the key fact and just threw away their contaminated cultures. Fleming benefited from chance, yes, but his observing mind was prepared for it.

### Formulating a hypothesis

If a problem is recognized only by a prepared mind, a hypothesis is generated only by a creative one. A *hypothesis* is a *tentative explanation* for observations or other known facts. In the early stages of investigation the scientist usually thinks of many such possible explanations. He then decides which, if any of them, he can and should subject to experimental test. To help him separate the very unlikely or even stupid hypotheses which may have occurred to him from those deserving further study, he can make use of several guidelines. A good hypothesis should be:

1  Consistent with all known facts, or more consistent with them than competing hypotheses.
2  Capable of being tested; that is, it should generate definite predictions, whether the results are positive or negative. Test results must also be repeatable by independent observers.
3  Simpler than competing hypotheses.

*"Sometimes I think you're a serious research-and-development man, Roberts, and sometimes I think you're just messing around."*

**FIGURE 2-6**
(Cartoon by Chon Day; © 1971, The New Yorker Magazine, Inc.)

***Consistency.*** A hypothesis must be consistent with all known facts, or must account for more of them better than competing views do. There are, however, certain reservations which may apply. Sometimes data appear to be in conflict with a hypothesis, but upon closer inspection we may find that the hypothesis can account for them. What is more, results can be erroneous. There are several examples in the history of science of men who persisted stubbornly in their views in the face of supposed "facts," only to be vindicated in the end. The so-called "facts" were shown to be erroneous. In biology, moreover, large individual differences in experimental plants or animals can produce occasional unexpected results that really do not affect the truth of the hypothesis being tested.

In most instances, however, the hypothesis which seems to explain the most is preferred. It is the one which is in accord with more of the facts, or explains them better than alternative explanations. For example, at one time the earth was held to be flat, but there was no accounting for the fact that tall ships' masts seemed to appear over the horizon before the ship itself could

**FIGURE 2-7**
An occasional inconsistent result does not necessarily discredit a hypothesis.

be seen. When it was proposed that the earth was round, at last this ancient observation could be reasonably explained.

***Predictability and testability.*** A hypothesis, if it is to have any practical value, must lead to one or more predictions. The hypothesis itself is not really subject to test, but one can test the predictions which follow from it. If the hypothesis is *true*, a number of consequences will follow deductively from it. On the other hand, if it is *false*, *other* consequences can be specifically predicted. For example, Einstein predicted, on the basis of his general theory of relativity, that light should be deflected by a strong field of gravity such as that possessed by the sun.

The only way such a hypothesis could be tested at the time was to measure precisely the position of a star whose image was close to the disk of the sun. The sun is so bright that this had never been attempted before,

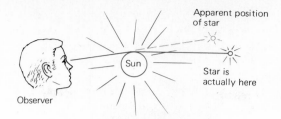

but there was a way in which it could be determined whether the prediction would come true: wait for an eclipse of the sun by the moon, and then measure the apparent positions of the stars near its disk. This was done, and the star images were indeed observed to be displaced to about the extent that Einstein had predicted.

In this example, a hypothesis led to a *definite prediction*. If it had not, there would have been no way to test the hypothesis, because there would have been no prediction to test. Unfortunately, sometimes hypotheses are formulated in such a way that one cannot predict the results if they should be false. By coincidence, a false hypothesis can sometimes lead to a true prediction. Just because we know nothing that would contradict a hypothesis does not make it automatically true. The absence of evidence cannot establish a hypothesis. Most often, this error takes the following form:

We know of no evidence for _____.
Therefore, _____ must be false or nonexistent.

Examples of this are numerous. There was a time when reputable scholars claimed that the biblical account of the Hittites was fictitious, because no extrabiblical evidence supported the existence of such people. Now, all competent scholars acknowledge the existence of that ancient nation, and considerable information about them is known. To make such a claim in the first place simply on the basis of negative evidence was obviously unscientific.

We must be able to design experiments that, if the hypotheses are false, will yield definite, tangible results to indicate that falsity.

*The unfalsifiable hypothesis.* When she was quite young, my daughter wondered what made the hot water drip from the top to the bottom of our coffee pot. Not wishing to explain the law of gravity just then, I told her that there was a little invisible man in the top of the coffee pot who stayed every morning just long enough to push the water down, and *he* was the one who made the faint dripping noises in the pot. Entangling myself still further, I explained that he could not be seen because he was invisible, and that the moment the coffee was made he flew away to another house to push the coffee down there.

Now as a matter of fact, there is no way that she, you, or anyone else can show me up! A mature mind will not accept my idea, however, because it is obviously not a "good" explanation. Why not?

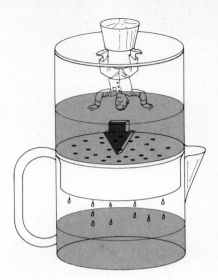

**FIGURE 2-9**
One kind of "bad" hypothesis is untestable because of the way in which it is proposed.

In the first place it cannot be tested. In fact, I have carefully proposed it in such a way that it cannot. Granted, one prediction could be made if it were true: the water would drip into the coffee pot. *However, if my theory were false, no unique consequences would result from its falsity.* The most obvious prediction would be that if my theory is false, the little man will not be there! But there is no way to test that. He is invisible, he can pass through the walls of the coffee pot, and he is weightless. Such a hypothesis is known as an unfalsifiable hypothesis. It cannot be disproved. Some widely accepted and much respected views fall into this category. However, it is important to remember that a proposal is not necessarily false because it is untestable. But if the proposal is not a working hypothesis, then it really is outside the scope of the scientific method.

*The untestable hypothesis.* There is a distinction between an unfalsifiable hypothesis where negative results are essentially impossible, and a hypothesis that is untestable. *An unfalsifiable hypothesis generates no negative predictions,* because no specific or concrete disproof seems conceivable. An untestable hypothesis does generate falsifying predictions. In the case of the untestable hypothesis these merely cannot be tested in our present state of knowledge.

In 1959 the Russian astronomer I. S. Shklovskii explained certain oddities in the orbits of the Martian moons by the hypothesis that they were hollow, and were in fact space stations launched by Martians! F. Zigel, another Russian, much taken by Shklovskii's imaginative suggestion, further proposed that the moons had been constructed between 1862 and 1877. At the time this hypothesis was untestable, but it did lead to a prediction: were one to travel to Mars he could examine the moons closely and determine that they were artificial.

Alas for the Martians! In 1972 the Mariner 9 spacecraft sent photographs

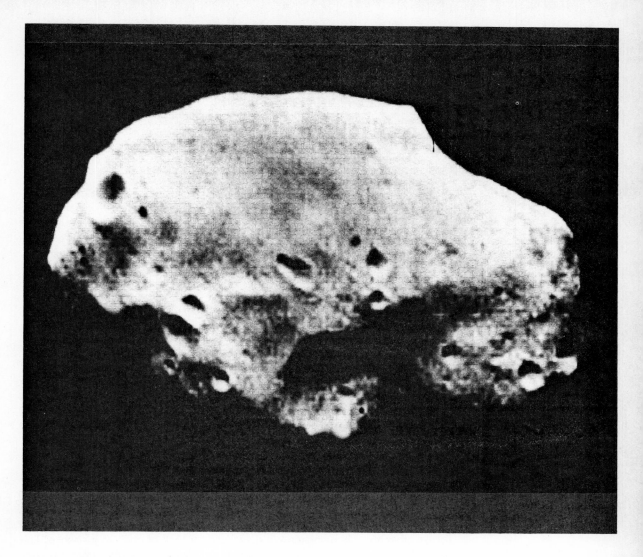

**FIGURE 2-10**
The most detailed image of
Phobos available to date is seen
in this computer-enhanced photograph taken by the Mariner 9
spacecraft during its thirty-fourth
orbit of Mars. It no longer seems
likely that Phobos is a Martian
space station. (NASA)

back to earth (Figure 2-10) which showed both moons to be irregular chunks
of rock about 20 miles across. Now we know Shklovskii's hypothesis to be
false. Still, it was a "good" hypothesis even though 13 years elapsed before
it could be tested.

*Disproof.* It is important to have an exact understanding of just what constitutes disproof of a hypothesis. Many people think that if a prediction fails to
occur, then the hypothesis has been disproved. That is not so.

To take an example, biologists have traditionally determined species in
quite a practical fashion. If animals can produce fertile offspring, they
belong to the same species. But one might put two mice of opposite sex
together in a cage and wait for some time for the blessed event to occur. A

negative result by itself would not necessarily show that they do not belong to the same species. They might, after all, just not particularly care for each other!

*Negative results do not disprove a hypothesis.* To discredit a hypothesis, an experiment must be designed which will give not a negative but a positive result if the hypothesis is false, as in the following example.

Green plants make sugar from carbon dioxide and water, using the energy of sunlight to do so, and giving off oxygen as a by-product.

$$6CO_2 \ + \ 6H_2O \longrightarrow C_6(H_2O)_6 \ + \ O_2$$

carbon dioxide    water              sugar       oxygen gas

From which compound did the oxygen gas, which is given off, originate? Since both carbon dioxide and water contain oxygen, either could have been its source. When radioactive oxygen became available after World War II, experimenters used it to discriminate between the two possibilities.

*Hypothesis:* The oxygen given off in the photosynthesis comes from the carbon dioxide.

*Prediction:* Radioactive oxygen incorporated in the carbon dioxide will be given off as oxygen gas if green plants are fed carbon dioxide containing radioactive oxygen.

$$6CO_2{}^* + 6H_2O \longrightarrow C_6(H_2O)_6 + O_2{}^*$$

*If the hypothesis is false:* radioactive oxygen incorporated in the carbon dioxide will appear in the *sugar.*

$$6CO_2{}^* + 6H_2O \longrightarrow C_6(H_2O^*)_6 + O_2$$

\* Radioactive.

The falsifying prediction came true. Therefore, the hypothesis was shown to be wrong. Now scientists realize that the oxygen originates from the water.

There is one qualification to the principle of negative evidence. In actual practice many hypotheses are discarded for lack of positive evidence. This is not as dangerous a practice as it may seem — if several *different* predictions of the same hypothesis fail to be borne out. In such a case it becomes *increasingly unlikely* (but not impossible) that the hypothesis is correct. One should not be hasty, however, to discard a hypothesis on the basis of a single unfulfilled prediction.

*Complete disproof of a general principle is rarely possible.* I am reminded of the old horror movie "The Fly." If I recall the plot correctly, a lady woke up one morning to discover that her husband had the head of a fly. This not being his normal state, she was alarmed. The producers of the movie offered a substantial reward to anyone who could prove this to be scientifically impossible. As far as I know the reward is still available and would be a great aid both to hungry students and to instructors. Surely you don't think anything like that could happen, do you? Claim the reward!

But how would you go about it? The best you could do would be to say that the idea is contrary to every established principle of biology and that it has never been done. The producer's lawyer would reply that history is full of outworn scientific ideas that have had to be discarded, and that just because it has not occurred before does not mean that it *cannot* happen. It is impossible, he would say, to prove anything scientifically impossible. Apparently so, for some time later there was another film called "The Return of the Fly." I have not seen it.

*Repeatability.* Another characteristic of a good hypothesis is that the experiment should be repeatable by the researcher, or by other scientists working independently of him, who should be able to obtain similar results. If it is not repeatable, we run into problems of a statistical nature (see Statistics, further on in this chapter). We may also be entitled to wonder whether the original researcher observed what he thought he did, or interpreted it correctly. Even more important, the principle of rejection of authority demands that we not accept blindly the conclusions or observations of even highly respected colleagues, particularly in our own field of study. We (or other scientists) must be able to repeat experiments and obtain reasonably close approximations of the findings of those who have worked before us.

There have, unfortunately, been several notable cases of scientific fraud which have retarded the progress of science in some fields for years or even decades. The only way to protect science from the accumulation of false ideas is to insist that it be self-correcting. The duplication of observations and experiments helps science to make needed corrections and to maintain whatever intellectual purity is necessary for its progress.

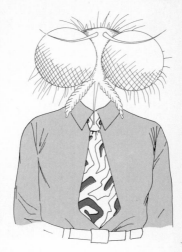

FIGURE 2-11
It is impossible to prove that anything is scientifically impossible.

### Testing a false hypothesis

As we shall see in Chapter 15, the "blending hypothesis" of inheritance holds that organisms tend to be intermediate in their characteristics between those exhibited by their parents. This hypothesis is based upon observation, for in a general sort of way many offspring *do* appear—at least superficially—to possess a blend of parental traits. The blending hypothesis is a "good" hypothesis, moreover, in that it gives rise to many specific predictions and suggests a number of experiments to test those predictions.

For instance, a common garden plant, *Nicotiana,* the four-o'clock, exists in several color varieties, of which we shall here consider only the red, white, and pink. Were we to cross red and white four-o'clocks, the blending hypothesis of inheritance would lead us to predict that the offspring would be of intermediate color, in this case, pink. That is exactly what is observed in such a cross. The hypothesis would be confirmed, so far.

But now let us see what would happen if we were to cross the pink-flowering plants with white ones. In that case, the blending hypothesis would predict that the offspring would be a very light shade of pink. What is actually observed, however, is that half of them are just the same shade of pink as the pink parent, and half of them are *white*. The blending hypothesis would now appear to be discredited.

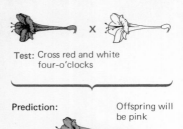

Test: Cross red and white
four-o'clocks

Prediction:                Offspring will
                           be pink

Observation:               Offspring are
                           pink

Hypothesis confirmed

Test: Cross pink and white
four-o'clocks

Prediction:                Offspring will be
                           very light pink

Observation:

Half the offspring    Half the offspring
are white             are parental shade
                      of pink

Hypothesis rejected

**FIGURE 2-12**
The blending hypothesis of
inheritance. Sometimes, by coincidence, a false hypothesis generates a true prediction.

The first prediction of the blending hypothesis chanced to be correct, but the second was not. The more predictions that one tests, the less likely it is that an erroneous hypothesis will be confirmed. In fairness to the blending hypothesis of inheritance, though it turned out to be false, we would remind you that it was a "good" hypothesis in that it gave rise to a number of testable predictions.

### Conducting experiments

Scientists do not usually wait around for natural confirmations of their predictions if they can help it. A total eclipse of the sun occurs rarely, but there was no other known way to test Einstein's prediction except to await the eclipse. When possible, it is desirable to set up situations which will produce observations or results relating to our hypotheses. These contrived situations are called experiments. They are devices for generating observations.

***Setting up the experiment.*** Farmer Brown wants to know whether he should use a particular brand of fertilizer on his farm. To test the idea, he adds fertilizer to a bean field. The crop of beans that year is greater than it has ever been before.

*Hypothesis:*   Fertilizer increases bean yield.
*Experiment:*   Add fertilizer to a bean field.
*Results:*   A bumper crop.
*His probable conclusion:*   Hypothesis is confirmed.

At this point, such a conclusion is not really justified. As any farmer knows, crop yields may vary from year to year as a result of variations in rainfall, insects, or other chance factors. Thus a bumper crop may not be due to the fertilizer he added.

***The control.*** To guard against that possibility, Farmer Brown should employ a *control,* that is, a second experiment performed *under the same conditions as the first,* except that the *one* factor he is attempting to test should be varied. In this case he would plow a second field (the control field) in the same way, using the same type and amount of seed, water, soil, insecticide, etc. However, he would not apply fertilizer to the control field. If both fields now show the same increase, the fertilizer could not be responsible for it. Also, if the untreated field yields more than the treated one, the fertilizer might even be shown to be worse than useless!

Even if the treated field shows greater yield than the untreated one, Farmer Brown is still not completely justified in saying that the fertilizer has produced the difference. It may have been due to chance again. Perhaps insects invaded one field, or grazing cattle trampled one crop. The possibility that a hypothesis might appear to be confirmed because of chance can be reduced but never completely eliminated. A result which seems to confirm a hypothesis is known as a *positive result*. If Farmer Brown performs the experiment many times and usually gets positive results, he can be increasingly sure that his hypothesis is true, but never completely sure. Put yourself in his place. If one time out of ten times the fertilizer apparently increased yield, you would probably attribute it to chance. If nine times out of ten increased yield, you would think that the fertilizer was responsible. But what about five times out of ten? Six?

*Statistics.* One consideration that would help you to decide in this instance is the size of the increase that is associated with the fertilizer. In general, the smaller the differences, the larger the number of observations which will be necessary to come to an assured conclusion. The science of *statistics* (the mathematical study of probability and chance) can help in such situations.

If Farmer Brown is convinced that fertilizer produces increases, he may wish to investigate *how much* increase a given quantity of fertilizer produces, or whether there is an upper limit to its effectiveness. He would probably add different amounts of fertilizer to different fields and observe the results. If he found that the more fertilizer he added, the more the crops grew, he would probably decide that fertilizer was responsible for the increase. This is called the method of *concomitant variation*. It is widely used.

Suppose, though, that Farmer Brown observes no consistent differences between his experimental and control plots. Is he justified in assuming that the fertilizer is without effect? First of all, he can make such an assumption only for himself. Other farmers with different soil, climate, or elevation might find it effective, although the likelihood of this is diminished by his experience. Still, such differences might make the fertilizer effective for someone else. Second, the fertilizer might increase yield by an indetectably small amount—a few ounces, perhaps. Since he weighs his crops in ton quantities, he would be unlikely to notice such small differences, and in any case, it would make no more practical difference to him than would the first consideration.

Farmer Brown may not be *absolutely sure* of his positive or negative results. However, that will not greatly concern him! Absolute certainty is not required before some practical decisions can be made.

### Theory, law, and truth

If a hypothesis seems to be conclusively confirmed, and gains general acceptance among scientists, it comes to be regarded as a *theory*. Theories are often somewhat broader in scope than the original hypothesis that gave rise to them. Theories may embody a number of related and generally accepted hypotheses. The atomic theory, for example, contains interrelated hypotheses from many areas of chemistry and physics.

To a large extent science is a social activity. Many hypotheses are proposed by different persons than those who eventually devise ways of testing them. The scientific process may involve several generations of accumulated observations in the eventual establishment of a major theory or law.

A *scientific law* may be defined as a theory which, over a long period of time, has yielded true predictions with unvarying uniformity. It thus becomes almost universally accepted. Occasionally, important exceptions even to laws are discovered. This usually does not result in the discarding of the law, but may result in its modification. Modifying a law can be like patching a garment. An old-fashioned housewife would patch dungarees for a long time. Only when they became more patch than pants would she discard them. Few laws of any age have escaped the patching process. Even Newton's law of gravity has been modified by Einstein's theory of relativity.

It is important to remember that there is no qualitative distinction between a hypothesis, a theory, and a law. No council meets (thank heaven!) to promote hypotheses to theories, or theories to laws. Hypotheses, theories, and laws are all approximations of the truth, differing only in that some are believed to be more tentative than others.

Science can be a reliable guide to nature, but it is not infallible and cannot be equated with absolute truth. Absolute truth is, in fact, outside the limits of science.

## THE ETHICS OF SCIENCE

Even though ethics is often believed to be outside the realm of scientific investigation, science has many ethical dimensions. Ethics may be applied to the scientific method itself, to the conduct of experiments, or to the public release and use of scientific findings.

Honesty and objectivity at every stage of the scientific method are indispensable if the results are to have any value. Deliberately or subconsciously doctored data may mislead scientists for generations. Scientists are human, and unfortunately some of them have been guilty of such doctoring.

Damage can also be done by withholding a portion of the results of a scientific investigation. Such a temptation is especially great among scientists who are directed to find "scientific" evidence which will support the claims made by their employers. In fact, a scientist's promotion or continued employment may even be dependent upon his willingness to suppress "unfavorable" information. Such scientists scarcely qualify as objective truth seekers, and their results are therefore more likely to be colored by their biases. Industries, public health experts, conservationists, and government agencies under attack by consumer advocates may hire scientists under such conditions. Fortunately, independent investigators frequently (but not always) uncover suppressed or doctored data on their own.

A particularly delicate area of concern involves the morality of whether certain investigations should be undertaken at all. Some investigations are by their nature morally illicit. The infamous "medical" experiments conducted by the Nazis upon concentration camp inmates will be almost universally acknowledged as immoral, regardless of the value of the information thus obtained (which was largely worthless).

There is, however, much more serious disagreement about the use of volunteers and other human subjects in medical research of a less serious nature. It may be that no one should be asked to volunteer for certain medical and psychologic experiments, regardless of his willingness to be subjected to such experiments.

Many also question whether experiments carried out upon living animals are moral. Generally, it is held that such experiments are permissible if every precaution is taken to avoid unnecessary suffering, for example, by the use of anesthesia. Sometimes suffering is a necessary feature of an experiment, particularly in psychology or stress physiology. It is our view that such studies should not be wantonly undertaken unless tremendous human benefit can convincingly be shown to be their likely result. We would also contend that it is wrong to endanger the survival of species of animals by their excessive use in experimentation. Both rhesus monkeys and chimpanzees are fast becoming scarce under the pressures exerted by some biomedical researchers who seem to care nothing for their conservation.

Sometimes investigations are avoided for fear of the social consequences of their possible results. The investigation of possible differences in intelligence among racial groups appears to have been delayed by such fears. Some consider certain types of military research illegitimate by virtue of their potential danger to humanity. On the other hand, so many projects have tremendous potential for both good and evil (such as the control of genetics and artificially lengthened life) that it is difficult to be dogmatic about the ethics of investigating such things.

We are inclined to feel that the social responsibility of the scientist extends more to the application made of his discoveries than to the choice of what he investigates. It is difficult, perhaps impossible, to determine the social relevance of a particular line of research in advance,[2] and it is doubtful that this should be a major criterion of its desirability. Rather, it would seem that the scientist has the responsibility, as Salvatore Luria puts it,

. . . not to be deterred by the widespread and increasing ignorance of the public, including governments, in scientific matters, but should try to break through that barrier.

Whether he is aware of it or not, the scientist's ethical decisions will influence every phase of his research from its inception to its conclusion.

---

[2] Recently an investigator made application for a patent which would permit the confinement of reacting gases in a thermonuclear power reactor. Were it to prove practical, much social good might result from it. However, the Atomic Energy Commission saw consequences of his discovery which he evidently did not, and immediately classified his patent as secret because of its potential military applications. This is an excellent example of what we mean.

**CHAPTER SUMMARY**

1 There is a method common to all science, though it is often practiced unconsciously.

2 All citizens need some knowledge of scientific method. It prepares them to understand contemporary issues, and to influence society in an intelligent way.

3 Logic is a means of drawing conclusions that are consistent with the evidence used to support them. Two varieties of thought are distinguished: deduction and induction.

4 Inductive logic suffers from several sources of error which must be carefully avoided, particularly sampling error and bias.

5 Inductive and deductive logic are both employed in the scientific method.

6 In using the scientific method, one must:
  (*a*) Recognize and state the problem.
  (*b*) Collect information or data bearing on the problem.
  (*c*) Formulate a creative hypothesis.
  (*d*) Make a deductive prediction on the basis of the hypothesis.
  (*e*) Make observations, or design and perform experiments to test the hypothesis.
  (*f*) Formulate a conclusion.

7 "Good" hypotheses generate definite predictions subject to test. Tests should be repeatable by independent workers. The simpler hypothesis is usually preferable.

8 Negative results do not necessarily constitute disproof, and complete disproof of a general principle is rarely possible.

9 A hypothesis which cannot be disproved even if it is false is known as an unfalsifiable hypothesis. A hypothesis which cannot be proved or disproved in the present state of knowledge, but which does generate falsifiable predictions is known as an untestable hypothesis.

10 The scientist uses controls and statistical data carefully in constructing valid experiments.

11 There is no qualitative difference between hypothesis, theory, and law. Hypotheses are regarded as more tentative than theories, and theories more tentative than laws, but all three are approximations of the truth.

12 Science is a tool for investigating nature, not a guide to absolute truth.

13 The practice of science entails a definite ethic, whether the practicing scientist is aware of it or not.

**FOR FURTHER READING**

Baker, Jeffrey, and Garland Allen: *Hypothesis, Prediction and Implication in Biology,* Addison-Wesley, Reading, Mass., 1968. A good source for the beginning student who wishes to delve more deeply into the subject but who does not wish to become buried in a tome.

Bettex, Albert: *The Discovery of Nature,* Simon and Schuster, New York, 1965. A fascinating pictorial history of science. Teachers will find it a rich source of examples for the scientific method in action.

Gardner, Eldon: *History of Biology,* Burgess, Minneapolis, 1960. A useful summary of the development of the scientific method as applied to the science of biology.

Luria, S. E.: "Modern Biology: A Terrifying Power," *The Nation,* Oct. 20, 1969, p. 406. One of the best recent considerations of the social potentials of the new approaches to biology.

Morris, Ian: "Is Science Really Scientific?" *Science Journal,* December 1966, p. 76. In our opinion, this is the finest short summary of modern scientific philosophy in print.

Riga, Peter J.: "Modern Science and Ethical Dimension," *The Catholic World,* August 1969, p. 213. A discussion of whether certain types of scientific investigation are ethical.

Ruby, Lionel: *Logic, an Introduction,* Lippincott, Philadelphia, 1960. An almost inexhaustible mine of clear thought.

Russell, George K.: "Vivisection and the True Aims of Biology," *The American Biology Teacher,* May 1972, p. 254. A notable summary of a controversial and little heard point of view.

# 3

# THE ORGANIZATION OF THE ORGANISM

**Chapter learning objective.** The student must be able to describe the organization of a complex multicellular organism with respect to its cells, tissues, organs, and organ systems.

*ILO 1*   Identify the levels of biologic organization and state why the cell is the basic biologic unit.

*ILO 2*   Describe the general characteristics of cells, e.g., size range and shape.

*ILO 3*   Identify methods by which scientists study cells.

*ILO 4*   Draw a diagram of a plant or animal cell, identifying each intracellular structure (as discussed in this chapter) and giving its function(s);
*or*
Describe, locate, and give the function(s) of each intracellular structure.

*ILO 5*   Distinguish between plant and animal cells in diagrams or by identifying three structural characteristics of plant cells that are not present in animal cells; identify the structural differences between prokaryotic and eukaryotic cells.

*ILO 6*   List factors that influence cells to reproduce by mitosis.

*ILO 7*   Describe the significance and function of mitosis with respect to maintaining chromosome constancy and efficient cell size.

*ILO 8*   Identify the five stages of a cell's life cycle, and describe the principal events characteristic of each.

*ILO 9*   Define the term *tissue*. Identify and give functions of each group of adult tissues.

*ILO 10*   Define the terms *organ* and *organ system*, and describe these levels of organization.

*ILO 11*   Define all new terms introduced in this chapter.

Order is the hallmark of life. Whether we study an individual organism or the world of life as a whole, we must marvel at its exquisite organization. Several levels of biologic organization may be identified (Figure 3-1). At the bottom of the organizational hierarchy[1] we find the minute and relatively simple atoms. These combine chemically to form molecules, which represent the next level of organization. Molecules, in turn, associate appropriately to form *organelles*, specialized structures within the cell.

Cells are the basic units of biologic organization. A living system may consist of only one cell or of several billion, but even the most complex organism begins life as a single cell, the fertilized egg. In most multicellular organisms, including man, cells associate to form *tissues* (e.g., muscle tissue), and various tissues are arranged into functional structures called *organs* (e.g., the heart). Each major biologic function is performed by a coordinated group of tissues and organs, an *organ system* (e.g., the circulatory system). Working together with far greater precision and complexity than the most complicated man-made machine, the organ systems make up the complex living organism. Organisms interact with one another and their environment, forming ecologic levels of organization known as populations, communities, ecosystems, and finally, the ecosphere. In this chapter we shall focus our attention upon the organization of a complex organism: its organelles, cells, tissues, organs, and organ systems.

## THE CELL—BUILDING BLOCK OF THE BODY

Beginning biology students sometimes wonder why so much time is devoted to the study of cells. An appreciation of the basic role of cells in living organisms and at least a general knowledge of their complex activities is essential to an understanding of life itself. Each new life begins with a fertilized egg and ends when cells cease to function. Like the bricks of a building, cells are the building blocks of the body. The *cell* is the smallest unit of living material capable of carrying on all the activities necessary for life. We might say that it is the smallest structure with a complete metabolism.

### General characteristics of cells

A cell consists of a tiny bit of jelly-like *cytoplasm* surrounded by a *cell membrane*. Almost all cells contain a *nucleus* and other internal structures designed to perform specific duties. Still, cells vary widely in their appearance, as a glance at Figure 3-2 will reveal. The size and shape of a cell are related to the specific job it must perform. Nerve cells possess long cytoplasmic fibers which are used to receive stimuli or to transmit impulses long distances through the body. Fibers of the sciatic nerve, for example, extend from spinal cord to foot. Although such a fiber may be more than a meter (39 inches) long, its diameter is so tiny that it cannot be seen without

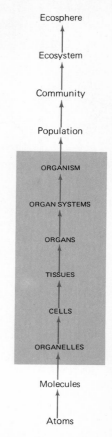

**FIGURE 3-1**
Levels of biologic organization. In this chapter our discussion will center upon the levels of organization shaded in the diagram.

---

[1] Actually even below this level we can point to organization among subatomic particles, the structural units of atoms.

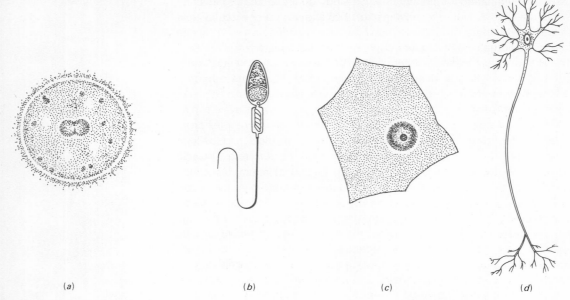

(a)          (b)          (c)          (d)

**FIGURE 3-2**
Cells come in many sizes and shapes. (a) Human fertilized egg cell. (b) Human sperm cell. (c) Liver cell. (d) Nerve cell.

the aid of a microscope. Certain white blood cells in the body resemble the unicellular ameba in their ability to change their shape as they flow along from one place to another destroying invading bacteria. Egg cells are round and large, while sperm cells are equipped with long, whiplike tails used to help propel them to the egg.

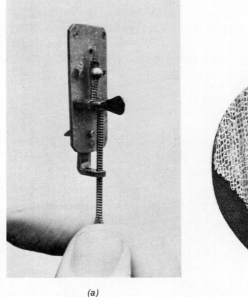

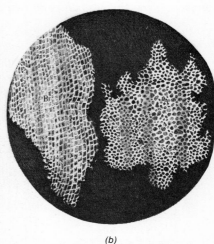

**FIGURE 3-3**
(a) A replica of a very early microscope. (b) The first drawing of cells as sketched by Hooke in 1665. Compared are honeycomb and cork cells.

(a)          (b)

The largest cell known is the ostrich egg, which contains a large quantity of yolk, nourishment for the developing bird. Neither the white of a bird's egg nor its shell is counted as part of the cell, because these structures are nonliving material secreted by the walls of the mother bird's oviduct. The smallest known cells are those of certain microorganisms. An "average" cell measures about 10 micrometers, or 1/2,500 inch, in diameter. That is, if you could line up about 2,500 typical cells end to end, the resulting cellular parade would measure only 1 inch.

### How cells are studied

Cells are so small that one might well wonder how we know so much about what goes on inside them. The biologist's most important tool for studying the internal structure of cells has been the *microscope*. In 1665 Robert Hooke examined a slice of cork with the aid of a crude homemade microscope (Figure 3-3). Because the tiny compartments he saw reminded him of the little rooms, or cells, of a monastery, Hooke called them *cells*. In 1839 two German scientists, Schleiden and Schwann, proposed that all living things are made up of cells and cell products. This became known as the *cell theory*. During the 1800s, scientists studied various cells and observed a variety of intracellular structures. These structures were collectively referred to as *organelles* (little organs) because investigators guessed that

**FIGURE 3-4**
An electron scanning microscope. (McGraw-Hill Encyclopedia of Science and Technology, 3d ed.)

they performed special jobs within the cell, just as our organs perform special jobs within our bodies. The gradual process of describing and understanding cellular structure was aided by the improvement of light microscopes and by the discovery of various methods for preparing and staining cells and their organelles.

Although most organelles were originally identified using an ordinary light microscope, the restricted magnifying and resolving power (ability to perceive fine detail, that is, to show two closely adjacent structures as two different structures, rather than one) of the light microscope hindered investigators in determining the fine structure of these tiny structures. A giant step was taken toward overcoming this limitation when the electron microscope (invented in 1931) became an important research tool in the late 1940s. This instrument has a resolving power more than 10,000 times that of the human eye. In comparison, the light microscope can resolve objects only 500 times better than the unaided human eye. The increased resolving and magnifying power of the electron microscope (Figure 3-4) enabled scientists to study the ultrastructure (fine detail) of each cellular organelle.

More recently organelles have been studied extensively by physical and chemical methods. Cells can be broken apart and then centrifuged (spun) at high speeds to separate the cellular components. Using a variety of such

**FIGURE 3-5**
Electron micrograph of a section through a cartilage cell. (×22,000) *N*, nucleus; *Nu*, nucleolus; *NM*, nuclear membrane; *CM*, cell membrane; *G*, Golgi apparatus; *M*, mitochondrion; *RE*, rough endoplasmic reticulum; *CV*, vacuole in cytoplasm. Black line indicates length of 1 micrometer. (Note: In this and other micrographs the stated magnifications have not been corrected for the reduction made in order to fit them into this book format. The stated values, in this case X22,000, are here consistently *greater than* the actual magnifications by anywhere from 5 to 10 percent.) (Lyle C. Dearden)

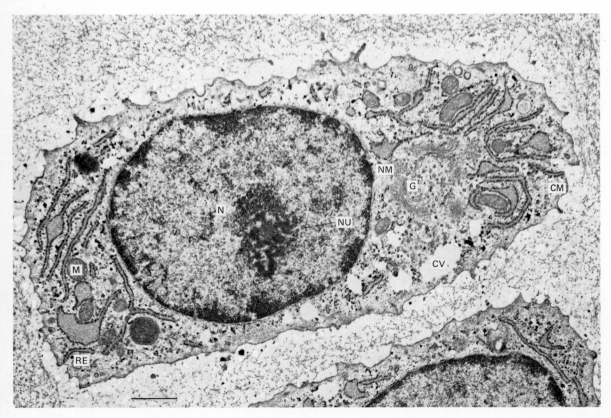

techniques, organelles can be separated from each other on the basis of size or density. Once separated from its surrounding cellular components, a given type of organelle can be analyzed biochemically to determine its composition. Other experiments can determine the function of an organelle by monitoring its workings in a test tube under controlled conditions.

## WITHIN THE CELL—STRUCTURE AND FUNCTION

A cell is a highly organized, amazingly complex structure (Figure 3-5). Every cell has its own control center, internal transportation system, power plants, factories for making needed molecules, packaging plants, and "self-destruct" system. In addition, many cells are specialized to perform still other functions.

### Cytoplasm

Cytoplasm is the jelly-like material of which cells are composed. It is a semisolid, gelatine-like solution of large molecules, notably proteins, and a variety of small molecules. Within the nucleus, the corresponding jelly-like material is referred to as *nucleoplasm*.

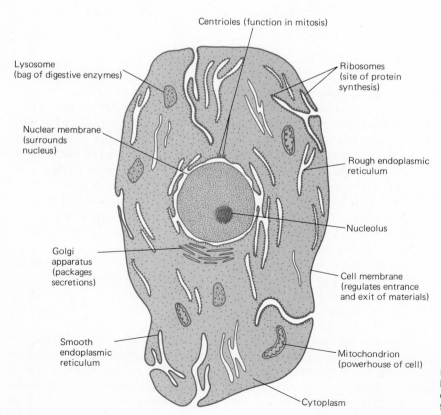

Centrioles (function in mitosis)

Lysosome
(bag of digestive enzymes)

Ribosomes
(site of protein
synthesis)

Nuclear membrane
(surrounds
nucleus)

Rough endoplasmic
reticulum

Nucleolus

Golgi
apparatus
(packages
secretions)

Cell membrane
(regulates entrance
and exit of materials)

Smooth
endoplasmic
reticulum

Mitochondrion
(powerhouse of cell)

Cytoplasm

**FIGURE 3-6**
Diagram of a generalized animal cell. Try to visualize the cell as a three-dimensional structure.

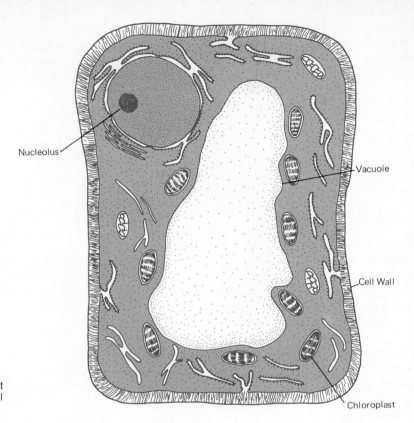

**FIGURE 3-7**
Diagram of a generalized plant cell. Compare with the animal cell in Figure 3-6.

Nucleolus

Vacuole

Cell Wall

Chloroplast

## The cell membrane

Every cell is surrounded by a delicate limiting membrane, the *cell membrane* (also called *plasma membrane*) (Figure 3-8). Its most obvious function is to prevent the contents of the cell from spilling out. It also regulates the passage of materials between the cell and its environment. Because it permits the passage of certain types of molecules while restricting the passage of others, the cell membrane is described as *selectively permeable.* Responding to varying conditions or needs, the cell membrane may present a barrier to a particular substance at one time and then actively promote passage of the same substance at another time. By regulating chemical traffic in this way the cell controls its own composition.

Viewed with an electron microscope the cell membrane appears to have three layers. Any membrane with such a three-layered structure is commonly referred to as a *unit membrane*. The precise arrangement of molecules in the cell membrane appears to be important in regulating the passage of materials into or out of the cell.

In animal cells the cell membrane is coated externally by a thin layer of protein and carbohydrate which is thought to be an adhesive that holds the cells of a tissue together. This cell coat helps cells recognize one another, and so determines which cells will associate with one another to form tissues. Plant cells are surrounded by a much thicker coating called the *cell wall.* Secreted by the cell itself, the cell wall is composed of a carbohydrate, most often cellulose. It forms a tough, protective covering that makes plant cells quite rigid. The cell wall does not present a barrier to materials passing in and out of the cell because it is studded with tiny pores.

## Endoplasmic reticulum and ribosomes

A system of internal membranes extends throughout the cytoplasm of many cells, forming an extensive complex of branching tubules, the endoplasmic reticulum (ER). The ER is continuous with the membrane which surrounds the nucleus and perhaps also with the cell membrane. Resembling an intricate tunnel system, the ER serves as a system for transporting materials from one part of the cell to another or from the cell to the environment. By dividing the cytoplasm into compartments, the ER may also separate one chemical activity from another and help each to maintain the proper relation to the other. It is also likely that expanded portions of ER may be temporary storage sacs for certain substances.

**FIGURE 3-8**
Electron micrograph of the cell membrane. (×135,000) Line indicates length of 1 micrometer. *CM*, cell membrane; *D*, desmosome. (A desmosome is a structural attachment between two adjoining cells.) (Dr. Lyle C. Dearden)

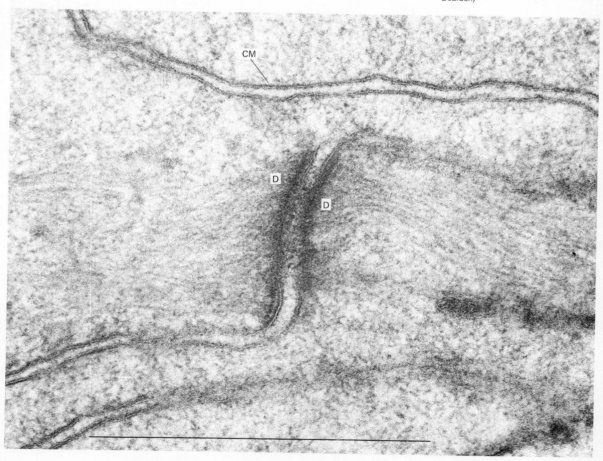

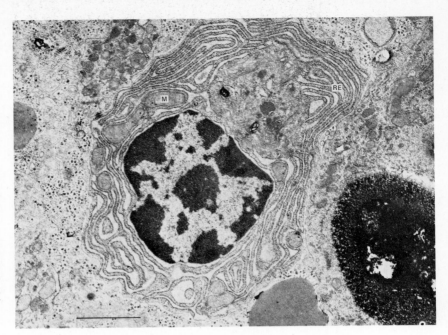

In an *electron micrograph*, a picture taken with an electron microscope, the ER may appear discontinuous (Figure 3-9). This is because such pictures are taken of a thin slice through the cell, while the ER is continuous in three dimensions.

There are two types of ER, smooth and rough. Smooth ER, so called because it lacks granules, may function specifically in the manufacture of certain hormones. Rough ER has a granular appearance because of tiny bodies called *ribosomes* along its outer walls. Ribosomes function as manufacturing plants where proteins are assembled. Recent studies have shown that rough ER is most extensive in cells that are actively engaged in protein synthesis. Rough and smooth ER are continuous, and each type can be converted into the other type according to the needs of the cell.

### The Golgi apparatus

Named after the Italian investigator Camillo Golgi who first described it in 1898, the *Golgi apparatus* is composed of stacks of platelike membranes that usually surround one end of the nucleus (Figure 3-10). Some investigators have claimed that the Golgi apparatus is a specialized portion of smooth ER; others argue that it is an independent organelle.

The Golgi apparatus functions as a packaging plant. Some proteins manufactured in the cell are sent to the Golgi apparatus, which adds a carbohydrate component, collects a large number of molecules, and then encloses them in a membrane. In cells which secrete substances, such as gland cells, these little packages pass out of the cell where they release their contents. Digestive enzymes are packaged and released in this manner by cells lining the intestine. In cells specialized to secrete substances, the Golgi apparatus is especially prominent, but this organelle performs an important function in nonsecreting cells as well, that of packaging intracellular digestive enzymes in the form of *lysosomes*.

## Lysosomes

Intracellular digestive enzymes are manufactured along ribosomes and transported by way of the endoplasmic reticulum to the Golgi apparatus. There, the enzymes are packaged, i.e., enclosed in membranous sacs which apparently bud off from the Golgi apparatus itself. These tiny bags of digestive enzymes, called *lysosomes*, are then dispersed throughout the cytoplasm. Lysosomes break down worn out organelles or other forms of cellular debris by means of these potent enzymes. When a white blood cell ingests a bac-

**FIGURE 3-10**
Electron micrograph showing the Golgi apparatus in a lymph cell. (×61,500) Black line indicates length of 1 micrometer. *GL*, Golgi lamellae; *GV*, Golgi vesicle. Can you identify a mitochondrion in this picture (cf. Figure 3-11)? Where is the nucleus? (Dr. Lyle C. Dearden)

terium, it is the lysosomes that provide the means for their destruction. Some cells such as amebas ingest food by packaging it with membrane pinched off from the cell membrane, thus forming a *food vacuole*. The food material is then digested by enzymes poured into the food vacuole by lysosomes.

When a cell dies, the lysosome membranes break, releasing lysosome enzymes into the cytoplasm. These enzymes actually digest the cell itself. This "self-destruct" system accounts for the rapid deterioration of cells following death.

Recent studies have suggested that cell damage and aging may be related to leaky lysosomes. Any digestive enzymes escaping from the lysosomes into the cytoplasm of the cell could injure intracellular structures. Damage done to cartilage cells by enzymes that have leaked out from lysosomes is thought to contribute to arthritis. Discovered in the 1950s, lysosomes are a relatively new structure to biologists; much research remains to be done before their whole story will be known.

### Mitochondria

The power plants of the cell are the mitochondria, the sites of cellular respiration (Figure 3-11). More than 1,000 mitochondria have been counted in a single liver cell, but the number varies among different cell types. In general they are more abundant in cells which are very active metabolically. Each mitochondrion is surrounded by two membranes. Folds of the inner membrane, called *cristae*, project into the interior of the mitochondrion. Cristae serve to increase the available surface, and some of the enzymes needed for cellular respiration are organized along these folds. Other enzymes are dissolved in the inner space (matrix) of the mitochondrion. Each mitochondrion has a small amount of DNA, enough to code for about 15 proteins, and also contains ribosomes. Some proteins are indeed synthesized within these organelles.

**FIGURE 3-11**
(*a*) Diagram of a mitochondrion, cut open so as to show the cristae. (*b*) Electron micrograph showing a cross section through a mitochondrion. (×96,000) Black line indicates length of 1 micrometer. *M*, mitochondrion; *C*, cristae. (Dr. Lyle C. Dearden)

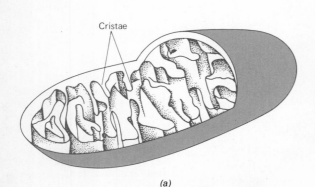

Cristae

(a)

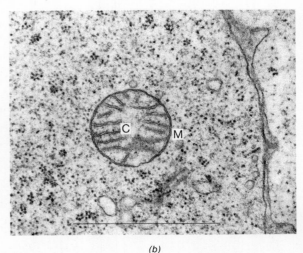

(b)

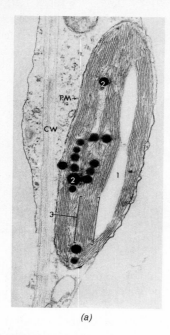

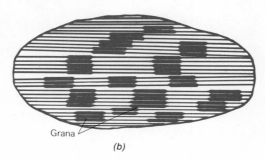

Grana

(a)                    (b)

**FIGURE 3-12**
(a) Electron micrograph of a
chloroplast from an orchid cell.
(×72,000) 1, starch grain; 2, droplets; 3, lamellae of grana; CW,
cell wall; PM, membrane of
plastid. (J. Arditti) (b) Diagram of a
chloroplast. (Don Moores)

## Plastids

Structures known as *plastids*, which produce or store food materials, are
found in almost all plant cells. *Chloroplasts* are the most important type, for
they contain the green pigment chlorophyll which traps light energy for photosynthesis. Structurally somewhat similar to mitochondria, each chloroplast
is surrounded by a double membrane and contains tiny stacks of membranes called *grana* (Figure 3-12). Enzymes needed for photosynthesis are
found between these membranes. Like mitochondria, chloroplasts contain
DNA and ribosomes and manufacture proteins. While a one-celled alga may
have only a single large chloroplast, cells of complex plants often possess
about thirty of these tiny factories. Some plastids contain pigments other
than chlorophyll, while one type (leukoplasts) which stores excess food in
the form of starch, contains no pigment.

## Other cytoplasmic structures

Most cells contain paired *centrioles* which lie in the cytoplasm close to the
nucleus. Each centriole is a cylinder whose wall is composed of nine
bundles of *microtubules*. Members of a pair are usually located at right
angles to each other, a characteristic which appears to be related to the
method by which centrioles duplicate themselves. Centrioles function in cell
reproduction, which will be described presently.

*Cilia* are tiny hairlike organelles which project from the surfaces of many
types of cells (Figure 3-13). An entire group of one-celled organisms (class
Ciliata) such as *Paramecium* swim about by rhythmically beating their cilia.
In complex organisms, cells lining passageways often possess cilia that
beat in a coordinated fashion to set up a current which moves materials.

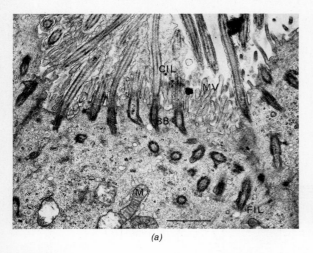

(a)

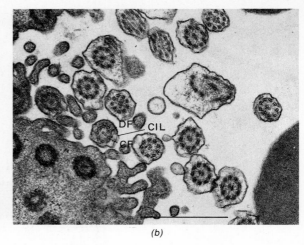

(b)

**FIGURE 3-13**

Electron micrographs of cilia. (a) Cilia of cells lining a respiratory passageway. (×37,400) Black line indicates length of 1 micrometer. *CIL,* cilium; *BB,* basal body; *M,* mitochondrion; *MV,* microvilli; *FIL,* filament. Microvilli are tiny projections of the cytoplasm found along the free surfaces of epithelial cells. (b) Cross section through several cilia. (×72,000) *CIL,* cilium; *CF,* central microtubules; *DF,* double microtubules. (Dr. Lyle C. Dearden)

Each of the ciliated cells lining the human trachea (windpipe) has about 270 of these tiny organelles. They move a film of mucus containing trapped dirt particles away from the lungs.

*Flagella* also function in locomotion. A flagellum is much longer than a cilium and moves with a wavy motion. Flagellates are one-celled organisms which depend upon flagella for swimming. Sperm cells of most organisms, including man, possess flagella which help propel them toward the egg.

At the base of each cilium or flagellum is a *basal body*, composed of microtubules, that is very similar to a centriole, and in fact is produced from a centriole in at least some types of cells. Each cilium or flagellum grows out from its basal body. The shaft of a cilium or flagellum consists of nine paired microtubules surrounding two central microtubules, a 9:2 arrangement of microtubules (Figure 3-13).

**FIGURE 3-14**

Formation of a large central vacuole in a plant cell. (a) Many small vacuoles in the cytoplasm are beginning to fuse to form (b), one large central vacuole in an immature cell from a developing root tip. (McGraw-Hill Encyclopedia of Science and Technology, 3d ed.)

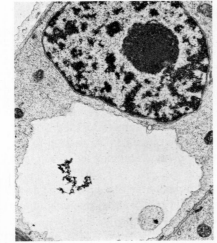

(a)

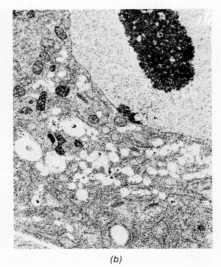

(b)

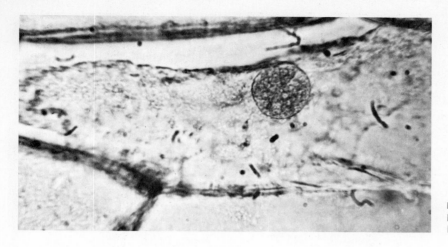

**FIGURE 3-15**
Photomicrograph of nucleus of
living onion cell.

A variety of other structures may be scattered throughout the cytoplasm. In plant cells large *water vacuoles*, membranous sacs containing water and dissolved substances, take up much of the volume of the cell (Figure 3-14). In some one-celled organisms water vacuoles are able to contract, expelling excess water from the cell.

## The nucleus

Of all the structures present within the cell the nucleus is most prominent (Figure 3-15). Perhaps for this reason investigators guessed that the nucleus served as the control center for the cell even before techniques to prove that hypothesis were available. During recent years all kinds of experiments have been performed which have confirmed the vital role of the nucleus. In one such experiment the investigator surgically removed the nucleus from a living ameba. The enucleated ameba was unable to eat or grow, and died after a few days. However, if after a day or two an enucleated ameba was given a new nucleus, it made a complete recovery. These and other experiments show that the nucleus is essential to the well-being of the cell.

The nucleus is surrounded by a double *nuclear envelope* which is continuous with the ER. Tiny pores in the nuclear membrane appear to permit the passage of molecules between the cytoplasm and the nucleus. Almost half the large molecules made in a cell are manufactured in the nucleus. Dispersed in the nucleoplasm are the *chromosomes* (Figure 3-16) and the *nucleolus*.

Except when cells are actually dividing, a threadlike network called *chromatin material* extends throughout the nucleoplasm. When a cell begins the process of reproduction (mitosis), the chromatin condenses into discrete rod-shaped bodies, the chromosomes. Each chromosome contains several hundred *genes* arranged in a specific linear order, while the genes, in turn, are composed of the nucleic acid *DNA*. Chemically coded within the DNA of the genes are instructions for producing all proteins needed by the cell. These proteins, more than anything else, determine what the cell will

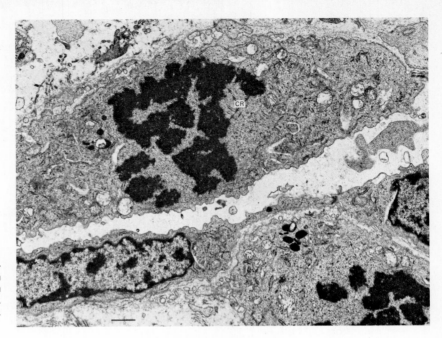

**FIGURE 3-16**
Chromatin material has condensed to form chromosomes in this epithelial cell, which is in the process of dividing. (×13,650) *CR*, chromosomes. (Dr. Lyle C. Dearden)

look like and what functions it will perform. The chromosomes serve as a chemical cookbook for the cell, while the genes might be compared to the individual recipes.

Each species has a characteristic number of chromosomes, and that number is present in every cell of the organism. In a man each body cell contains an identical set of 46 chromosomes containing thousands of genes. It is important to understand that each cell of an individual contains the same set of genetic instructions (DNA). Before cell division can take place DNA duplicates itself and two complete sets of chromosomes are formed. During cell division one set of chromosomes is distributed to each new cell.

When the cell is not dividing the chromatin stretches out in a set of long, very thin threads. Why is this so? Only when chromosomes are greatly elongated can genes actively function. We may compare the condensed chromosome to a closed cookbook. The recipes are all inside but the pages cannot be read. When the book is open the instructions can be followed.

The nucleus may contain one or more large, prominent *nucleoli* (singular, *nucleolus*). Each nucleolus contains a large quantity of the nucleic acid RNA. Recently studies have indicated that nucleoli are the factories where the RNA components of the ribosomes are assembled.

## DIFFERENCES IN SOME MAJOR CELL TYPES

Even in their diversity the thousands of different kinds of cells share many common features of structure and function. There are, however, some major differences which we should note.

## Plant and animal cells

At several points in our discussion of cell structure you may have noticed that plant and animal cells differ in certain features (Figures 3-6 and 3-7). First, while all cells are limited by cell membranes, plant cells are surrounded by rigid cell walls. Second, plant cells contain plastids, such as chloroplasts, which are absent in animal cells. Third, plant cells have large water vacuoles, and fourth, centrioles are almost always absent in the cells of complex plants.

## Prokaryotic and eukaryotic cells

Our discussion of cellular structure has centered upon cells characteristic of multicellular organisms and of some single-celled ones. Distinguished by the presence of a definite nucleus, they are called *eukaryotic cells*. In contrast the *prokaryotic* cells of blue-green algae and bacteria do not have a clearly defined nucleus. Prokaryotic cells do have DNA, which sometimes is condensed in the cell center, giving it the appearance of a nucleus, but no nuclear membrane is present. Prokaryotic cells also lack complex internal membranous organelles such as mitochondria and endoplasmic reticulum. The functions of these organelles may be assumed by relatively simple infoldings of the cell membrane. In prokaryotic cells, flagella are single simple strands, unlike the 9:2 organization of microtubules characteristic of eukaryotic flagella.

## CELLULAR REPRODUCTION—MITOSIS

Mitosis is the process by which most cells multiply. Although a human being begins life as a single cell, by the time of birth he is composed of billions of cells. Before we discuss how this happens, perhaps we should ask why.

## Why cells divide

Why is a man or an elephant made of billions of cells? Why doesn't a fertilized egg simply grow in size to produce a man composed of one giant cell? Why are one-celled organisms small? Have you ever stopped to wonder why we do not see amebas as large as whales slithering around? In fact amebas are so small that even the larger ones are barely visible to the unaided eye.

Probably the main reason that cells do not get large is that it is inefficient for them to do so. All materials must pass in or out of the cell through the cell membrane, so its size in comparison with the size of the rest of the cell is critical. As a cell increases in size its volume increases at a greater rate than its surface. (The surface of a sphere increases as the square of its radius, while the volume increases as the cube of its radius.) Thus as a cell grows, its cell membrane (the cell surface) becomes unable to provide sufficient oxygen and nutrients for all regions of the cell. Wastes produced within the cell must move longer distances to reach the cell membrane and exit from the cell. Too great an increase would threaten the well-being of the cell. There seems also to be a limit to the amount of cytoplasm that a nucleus can control. In fact very large cells may have more than one nucleus.

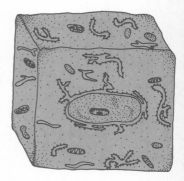

**FIGURE 3-17**
Four small cells possess a much greater surface area (cell membrane) in relationship to their total volume than one large cell. If this concept is difficult for you to understand think of it in terms of milk cartons. Although a gallon carton contains the same amount of milk as four quart cartons, its surface area is far less.

When its size approaches the limits of efficiency, a cell divides to form two cells. Each new cell is half the size of the mother cell, but the relative size of both the cell and nuclear membranes is greatly increased in proportion to the volume of the cell.

In unicellular organisms cell division results in the production of two new individuals, while in multicellular organisms the two new cells may remain associated to form a part of the whole organism. The number of cells, not their individual size, is responsible for the different sizes of various organisms. The cells of an earthworm, a man, or an elephant correspond in size; the elephant is larger because its genes are programmed to provide for a larger number of cells.

In one-celled organisms the single cell must carry on all life activities. An organism composed of many cells can assign specific tasks to different cells. Its cells become specialized to perform certain functions, so that the labor is divided among different groups of cells. Such an organism can be highly proficient at performing a wide variety of activities.

Still other factors are pertinent to cell reproduction. Instructions in the genes of the cell probably determine the normal size, as well as shape, of any given cell. Certain hormones stimulate the rate of cellular metabolism, which in turn influences growth. In many organisms cell reproduction proceeds rapidly during the early years of life, the growth period. Later the process is restricted to replacement of worn-out or damaged cells. In adult human beings cell reproduction is greatest among cells that produce blood cells and among cells lining the intestine.

### The process of mitosis

We have established that development and growth of an organism require the production of new cells. But how do these cells come to be? In 1855 the biologist Rudolph Virchow suggested that cells arise only from already existing cells. This was an important extension of the cell theory, and the beginning of the realization that life is a continuous process. Cells do not arise spontaneously from nonliving substances, but are produced only as offspring of living cells. Every living cell, then, has a lengthy family tree through which its cellular ancestors may be traced back through the ages. Every cell in your body can be traced back to the zygote, which was produced by the union of parental egg and sperm cells. Each reproductive cell can in turn be traced back to the zygote of its producer. And so on, back through generations to the ultimate origin of living things.

The process by which most cells divide to form two new cells is called *mitosis*. The most significant feature of mitosis is that it provides for the precise duplication and distribution of chromosomes so that each new cell contains the identical number and kinds of chromosomes present in the original mother cell. This means that when a zygote divides to form two cells *each* new cell contains all the genetic information present in the zygote. After hundreds of mitoses every cell in the body contains a copy of the same genetic information.

The life history of the cell may be divided into five stages, or *phases*.

These phases are not completely distinct from one another, but blend almost imperceptibly from one into the next. (As you read the following detailed description of mitosis, refer to Figures 3-18 and 3-19.)

*Interphase.* Most of the life of the cell is spent in interphase actively synthesizing needed materials and growing. The term *interphase* means "between phases" and describes the state of the cell between reproductive events. During interphase the chromosomes are greatly elongated, a condition rendering them so thin that they are not visible with the ordinary light microscope. Only in places where they are coiled can they be seen as the dark granular patches referred to as chromatin material.

Two important events take place during interphase in preparation for mitosis: (1) duplication of the centrioles (in animal cells) and (2) duplication of the chromosomes. Chromosome duplication begins with the precise synthesis of matching strands of DNA from raw materials in the cell. Next, each chromosome separates from its new twin to form a pair of chromatids. The

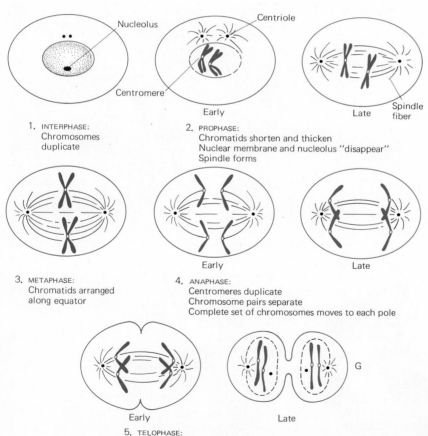

1. INTERPHASE:
   Chromosomes
   duplicate

2. PROPHASE:
   Chromatids shorten and thicken
   Nuclear membrane and nucleolus "disappear"
   Spindle forms

3. METAPHASE:
   Chromatids arranged
   along equator

4. ANAPHASE:
   Centromeres duplicate
   Chromosome pairs separate
   Complete set of chromosomes moves to each pole

5. TELOPHASE:
   Set of chromosomes at each end of cell
   Nucleus reorganizes
   Cytoplasm cleaves

**FIGURE 3-18**
Stages in the life cycle of a cell. The cell shown has a normal chromosome number of two. This diagram depicts animal mitosis. Compare with the photographs in Figure 3-19 showing plant cell mitosis.

term *chromatid* is used because the separation is not yet complete; twin chromatids remain attached by a special structure called the *centromere*. Chromatids begin to condense by coiling, and by the time mitosis actually begins they have become so much shorter and thicker that the light microscope reveals them as discrete rod-shaped bodies.

*Prophase.* During this first stage of mitosis each chromatid becomes tightly coiled, forming a helix which is visible as a dark rod-shaped body under the light microscope. Several other dramatic events take place. The nuclear membrane dissolves so that the nuclear contents are free to mingle with the cytoplasm. The nucleolus becomes disorganized. At about the same time, the centrioles begin to migrate toward opposite ends of the cell. Microtubules form and appear to radiate from the centrioles into the former nuclear region where the chromosomes remain gathered.

*Metaphase.* The second phase of mitosis, metaphase, is reached when one pair of centrioles is present at each end, or pole, of the cell. Centromeres of the chromatids attach to microtubules radiating from the centrioles. The overall appearance is that of a *spindle* made of fibers and marked centrally

**FIGURE 3-19**

Photomicrographs of cells in an onion root tip. (a) In the tissue shown many cells are undergoing mitosis. Can you identify each stage? (×500) (b) Cell in center of photo is in late prophase. (×1,600) (c) Cell in center of photo is in metaphase. (×1,600) (d) Cell in center of photo is in anaphase. (×1,600) (e) Cell in center of photo is in early telophase. (Roy M. Allen)

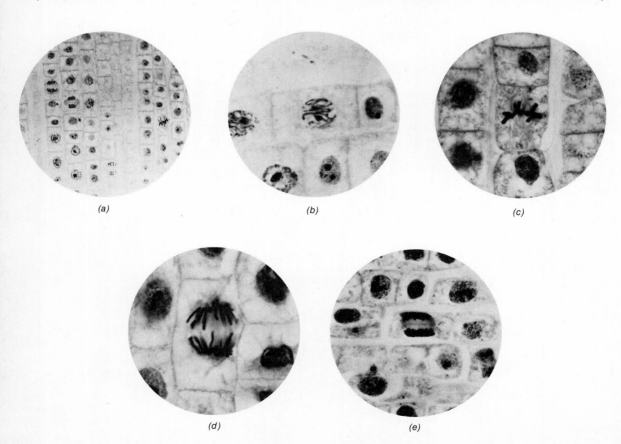

(a)

(b)

(c)

(d)

(e)

by the chromatids. Indeed the chromatids are seen to have arranged themselves along the equator of the cell. Each chromatid is now completely coiled so that it appears quite thick and discrete. In fact, chromosomes in metaphase can be seen more clearly than those at any other stage; they are sometimes photographed and studied medically to determine possible chromosome abnormality.

*Anaphase.* During the first part of anaphase the chromatids separate at the centromeres, so that each chromatid becomes a complete, independent chromosome. The next important process is that of equally distributing the two sets of chromosomes. Immediately upon becoming independent of each other, the two chromosomes that had been paired chromatids seem to repel each other and begin to move apart. One theory holds that the spindle fiber to which each is attached in some way pulls the chromosome toward its centriole at the pole of the cell. The overall result is that one complete set of chromosomes begins to move toward one end of the cell while the other complete set moves towards the opposite pole.

*Telophase.* With the arrival of a complete set of chromosomes at each end of the cell, telophase begins. At this time the cell begins to constrict, generally around the center. Slowly the cell becomes more and more constricted at its center until it has completely divided to form two daughter cells. As the cell constricts, or *cleaves*, the events taking place in the region of each set of chromosomes reestablish interphase conditions. The chromosomes begin to elongate by uncoiling. As they become longer, they become thinner and less visible. Around each set of chromosomes a nuclear membrane forms, and in association with specific regions of certain chromosomes, nucleoli are organized. Spindle fibers are no longer visible. Mitosis is complete, and each new cell enters a period of interphase, thus beginning a new life cycle.

Mitosis in the cells of complex plants differs from the above description in two ways: (1) the spindle forms in prophase without the presence of centrioles; (2) the actual division of the cell in telophase begins with the formation of a cellulose partition, the *cell plate*, in the middle of the cell. The beginning of the new cell wall, the cell plate grows outward to join with the old cell wall, thus separating the daughter cells from each other.

## Main events of mitosis

In summary, the main events of mitosis are (1) the duplication of the chromosomes, (2) disappearance of nucleus and nucleolus, (3) formation of the spindle, (4) separation and migration of the chromosome sets to opposite ends of the cell, (5) restoration of the original condition of the nucleus, and (6) cleavage of the cell into two complete cells. Though mitosis generally includes the division of the cytoplasm so that two cells are formed, there are instances when mitosis takes place without accompanying cytoplasmic division. For this reason some biologists restrict the meaning of mitosis to nuclear division and use the term *cytokinesis* to refer to cytoplasmic division.

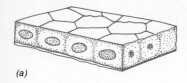
*(a)*

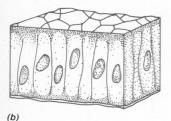
*(b)*

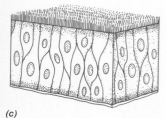
*(c)*

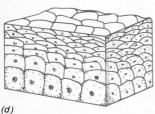
*(d)*

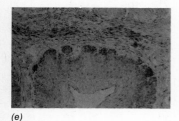
*(e)*

**FIGURE 3-20**
Epithelial tissue. (*a*) Cuboidal epithelium (as found lining the kidney tubules). (*b*) Columnar epithelium (the type that lines the intestine). (*c*) Pseudostratified columnar epithelium with cilia (the type that lines the trachea). (*d*) Stratified squamous epithelium (skin). (*e*) Photomicrograph of stratified squamous epithelium. (Carolina Biological Supply Company)

### Timing of mitosis

In certain cells mitosis takes only a few minutes; in others several hours may be required. Prophase and telophase are the longest stages of cell division, but interphase represents the longest period in the life of most cells.

Though certain cells are capable of dividing every 20 minutes, other cells never divide at all. The highly specialized nerve cells and muscle cells of the human body are unable to reproduce. We are born with all the nerve cells and muscle cells we will ever have. As they wear out or are injured or destroyed, we are left with fewer and fewer. At the other extreme, it has been estimated that as many as 10 million red blood cells can be produced in the body by mitosis every second.

## FROM SINGLE CELL TO COMPLEX ORGANISM

During development the fertilized egg divides by mitosis again and again, giving rise to the billions of cells that make up a complex organism. Cells associate with one another to form functional groups. They also become different from one another (cellular differentiation), specializing to perform specific functions.

### Tissues

A tissue is a group of similar cells associated for the purpose of carrying on specific functions. In man and other complex animals four main groups of differentiated, or specialized, tissues are recognized: *epithelial tissue*, *connective tissue*, *muscle tissue*, and *nerve tissue*.

*Epithelial tissue.* Epithelial tissue (Figure 3-20) consists of cells fitted closely together and arranged in one or more layers. This tissue is specialized to cover body surfaces, so one of its surfaces is always free (that is, not covered by other tissue). As skin, epithelial tissue covers the outside of the body; as lining for body tubes, it covers the inside surfaces of digestive and respiratory tubes and blood vessels. Several different types of epithelial tissue can be distinguished by the shape of component cells, their arrangement, and the number of cell layers in the tissue.

As a covering or lining, epithelial tissue protects the body. Epithelial tissue of the skin protects us from mechanical injury, from invading bacteria, and from excessive water loss. Epithelial tissue may also be specialized to perform the important functions of *secretion*, *excretion*, and *absorption*. In glands epithelial tissue is specialized to secrete specific substances, such as lubricating substances, enzymes, or hormones (substances that regulate metabolic activities). Epithelial tissues in the kidney excrete waste materials from the body and absorb needed materials from the urine. Certain of the epithelial cells lining the intestine absorb nutrients.

Because many epithelial tissues bear the brunt of the constant wear and tear to which the body is subjected, their cells are continuously damaged. As cells die and slough off they are replaced by new cells produced by mitosis.

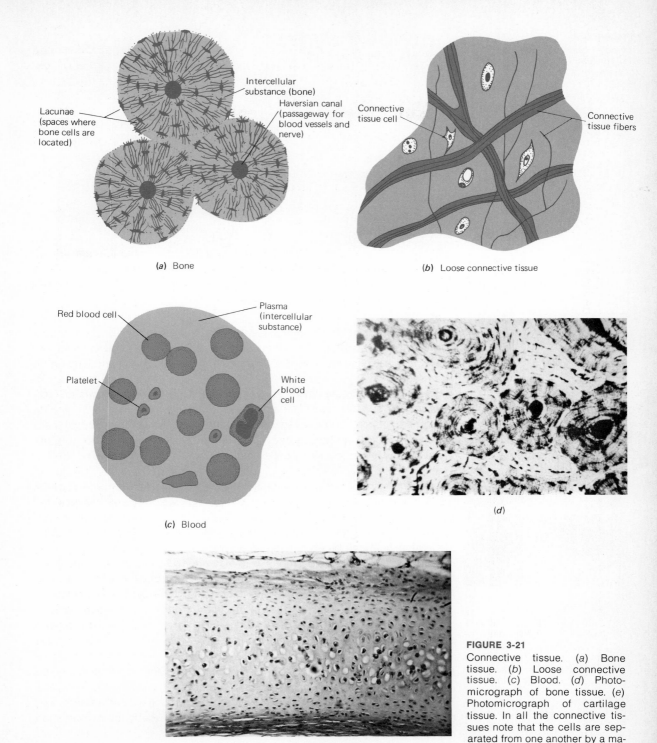

(a) Bone

(b) Loose connective tissue

Intercellular substance (bone)

Haversian canal (passageway for blood vessels and nerve)

Lacunae (spaces where bone cells are located)

Connective tissue cell

Connective tissue fibers

(c) Blood

Red blood cell

Plasma (intercellular substance)

Platelet

White blood cell

(d)

(e)

**FIGURE 3-21**
Connective tissue. (a) Bone tissue. (b) Loose connective tissue. (c) Blood. (d) Photomicrograph of bone tissue. (e) Photomicrograph of cartilage tissue. In all the connective tissues note that the cells are separated from one another by a matrix. Compare to epithelial tissue.

**FIGURE 3-22**
Muscle tissue. (*a*) Striated muscle tissue. (*b*) Smooth muscle
tissue. (*c*) Cardiac muscle tissue. (*d*) Photomicrograph of
striated muscle tissue.

(*a*) Striated muscle tissue

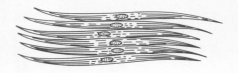

(*b*) Smooth muscle tissue

(*c*) Cardiac muscle tissue

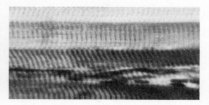

(*d*) Striated muscle

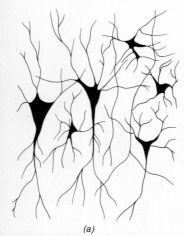

(*a*)

**FIGURE 3-23**
Nerve tissue. (*a*) Diagram of
nerve tissue. (*b*) Photomicrograph of nerve tissue.

(*b*)

*Connective tissue.* Connective tissue (Figure 3-21) supports and connects
all the structures of the body. Bone, cartilage, dense and loose connective
tissue, and fluid tissues (blood and lymph) are some of the types of connective tissue.

Certain cells of connective tissue produce nonliving substances called
*intercellular substances* which lie between adjacent cells. The intercellular
substances of bone and cartilage confer upon these tissues the strength that
enables them to support the body. The cell products make up the bulk of
these tissues, while the living cells comprise only a minor portion.

*Muscle tissue.* In most animals muscle is the most abundant tissue (Figure
3-22). It accounts for nearly two-thirds of the body weight in man. Muscle is
specialized for contraction and is the basis for almost all movement in
animals. There are three types of muscle: *skeletal*, *cardiac*, and *smooth*.
Skeletal muscle can be contracted voluntarily. Characterized by a pattern of
light and dark stripes, or striations, it is also referred to as *striated muscle*.
*Cardiac muscle*, the main tissue of the heart, is a kind of striated muscle that
is not usually under voluntary control. The third type of muscle, *smooth
muscle*, is also involuntary. Found in the body organs, its contractions are
responsible for such internal movements as the waves of contraction called
*peristalsis* that move food through the digestive tract.

*Nerve tissue.* Specialized for receiving stimuli and for transmitting nerve
impulses, nerve tissue (Figure 3-23) is perhaps the most highly specialized tissue of the body. Each nerve cell, or *neuron*, consists of a *cell body*
containing the nucleus, and elongated extensions of the cytoplasm called
*nerve fibers*.

## Organs

Different types of tissues cooperating with one another to perform a particular biologic function constitute an *organ*. The brain, heart, stomach, and eye are examples. Although an organ may be composed primarily of one type of tissue, other types are needed to provide support, protection, blood supply, and conduction of nerve impulses. For example, the brain is composed mainly of nerve tissue, but epithelial and connective tissues protect it from injury and transport vital nutrients and oxygen to it. The intestine is lined with epithelium that secretes digestive enzymes and absorbs nutrients. Layers of muscle make up the bulk of its wall and contract in waves, moving food along the digestive tube. Nerve tissue places the digestive tube in communication with other parts of the body such as the brain. Connective tissue supplies the intestine with blood and holds all its tissues together, as well as holding the tube in place in the body.

## Organ systems and organisms

Various tissues and organs coordinate their activities to perform a specialized set of functions. Such an organized complex of structures is termed an *organ system.* In the human body, as in other complex animals, we can identify 10 organ systems (Figure 3-24), each responsible for a specific group of activities. Working together, these 10 organ systems make up the complex *organism.*

**CHAPTER SUMMARY**

1  The cell is the basic unit of biologic organization.
2  Biologists have learned much about cellular structure by studying cells with light and electron microscopes, and with chemical techniques.
3  Most cells have elaborate organelles which perform specific intracellular duties.
   (*a*) The cell membrane regulates molecular traffic entering and leaving the cell.
   (*b*) The endoplasmic reticulum (ER) plays a role in the transport and storage of materials within the cell.
   (*c*) Ribosomes serve as factories where proteins are manufactured.
   (*d*) The Golgi apparatus is a packaging plant for lysosomes and for cellular secretions.
   (*e*) Lysosomes function in intracellular digestion, and form the "self-destruct" system of the cell.
   (*f*) Mitochondria are the power plants of the cell.
   (*g*) The nucleus is the control center for the cell, and houses the chromosomes and nucleolus.
   (*h*) Other common intracellular structures are plastids, cilia and flagella, centrioles, and vacuoles.
4  Plant cells differ from animal cells in that they possess a cell wall, plastids, and large water vacuoles. Prokaryotic cells of bacteria and blue-green algae lack a nuclear membrane and other membranous organelles.
5  Eukaryotic cells multiply by mitosis, a process ensuring that each new cell will contain a chromosome complement identical to that of the mother cell.

**6** When cells grow to a certain characteristic size determined by their genes, they reproduce by mitosis. The cell membrane to cell volume ratio is an important factor in limiting cell size.

**7** Mitosis may be divided into four stages: prophase, metaphase, anaphase, and telophase.

**8** Similar cells become organized into specialized groups called tissues. Various tissues become organized to form the organs of the body.

**9** Ten organ systems may be identified in the human body. Working together, these constitute the complete organism.

**FOR FURTHER READING**   Asimov, Isaac: *Human Body*, New American Library, New York, 1964. A popular introduction to human structure and function.

Barry, J. M.: *Molecular Biology*: *Genes and the Chemical Control of Living Cells*, Concepts of Modern Biology Series, Prentice-Hall, Englewood Cliffs, N.J., 1964. For the

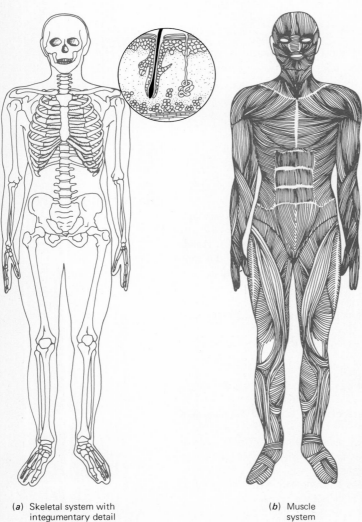

(*a*) Skeletal system with integumentary detail

(*b*) Muscle system

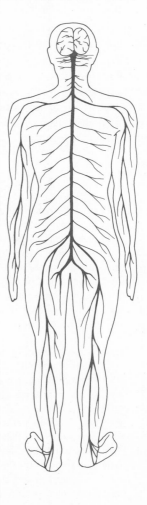

(*c*) Nervous system

student who wants to know more about genes and how they control the activities of the cell.

Gillie, Oliver: *The Living Cell,* Funk & Wagnalls, New York, 1971. An excellent and readable introduction to cells and cell function.

*The Living Cell,* readings from *Scientific American,* Freeman, San Francisco, 1965. A collection of 24 articles on the cell.

McElroy, William D., and Carl P. Swanson: *Modern Cell Biology,* Foundations of Biology Program, part 2, Prentice-Hall, Englewood Cliffs, N.J., 1968. An easily comprehended and excellent introduction to the cell and to biology.

Moner, John G.: *Cells, Their Structure and Function,* Concepts of Biology Series, Wm. C. Brown Company Publishers, Dubuque, Iowa, 1972. A paperback which delves into the structure and chemistry of cells and cell processes.

"The Worlds within Us," *Life,* Jan. 9, 1970. An impressive group of micrographs of various cells and tissues of the human body.

**FIGURE 3-24**
The organ systems of man. (Only the female reproductive system is shown in part *e.* See Chapter 12 for more detailed diagrams of both male and female systems.)

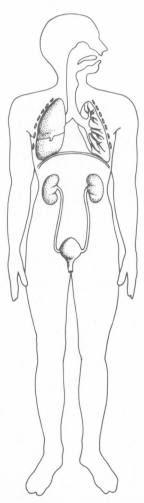

(*d*) Respiratory and excretory systems

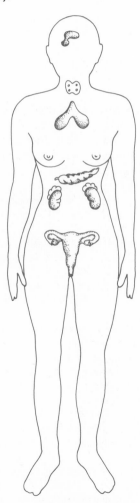

(*e*) Reproductive and endocrine systems

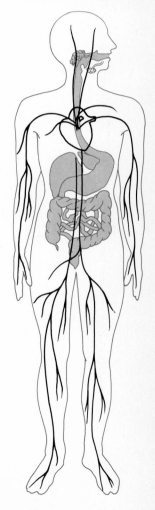

(*f*) Digestive and circulatory systems

# 4 NUTRITION

**Chapter learning objective.** The student must be able to discuss the problems involved in attaining an adequate nutritional balance for all human beings, with respect to nutrients needed, effects of specific deficiencies, patterns of energy transfer through food chains, and population problems and solutions.

*ILO 1*  Identify each of the basic nutrients; describe its general chemical structure or composition and its source. Give its role in the body and the effect of a deficiency in the diet.

*ILO 2*  Write the laws of thermodynamics, and apply them to living systems.

*ILO 3*  Identify the various modes of nutrition encountered among living organisms, and give examples of each type.

*ILO 4*  Describe the principal steps of the photosynthetic process, and be able to write a summary equation for photosynthesis, explaining the origin and fate of each substance involved.

*ILO 5*  Trace the transfer of energy through the biologic world beginning with photosynthesis, and describing the pattern of energy use by an individual organism, as well as its transfer up the trophic levels of a food pyramid.

*ILO 6*  Diagram a simple food chain, or analyze one already drawn. Identify levels of greatest biomass and energy, and identify the types of organisms found at each level (e.g., herbivores, producers, carnivores, etc.).

*ILO 7*  Discuss the problem of protein poverty, comparing vegetable and animal protein and describing the consequences of protein deficiency.

*ILO 8*  Discuss malnutrition resulting from nutrient deficiencies other than protein; e.g., vitamin A and iron.

*ILO 9*  Discuss the relationship between population and food supply—current as well as future.

*ILO 10*  Discuss the solutions to problems of inadequate food supply given in the chapter, citing advantages, disadvantages, and examples of each.

*ILO 11*  Define each new term introduced in this chapter.

Obtaining food is of such vital importance that both individual organisms and ecosystems are built around the central theme of nutrition, the process by which an organism takes in and assimilates food. An organism's body plan, as well as its life style, is adapted to its particular method of procuring food (Figure 4-1). In plants, leaf cells possess chlorophyll for trapping energy from sunlight, while extensive root systems are designed for absorbing water and minerals from the soil. The sharp teeth and claws of lions, as well as their long, quick legs, enable them to hunt and kill smaller animals. Clams, on the other hand, are equipped for an entirely different nutritional regimen. They are adapted to filter plankton (tiny plants and animals) from the water, and need only sit and wait for their dinner. Parasites such as lice may possess elaborate adaptations which help them maintain their own peculiar nutritional mode.

Organisms require food both as a source of energy to run the machinery of the body and as raw material for growth and repair of tissues. *Producers* manufacture complex food molecules out of simple nutrients. *Consumers* lack the chemical ability to produce their own food, and are dependent upon producers for their sustenance. *Decomposers* manage on the leftovers, the wastes and corpses of producers and consumers.

Man has managed to increase his food supply greatly by developing scientific methods of agriculture. Still, human population has multiplied faster than the food needed to nourish it, and nutrition is today one of the world's most serious problems. Between 40 and 60 percent of the people on earth suffer from malnutrition. More than 10 million of these human beings, most of them children, die each year as a result of starvation. Even in affluent America, hunger is a daily problem for more than 14 million people, nearly half of them children. More than 20 percent of American families are eating diets rated as poor by the U.S. Department of Agriculture, and only half the families in the United States have diets rated as good.

**FIGURE 4-1**
Organisms are adapted to their particular mode of procuring food. (*a*) The body and behavior of the lion are especially suited to its carnivorous life style. (Patricia Gill) (*b*) The robber fly is equipped with a beak for piercing the exoskeleton of its insect prey. (*c*) The claws of the human body louse are an adaptation for hanging onto hairs of its host's body.

(a)

(b)

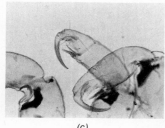
(c)

## THE BASIC NUTRIENTS

From about twenty chemical elements, absorbed in the form of simple compounds and salts, green plants are able to produce all the different kinds of organic molecules that they need. Animals are not such sophisticated chemists. Although they require approximately the same 20 chemical elements, these elements must already be put together in the form of about 40 chemical substances, many of them organic compounds. These essential nutrients are provided by a balanced diet consisting of proteins, carbohydrates, lipids, vitamins, minerals, and water (see Tables 4-1 and 4-3). With only slight variation, animals all require the same basic nutrients, but our discussion will primarily center upon human nutritional needs.

### Water

A daily dietary intake of about 2.5 quarts of water is required by an average adult. Because plant and animal tissues are composed largely of water, much of this daily requirement is met by eating food. For example, a raw apple is composed of almost 85 percent water; a slice of roast beef is 60 percent water. We drink the remaining required quart or so directly as water or other beverages.

The human body is about 65 percent water by weight, but the amount in specific organs varies. Skin is about 70 percent water, while bone is only 25 percent. Most surprising, perhaps, is that the human brain is about 80 percent water.

Water performs an impressive variety of functions in the body. All the chemical reactions essential to life take place in a watery medium. Water is an almost universal solvent because so many chemical compounds can be dissolved in it. The molecular motion of these dissolved substances lets them contact and interact with one another. In many chemical reactions water serves as an active ingredient. Water transports materials both intracellularly and from one location in the body to another. Blood plasma, which is 90 percent water, transports nutrients, hormones, and other substances throughout the body. Wastes are carried from the body in urine and sweat (both 99 percent water). Because of its ability to absorb considerable heat without rapidly changing its own temperature, water helps to maintain the constant temperature of the body. About 25 percent of body heat is lost by water evaporation from skin and lungs.

Chemically water is composed of two atoms of hydrogen attached to one atom of oxygen, and has the chemical formula $H_2O$ (Table 4-1).[1] Structurally it may be represented H—O—H.

### Minerals

Minerals are inorganic nutrients generally ingested as salts dissolved in food and water. Essential minerals include sodium, chlorine, potassium,

---

[1] For explanation of chemical symbols and formulas see Appendix A, Basic Principles of Chemistry.

**Table 4-1. Major groups of nutrients**

| Type | Chemical components | Function |
|---|---|---|
| Water | $H_2O$ | Structural component of cytoplasm; solvent; transports materials. |
| Vitamins: Fat-soluble: A, D, E, K Water-soluble: B complex and C | Small organic molecules | Most of them serve as coenzymes in metabolic reactions. |
| Minerals: Calcium, phosphorus, sodium, chlorine, magnesium, sulfur, potassium, and trace elements | Usually in form of salts | Many are structural components of important molecules; others serve to activate certain enzymes. |
| Carbohydrates: Monosaccharide, e.g., glucose Oligosaccharide, e.g., fructose Polysaccharide, e.g., glycogen | $C_x(H_2O)_y$ | Serve as principal fuel molecules for cellular respiration. |
| Lipids: At least two fatty acids are essential | C, H, O arranged into glycerol and fatty acid subunits | Storage form for fuel in the body; structural component of cellular membranes. |
| Proteins: 8 amino acids are essential (10 in children) | C, H, O, N, S, and sometimes P arranged into amino acid subunits | Provide amino acids essential for building body proteins; function as enzymes. |

magnesium, calcium, sulfur, and phosphorus. About thirteen others, such as iron, iodine, and fluorine, are referred to as *trace elements* because they are required in such small quantities.

Minerals are needed as components of all body tissues and fluids. Salt content (about 0.9 percent) is vital in maintaining the fluid balance of the body, and as they are lost from the body daily in sweat, urine, and feces, salts must be replaced by dietary intake. Sodium chloride (common table salt) is the salt needed in largest quantity in blood and other body fluids. A deficiency in sodium chloride results in dehydration.

Phosphorus and calcium are important structural components of bones and teeth. Phosphorus is also a component of nucleic acids, the high-energy molecule ATP (adenosine triphosphate), and several other important organic compounds. Calcium is required for biochemical reactions essential to blood clotting, muscle contraction, and transmission of nerve impulses.

Iron, the mineral most likely to be deficient in the diet, is an essential part of the hemoglobin molecule, the compound in red blood cells that transports oxygen. When iron intake is inadequate, the body cannot synthesize enough hemoglobin, and there is a decrease in the number of red blood cells. As a result, the capacity of the blood to transport oxygen to the cells is reduced, and cellular metabolism is slowed. People with this condition, known as iron-deficiency anemia, lack energy and are easily fatigued. In the United States a high percentage of infants of all races and income levels have iron deficiencies, especially during the last half of their first year. Milk alone does not provide sufficient iron for the needs of a growing baby at this time. An estimated 13 percent of adult women and 58 percent of pregnant women are also iron-deficient; because iron is lost during menstruation, and because iron is required by a developing fetus, increased iron intake is needed during a woman's reproductive years, particularly during pregnancy. In many other parts of the world iron-deficiency anemia is even more prevalent and more severe.

Perhaps iron-deficiency anemia is common because many foods that are rich in iron, such as steak or nuts, are expensive, while others, such as liver, are not popular. Flour, bread, and some cereals are now fortified with iron as a nutritional measure aimed at helping people ingest adequate amounts of this mineral.

Iodine, another trace element, is needed by the thyroid gland for making its hormone, thyroxine. Deficiency of iodine results in goiter, abnormal enlargement of the thyroid gland. In recent years the availability of iodized salt (table salt to which iodine has been added) has dramatically reduced the incidence of goiter. Seafood is a rich natural source of this mineral.

Of all minerals, perhaps fluorine has generated the most controversy in recent years. A great deal of evidence indicates that this mineral nutrient is necessary for dental health. Small amounts of fluoride (salt of fluorine) included in the diet during the years of tooth development affect the crystalline structure of teeth, making them resistant to decay. For more than 30 years, it has been known that persons living in communities in which the water supply contains fluoride [about 1 part per million (ppm)] have a much lower incidence of tooth decay than those living in communities where it is lacking. Since 1945, some communities lacking natural fluoride have fluoridated their water supply by adding this mineral. Studies show that fluoridation has reduced dental decay by about 60 percent in people who have drunk fluoridated water during early childhood (the period of tooth development). More than 3,000 communities in the United States, including Chicago and New York City, currently fluoridate their municipal water supply. Although the evidence at this point seems to weigh heavily in favor of fluoridation, many communities are still engaged in heated debate over the prospect of adding fluoride to their water supplies. Some persons cite dangers of fluoride overdose. Indeed, when water containing large amounts of fluoride (4 to 6 parts per million) is continually ingested during the period of tooth development, the enamel may become mottled, and in very large amounts fluoride is harmful. But then, so is chlorine, which is added to all water supplies to kill bacteria. Clearly, the amount of fluoride added to

FIGURE 4-2
Mineral deficiencies in plants result in characteristic plant diseases. (a) Potassium-deficiency disease of cabbage. (b) Iron-deficiency disease of cabbage. (c) Magnesium deficiency disease in bean plant. (McGraw-Hill Encyclopedia of Science and Technology, 3d ed.)

water must be regulated carefully, just as the amount of chlorine added is monitored.

Other minerals will be discussed throughout the text in relation to their specific physiologic activities. Effects of some mineral deficiencies in plants are shown in Figure 4-2.

## Vitamins

Vitamins are organic compounds required by the body for biochemical processes. They function primarily as *coenzymes*, substances which activate needed enzymes. Very small amounts are required in comparison to other dietary constituents. Vitamins may be divided into two main groups. *Fat-soluble vitamins* are those that can be dissolved in fat and include vitamins A, D, E, and K. *Water-soluble vitamins* are the B and C vitamins. Table 4-2 gives the sources, functions, and consequences of deficiency for most of the vitamins.

While millions of people gulp down their daily vitamin pills with great fervor, two opposing schools of thought debate the need for vitamin supplements. One group, which includes a large number of physicians, claims that people who eat a nutritionally balanced diet have no need for vitamin pills. The other group argues that most of us do not eat a balanced diet and therefore are likely to suffer from vitamin deficiencies. Debates also rage over the advisability of taking large amounts of certain specific vitamins, such as vitamin C for colds or vitamin E as a protection against vascular disease. To date there is no compelling evidence to support claims that large quantities of any vitamin are beneficial, and in fact more conclusive evidence is needed before we will fully understand all the biochemical roles played by vitamins or the interactions between various vitamins and other nutrients. Meanwhile the vitamin supplement industry is doing a booming business.

**Table 4-2. The vitamins**

| Name of Vitamin | Function | Effect of deficiency | Primary sources | Minimum* daily adult requirement |
|---|---|---|---|---|
| A | Promotes growth of bones and teeth. Essential for normal growth and maintenance of epithelial cells. Component of retinal pigments necessary for vision in dim light | Nightblindness; blindness; dry, rough skin; retarded growth | Green and yellow vegetables, fruit | 5000 IU† |
| D | Promotes intestinal absorption of calcium. Helps maintain normal calcium and phosphate levels | Bone deformities (called rickets in children) | Fish, meat, fortified milk | 400 I.U. |
| E | Inhibits oxidation of unsaturated fatty acids. Other functions not clear | Fragility of red blood cells (more research needed) | Plant oils | 30 I.U. |
| K | Essential for blood clotting | Prolonged blood clotting time | Normally supplied by intestinal bacteria; (green vegetables) | About 1 mg‡ |
| *B complex* | | | | |
| $B_1$ (thiamine) | Coenzyme in carbohydrate metabolism | Beriberi (malfunctioning circulatory and nervous system) | | 1.0 mg |
| $B_2$ (riboflavin) | Coenzyme in cellular metabolism | Blurred vision, poor growth | | 1.6 mg |
| Niacin | Coenzyme in cellular metabolism | Pellagra; early symptoms are fatigue, weight loss; associated with protein deficiency | Liver, egg, milk, meat, whole or enriched grains | 18 mg |
| $B_6$ (pyridoxine) | Coenzyme needed for normal amino acid and fatty acid metabolism | Dermatitis; neural malfunctions | | 2 mg |
| Pantothenic acid | Coenzyme for important reaction in cellular respiration | Mental symptoms, dizziness | | 10 mg |
| $B_{12}$ | Coenzyme for synthesis of nucleic acids; essential for maturation of red blood cells; functions in nervous tissue metabolism | Pernicious anemia in individuals with faulty absorption | | 5 mg |

* Refers to the minimum needs of an "average" person.
† I.U., abbreviation for "International Units," is a standard unit for measuring certain vitamins.
‡ mg (milligram) = 1/1,000 of a gram (28.4 grams = 1 ounce).

**Table 4-2.** (*Cont.*)

| Name of Vitamin | Function | Effect of deficiency | Primary sources | Minimum* daily adult requirement |
|---|---|---|---|---|
| *B complex* (*Cont.*) | | | | |
| Biotin | Coenzyme needed for cellular metabolism. | Skin lesions; muscle pain; sleeplessness | Produced by intestinal bacteria | About 200 mg |
| Folic acid | Needed for synthesis of nucleic acids | Certain types of anemia | | 0.4 mg |
| C (ascorbic acid) | Needed for amino acid metabolism. Promotes formation of strong teeth and bones; promotes firm gums | Scurvy, weakness, joint pain, anemia | Citrus fruit, tomatoes | 60 mg |

Sometimes people suffer from an *excess* of certain vitamins. Some people think that if one vitamin capsule daily is healthy, four or five might be even better. Actually, such a daily overdose may be quite harmful. While the body appears able to handle a moderate overdose of the B and C vitamins, surpluses of the fat-soluble vitamins are not easily excreted. An excess of vitamin D can cause weight loss, nausea, diarrhea, and, eventually, mineral loss from the bones and calcification of soft tissues, including heart, blood vessels, and kidney tubules. One common result is renal disease. In children 2,000 units or more a day results in growth retardation, and high doses of vitamin D taken by pregnant women have been linked to mental retardation in the developing child. Overdosage of vitamin A results in skin ailments, slow growth, enlargement of liver and spleen, and painful swelling of long bones.

## Carbohydrates

About 50 percent of the calories[2] that we ingest daily are in the form of carbohydrates, familiarly known as sugars and starches. Carbohydrates are not considered essential because when they are not included in the diet the body can synthesize them from other nutrients. Carbohydrates are used as fuel molecules for cellular respiration. The carbohydrate *glucose*, the most common fuel molecule in living systems, is stored as glycogen in the liver and in muscles, and can be rapidly mobilized as needed. An excess of carbohydrate is converted to lipids by the liver and stored in the body as fat tissue. Besides functioning in many types of metabolic reactions, carbohydrates serve as structural components of some tissues. In plants carbohydrates are an integral part of the wall surrounding each cell.

Each carbohydrate molecule is composed of carbon, hydrogen, and ox-

---

[2] The energy value of foods is measured in Calories per gram of food. A Calorie (actually a kilocalorie; customarily shown with a lowercase c, calorie, in popular usage) is defined as the amount of heat required to raise the temperature of a kilogram of water from 15 to 16 degrees Celsius (centigrade).

ygen. The term carbohydrate, hydrate (water) of carbon, stems from the fact that there are twice as many hydrogen atoms as there are oxygen, the same ratio found in water ($H_2O$). Carbohydrates may be divided into three groups based on molecular size. The smallest, called *monosaccharides* ("one sugar"), are the simple sugars. Glucose, the most common monosaccharide in living organisms, has the chemical formula $C_6H_{12}O_6$. Its structural form is shown below. *Fructose*, fruit sugar, has the same chemical formula as glucose, but its atoms are arranged differently so that its chemical properties are distinct.

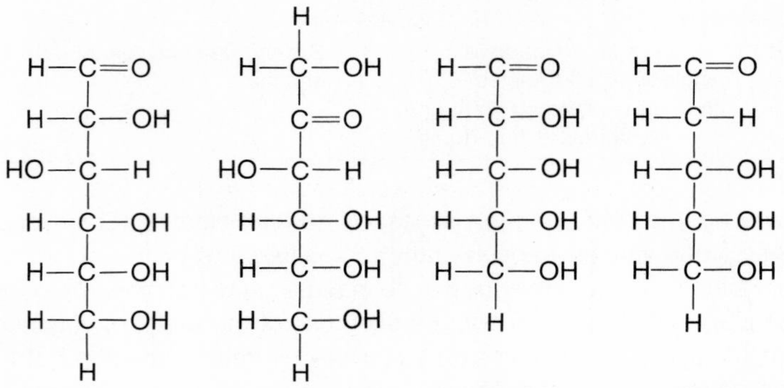

glucose          fructose          ribose          deoxyribose

Ribose and deoxyribose (containing one less oxygen) are five-carbon monosaccharides which are components of the nucleic acids, RNA (*ribo*nucleic acid) and DNA (*deoxyribo*nucleic acid).

Actually these compounds usually form rings, rather than remaining in the straight chains shown above. The ring structures of glucose and fructose may be drawn as follows:

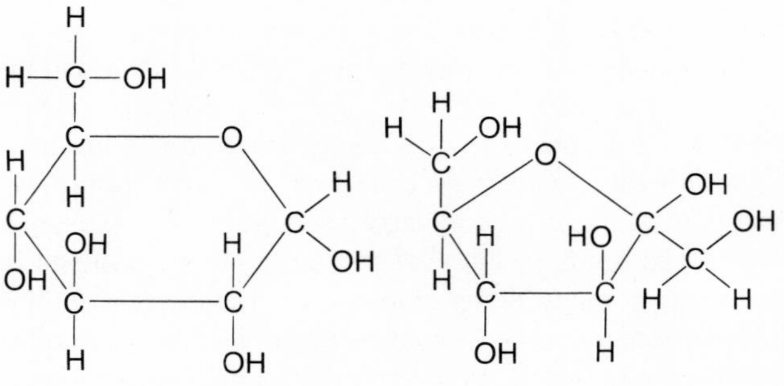

glucose          fructose

The *oligosaccharides* ("few sugars") consist of two or three monosaccharides chemically bonded to form double or triple-size molecules. *Sucrose*, the sugar we use to sweeten our foods, consists of a molecule of glucose chemically combined with a molecule of fructose. Other common oligosaccharides are lactose (the sugar in milk) and maltose (the sugar in malt).

sucrose

Most carbohydrates are *polysaccharides* ("many sugars"). Each polysaccharide molecule consists of repeating units of a simple sugar. *Glycogen,* the storage form of glucose, actually consists of great numbers of glucose molecules chemically bonded together. In plants, carbohydrates are stored in the form of *starch,* large molecules consisting of repeating units of maltose. *Cellulose,* the polysaccharide responsible for the rigidity of the plant-cell wall, is the most plentiful organic compound on earth. Cotton is composed of about 90 percent cellulose. Although man does not possess the enzymes needed to digest it, cellulose serves as bulk within the intestine, mechanically stimulating elimination of undigested food.

When two simple sugars combine chemically, a molecule of water is produced. This type of chemical reaction is called a *dehydration* reaction.

Glucose + fructose $\longrightarrow$ sucrose + water

On the other hand when sucrose is broken down, as in digestion, glucose and fructose form, and water must be added chemically.

Sucrose + water $\longrightarrow$ glucose + fructose

This type of reaction is referred to as *hydrolysis* and is important in digestion.

## Lipids

Lipids account for about 40 percent of the calories in an average American diet. The most common lipids are the fats. Lipids are important components of cell membranes and other membranous organelles and are also used as fuel in the body. In fact, they yield the most fuel energy of any food substance (9 calories per gram), more than twice as much as carbohydrate (only 4 calories per gram). For this reason, it is efficient for the body to store fuel in the form of lipid molecules. They are stored primarily in the cells of adipose (fat) tissue. Too much stored lipid results in obesity, a health problem which will be discussed in Chapter 5. During the semistarvation which we call dieting, as during actual starvation, the body utilizes its lipid reserves for energy.

Like carbohydrates, lipids are composed of carbon, hydrogen, and oxygen. Lipids, however, contain much less oxygen per molecule. Of the several types of lipids, only two, the neutral fats and the steroids, will be discussed.

The lipids found most commonly in living systems are the *neutral fats*. Each molecule of neutral fat consists of a *glycerol* molecule to which one, two, or three *fatty acids* are attached. Each time a fatty acid chemically combines with glycerol, a molecule of water is split off. An example of this dehydration reaction is shown below. Neutral fats are called *mono-*, *di-*, and

glycerol      +      fatty acid (stearic acid)      ⟶      fat (monoglyceride)      + water

*triglycerides*, which have one, two, or three molecules of fatty acid on each glycerol molecule, respectively.

Besides differing in the number of fatty acids present, fats contain different types of fatty acids. About thirty different kinds of fatty acids are commonly found in animal lipids. All fatty acids have a characteristic end group known as a carboxyl group,

$$-\overset{\displaystyle O}{\overset{\|}{C}}-OH$$

but differ from each other in the number and arrangement of carbon atoms in the remainder of their structure. *Saturated* fatty acids contain the maximum number of hydrogen atoms chemically possible, while *unsaturated* fatty acids contain carbon atoms that are doubly bonded to one another, rather than fully saturated with hydrogens (Figure 4-3). These unsaturated fatty acids are the oils, and most of them are liquid at room temperature. At least two fatty acids (linoleic and arachidonic acids) are essential nutrients, which must be included in the diet of children. Whether they are essential in the adult diet has not been established.

There has been considerable controversy concerning the role of fats in heart disease and in the development of *atherosclerosis* (hardening of the arteries). Current research indicates that atherosclerosis may be promoted by a relative deficiency in polyunsaturated fatty acids in the diet as compared with saturated fats. Safflower, corn, peanut, and cottonseed oils are rich sources of polyunsaturated fatty acids; animal fats are predominantly saturated. For these reasons, there has been a recent shift (at least in some circles) away from butter, fatty meats, and other animal fats in favor of vegetable oils.

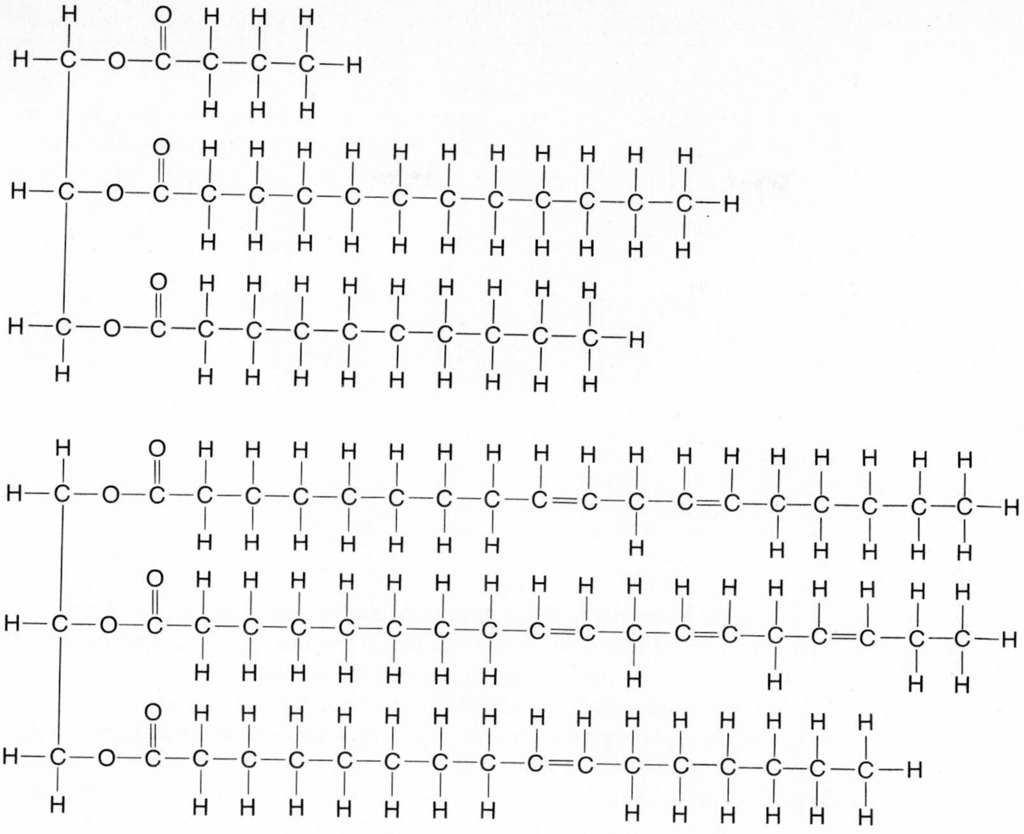

*Steroids* are another group of lipids that are of biologic importance. Both male and female reproductive glands, as well as the adrenal gland, synthesize steroid hormones. Cholesterol, which is characteristic of animal but not of plant tissues, is found in all human tissues and body fluids. Another steroid, ergosterol, is used by the body to make vitamin D.

Structurally, steroids differ markedly from neutral fats. Their carbon atoms are arranged in four rings, from which extend characteristic side chains of carbon atoms. The length and structure of these side chains distinguish one type of steroid from another.

**FIGURE 4-3**

Comparison of a saturated and an unsaturated fat. (*a*) A fat containing three saturated fatty acids. (*b*) A fat containing three unsaturated fatty acids.

### Proteins

High-quality protein is the most expensive and least available of all the nutrients. It is also the most important. The protein content of a person's diet is often a measure of his economic status. Protein consumption is even an index of a country's nutritional standard, and reflects agricultural and economic conditions.

**Table 4-3. Basic food groups for balanced diet**

| Group, with representative foods | Number of servings daily | Calories | Protein | Calcium | Iron | A | D | E | B | C |
|---|---|---|---|---|---|---|---|---|---|---|
| | | These foods rich in: | | | | Vitamins | | | | |
| Meat group: meat, fish, poultry, eggs | 2 or more | + | + | + | + | − | − | + | + | − |
| Milk group: cheese, milk, and other dairy products | Children, 3 cups; nonpregnant adults, 2 cups | − | + | + | − | − | + | − | + | − |
| Fruit group | 2 or more (should include one serving of citrus fruit or tomatoes for vitamin C) | − | − | − | + | + | − | − | − | + |
| Vegetable group | 2 or more | + | + | + | + | + | − | − | + | − |
| Grains and starchy vegetables: bread, cereal, rice, corn | 2 or more | + | − | − | − | − | − | + | + | − |

In order to appreciate the importance of protein, we must understand its functions in the body. Many cellular structures are composed, at least partly, of protein, so that growth and repair as well as maintenance of body tissues depend upon an adequate supply of this foodstuff in the diet. Some proteins also serve as *enzymes*, catalysts that regulate the thousands of different chemical reactions that take place in a living system. A specific kind of enzyme is required for each type of biochemical reaction.

The protein constituents of a cell are the clue to its life style. Each cell type has characteristic types, distributions, and amounts of protein that determine what the cell looks like and how it functions. A muscle cell is different from other cell types by virtue of its large content of the proteins *myosin* and *actin*, which are largely responsible for its appearance and its ability to contract. The protein hemoglobin of red blood cells is responsible for the specialized function of oxygen transport. Most proteins are species-specific—that is, they vary slightly in each species so that protein complement (as determined by the instructions in the genes) is also largely responsible for differences among species. Thus a dog is different from a fox or a coyote because its cells have the types and distribution of proteins which constitute dog cells.

A basic knowledge of protein chemistry is essential for understanding nutrition, as well as other aspects of metabolism. Proteins are composed of carbon, hydrogen, oxygen, nitrogen, and usually sulfur. Atoms of these elements are arranged into molecular subunits called *amino acids*. About twenty different amino acids are found in proteins. Every amino acid has a characteristic carboxyl group (—COOH) and an amino group (—NH$_2$), but each is distinct from other types of amino acids in the number and arrangement of its remaining atoms. Three examples follow.

amino group    carboxyl group

glycine
(the simplest amino acid)    alanine    lysine

Amino acids combine chemically with each other by bonding the carboxyl group of one molecule to the amino group of another. This is a dehydration reaction, since water splits off as the molecules combine. The chemical bond linking two amino acids together is called a *peptide bond*.

glycine    alanine    dipeptide

When two amino acids combine, a *dipeptide* is formed; a longer chain of amino acids is a *polypeptide*. A protein consists of one or more long polypeptides, each containing from 20 to several hundred amino acids. Polypeptides characteristically fold into specific coiled or pleated structures and then arrange themselves into fibers, crystals, or specific layers to produce the final structure of a completed protein.

A relatively small protein may be composed of five or six hundred amino acids. Hemoglobin consists of 574 amino acids arranged in four polypeptide chains. Its chemical formula is $C_{3,032}H_{4,816}O_{872}N_{780}S_8Fe_4$. Most proteins are much larger. A few small proteins, such as the hormone insulin, have been made artificially.

It should be clear that the various proteins differ from one another with respect to the number and types of amino acids that they contain, as well as in the arrangement of these amino acids. The 20 amino acids found in biologic proteins may be thought of as letters of a protein alphabet. Each protein is a word made up of certain types, numbers, and arrangements of amino acid letters. Thus, not all kinds of amino acids must be present in any particular protein.

Of the 20 amino acids found in man, the human body is able to synthesize 12 when provided with the appropriate molecular components (e.g., certain small organic molecules, other amino acids, etc.). The remaining 8 (10 in children) must be included in the diet because they cannot be synthesized in the body. These are referred to as the *essential amino acids*. A balanced diet containing all the essential amino acids in adequate amounts is necessary for proper growth and metabolism.

## TRANSFER OF ENERGY THROUGH THE WORLD OF LIFE

All living systems depend upon energy; without it there could be no life. Producers manufacture food using radiant energy from the sun. Then energy is transferred from one organism to another by consumers and decomposers that eat one another.

Energy is usually defined as the capacity to do work. Here we are concerned with the capacity of living systems to do biologic work. To stay alive, cells must continuously process nutrients, synthesize molecules, make cells or cellular structures, move, and transport materials. All these activities require an expenditure of energy by cells.

### The laws of thermodynamics

All changes that take place in the universe may be described in terms of the laws of thermodynamics. The *first law of thermodynamics,* known also as the *law of conservation of energy*, states that during ordinary chemical or physical processes energy cannot be created or destroyed, but can be transformed from one form to another. Green plants transform radiant energy into chemical energy, and chemical energy may be converted to heat energy. Although energy is constantly lost from biologic systems in the form of heat, it can be accounted for in the surrounding physical environment.

During biologic activities a great deal of valuable energy flows to the environment as heat. These energy losses are in accordance with the *second law of thermodynamics*, which explains that an activity involving an energy transaction can occur only when some of the energy is degraded, that is, changed from a concentrated form to a random, disorganized form. With every biologic activity, energy is lost from the organism, so that the amount of useful energy is decreased. Cellular respiration is only about 55 percent efficient, the other 45 percent of the energy being lost as heat. (Compared to most man-made machines, however, such biologic processes are quite efficient.) Because they are highly organized, living systems are very unstable, and in a sense life is a constant struggle against the second law of thermodynamics. Continuous energy input is essential both for individual organisms and for ecosystems. Thus consumers and decomposers must eat and producers must carry on photosynthesis.

### Modes of nutrition

Organisms may be distinguished according to their eating habits. *Autotrophs* ("self-nourishers") are organisms that make their own food from simple inorganic raw materials. Most autotrophs are *photosynthetic autotrophs.* All green plants and some purple and green bacteria which use sunlight as an energy source for food production are included in this group of producers. However, there are other types of producers. A few types of

*(a)*

*(b)*

**FIGURE 4-4**
(a) The ant lion, larva of the lacewing fly, is a predator which lives in a crater which it digs in the sand. When an ant wanders into the crater, the ant lion throws sand at it until the ant falls to the bottom. The ant lion then pierces the ant's body with its piercing and sucking mouth parts and sucks the ant's body fluids. (b) The rhinoceros is an herbivore. (Busch Gardens)

**Table 4-4. Modes of nutrition**

| Types of organisms according to nutritional mode | Examples |
| --- | --- |
| I. Autotrophs | |
|   A. Photosynthetic | Green plants |
| | Purple bacteria |
|   B. Chemosynthetic | Nitrite bacteria |
| | Nitrate bacteria |
| II. Heterotrophs | |
|   A. Consumers | |
|     1. Herbivores | Cow |
|     2. Carnivores | Lion |
|     3. Omnivores | Man, bear |
|     4. Parasitoids | Brachonid wasp |
|   B. Decomposers | Fungi |
| | Bacteria |
| | Some animals (maggots) |
|   C. Symbionts | |
|     1. Parasites | Tapeworms, fleas |
|     2. Commensals | Certain bacteria in human intestine |
|     3. Mutualistic partners | Flagellate in intestine of termite; certain bacteria in human intestine |

**FIGURE 4-5**
The tomato hornworm, larval stage of the hawkmoth, is almost covered by parasitoids, pupal-stage cocoons of the Brachonid wasp. The adult wasp lays her eggs under the skin of the hornworm. When the eggs hatch, the larvae feed on the body fluids of the worm until ready to spin cocoons. Then they emerge from the skin and spin their cocoons on the worm's body. The hornworm has little chance of developing into an adult moth. (Ralph B. Colson)

bacteria obtain energy for food production by "burning" certain inorganic compounds. These are *chemosynthetic autotrophs.* In a later chapter we shall discuss in detail the activities of the nitrate bacteria which obtain their energy by converting nitrites to nitrates, and the nitrite bacteria, which convert ammonia to nitrites. These chemosynthetic bacteria are vital links in any ecosystem because they chemically transform nitrogen compounds into forms needed by green plants.

All other organisms are *heterotrophs,* which get their food from some other source. We may divide heterotrophs into three groups: *consumers, decomposers,* and *symbionts.* Consumers may be variously subdivided. Those which live directly upon plant matter are *primary consumers*, or *herbivores.* Primary consumers are themselves consumed by various *predators,* sometimes called *carnivores,* which may also consume one another. The demarcation between primary consumer and predator is not always sharp, for some animals, called *omnivores,* include both plant and animal material in their diets. Man is an omnivore. The "top" or ultimate predators in an ecosystem are animals that have few or no natural enemies. Lions, eagles, or killer whales are examples of predators not preyed upon by anything else.

A most unusual form of predation is practiced by *parasitoids* (Figure 4-5). These are predators, not parasites, but they differ from more typical predators in that they consume their prey alive a little at a time. Many solitary

**FIGURE 4-6**
Decomposers. The fruiting bodies (mushrooms) of a fungus.
(Lynn Nieman)

wasps, for example, lay their eggs upon living prey. When the larvae hatch they eat the tissues of their victim, leaving the meat alive and fresh until the very end.

While most consumers must find, eat, digest, and absorb food, decomposers (Figure 4-6) make their living by breaking down the wastes and corpses of other organisms. All the nutrients they require are absorbed directly through the cell membrane.

Symbionts (Figure 4-7) are organisms which live in intimate (symbiotic) association with other types of organisms. Three types are *parasites, commensals,* and *mutualistic partners.* Parasites live at the expense of their hosts' well-being, but an effective parasite does not kill its victim. It has been said that while predators and parasitoids live upon capital, parasites live upon interest. Ectoparasites, such as fleas and ticks, live outside the host's body, while endoparasites, such as tapeworms and hookworms, live inside. Occasional parasites such as mosquitoes or bedbugs prey upon hosts as the need arises or as opportunity is afforded. Whether a parasite nourishes itself from food ingested by its host or by sucking blood, it is purely a freeloader.

Commensals also live at the expense of the host, but the host does not seem to mind, since it is not harmed by the relationship. Some types of bacteria live as commensals in the human intestine, nourishing themselves upon wastes that would only be eliminated anyway. Certain mites live quietly in the pores of most adults' faces without producing any known damage.

Mutualistic partners are two associated organisms of different species which live together and are beneficial to each other. They may also be interdependent. For example, a one-celled flagellate living in the intestine of the termite digests the celluose in the wood which the termite eats. Without this mutualistic partner the termite, though it might eat great quantities of wood, would starve, because it does not have the necessary enzymes for digesting its meal. The flagellate is nourished by the digested wood, and yet could not chew the wood for itself, so the arrangement is a happy one for both.

Symbionts may be classified in more than one nutritional group. For example, the termite could also be considered a consumer. See Table 4-4 for a summary of the various classifications.

### Energy capture and utilization

The flow of energy through living organisms may be divided into three major stages: photosynthesis, cellular respiration, and expenditure in biologic work. All these stages can take place within the producer, or energy can be transferred to other organisms through *food chains.*

*Photosynthesis.* All biologic energy originally comes from the sun. During photosynthesis green plants convert radiant energy from the sun into *chemical energy,* the only form of energy cells can use to carry on metabolic activities. Chemical energy is the energy found in chemical bonds—that is, the

**FIGURE 4-7**
Symbionts. (*a*) The flea is an ectoparasite. Its body shape is adapted for slipping through fur. Note also its long hind legs built for jumping. (*b*) Aphids are plant ectoparasites. (*c*) The dodder is a plant ectoparasite. Note penetration of parasitic vascular tissue into body of host. (*d*) Mistletoe in oak tree. Mistletoe is only partially parasitic, since its green leaves do carry on photosynthesis. (*e*) Commensalism. In this rather unusual partnership, barnacles have formed a commensal relationship with a diamond back terrapin. (Roy Lewis) (*f*) Commensalism. This bromeliad, closely related to the pineapple plant, is an epiphyte (air plant). It does not harm the host.

(*a*)

(*b*)

(*c*)

(*d*)

(*e*)

(*f*)

energy that holds atoms together. A considerable amount of such energy is contained in the bonds of organic molecules. When these bonds are broken, energy is released and may be used by the cell for its activities, such as putting together atoms in the synthesis of needed molecules. Chemical energy is the principal energy currency in the biologic world.

The principal raw materials for photosynthesis are water taken in through the roots of a plant and carbon dioxide in the air which diffuses into the leaves. With radiant energy as their energy source, plants are able to use these ingredients to produce carbohydrates. Oxygen is produced during the process. In fact, photosynthesis is really the only source of atmospheric oxygen. A summary equation for photosynthesis follows.

$$6CO_2 + 6H_2O + \text{light energy} \xrightarrow{\text{in presence of chlorophyll}} C_6H_{12}O_6 + 6O_2$$

carbon    water                                        glucose   oxygen
dioxide

This equation tells us what happens during photosynthesis, but not how it happens. The "how" is much more complex and involves a long series of chemical reactions. Although a detailed examination of photosynthesis is beyond the scope of the beginning biology student, we will describe the principal steps.

*The light phase.* Reactions of the light phase can take place only in the light. We can identify four steps.

1  The green pigment chlorophyll is energized by sunlight. Chlorophyll is an organic molecule which contains a central atom of magnesium. Together with other necessary chemical components and enzymes, chlorophyll is arranged along the grana of the chloroplasts. In complex plants, chlorophyll is concentrated primarily in the leaf cells. Chlorophyll molecules absorb and utilize light in the red and blue range of the spectrum. Plants are green because green light is *reflected* rather than absorbed. Of the light that falls upon a plant, only about 2 percent is actually absorbed by chlorophyll. The rest is converted into heat or reflected back into the atmosphere. When light energy is absorbed, chlorophyll becomes temporarily energy-rich, or *activated.*

2  Some of the energy of the activated chlorophyll is packaged as chemical energy in the special energy molecule, *ATP*. ATP is an organic molecule with three phosphate groups.[3] The second and third phosphate groups are attached to the molecule with larger-than-average quantities of energy. The most common way to store energy temporarily in cells is to package it in the third phosphate bond by making ATP from ADP, which has only *two* phosphates.

$$\text{ADP} + \text{P} + \text{energy} \longrightarrow \text{ADP} \sim \text{P}$$
$$\text{(ATP)}$$

The wavy line holding the last phosphate onto the ADP signifies a high-energy bond.

---

[3] See chemical structure of ATP as diagrammed in Chapter 8.

**3** Water is split. Much of the energy trapped by the chlorophyll is used to split water into hydrogen (H) and hydroxyl (OH) groups. Some of the hydroxyl groups combine to re-form water, but during the process the extra oxygen atoms combine to form molecular oxygen. While some of this oxygen is utilized by the plant for cellular respiration, most of it is released into the atmosphere.

**4** The hydrogen from the water is picked up by a hydrogen carrier molecule called NADP:

$$NADP + H_2 \longrightarrow NADPH_2$$

*The dark phase.* Dark-phase reactions use the energy of ATP and hydrogen from the $NADPH_2$ to manufacture carbohydrate from the carbon dioxide of the air. The first carbohydrate produced is RDP (ribulose diphosphate), a five-carbon compound. It combines with carbon dioxide to form a six-carbon compound, which immediately splits to form two three-carbon compounds, each called PGA (phosphoglyceric acid). ATP and $NADPH_2$ react with PGA in a series of reactions that results in a three-carbon compound called PGAL (phosphoglyceraldehyde). PGAL can be used by plants to make glucose, more RDP, and several other organic molecules.

Reactions of the light and dark phases of photosynthesis may be outlined as shown in the accompanying summary.

**Reactions of photosynthesis**

---

*LIGHT PHASE*
  *Summary:*


**$H_2O$ + ADP + P + NADP + energized chlorophyll $\longrightarrow$ $O_2$ + ATP**
**+ NADP-$H_2$ + chlorophyll**

  Actually a series of reactions:
  Chlorophyll + light energy $\longrightarrow$ energized chlorophyll
  ADP + P + E (from energized chlorophyll) $\longrightarrow$ ATP + chlorophyll
  $4H_2O$ + E (from energized chlorophyll) $\longrightarrow$ 4H + 4OH
  NADP + $H_2$ $\longrightarrow$ NADP-$H_2$
  4OH $\longrightarrow$ $2H_2O$ + $O_2$

*DARK PHASE*
  *Summary:*


**$CO_2$ + NADP-$H_2$ + ATP $\longrightarrow$ carbohydrate + ADP + P + NADP**
  Actually a series of reactions:

  RDP + $CO_2$ $\longrightarrow$ 6-C compound $\xrightarrow{\text{splits}}$ 2PGA
  PGA + NADP-$H_2$ + ATP $\longrightarrow$ PGAL + $H_2O$ + NADP + ADP + P
  PGAL $\longrightarrow$ used to make glucose and other organic compounds needed by plant

---

*Cellular respiration.* Photosynthesis provides the plant with the basic organic compound, carbohydrate. All the remaining useful energy captured by the chlorophyll molecules has been converted into the chemical energy holding together the atoms of the carbohydrate molecule. The rest of the captured energy has been lost to the plant as heat. Thus, the plant's available energy is now in the form of carbohydrate molecules. Some of these are used as fuel in the process of cellular respiration. Each carbohydrate molecule is broken apart bit by bit, and as each chemical bond is broken, the energy released is efficiently processed. Ultimately the energy from the carbohydrate is more conveniently packaged in the form of ATP, whose high-energy bonds are temporary storage places for energy that can be released instantly whenever needed by the cell for performing work.

*Expenditure of energy in biologic work.* Cellular respiration provides ATP energy needed by the cell for all types of biologic activities. *Synthesis*, the process of manufacturing the thousands of different types of molecules needed by cells, is one form of biologic work. Energy is required to bond together chemically the component atoms of new molecules. Proteins and nucleic acids are just two kinds of molecules which must be continually produced in each cell. In addition some synthesized molecules are put together to build cellular structures. During growth the synthetic demands of the system are especially great, but new molecules and cell parts are needed continually in all cells for maintenance and repair. Another type of work performed constantly by living systems is transporting materials within cells and from one cell to another. An example is the absorption of nutrients from the intestine, a process termed *active transport*. Transmission of nerve impulses depends upon one form of active transport in which sodium ions are constantly pumped out of the nerve cell and potassium ions are pumped in, moving in a direction opposite to that in which they would normally move. Undoubtedly the most common type of biologic work is that of movement. During muscle contraction the chemical energy of ATP is converted into mechanical energy which is expended in the mechanical work of movement.

As an organism carries on each of these activities, a great deal of energy is converted to heat. Some heat energy is put to good use by warm-blooded organisms in maintaining body temperature, but most of it flows to the environment, becoming useless to the biologic world. Thus there is a one-way flow of energy through living systems, requiring continuous input of radiant energy from the sun. Only producers can perform this input function by their photosynthetic activities. All other organisms must obtain their energy by eating the organic molecules produced by plants, or by eating organisms that have eaten plants.

### Transfer of energy through food chains

A food chain is a hierarchy of organisms that eat, and in turn are eaten by, one another. A very simple food chain might consist of grass being eaten by a cow, which in turn is eaten by man.

In all food chains the producer, in this case the grass, occupies the first *trophic* (nourishment) *level*. When the cow, the primary consumer, eats the grass, some of the chemical energy of the grass's organic molecules is

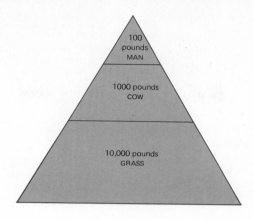

**FIGURE 4-8**
A simple food pyramid.

transferred to the second trophic level. The cow now enjoys the benefits of the photosynthetic activities of the grass. Man in this case is a secondary consumer. When he eats the cow he gains some of the chemical energy originally captured by the producer but now stored in the bonds of the cow's own organic molecules. Thus in a food chain energy is transferred from the eaten to the eater.

As a consequence of the second law of thermodynamics, a large amount of useful energy is lost as heat when it is transferred from one level of a food chain to the next. Part of the energy loss between trophic levels is due to the energy that each organism uses for its biologic activities. An additional loss is due to undigested food in feces and the organism's ultimate fate as a corpse. Both feces and dead organisms represent energy and furnish subsistence to decomposers. Complex subsidiary food chains originate with these decomposing organisms.

In studying energy transfer it is convenient to think of a food chain as a *food pyramid* (Figure 4-8) which shows diagrammatically the relative size decrease of each level. The first trophic level is always the largest, containing the greatest *biomass*, or total weight of the organisms, as well as the greatest amount of energy. Only about 10 percent of the mass can be transferred to the second trophic level, so that its biomass is only 10 percent as great as the producer level upon which it feeds. Only 10 percent of the biomass at the second trophic level can be supported on the third level, and so on up the chain. In our grass-cow-man food pyramid, then, if we begin with 10,000 pounds of grass we can roughly predict that over a given time period, this amount of grass can support 1,000 pounds of cow which in turn can support 100 pounds of man (Figure 4-8). Another type of terrestrial food chain is compared with an aquatic food chain in Figure 4-9. Note that an organism can and frequently does change its position in a food chain. In nature, organisms relate to one another nutritionally in a pattern called a *food web*, which is more realistically complex. In the aquatic food chain shown in Figure 4-9, man is shown on the sixth level, but man can, and often does, eat producers directly, which allows him to occupy the second trophic level as

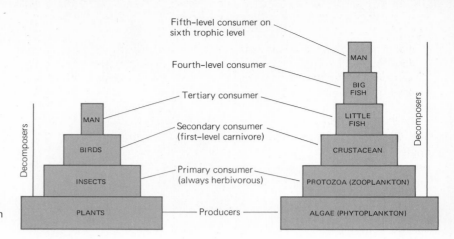

**FIGURE 4-9**
Comparison of a terrestrial with an aquatic food chain.

well. Because of the great quantities of energy lost as it is transferred up through the food chain, it is usually ecologically most efficient, as well as economically necessary, for people to eat the lower components of a food chain directly. It should be clear that the higher the food-chain level, the lower the amount of biomass that can be supported.

We can indicate the decomposers by a line extending from the producers all the way up to the top of the pyramid. The decomposers make use of the energy in feces and corpses, energy that is not generally available to the organisms making up the main body of the food chain. A complex side chain based upon decomposers actually branches off at each level. An example of such a subsidiary food chain is shown in Figure 4-10.

The trophic relationships which exist within any ecosystem link organisms together in an intricate network of chemical and energy flow. Many nonnutritional aspects of an ecosystem are related to the pattern of what eats which and how much.

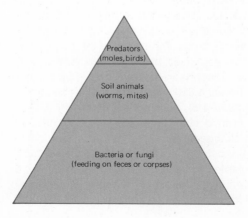

**FIGURE 4-10**
Subsidiary food chain based upon decomposers.

# THE PROBLEM OF WORLD FOOD SUPPLY

If all food in the world were equally distributed and each person received identical quantities, we would all be malnourished. If the entire world's food supply were parceled out at the dietary level of the United States, it would feed only about one-third of the human race.[4]

Of the earth's 3.7 billion people, more than 2 billion suffer from malnutrition. Our rapidly mushrooming human population promises to intensify the crisis. Despite recent advances in agriculture, an ever greater number of people suffer from malnutrition each year. People are multiplying at a much greater rate than their food supply is.

In discussing food shortages, it is helpful to understand the distinction between *undernutrition* and *malnutrition*. People suffering from undernutrition do not have enough food to fill their stomachs. Obviously their daily caloric intake is also inadequate. On the other hand, people can have enough food so that they do not feel hungry, yet still be starving. These people suffer from malnutrition; they lack one or more of the essential nutrients in their diet, usually essential amino acids found in some forms of protein.

People whose bodies are weakened as a result of malnutrition or undernutrition often die from illnesses which would be only minor in well-nourished persons. Common childhood diseases such as measles, chicken-pox, and whooping cough are often fatal to malnourished people. A recent study showed that in undeveloped countries the death rate from measles may be more than 300 times greater than in North America.

## Protein poverty

Protein poverty is now the world's major health problem. At least one of every six human deaths can be attributed to poor nutrition, usually protein deficiency. Of all required nutrients, the essential amino acids are the ones most often absent or deficient in the diet.

*Protein quality.*　Not all proteins contain the same types or quantities of amino acids, and many proteins do not contain all the essential amino acids. The highest-quality proteins, those that contain the most appropriate distribution of essential amino acids for human nutrition, are found in animal tissues. Such foods as meat, eggs, milk, and fish contain *high-quality* proteins. Other foods such as gelatin or soybeans may be almost 100 percent protein, yet they do not contain all the essential amino acids or do not contain them in the right proportions. Thus a breakfast containing eggs and milk, for example, is better nutritionally than a cereal product which may contain the same amount of protein. Unfortunately, because they are expensive, animal foods account for only 10 percent of the total caloric intake of the

---

[4] Georg A. Borgstrom, "The Dual Challenge of Health and Hunger: A Global Crisis," *Bulletin of the Atomic Scientists*, October 1970, p. 42.

(a)

(b)

**FIGURE 4-11**
Primitive methods of agriculture still prevail over most of the world. (*a*) Here we see Burmese women transplanting rice. (I. W. Kelton) (*b*) Irrigation system in Madras, India. Some of the channels date back 1800 years. (United Nations)

human population. Furthermore, even this figure is misleading because in the developed countries a much greater proportion of animal foods is eaten than in the underdeveloped countries. It has been estimated that in the developed countries an average of 50 pounds of animal food is consumed per person each year (actually, more protein than necessary), while in the underdeveloped countries the average falls to only 2 pounds of animal food per person per year. And again, this is only an average; not every person gets his annual 2-pound share, since in wealthier families each person might consume 10 pounds or more annually. Indeed, the diets of many consist totally of plant foods.

It should be clear that the vast majority of people rely upon vegetable protein for their essential amino acids. As a rule, vegetable protein is inferior to animal protein in amino acid distribution. Most vegetable protein is deficient in at least one essential amino acid, often lysine, tryptophan, or threonine. Another problem is that protein is less concentrated in plant tissues than in animal tissues. The new high-yield grains contain from 5 to 13 percent protein; meat contains about 25 percent. Furthermore, because plant cells are surrounded by cellulose cell walls for which man has no digestive enzymes, vegetable protein is considerably less digestible than animal protein; even the protein that is present cannot be utilized entirely.

The staple food for most humans is cereal grain—usually rice, wheat, or corn. None of these foods provides an adequate distribution of essential amino acids for growing children. Wheat contains less than 1 percent lysine, which is not adequate to support growth. (An adult, however, can get along with wheat as his sole source of protein.) Ordinary corn lacks the amino acids lysine and tryptophan and is also low in cystine. In some underdeveloped countries people must rely upon starchy crops such as sweet potatoes or cassava as their staple food. The total protein content of these foods is less than 2 percent, far too low to provide the body with needed raw materials.

Peanuts, soybeans, and other legumes have more than twice the protein content of the cereal grains, and thus are important in human nutrition. However, they are not produced in nearly the quantities of the grains, and a large percentage of these products is used as feed for domestic animals.

With care, a purely vegetarian diet can be planned to provide adequate nutrition. In such a diet several vegetable proteins that are known to complement one another with respect to their amino acid content must be included. Vegetable proteins may have different deficiencies so that a variety in which one cancels out the drawbacks of others must be selected. Because man cannot store amino acids in his body, *all the essential amino acids should be provided in each meal*. In order to synthesize needed protein molecules the body must have all the amino acid components available at the same time. Such nutritional planning requires a certain amount of sophistication in nutritional principles, as well as the availability of the needed varieties of vegetable foods.

**Effects of protein deficiency.** It has been suggested that the problem of protein poverty is a major factor holding back the underdeveloped countries. Millions of persons suffer poor health, retarded physical and mental development, and a lowered resistance to disease because of protein deficiency.

Recent studies have shown a striking relationship between adequate diet and normal growth and development in young children. One study of nutrition in poor families in the United States showed that one-third the children up to age 6 in urban ghettos and rural slums have suffered from growth retardation. Intelligence levels have also been correlated with proper nutrition. Human brain cells multiply rapidly during embryonic development and early infancy. At about 6 months of age, the human brain stops proliferating new cells, and from that time no new brain cells can ever be made. Growth continues only by cells increasing in size. By age 3, the human brain has attained 80 percent of its adult size. If the diet of a pregnant mother does not provide the growing embryo with needed nutrient materials for growth, or if the diet of an infant is inadequate, the developing brain will be unable to grow and form new cells. A recent analysis of the brains of babies who had died of malnutrition revealed that the number of brain cells present was up to 60 percent below normal. Furthermore, when necessary nutrients are not provided during the critical period of brain-cell proliferation, the resulting retardation is permanent. A child who has been a victim of severe malnutrition before birth or during the first months of life cannot make up the critical brain growth even if he is later provided with the best of diets. Yet in most underdeveloped countries two-thirds of the children are malnourished (Figure 4-12).

Severe nutritional deficiency often results in the diseases *marasmus* and *kwashiorkor*. All too common in poorer nations, these diseases have also been identified in the United States. Marasmus is a condition of inadequate total caloric intake associated with severe protein deficiency. Children with this disease do not have enough to eat, and whatever is available is certainly not rich in protein. Marasmus is most common among babies who are weaned early and placed on inferior vegetable (cereal) diets. They become emaciated, and if not treated, they die.

Kwashiorkor is a consequence of severe protein deficiency. Children with

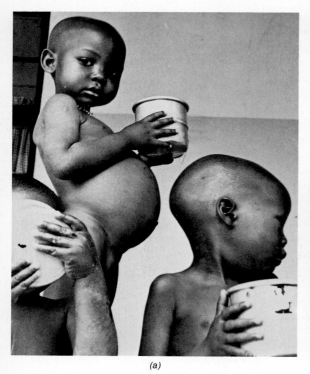
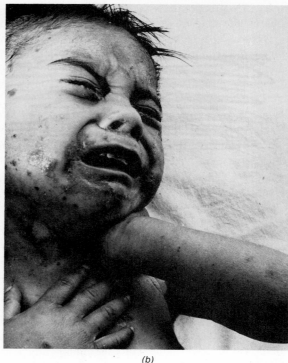

**FIGURE 4-12**
Children with deficiency diseases. (a) Kwashiorkor; (b) marasmus. (OMIKRON)

this condition may have plenty to eat, but their diets do not provide adequate amounts of the essential amino acids. The term kwashiorkor is an African word that means "displacement sickness," and refers to the fact that when a child is displaced from his mother's breast by the arrival of a new sibling he often develops symptoms of this condition. The vulnerable child is usually on a diet consisting almost exclusively of corn, cassava, or plantain. Children with kwashiorkor have stunted mental and physical growth, anemia, and impaired metabolism. They often suffer from loss of appetite and apathy, and they may display swollen bellies. A decrease in the amount and activity of important enzymes can be measured. Because the digestive enzymes required to break down ingested protein are themselves proteins, they are often not synthesized by a malnourished system. The necessary component amino acids are just not available to the cells. In the resulting unfortunate cycle, the small amounts of dietary protein that are provided by the diet cannot be digested, and are consequently wasted.

Children with these deficiency conditions are highly susceptible to parasites and other disease organisms. Their bodies are not able to fight off even common childhood diseases that would not prove dangerous to a healthy child. When untreated, kwashiorkor brings death to as many as 90 percent of its victims. In malnourished children, the final death blow is often dealt by diarrhea and dehydration. Malnutrition intensifies diarrhea, and diarrhea lessens the ability of the intestine to absorb nutrients, setting up a vicious cycle.

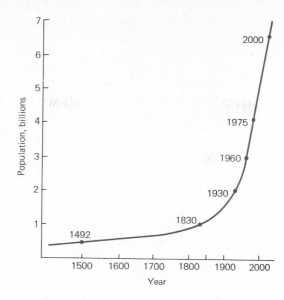

FIGURE 4-13
Human population explosion.

## Deficiencies of other nutrients

We have emphasized protein deficiency and its consequences because essential amino acids are the nutrients most often absent in the diet, and also because the problem of protein poverty is the most difficult to solve. The effects of various vitamin deficiencies have been indicated in Table 4-2. One of the most common is vitamin A deficiency, which often results in blindness. Some 11 million school-age children in the underdeveloped countries suffer from extreme vitamin A deficiency. An estimated 1 million persons in India alone are blind as a result of inadequate intake of this nutrient. Iron and calcium are the most common deficiencies among the mineral nutrients. Vitamin and mineral deficiencies can be prevented by adding them to various foods, and indeed many foods are fortified with nutrients.

Millions of people suffering from mild malnutrition do not exhibit obvious disease symptoms. However, they are not able to function efficiently because they are weak, easily fatigued, and highly susceptible to infection.

## Population and food supply

To give our current population a minimally sound diet would require the immediate doubling of world food production.[5]

Yet the population continues to increase at a frightening rate. Figure 4-13 illustrates the quickening pace of human reproduction. This graph shows that it took from the time of the first human beings on earth until the year 1830 for the human population to reach 1 billion. But only 100 years were required, from 1830 to 1930, to expand the population to 2 billion. In only 30 years (by 1960) the third billion was added. By 1975 the human population

---

[5] Ibid.

will probably exceed 4 billion. The doubling time for world population is now only 37 years. Thus, by the year 2000 (less than 30 years away) about 7 billion persons will be crowded onto our planet.

Professor K. E. F. Watt[6] of the University of California, estimates that by the year 2000 the United States will have only 84 percent of the land necessary to produce food for its own citizens (more than 300 million). The rest of the land will be covered by housing. Every day 300 acres of agricultural land in California alone are lost to concrete highways and buildings. The U.S. Department of Agriculture lends support to Watt's estimate by reporting that the United States can continue to help feed the developing countries only until 1984. After that we will need all the food we produce to feed our own citizens.

## What are the solutions?

Is it possible to multiply our supply of food sufficiently to meet the demands of a hungry world? We shall discuss the following three possibilities:

1 Improving conventional agriculture (including the Green Revolution)
2 Farming the sea
3 Developing various exotic food sources

*Improving conventional agriculture.* Food production could be increased either by cultivating more land or by improving crop yields on land already being farmed.

*Cultivating more land.* Although optimists are quick to point out that there are millions of potentially farmable acres of land, the truth is that much of it is of only marginal quality. In fact, "almost all the land that can be cultivated under today's economic circumstances is now under cultivation."[7] In many regions the soil is so poor that it is not practical to attempt to cultivate it. When a tropical forest is cleared, rain, sun, and oxygen make the soil unfit for agriculture within a few years. Many such soils turn into *laterite*, a rocklike material so tough that temples built from it almost a thousand years ago still stand in Cambodia.

> The hungry nations have been and are hungry because they have a poor piece of real estate. The soils are too dry, too wet, too rocky, too thin, or too mountainous to fulfill adequately the agricultural needs of the country.[8]

Some areas possess fertile soil but little rainfall. In these regions farming is dependent upon artificial irrigation. Unfortunately, the water demands of an expanding population, as well as the consequent pollution of waterways

---

[6] Walter Howard, "The Population Crisis Is Here Now," *Bioscience*, September 1969, p. 782.
[7] Paul R. Ehrlich and Anne H. Ehrlich, *Population, Resources, Environment*, Freeman, San Francisco, 1970, p. 91.
[8] William C. Paddock, "How Green Is the Green Revolution?" *Bioscience*, Aug. 15, 1970, p. 898.

are steadily reducing the amount of fresh water available for irrigation. There has been much talk recently about desalting seawater for use in irrigation systems. However, desalting is such an expensive procedure that it is not a practical solution to current problems.

*Improving crop yields—the Green Revolution.* The most practical solution, and the one with the most potential for alleviating the current hunger crisis, is the application of new technologic advances for improving crop yields on land already under cultivation. More intensive and sophisticated methods of agriculture have contributed to recent significant increases in crop yields. Some agriculturists are so hopeful of continued progress in multiplying food supply by these methods that they have suggested that we are in the midst of a Green Revolution. This agricultural revolution is based primarily upon the development of new "miracle grains" of rice and wheat, more intensive use of fertilizers, pesticides, and irrigation, and recent breakthroughs in the development of high-protein strains of cereals.

What are "miracle grains"? These are new genetic varieties of wheat and rice that have been especially bred to produce greater yields. They are short-stemmed to permit a heavier head of grain, because older, long, thin-stemmed varieties could not support the heavy weight. These new grains are adaptable to a wide range of climatic conditions, and they mature quickly so that two or three crops can be grown per year. But what of their drawbacks? Their chief disadvantage is that their success depends largely upon adding great quantities of fertilizer to the soil. Most countries do not have the natural resources for producing fertilizer, so it must be imported. At present, many farmers must rely on government subsidies in order to buy imported fertilizer, and in underdeveloped countries capital for such purposes is in short supply. Moreover, intensive use of chemical fertilizers destroys the natural ecology of the soil. They create conditions that destroy the bacteria necessary for normal cycling of nitrogen, while at the same time favoring bacterial denitrification (so that large quantities of the inorganic nitrogen added in the fertilizer become useless to the plants). The result is that the soil soon becomes unable to maintain itself, becomes "addicted" to fertilizer, and fertilizers must be added continuously forever after. It has been calculated that nature requires about 500 years to produce 1 inch of good top soil. But man can destroy it in just a few years. The Green Revolution depends upon fertilizer input, liberal use of pesticides, irrigation, government subsidies, and the sophistication and know-how for using these techniques.

We have seen that most people must depend upon vegetable protein as the sole source of their essential amino acids. At this time many of the miracle grains are deficient in essential amino acids. However, another recent breakthrough in the applied genetics of cereal grains holds out new promise for improving the quality of vegetable protein. Breeders have discovered that a strain of corn possessing the gene called opaque-2 has an unusually high amount of the essential amino acid lysine which is normally deficient in corn (Figure 4-14). When this gene is transferred to conventional varieties, not only is their lysine content increased, but their balance of other amino acids is also improved. This high-protein corn has been named *opaque-2* after the

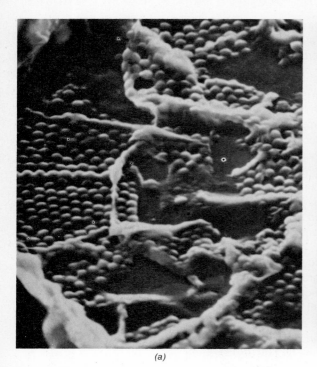

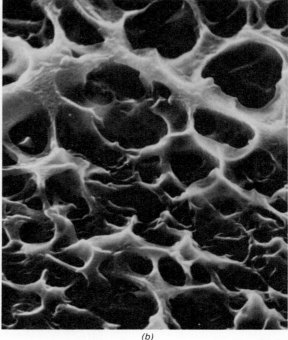

(a)                                                    (b)

**FIGURE 4-14**
Scanning electron micrographs of sections through corn endosperm (kernel) from which starch has been removed. (*a*) Normal corn. The protein is stored in spherical bodies and is of relatively poor quality. (*b*) High-lysine corn. Much of the protein is located in a matrix of interstitial material. The protein has a higher nutritional value. (McGraw Encyclopedia of Science and Technology, 1972 Yearbook and USDA)

gene. The search is now on for high-protein varieties of rice and other grains. These may then be combined with high-yield strains to produce supergrains that are high-yield and high-protein (Figure 4-15).

Man has managed to breed certain strains of grain that are resistant to various plant pests and diseases. Although such disease-resistant crops may appear to be an answer to the farmers' prayers, they are not without genetic hazard. If a few supergrains with the same disease-resistant characteristics are planted worldwide, a new or different plant disease or insect could destroy large portions of the world food supply with a single disastrous blow. And new varieties of insects and disease organisms do appear in nature constantly. As we shall learn in our study of genetics, genetic variability is the key to survival. When man imposes an artificial homogeneity upon the genetic complements of his crops, he makes them vulnerable to extinction. Food production could be devastated before breeders had time to develop new strains. Several times over the past few decades man's simplification of nature has caused such disaster. In 1946, for example, oats resistant to a troublesome type of rust were planted over 30 million acres. This amounted to two-thirds of the United States oat crop. A new disease strain attacked these crops, causing such destruction that in only 2 years these oats disappeared from American soil.[9] In 1970 a corn blight destroyed

---

[9] Ibid., p. 900.

15 percent of the United States corn crop. The problem was traced directly to the genetic uniformity of the corn (Figure 4-16). Similar crop disasters have been reported in other countries using new strains of rice and other crops. A recent study of major crops in the United States revealed that because consumers and farmers both prefer uniform produce, crops are genetically uniform and, consequently, very vulnerable. For example, 96 percent of the pea crop consists of only two genetic varieties, and 60 percent of the dry bean crop is composed of two varieties.

For his work in initiating the Green Revolution, Dr. Norman Borlaug was awarded the 1970 Nobel Peace Prize. Dr. Borlaug himself believes that the Green Revolution can buy only 20 more years unless we solve the population problem. "The unrealistic attitudes of the world towards unreasonable population growth is the biggest problem we face now."[10] He goes on to say that we should not produce more human beings than we are able to feed. While the Green Revolution can be expected to help alleviate existing hunger, it cannot multiply food at anywhere near the rate that we are multiplying hungry mouths.

FIGURE 4-15
An agronomist examines a new wheat variety at the Nebraska Experiment Station, where efforts are being made to increase the lysine percentage in wheat. (USDA)

*Farming the sea.* There is a popular myth among unenlightened optimists that the seas are filled with food resources just waiting to be harvested. The truth is that overfishing already has caused the decline of many marine species. Because most fish are near the top of long food chains, they are relatively few in number. Fish are actually most plentiful not in the open sea, but in coastal areas and estuaries where an abundance of minerals supports food chains. Unfortunately these are also the areas closest to dense human populations, and therefore they are becoming increasingly polluted by man's wastes. If man were to harvest algae on a large scale, he would also deplete fish populations by depriving them of their food-chain base. Furthermore, technology is not yet available for harvesting algae economically or for converting it into edible food once it is harvested. Practical methods for raising marine animals on sea farms have not yet been developed either.

*Exotic food sources.* Various methods have been proposed for solving the problem of protein poverty. One such proposal involves fortifying cereal foods with protein obtained from soybeans or other sources. Incaparina, a product developed by the Institute of Nutrition for Central America and Panama, consists of corn and cottonseed meal enriched with vitamins A and B. Although it has been available in Central America for several years, it has not been accepted by the people and therefore has not done much good. Apparently its bland texture and taste are not appealing.

Bacteria, yeast, and algae can be cultured on sewage. These single-celled organisms produce great quantities of protein, and can be harvested for animal feeds, if not eaten directly by human beings. Development of such

[10] Norman Borlaug, press conference as reported by Associated Press, Oct. 22, 1970.

**FIGURE 4-16**
Corn ears damaged by the Southern corn leaf blight that recently spread through the United States. (USDA)

animal feeds would free millions of tons of grain and legumes (now being fed to animals) for human use.

One promising exotic food source is the culture of yeast and bacteria on petroleum. These organisms synthesize protein several thousand times faster than domestic animals. Such single-celled organisms can be harvested, dried, and purified. The resulting white powder can be added to various foods. So far, protein produced in this way is being used only for livestock feed, but eventually it will be processed for human consumption. Although protein from petroleum offers some promise, it is not a long-term solution because the amount of petroleum on earth is limited and will eventually run out.

***Controlling population.*** The only long-term solution to the problem of world food supply is population control. Without a halt to present rates of population growth, worldwide famines may be expected before the end of this century. If a great deal of money, research, and planning, as well as international cooperation, is put into the technology of the Green Revolution, and if other possible solutions are explored, we may be able to buy the time needed to halt population growth before complete disaster occurs.

1 Required nutrients are obtained by eating a balanced diet consisting of proteins, lipids, carbohydrates, vitamins, minerals, and water.

2 Each essential nutrient performs a specific function or group of functions. A deficiency of any nutrient in the diet results in undesirable physiologic consequences.

3 The laws of thermodynamics govern the passage and utilization of energy in the biologic world as well as in the physical world.

4 On the basis of their mode of nutrition, organisms may be classified as autotrophs or heterotrophs, and into subcategories.

5 In the process of photosynthesis, green plants convert water and carbon dioxide into carbohydrate molecules. Oxygen is produced as water is chemically split.

6 After radiant energy is converted into the the chemical-bond energy of carbohydrates during photosynthesis, plants may use carbohydrates as fuel for cellular respiration.

7 Only about 10 percent of the total amount of the solar energy captured by a plant during photosynthesis can be transferred to the consumer that eats it. The longer a food chain, the less there is on the top level to show for the energy captured by the producers.

8 Much of the world suffers from protein poverty and its accompanying disabilities. Many persons also suffer from various vitamin and mineral deficiencies.

9 Possible methods for increasing food production include improving conventional agriculture, developing practical methods for farming the sea, and developing various exotic food sources such as growing algae upon sewage.

10 In order to solve the problem of world hunger, it is necessary to solve the problem of overpopulation. Food production is not presently keeping up with population growth, and is even less likely to in the future.

**FOR FURTHER READING**

Borgstrom, Georg: *The Hungry Planet,* Macmillan, New York, 1967. An in-depth report of the world food crisis.

Dumont, Réné, and Bernard Rosier: *The Hungry Future,* Praeger, New York, 1969. A detailed study of the world food problem.

Ehrlich, Paul R.: *The Population Bomb,* Ballantine Books, New York, 1972. This best-selling paperback is must reading because it strikingly relates the problems of hunger and overpopulation, showing how the environment is damaged as we struggle to provide more food and other necessities for "too many people."

Lehninger, Albert L.: *Bioenergetics,* W. A. Benjamin, Inc., New York, 1965. An excellent paperback for those interested in knowing more about energy flow and utilization through the biologic world.

Paddock, William C.: "How Green Is the Green Revolution?" *Bioscience,* Aug. 15, 1970, p. 897. An excellent article summarizing the facts and fictions of the Green Revolution.

# 5

# PHYSIOLOGY OF NUTRITION

**Chapter learning objective.** The student must be able to describe the steps involved in processing ingested food and the utilization of digested nutrients by the body, as well as discuss questions about obesity, food additives, and health foods.

*ILO 1* On a diagram or model identify each of the digestive structures discussed in this chapter, giving the function of each structure.

*ILO 2* Trace the pathway traveled by an ingested meal, describing each of the changes that takes place en route.

*ILO 3* Define the term *enzyme*, describe its function, and give four factors that influence enzymatic activity.

*ILO 4* Describe the process of food absorption.

*ILO 5* Trace the route and fate of each major group of nutrients after absorption from the digestive tube.

*ILO 6* Describe obesity, giving its causes and best remedies.

*ILO 7* List five types of food additives of current importance, give at least three reasons for the use of food additives, and produce arguments in favor of their strict regulation.

*ILO 8* Give two arguments for and one against the use of health foods.

*ILO 9* Define each new term introduced in this chapter.

A plant has no need for a digestive system because most of the chemically simple nutrients it "eats" are already in the form required for photosynthesis. Animals eat chemically complex foods which must be broken down into simpler forms before they can be utilized by individual cells. In very simple organisms such as amebas digestion takes place within each cell and depends upon the action of lysosomes, but in more complex forms food is broken down by specialized digestive systems. Like most complex animals, man has a complete digestive tube (known more formally as the *alimentary canal*), extending from *mouth,* the opening for food intake, to *anus,* the opening for elimination of unused food (Figure 5-1). An organism with this type of digestive system possesses a tube-within-a-tube body plan. Food is processed within the digestive tube, but needed nutrients must be moved out of the tube and distributed to each of the billions of cells of the body.

The process of nutrition begins with *ingestion,* in which food is taken into the digestive tube. During *digestion* food is mechanically and chemically broken down into molecules small enough to pass through the wall of the digestive tube. By means of *absorption* these small nutrient molecules pass through the cells lining the digestive tube, through the cells which line the blood capillaries, and into the blood. Nutrients are *transported* by the circulatory system to each individual cell of the body. The process of nutrition is not complete until the required nutrients are delivered to each body cell. Undigested food is *eliminated* from the body through the anus.

## PROCESSING FOOD

The digestive tube consists of mouth, pharynx, esophagus, stomach, small intestine, large intestine, and anus (see Table 5-1 and Figure 5-2). Food materials are processed as they pass through each of these regions of the tube in sequence. In addition, three types of accessory digestive glands—the salivary glands, the pancreas, and the liver—secrete digestive juices into the tube. These secretions function in breaking down food into its component molecules. Each type of food is chemically digested into its

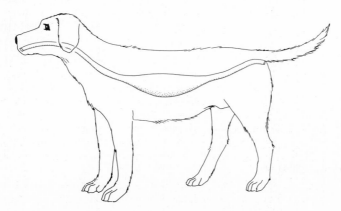

FIGURE 5-1
A tube-within-a-tube body plan is characteristic of all complex animals. In this diagram the digestive tube has been shortened and straightened to illustrate that it is indeed a tube which extends from mouth to anus.

molecular components by means of an orderly sequence of chemical reactions dependent upon specific enzymes.

**Table 5-1. Organs of digestion**

| | |
|---|---|
| *Digestive tube:* | |
| Mouth | Teeth bite and chew food, mechanically breaking it down into small pieces. |
| | Salivary amylase, an enzyme in saliva, initiates chemical digestion of carbohydrates. |
| Pharynx | Throat. Opens into esophagus. |
| Esophagus | Muscular tube leading from pharynx to stomach. |
| | Food is pushed along esophagus by a wave of muscular contraction called *peristalsis* (Figure 5-4). |
| Stomach | An elastic muscular sac which can stretch to a capacity of 1–2 qt as it receives food. |
| | Millions of gastric glands in stomach wall secrete gastric juice needed for chemical digestion. |
| | Churning movements of the stomach help reduce the food to a soupy mixture called *chyme*. |
| | Chyme leaves the stomach through its narrow exit, the *pylorus*. |
| Small intestine | May be up to 20 ft. long, but is only 1 in. in diameter. Most chemical digestion takes place in the *duodenum*, the first 10 in. or so of the small intestine. Small glands in lining secrete digestive juices containing enzymes needed for chemical digestion. Liver and pancreas both empty their secretions into the duodenum. |
| | Nutrients are absorbed through *villi* (Figure 5-6), tiny folds in lining of small intestine. |
| Large intestine (colon) | Shorter but with greater diameter, than small intestine. First section called *cecum*. The vermiform *appendix* hangs down from the cecum. |
| | Food which has not been absorbed in the small intestine passes into the large intestine in a semisolid state. Water is absorbed from these wastes so that they are solidified into the consistency of normal feces. Bacteria which thrive in the large intestine ingest the last remnant of man's meal. Some of these bacteria synthesize vitamin K and certain of the vitamin B complex, which are then absorbed and utilized by the human host. |
| *Accessory digestive glands:* | |
| Salivary glands | Three pairs located near the mouth cavity secrete more than a quart of saliva each day. Saliva initiates carbohydrate digestion, moistens the mouth, dissolves some of the food, and helps lubricate the throat. |
| Liver | Largest internal organ. Secretes bile, which mechanically promotes digestion and absorption of lipids and fat-soluble vitamins. Bile is stored in the gallbladder, then delivered into the duodenum as needed. |
| Pancreas | Secretes pancreatic juice containing several digestive enzymes. |

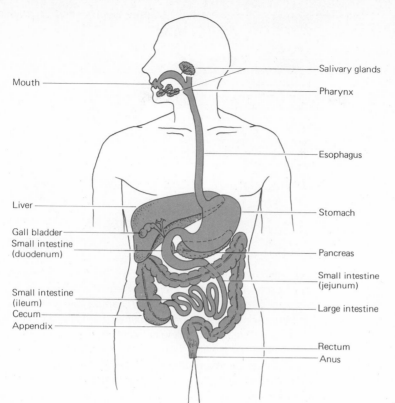

Mouth

Liver

Gall bladder
Small intestine
(duodenum)

Small intestine
(ileum)
Cecum
Appendix

Salivary glands

Pharynx

Esophagus

Stomach

Pancreas

Small intestine
(jejunum)

Large intestine

Rectum
Anus

**FIGURE 5-2**
The human digestive system. Note that the salivary glands, liver, and pancreas are not actually a part of the digestive tube. They are accessory glands which empty their contents into the digestive tube.

## Enzymes

Enzymes are organic catalysts that regulate chemical reactions. They are proteins, and are highly specific. That is, an enzyme can become involved only in one chemical reaction or in a small number of similar chemical reactions. There are thousands of different types of enzymes, each designed to regulate a particular type of chemical reaction. The substance that an enzyme acts upon is called the *substrate.* Although many biochemical reactions can proceed without the help of an enzyme, the rate of the reaction is greatly increased by enzymatic action. In certain reactions, one molecule of an enzyme can catalyze (speed the transformation of) as many as 1 million molecules of substrate per minute. An enzyme is never altered by the reaction, and thus is free to function again and again. Enzymes function by providing a convenient meeting place for reactants. The shape of a particular enzyme molecule permits only substrate molecules of a certain shape and composition to come into close contact with it and ultimately with each other (Figure 5-5). We may think of an enzyme as a molecular lock into which only specifically shaped molecular keys (the reactants) can fit. Before they can perform their job, many enzymes must be assisted by a *coenzyme,* often a vitamin or substance derived from a vitamin. A coenzyme generally accepts atoms from the substrate or donates them to it.

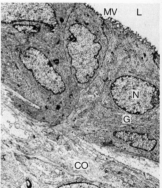

MV  L

N

G

CO

**FIGURE 5-3**
Electron micrograph of epithelial cells lining the digestive tube. Note the microvilli (*MV*), minute outpocketings of the cytoplasm, which greatly increase the absorptive surface of each cell. *L,* lumen (inside space of tube); *N,* nucleus; *G,* Golgi apparatus; *CO,* collagen (a protein characteristic of loose connective tissue) fibers. (×10,000) (Dr. Lyle C. Dearden)

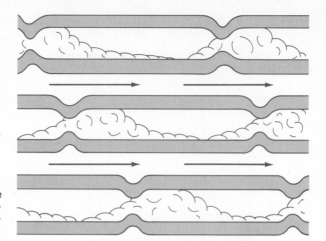

**FIGURE 5-4**
Food is moved along through the digestive tube by waves of muscular contraction called *peristalsis*.

Enzymes are usually named by adding the suffix *-ase* to part of the substrate name. For example, lipase is the enzyme that acts upon a lipid. Some enzymes, such as pepsin, discovered and named before this system of nomenclature had been adopted, have retained their original names. However, any chemical substance bearing a name that ends in -ase is an enzyme.

Enzyme activity may be influenced by a number of factors. (1) The rate of enzymatic activity is influenced by the concentration of substrate and by the concentration of enzyme. (2) Enzymes have an optimum (best) temperature for their activity. A rise in temperature generally increases the rate of enzymatic activity, but above a certain temperature the enzyme is inactivated. Like all proteins, enzymes may be denatured by too much heat. (3) Each en-

**FIGURE 5-5**
How an enzyme works. An enzyme may be thought of as a molecular lock. Only certain molecular keys can fit into the lock, and these are the reactants. In this illustration molecules *A* and *B*, but not *C*, fit the lock and are thus brought into contact by the enzyme. In this reaction *A* and *B* join to form compound *A-B*. This new compound then leaves the enzyme, which is thus free to bring other molecules of *A* and *B* together.

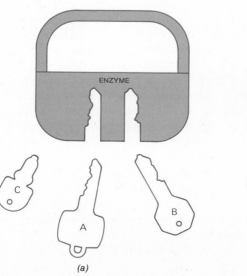

*(a)*

*(b)*

zyme has an optimum pH (level of acidity). Pepsin in the stomach is activated by hydrochloric acid in gastric juice and functions best in a highly acidic environment. When subjected to the alkaline conditions of the duodenum it ceases to function. On the other hand, the pancreatic and intestinal enzymes find such an alkaline medium just right for their activity.

Certain compounds inhibit the normal activity of an enzyme. Knowledge of enzyme inhibition has been put to practical use in medicine, as, for example, in the use of sulfa drugs, which function by inhibiting a necessary biochemical reaction within the infecting bacteria. Sulfa actually confuses the enzyme system because it is chemically similar to the normal substrate. Some inhibitors permanently inactivate, or poison, enzymes, for example, cyanide, which irreversibly poisons certain respiratory enzymes, resulting in death.

## Details of digestion

Table 5-2 provides the details of the step-by-step breakdown of each major group of organic foods. Note that in each case a large molecule is separated into its component molecular subunits. Polysaccharides are digested into simple sugars, lipids into glycerol and fatty acids, and proteins into their component amino acids. Nucleic acids are digested into individual nucleotides. These molecular products of digestion are small enough to be absorbed through the intestinal villi and into the circulatory system.

Regulation and coordination of digestive events are functions of both the nervous and the endocrine systems. These systems act and interact in response to chemical, mechanical, or psychic stimuli imposed by food.

**Table 5-2. Digestion**

| Location of activity | Source of enzyme | Digestive process |
|---|---|---|
| Mouth | Salivary glands | *Carbohydrate:* <br> Polysaccharides $\xrightarrow{\text{salivary amylase*}}$ disaccharides |
| Small intestine | Pancreas | Polysaccharides $\xrightarrow{\text{pancreatic amylase}}$ disaccharides |
| | Intestinal glands | Disaccharide $\xrightarrow{\text{specific disaccharidase}}$ monosaccharide |
| | | For example: sucrose $\xrightarrow{\text{sucrase}}$ glucose + fructose <br> (a disaccharide)  (monosaccharides) |
| Small intestine | Liver | *Lipid:* <br> Fat $\xrightarrow{\text{bile salts}}$ tiny fat droplets |
| | Pancreas | Fat droplets $\xrightarrow{\text{lipase}}$ glycerol + fatty acids |
| Stomach | Stomach (gastric glands) | *Protein:* <br> Protein $\xrightarrow{\text{pepsin}}$ polypeptides |
| Small intestine | Pancreas | Protein $\xrightarrow[\text{chymotrypsin}]{\text{trypsin}}$ polypeptides |
| | Pancreas | Polypeptides $\xrightarrow{\text{peptidases}}$ dipeptides |
| | Intestine | Dipeptide $\xrightarrow{\text{dipeptidase}}$ free amino acids |

* The enzyme required for each reaction is indicated above the arrow.

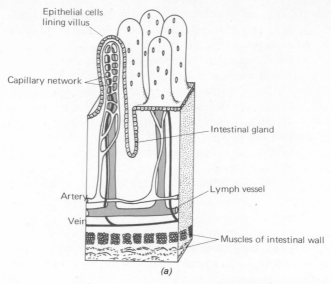

Epithelial cells
lining villus

Capillary network

Intestinal gland

Lymph vessel

Artery

Vein

Muscles of intestinal wall

*(a)*

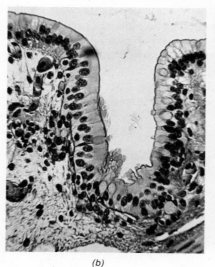

*(b)*

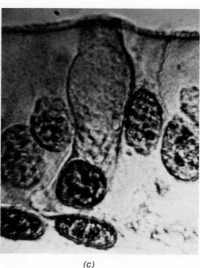

*(c)*

**FIGURE 5-6**
(*a*) Intestinal villi. One villus is sectioned to show the blood vessels and lymph vessel within it. (*b*) Photomicrograph showing parts of two adjacent villi. The clear space is the lumen (inside space) of the tube. (*c*) Goblet cells such as the one shown in the center of this photomicrograph secrete mucus. They are found among the epithelial cells lining the digestive tube.

## Absorption and utilization of nutrients

The molecular end products of digestion are small enough to be absorbed into the blood, which transports them to the individual cells of the body. Only a few substances—water, simple sugars, salts, alcohol, and certain drugs—pass through the wall of the stomach. Most absorption takes place through the lining of the small intestine.

Absorption is the job of the estimated 5 million intestinal villi. As illustrated in Figure 5-6, each villus contains a lymph vessel and a network of capillaries (tiny blood vessels). Nutrients must pass through the villus wall and through the single layer of cells lining the capillary in order to reach the blood.

Some nutrients are absorbed by simple *diffusion.* Molecules tend to distribute themselves evenly throughout a given space, diffusing from a region where they are highly concentrated to a region where they are less concentrated. A familiar example is the diffusion of molecules around the room from a heavily perfumed woman.

Water moves in and out of cells by *osmosis,*[1] the passage of water through a semipermeable membrane (such as the cell membrane). In osmosis, water always moves from a region where dissolved substances are less concentrated to a region where they are more concentrated.

Most nutrients are probably absorbed by the process of *active transport,* a form of work performed by the cell membrane. Energy is required for moving materials across membranes in this way, though the complex details of passage mechanisms are not yet known.

All digestive nutrients are absorbed directly into the blood except the lipids. Lipid components (fatty acids and glycerol) are absorbed through the villi but enter the lymph vessels. They are carried by the lymph system to the region of the left shoulder, where the lymph fluid and its contents enter the blood circulatory system.

Once absorbed, nutrients in the blood are transported from the tiny capillaries of the intestine to the large *hepatic portal vein.* This vein enters the liver, where it divides into a capillary network that enables materials to be exchanged between blood and liver. As blood courses slowly through these capillaries, liver cells inventory the nutrient content of the blood. If a surplus of any type of nutrient is present, the excess is absorbed into the liver cells. Here it is either stored or converted into other materials. Under normal circumstances, blood leaves the liver carrying sufficient nutrients to meet the requirements of all the cells of the body. The blood has been described as a traveling smorgasbord from which each cell selects whichever nutrients it needs.

See Table 5-3 for a list of the more common disorders of the digestive system.

**Fate of carbohydrates.** An important function of the liver is to help maintain a constant blood-sugar (glucose) level. Cells require a constant supply of glucose for use as fuel in cellular respiration. After a meal, when surplus sugar reaches the liver, the excess is converted into a large storage molecule, *glycogen* (Figure 5-7). As the body cells deplete the circulating glucose, glycogen is converted back into glucose and released into the blood. Sugar surpluses exceeding the amount normally stored as glycogen are converted into fat and stored in adipose (fat) cells located throughout the body.

**Fate of amino acids.** Most excess amino acids are removed from circulation by the liver. In the liver cells these are *deaminated,* that is, the amine

---

[1] See Appendix A. Basic Principles of Chemistry, for a more detailed discussion of osmosis.

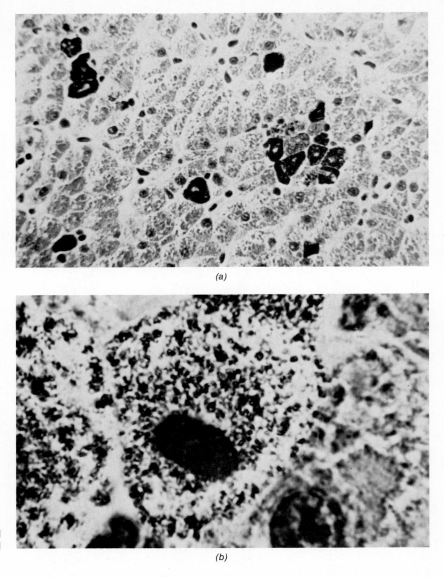

**FIGURE 5-7**
(a) Glycogen granules stored within liver. (b) One liver cell with glycogen.

(a)

(b)

group is removed. Amine groups are waste products and are converted to *urea* for excretion from the body. The remaining carbon-hydrogen-oxygen chain of the amino acid may be converted into carbohydrate or lipid and used as fuel or stored. Thus, even a person who eats protein almost exclusively can become fat if he eats too much. Amino acids circulating in the blood are removed as needed by individual body cells and used primarily for synthesis of proteins (see Table 5-4).

*Fate of lipids.* Although they take a circular route through the lymphatic

**Table 5-3. Common disorders of the digestive system**

| Disorder | Description |
| --- | --- |
| Heartburn | A painful sensation resulting from small amounts of gastric juice being regurgitated from the stomach into the esophagus. |
| Vomiting | Reflex action for rapidly emptying stomach contents through mouth. May occur as a result of ingestion of substances irritating to the stomach, mechanical stimulation of the pharynx, irritation of certain organs such as the inner ear (motion sickness), or even conditioned psychic stimuli. |
| Ulcers | A sore that occurs when a small area of the lining of the digestive tube is digested. Mechanical injury and nervous stress appear to contribute to failure of normal protective mechanisms. |
| Appendicitis | Infection of appendix. |
| Constipation | Condition produced when food moves through the digestive tube too slowly, too much water is removed, and dry, hard feces result. |
| Diarrhea | Condition produced when food moves through the digestive tube too quickly, insufficient water is removed, and watery feces, often accompanied by frequent defecation, result. |
| Gallstones | Condition produced when cholesterol, a component of bile, precipitates (comes out of solution), forming crystals which build up into stones. |
| Cirrhosis of the liver | Condition in which liver has been damaged and nonfunctional tissue replaces normal liver cells. Common in alcoholics and in those who have suffered from hepatitis. |

**Table 5-4. Utilization of nutrients**

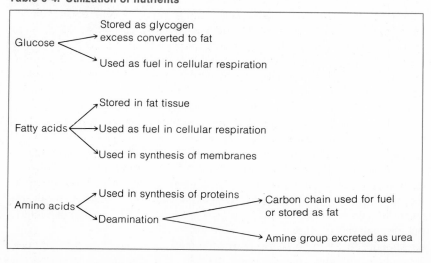

system, excess lipid nutrients do eventually reach the liver. Some are converted by the liver into cholesterol and lipids needed in the production of membranous organelles. Body cells can also utilize lipid components as fuel for cellular respiration. Excess lipids are stored in adipose tissue (Figure 5-9).

## THE PROBLEM OF OBESITY

It is stated in *The Talmud,* "In eating, a third of the stomach should be filled with food, a third with drink, and the rest left empty."

While 14 million Americans do not have enough to eat, 40 million others eat too much. Obesity, the excess accumulation of body fat, is a serious form of malnutrition, and in our affluent society it has become a problem of epidemic proportions. An overweight person places an extra burden upon his heart and, because he is more susceptible to heart disease and other ailments, tends to die at a younger age than persons of normal weight. According to insurance statistics, men who are 20 percent or more overweight bear a 43 percent greater risk of dying from heart disease, a 53 percent greater risk of dying from cerebral hemorrhage, and a 133 percent greater risk of dying as a result of diabetes than men of normal weight. A man who is 20 percent overweight is 30 percent more likely to die before retirement age than if his weight were normal. Still, one-third of our working population is 25 percent or more overweight.

Why are so many persons overweight? Recent studies show that in some cases the problem stems from early childhood. The number of fat cells in the body is apparently determined early in life. When babies or small children are overfed, abnormally large numbers of fat cells are formed. Later in life these fat cells may be fully stocked with excess lipids or may be shrunken, but they are always there. People with such increased numbers of fat cells are thought to be more liable to obesity than those with normal numbers. Thus parents, in their concern for weight gain in their babies, may do them a lifetime disservice by overstuffing them—and making it more difficult for them to control their weight throughout life. Recent research also indicates that there may be individual differences in the efficiency with which we metabolize our food. Most people, it seems, can adapt to changes in the amount of food intake without significant fluctuation in weight. In some cases metabolic disorders may influence the distribution of fat or increase in appetite, but far too many persons blame their obesity upon "glandular problems."

Whatever the underlying causes, the actual process of becoming fat results from overeating. Often obesity can be traced to a combination of poor eating habits and psychologic problems. Some persons compensate for their frustrations or lack of satisfaction in life by overindulging themselves with food. In many cases such attendant psychologic problems must be resolved before the problem of obesity can readily be solved.

To maintain normal body weight, we must maintain a balance between the amount of energy taken into the body and the amount of energy used in activity. Body weight can remain constant only if calorie intake equals calorie expenditure. When we eat more fuel than our bodies need for energy, the excess is stored as fat. Most of us get little physical exercise and so require less food than if we were engaged in heavy physical labor. Our dietary patterns, however, still reflect those of our great grandparents, whose lives required substantial physical activity. Many persons become overweight in middle age as they decrease their physical activities without also decreasing their food intake. Any change in habits of energy expenditure must be accompaniend by new eating habits (see Table 5-5).

People who want to lose weight should reduce their daily intake of calories and increase the energy output by a program of exercise, so that energy output is greater than energy input. The body then draws on its fat reserves for the portion of energy not supplied by the diet. In addition to decreasing the *total* number of calories consumed, the typical reducing diet decreases the proportion of carbohydrates (see Table 5-6). Sufficient high-quality protein must be included to ensure that important body tissues will not be broken down for their amino acid content.

Often a dieting individual becomes discouraged if he does not begin to lose weight immediately. When tissue is metabolized (broken down) for use as fuel, water is temporarily retained to replace it. After 2 or 3 weeks this excess water is excreted, resulting in rapid weight loss at that time.

Because there are so many overweight persons, dieting has generated a multimillion-dollar industry embracing diet foods, formulas, pills, books,

**FIGURE 5-9**
(*a*) Accumulation of fat in adipose tissue. In the first group of fat cells shown, only a few small fat droplets are present. The second group of cells has synthesized more fat, and some of the fat droplets have fused to form larger droplets. The third group of cells are typical fat cells. So much fat is being stored that the fat droplets have fused to form one giant fat droplet, and the cytoplasm has been reduced to a small area around the droplet. (*b*) Photomicrograph of fat tissue. Fat droplets are seen as large black spots.

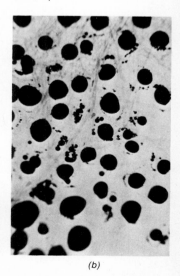

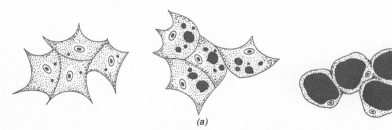

(*a*)

(*b*)

**Table 5-5. Approximate energy expenditure with various activities***

| Activity | Calories per hour |
|---|---|
| Sitting at rest | 15 |
| Walking | 165 |
| Running | 700 |
| Mental work† | 8 |
| Writing | 20 |
| Swimming | 200–700 |
| Bicycling (moderate speed) | 240 |

* Values refer to energy expended above that required for basic metabolism. Energy value computed for an "average" adult.
† Mental effort causes very little increase in metabolic rate. Solving mathematic problems, for example, increases energy expenditure by only 3 or 4 percent.

clubs, and slenderizing devices. Gulping down diet pills has become somewhat of a fad. Amphetamines and amphetamine-like preparations reduce appetite so that the dieter does not feel like eating as much as usual. These drugs may act upon the appetite-control mechanisms in the brain, but their precise mode of action is not known. A tolerance to amphetamines is often developed within a few weeks so that the prescribed dosage no longer causes the intended effect. When this occurs the drug should be discontinued, but instead the dosage often is increased. In fact, may persons become psychologically addicted to amphetamines. They would do better to muster the motivation necessary to reduce food intake without the help of diet pills.

Eating foods of the proper quality is as important as cutting down the quantity of food. Fad diets should be avoided. When carefully analyzed for nutritional balance, effective reducing diets are basically quite similar. Yet every few months a diet put forth in a book or magazine article becomes popular because someone has thought up a new and exciting title. Extreme diets should never be undertaken without the supervision of a physician. In their frantic attempts to lose weight suddenly, people sometimes reduce food intake drastically and rely almost solely on vitamin pills for nourishment, but these supply only vitamins, not other essential organic nutrients. Even those of normal weight often think they can get by nutritionally by eating what they like (such foods as potato chips, desserts, and carbonated beverages) and maintaining a balanced diet through vitamin and mineral supplements. Such supplements are no substitute for proper food and do not supply needed protein, fats, and bulk. When deprived of dietary protein, the body is forced to metabolize needed tissue in order to obtain amino acids. Some vitamin pills also contain amino acid supplements, but these should be used only on the advice of a physician. Parents should be especially careful not to subject their children to fad diets. Because they are growing, children need exceptionally well-balanced diets and can be permanently harmed by their parents' passing nutritional whims.

Another unsatisfactory practice is the reduction of salt and/or water intake. Weight loss does occur, but from dehydration rather than from metabolizing fat tissue. Crash diets cannot be maintained for long periods of time. After a few days or weeks, the dieter reverts to his former eating habits and quickly provides his shrunken fat cells with new storage supplies. The best reducing diets are those which establish new eating habits that can be maintained permanently.

**Table 5-6. The average daily American diet**

| | Calories | Percent of total calories | Grams | Calories per gram |
|---|---|---|---|---|
| Total calories* | 3257 | | | |
| Protein | 376 | 12 | 94 | 4.1 |
| Fat | 1269 | 39 | 141 | 9.3 |
| Carbohydrate | 1612 | 49 | 403 | 4.1 |

* Note that the total calories are in excess of the recommended total, which is about 2800 calories for an adult male and 2000 for an adult female.

Carefully developed diets which contain all essential nutrients and require that certain foods be eaten (as well as others avoided) go a long way toward staving off hunger and thus overcoming one of the largest barriers to successful dieting. Many persons on the famous "Weight-watchers' Diet," for example, claim that they are less hungry and more satisfied than when they were consuming unbalanced diets containing far more calories.

## FOOD ADDITIVES

Next time you are in a supermarket take the time to read the ingredients on a box of cookies, instant potatoes, or other prepared foods. Chances are you will be mystified, if not appalled, at the list of food additives such as artificial flavoring, emulsifiers, artificial coloring, preservatives such as BHA and BHT, and so on. Each of us consumes an estimated 5 pounds per year of chemicals that have been deliberately added to our foods. Food additives are big business. The food industries buy about $500 million worth annually. About 2,000 different additives are used to facilitate processing, prevent spoiling, render food enticingly colored and flavored, and keep it properly moist and appealing for the American consumer. An estimated 60 percent of additives are used to make food more attractive—that is, improve color, flavor, or texture. Besides these intentional additives, a wide array of pesticides, hormones, and antibiotics used in food production also find their way onto our dinner tables.

Some additives are natural substances such as sugar and salt, but many others are synthetic chemicals foreign to the body. Just how safe are these chemicals? No one seems to know for sure, but increasing numbers of journal and newspaper articles are raising alarming questions about some commonly used additives, the safety of which everyone has taken for granted. The recent furor over artificial sweeteners, the debate over monosodium glutamate added to baby foods, the evidence that nitrite, a commonly used preservative (read the ingredients on your frankfurter or luncheon meat package), causes cancer in mice, and the mounting evidence that DDT continues to build up in our food supply, are examples of the uncertainty that surrounds the issue of safe food. Moreover, the export of American oranges has been endangered because the Food and Drug Administration allows 40 percent more decay retardant (biphenyl) than is permitted by the European Common Market. And in 1970 Sweden banned imports of United States beef and liver for fear that the hormones used to increase beef yield may present a hazard to consumers.

We might rightfully wonder, who is watching the store? The Food and Drug Administration (FDA) of the U.S. Department of Health, Education, and Welfare is the agency primarily responsible for safeguarding the American food supply. In a 1970 report written by FDA officials themselves, inadequate legislation, money, and manpower are cited as factors preventing the FDA from dealing adequately with the problems. Powerful lobbies representing the food industry seem to strive to keep the FDA ineffective. On the bright side, consumer advocate Ralph Nader and others have recently done a great deal to bring matters of food safety to public attention.

The cyclamate story serves as warning to the American consumer that each of us must keep informed and ready to exert consumer pressure. Many food additives are legally classified according to a 1958 law[2] as "Generally Regarded as Safe" (GRAS). Many of these additives, however, have not been subjected to the scientific testing now required for other food additives. Cyclamates, used as artificial sweetners, were placed on the original 1958 GRAS list of almost 200 additives despite a recommendation by the National Academy of Sciences that distribution of foods containing cyclamates should be controlled. By 1968 three out of four American households were using artificially sweetened foods and drinks, and in that year 17 million pounds of cyclamates were produced. These chemicals were used to sweeten children's vitamins, to cure bacon and ham, and in other food and drug processing, with no requirement that they be listed as an added ingredient! At that time it would not even have been possible for a consumer to estimate his daily intake of cyclamates. While cyclamates were rising swiftly in popularity both among food manufacturers and among the figure-conscious public, an increasing number of scientists were discovering that cyclamates were not really as safe as the FDA had thought. Over a 15-year period it was shown that cyclamates cause bladder cancer in rats, cause birth defects in chick embryos, and break chromosomes, and that during processing and in the human body cyclamates are converted into a toxic substance called CHA. Finally in 1968 the FDA moved to warn Americans that artificially sweetened diet drinks should not be consumed by the general public. Ironically, studies comparing the effects of cyclamates and sugar on body weight showed no significant difference. Because cyclamates lower blood-sugar levels, they lead to an increase in food intake and ultimately an increase in body weight. For all the diet drinks consumed, the battle of the bulge was still not being won.

Another interesting case is that of DES (diethylstilbestrol), a synthetic chemical which mimics the action of natural female sex hormones. When added to cattle feed it stimulates growth, fattening the animals more quickly and on less grain (a savings of millions of dollars each year to the cattle industry). DES was first approved for use in cattle in 1954. The problem? DES ends up in the beef we eat in concentrations larger than those shown to cause cancer in mice.[3] According to the Delaney clause in the Food Additive Amendment of 1958, no substance shown to cause cancer in man or animals may be added to food, and at the time of this writing a legal battle is being waged over the use of DES. Since DES is fed to about 75 percent of the cattle slaughtered each year in the United States, banning its use is likely to cause an estimated 4 cents per pound rise in the price of beef.

---

[2] The 1958 Food Additives Amendment to the Food, Drug, and Cosmetic Act (1938).

[3] In 1971 eight women were admitted to a Massachusetts hospital with a rare form of cancer which was traced to DES exposure each had experienced as a fetus. Their mothers had been treated with this hormone as an abortion preventive. The ability of DES actually to prevent abortion, incidentally, is questionable. Since this 1971 study many additional cases of DES-related cancer have been identified.

Yet another subject for debate is the use of sodium nitrite (or sodium nitrate, which through bacterial action is converted to sodium nitrite). Sodium nitrite is added to almost all canned meat and cured-meat products such as hot dogs, lunch meats, bacon, ham, and sausages, as well as to smoked fish. Though it is used primarily as a preservative, sodium nitrite also gives these foods their red color and contributes to their flavor. Researchers have recently found that nitrite may react with certain other chemicals (secondary and tertiary amines) in these foods to form cancer-producing *nitrosamines.* About 80 percent of nitrosamines tested have been shown to cause cancer in laboratory animals. All types of animals, including monkeys, thus far tested have been susceptible to the carcinogenic (cancer-producing) action of these chemicals. Although long-term studies have not yet been carried out on human beings, nitrosamines are almost certain to act as carcinogens in man also. Some scientists feel that any amount of carcinogen in food is too much, and that sodium nitrite should no longer be used. The meat industry is understandably worried. Seventy percent of all pork sold in the United States, for example, is either cured or canned. And if a satisfactory substitute for sodium nitrite is not found the all-American hot dog may well become a thing of the past.

A class of unintended additives which we should mention is filth. In 1972 the FDA released an eight-page set of guidelines describing the amount of filth it considers as "natural and unavoidable" in certain foods. For example, one rat or mouse pellet per pint of wheat is permissible, as is mold or insect damage in 10 percent of coffee beans.

The problem of food-additive safety is extremely complex. It is not a simple matter to prove that a particular chemical is harmful. First, since we consume hundreds of additives, it is difficult to demonstrate a cause-and-effect relationship between a single chemical and a specific physiologic effect. And second, it may take years, even generations, before the effects of additives become apparent. For example, human bladder cancer has a latent period averaging 15 years. Should we continue using cyclamates for this 15-year waiting period in order to find out for certain that these chemicals cause human, as well as mouse, bladder cancer? Genetic damage may not manifest itself until one attempts to produce offspring. Even then, the first generation, or even the second may be fortunate and not exhibit outwardly the inner damage of mutated genes. But genetic damage is inherited and passed on to future generations. Do we want to take this risk? Or should we refrain from adding chemicals to our foods until substantial independent research establishes that they are safe?

## HEALTH FOODS

Concern over food additives has sparked new interest in "health foods," a term which encompasses a wide variety of products, including foods processed without harmful additives, "meatless meats," and organic foods (produce grown without chemical fertilizers or pesticides). Most health food stores also carry a vast array of vitamin supplements, many of dubious value.

Because health foods are presently more expensive to produce, their cost to the consumer is higher than conventionally processed foodstuffs. Are they really worth their price? The answer depends upon the particular food in question and upon its legitimacy as a naturally prepared food. Produce which is really grown organically may indeed be healthier, and certainly does not take the toll on the environment exacted by conventionally grown foods. Peanut butter, cookies, and other products prepared without additives no doubt are worth the price difference. "Meatless meats," foods that promise to become increasingly popular, are made of textured vegetable (mainly soybean) protein. Designed to look and taste like steak, meatloaf, or other common meat dishes, these vegetables are actually less expensive than the real thing. Nutritionally these synthetic meats may be superior, since they contain no cholesterol and are low in saturated fats.

Like all growing industries the health-food business has attracted some unscrupulous promoters whose exaggerated or false claims may deceive unwary consumers. The label "health food" does not guarantee the cleanliness of a product, nor does it ensure that a particular food was actually organically grown. Consumers should beware claims that certain foods or diets can cure cancer, epilepsy, sexual inadequacy, or any other medical condition. Homely remedies are sometimes substituted for rational medical treatment with disastrous results. Those interested in health foods would do well to study nutrition and investigate local dealers sufficiently to know just what benefits may be gained from their investment.

1 As food passes through the digestive tube it is mechanically and chemically broken down into small molecules.
2 Three accessory digestive organs—the salivary glands, the liver, and the pancreas—pour their secretions into the digestive tube.
3 The chemical reactions of digestion depend upon the activity of specific enzymes.
4 Nutrients are absorbed through the intestinal villi into the circulatory system, which transports them to the various cells of the body.
5 The liver is responsible for processing absorbed nutrients. It stores excess carbohydrate, deaminates excess amino acids, and converts surplus nutrients to fat for storage in adipose tissue.
6 Obesity is a consequence of imbalance between fuel intake and energy output. Eating habits are important both in causing the condition and in solving the problem.
7 Many common food additives have not been adequately tested to establish that they are safe.
8 "Health foods," including organically grown produce, meatless "meats," and foods prepared without food additives, are becoming increasingly popular.

Griffin, Donald R., and Alvin Novick: *Animal Structure and Function,* Holt, New York, 1970, chap. 4. A discussion of the types of digestive systems found in various organisms.

Jacobson, Michael F.: *Eater's Digest, The Consumers Factbook of Food Additives,* Doubleday, New York, 1972. An in-depth discussion of some commonly used and potentially hazardous additives.

Turner, James S.: *The Chemical Feast,* Grossman, New York, 1970. An indictment of the FDA by a member of Ralph Nader's Study Group on the FDA and food additives.

Wade, Nicholas: "DES: A Case Study of Regulatory Abdication," *Science,* July 28, 1972, p. 335. Documents the use and abuse of DES over the past several years.

# 6

# INTERNAL TRANSPORT

*Chapter learning objective.* The student must be able to describe the structures and functions (including immunity) of the circulatory system and discuss its common disorders.

*ILO 1*  List seven functions of the circulatory system and explain each.

*ILO 2*  Describe the composition of blood and give the function for each component.

*ILO 3*  Describe the structure and give the function of each part of the heart.

*ILO 4*  Trace a drop of blood through the pulmonary and systemic circulatory systems, naming each vessel through which it passes in proper sequence.

*ILO 5*  Describe and give the causes of the common disorders of the circulatory system such as atherosclerosis, anemia, leukemia, coronary thrombosis.

*ILO 6*  Describe the structure and functions of the lymph system.

*ILO 7*  Explain immunity in cellular and molecular terms, and describe the immunologic basis of vaccines; distinguish between active and passive immunity.

*ILO 8*  Review the major problems encountered in organ transplants and the attempts which are being made to solve them.

*ILO 9*  Define each new term introduced in this chapter.

Each cell of a living system requires a constant supply of oxygen and nutrients and some means of getting rid of metabolic wastes. Very small organisms have few cells and each is in close contact with the surrounding environment. Gas exchange in such organisms depends upon simple diffusion, and each cell is able to take care of its own nutritional needs. However, diffusion of gases occurs more than 10,000 times faster through air than through tissues. As long as an organism is less than 1 millimeter thick, diffusion is adequate, but if an organism is to grow any larger it must possess some other mechanism for transporting materials to all its cells.

Complex plants get around this diffusion efficiency problem with special systems of air spaces between cells which extend throughout the plant. Air enters the plant body through special openings and passes through the system of air spaces to the innermost cells. Nutrients are distributed by a specialized vascular system which conducts water and minerals from the roots to the upper tissues, and transports carbohydrates from the leaves to other cells.

In large multicellular animals the problem of internal transport is most efficiently solved by the development of a specialized circulatory system. In man, and other vertebrates, the circulatory system consists of a heart, blood vessels, blood, lymph, lymph vessels, and associated organs such as the thymus, spleen, and liver. Some of the vital functions performed by the circulatory system are:

1 Transport of nutrients to each cell of the body
2 Transport of oxygen to each cell of the body
3 Transport of wastes from each cell to excretory organs
4 Transport of hormones
5 Regulation of inorganic substances and fluid balance
6 Protection of the body against invading microorganisms
7 Regulation of body temperature


## THE BLOOD

An adult man possesses about 5 quarts of blood. Whole blood consists of a fluid known as *plasma* in which several types of cellular components are suspended.


### Plasma

When whole blood is prevented from clotting and allowed to stand for a few hours, the cellular components sink gradually to the bottom of the test tube, and thereby become separated from the plasma. Plasma consists of water (92 percent) and a variety of dissolved proteins and other chemical substances, including nutrients such as glucose, dissolved gases, salts, hormones, and wastes. Plasma is yellowish, with a slightly alkaline pH.

Three types of protein found in plasma are *albumins, globulins,* and *fibrinogen.* Plasma proteins, especially the albumins, regulate water balance between the blood and the surrounding tissue fluids. Albumins also

**Components of whole blood**

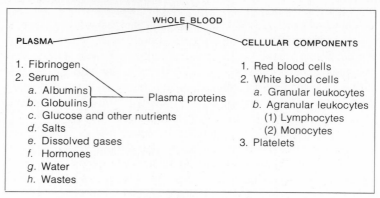

transport molecules such as fatty acids which are not very soluble in blood. Fibrinogen functions in the clotting process (see Platelets, further on in this chapter). When fibrinogen is removed from plasma, the remaining liquid is called *serum*. Many of the globulins are *antibodies,* substances which provide immunity against invading disease organisms or other foreign substances.

### Red blood cells

Each of us has about 23 trillion ($23 \times 10^{12}$) red blood cells, or *erythrocytes,* suspended in our plasma. These cells are so tiny that about 3,000 of them lined up end-to-end would measure only 1 inch. Red blood cells transport oxygen.

Red blood cells develop inside certain bones in special tissue called red bone marrow (Figure 6-1). As each cell differentiates, it manufactures great quantities of the protein *hemoglobin,* the pigment which gives blood its characteristic red color. Each hemoglobin molecule contains four atoms of iron, each of which can combine with one molecule of oxygen. Thus, each hemoglobin molecule can combine chemically with four molecules of oxygen to form the compound *oxyhemoglobin*. It is in this molecular condition that oxygen is transported throughout the body.

The more hemoglobin contained within a red blood cell, the more oxygen

**FIGURE 6-1**
(*a*) Photomicrograph of normal blood. (*b*) Photomicrograph of bone marrow tissue. (*c*) Blood of patient with sickle-cell anemia. (Dr. Sorrell Wolfson)

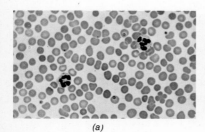

(*a*)

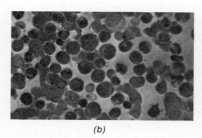

(*b*)

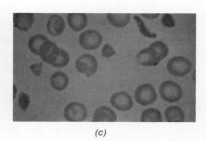

(*c*)

it can transport. Perhaps this is why during the differentiation of a red blood cell most of its organelles—nucleus, nucleolus, mitochondria, ER, and ribosomes—are discarded, providing more space for hemoglobin molecules. The spherical shape of the immature cell gradually changes into a double concave, doughnut-like affair, a shape wonderfully suited to its function of absorbing and releasing oxygen. Thus the mature red blood cell is a tiny sac of hemoglobin, completely lacking a nucleus, and shaped with a considerable surface area for gas exchange.

Having no nucleus, a mature red blood cell is severely limited in its ability to carry on normal cellular functions. Probably for this reason, the red blood cell has a short life-span—about 120 days. As blood circulates through the liver and spleen, worn-out red blood cells are removed from circulation and destroyed. Their hemoglobin molecules are taken apart so that some of the components such as iron can be sent back to the red bone marrow for reuse. It is interesting to contemplate that about 2.4 million red cells are destroyed each second. These are immediately replaced by new red blood cells from the bone marrow. The body is indeed a highly productive machine.

The inability to carry sufficient oxygen to meet the demands of the cells is called *anemia*. This condition results from too few or abnormal red blood cells. In the final analysis anemia can be traced to an inadequate amount of hemoglobin. The most common cause is a deficiency of iron in the diet. Without iron, the body cannot manufacture hemoglobin. Pregnant women, who require especially high levels of iron in the diet, and babies on milk diets are highly susceptible to this type of anemia.

An inadequate amount of vitamin $B_{12}$ is another cause of anemia, since this vitamin is necessary for normal maturation of red blood cells. Some persons have difficulty in absorbing this vitamin. Anemia may also be caused by loss of blood, as in hemorrhage, which decreases the amount of circulating hemoglobin as well as the total blood volume. Hookworm infestation can cause a chronic loss of blood, which also results in anemia. *Hemolytic anemia* results from disease which destroys red blood cells more quickly than they can be replaced. An example of a hemolytic agent is the malaria parasite. Another example of hemolytic anemia, sickle-cell anemia, is an inherited condition in which hemoglobin molecules are not synthesized accurately. The resulting abnormally shaped red blood cells are fragile and are easily destroyed during circulation.

### White blood cells

White blood cells, known as *leukocytes,* protect the body against invading bacteria and other foreign substances. These cells are able to leave the blood, passing out through the walls of the capillaries. Capable of independent locomotion similar to that of an ameba, they wander through the connective tissues of the body. Great numbers are present in the lungs and intestinal walls, areas where air and food intake bring foreign matter in contact with body tissues. Many white blood cells wander into the large intestine or urinary system and are expelled from the body with wastes. The average life-span of a granular leukocyte is only 5 to 10 days.

There are five different types of white blood cells. Three of them are referred to as *granular leukocytes* because they contain large characteristic granules in their cytoplasm. The remaining two varieties lack specific granules and are classified as *agranular leukocytes*.

Like red blood cells, granular leukocytes are formed in the red bone marrow. Their job is to combat infection. They serve as a barrier to spreading bacteria and provide digestive enzymes which immobilize or destroy bacteria. Some white cells actually ingest bacteria and kill them with lysosomal enzymes. Eventually the strong enzymes damage the leukocytes, as well as the bacteria they have ingested, and leukocytes die in great numbers as a consequence of their activities. Honor your leukocytes—they fight and die for your well-being.

The two types of agranular leukocytes are *monocytes* and *lymphocytes*. These cells are produced in the lymph glands, spleen, and thymus. Whenever needed, monocytes leave the blood and enter the connective tissues. Here they develop into cells called *macrophages*. The term macrophage means "big eater," and these cells ingest certain types of bacteria and all sorts of debris resulting from cellular breakdown. Macrophages are the janitors of the body. Lymphocytes are responsible for the production of antibodies, as will be discussed presently.

Normally there are only 6,000 to 8,000 white blood cells per milliliter of blood. In contrast, the more numerous red blood cells may number as many as 5 million per milliliter, so there are about 1,000 red cells for every white cell. When bacteria infect any part of the body a chemical message is sent to the red bone marrow, which responds by sending great numbers of granular leukocytes into the circulatory system. In a short time the number of white blood cells per milliliter may reach as many as 30,000. For this reason, a white blood cell count is standard procedure in diagnosing the presence of a bacterial infection. During viral infections, the white blood cell count usually decreases, but the reason for this is not known. During some parasitic infections a specific type of granular leukocyte may increase in number. This can aid a physician in making a diagnosis.

*Leukemia* is a blood cancer characterized by an uncontrolled multiplication of the cells that normally give rise to white blood cells. The white blood cell count may rise to 1 million per milliliter of blood. These white blood cells are abnormal and cannot function effectively in destroying bacteria, so that the victim often dies of infection. In addition, the tremendous numbers of white blood cells forming in the bone marrow crowd out developing red blood cells, which results in severe anemia (Figure 6-2). Disorders also arise in the clotting mechanism because the large cells that give rise to platelets are unable to develop properly. Such interference with normal body function leads to eventual death. The course of the disease may be slowed by administering drugs that inhibit mitosis, thus slowing the production of white blood cells. By the judicious combination and precise timing of drug administration, physicians have been able to induce remissions for as long as 5 years in some cases. Whether these are actually "cures" is not yet certain. At present a possible vaccine against leukemia is being tested.

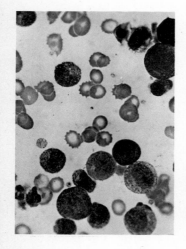

**FIGURE 6-2**
Chronic granulocytic leukemia. Abnormal numbers of mature granular leukocytes, as well as immature ones, are seen in the blood (large, dark, granular cells). Immature red blood cells bearing nuclei may also be seen.
(Carolina Biological Supply Company)

Leukemia is primarily a disease of childhood. Its cause is not known, but there is some evidence that it results from a combination of factors, including a genetic tendency coupled with infection by a virus.

## Platelets

Certain large cells in the bone marrow bud off small fragments of their cytoplasm, called *platelets*. A platelet is not actually a whole cell, but a bit of cytoplasm enclosed by a cell membrane (Figure 6-3). It contains whatever cytoplasmic organelles were present when its cytoplasm was part of the mother cell, but entirely lacks a nucleus.

Platelets patch or plug damaged blood vessels so that blood will not spill out. When the wall of a blood vessel is injured, as when you cut your finger, platelets seal the break by adhering to the wall in large numbers. This process is aided by a complex series of chemical reactions which produce tiny fibers. The fibers reinforce the platelets, forming a strong clot. A summary of this process follows:

Injured tissue or platelets release thromboplastin, an enzyme that triggers the following sequence of events:

Thromboplastin activates a plasma protein called prothrombin.*

$$\text{Prothrombin (a plasma protein)} \xrightarrow[\text{calcium}]{\text{thromboplastin}} \text{thrombin (active form of prothrombin)}$$

Note that calcium is required for the above reaction. The thrombin serves as an enzyme, activating the plasma protein, fibrinogen.

$$\text{Fibrinogen} \xrightarrow{\text{thrombin}} \text{fibrin}$$

Fibrin consists of tiny threads, which serve to reinforce the platelet clot and also entrap red and white blood cells.

Certain globulins, referred to collectively as the antihemophilic factor, also participate in the clotting process. Their absence is associated with bleeder's disease, hemophilia.

* Vitamin K is necessary for the production of prothrombin.

## Blood transfusion[1]

When extensive quantities of blood have been lost through accident, surgery, or chronic disease, the blood transfusion is a lifesaving tool of medicine. Not even plasma is an adequate substitute for whole blood in many cases.

Yet despite the great demand for whole blood, only about 3 percent of the adult population of the United States donates blood regularly. With a shelf life under refrigeration of only about 2 weeks, blood is also a very perishable

---

[1] Blood types and their inheritance will be discussed in Chapter 16.

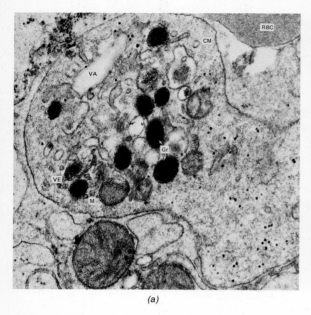

(a)

(b)

**FIGURE 6-3**
(a) Electron micrograph of a platelet. (×74,000) Note the absence of a nucleus. Part of a red blood cell (RBC) may be seen above platelet at right of photograph. (*CM,* cell membrane; *VA,* vacuole; *Gr,* granules; *VE,* vesicles; *M,* mitochondrion.) (Dr. Lyle C. Dearden) (b) Scanning electron micrograph of red blood cell enmeshed in fibrin. (Emil Bernstein and *Science*)

commodity. It is believed that if as few as 5 percent of the populace were to donate blood regularly, both the demand for blood and the problems created by its perishability would be met.

One may donate a pint of blood as frequently as once every 2 months with no known ill effects. The donor is usually made comfortable on a couch or special chair, a needle is inserted in the antecubital vein of the arm, and the blood is drawn off by gentle suction while the donor reads or relaxes. Perhaps because of fears arising from inadequate knowledge, many persons are reluctant to donate blood, but the procedure is as safe as any known to medicine.

## THE ORGANS OF CIRCULATION

The structures of the circulatory system include the blood vessels, heart, liver, and spleen. The lymph structures, though also circulatory organs, form a specialized subsystem and will be discussed separately.

### Blood vessels
Blood flows throughout the body within a continuous system of tubes, the blood vessels. Since blood does not pour freely into any of the body cavities, man and other vertebrates are said to have a *closed circulatory system.* It differs from the open systems found in some animal groups such as the insects.

There are three main types of blood vessels: *arteries, capillaries,* and *veins.* An artery is a blood vessel that carries blood away from the heart, toward other tissues. As an artery approaches or enters an organ, it divides into many smaller branches. Each branch carries blood to a different part of

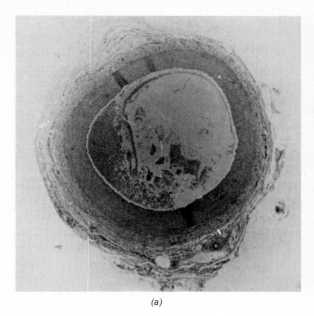

(a)

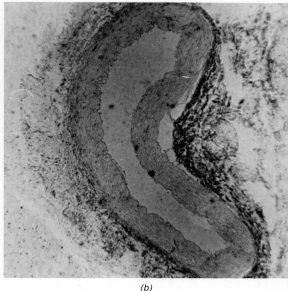

(b)

**FIGURE 6-4**
(a) Section through an artery. (b) Section through a vein.

the organ. Within the tissues of the organ the artery branches divide into smaller and smaller vessels until they are so tiny that their walls are only a single cell thick. These tiny vessels are called *capillaries*. After coursing through an organ, capillaries eventually merge to form larger and larger vessels that transport the blood back toward the heart. These are known as *veins*.

The walls of arteries and veins are thick, which prevents gases and nutrients from passing through them (Figure 6-4). The exchange of these materials between the blood and individual cells of the body tissues takes place through the thin walls of the capillaries (Figure 6-5). The network of capillaries is so extensive that at least one of these tiny vessels is located close to almost every cell in the body. It has been estimated that the total length of all capillaries in the body is more than 60,000 miles!

## Heart

The blood circulates through the body as a result of the heart's pumping action. Located in the chest cavity, the heart is a muscular organ that consists of four chambers (Figure 6-6). Each side of the heart has a receiving chamber, the *atrium,* and a pumping chamber, the *ventricle.* The wall of the heart is composed of cardiac muscle covered by a tough connective tissue membrane. Another membrane, the *pericardium,* surrounds the entire heart but is separated from it by a space known as the *pericardial cavity.* This cavity contains lubricating fluids for reducing friction. Inside, the heart has four *valves,* flaplike structures which prevent blood from flowing backwards. A valve is located between each atrium and its respective ventricle and between each ventricle and its associated artery. Each valve closes or opens at the appropriate time in response to pressure and muscular activity.

**FIGURE 6-5**

Electron micrograph of a section through a capillary. The capillary has been sliced at an angle so that it is larger than a cross section. The red blood cell (RBC) shown was also sliced so that just the tip of it is visible. Note how thin the capillary wall is. The black line represents the length of 1 micrometer. (×41,200) *L*, lumen (space inside capillary); *BM*, basement membrane (thin layer of nonliving material which binds epithelial to underlying connective tissue); *CAP*, capillary; *PV*, small droplets of fluid which have been taken in by the cell membrane. This capillary was from the heart muscle of the right ventricle. Contrast the thickness of the capillary wall with the walls of the artery and vein in Figure 6-4. (Dr. Lyle C. Dearden)

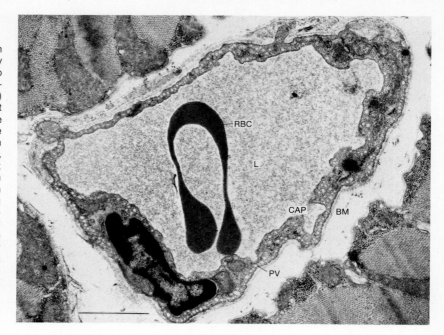

***The heartbeat.*** When a physician listens to the heartbeat (usually with a stethoscope) he is able to distinguish two distinct sounds that occur in repeating rhythm. The first sound, called a "lubb," is heard as first the atria and then the ventricles contract. This sound is primarily caused by the closing of the valves between the atria and ventricles. The second sound is a short, sharp "dupp," heard as the heart begins to relax; it is mainly a result of the valves snapping shut at the entrance of the two large arteries leading from the heart. The phase of the heart's cycle when the ventricles contract is called *systole,* and the phase of relaxation is known as *diastole.* In an average lifetime, the heart beats an estimated 2.5 billion times.

Even when removed from the body, the heart continues to beat and will function for some time if kept in an appropriate liquid medium. Each heartbeat begins in a specialized node of tissue called the *pacemaker,* which is located in the right wall of the atrium. From the pacemaker impulses are transmitted through the muscle fibers of the atria to a second node, located in the wall between the atria just above the ventricles. From here, impulses sweep through specialized fibers to all parts of the ventricles.

Two separate nerves control the rate at which impulses for contraction develop in the pacemaker. Branches of the *vagus nerve* coming from the brain slow the heartbeat, while a second nerve coming from the spinal cord increases the rate of the heartbeat. These nerves are part of an elaborate system of self-regulation controlled by the brain, which receives various messages concerning the changing needs of the body for blood. When the body is under stress the hormone epinephrine stimulates the frequency and force of the heartbeat.

The normal heart rate is about 70 beats per minute, and the output of blood during this time is about 5 quarts (an amount equivalent to the total volume of blood in the body). During vigorous exercise the heart may beat as many as 200 times per minute and its output may increase to 30 quarts. (This means that an amount equal to the total blood volume in the body would be pumped through the heart every 12 seconds.) The heart of an athlete is conditioned to pump a greater quantity of blood per beat, and is thus very efficient. The athlete's heart does not have to beat as often as the heart of a person who is not in good physical condition to distribute the same quantity of blood.

Each wave of heart contraction is accompanied by a characteristic pattern of electrical activity. This can be measured and recorded so that the electrical activity of the heart can be analyzed. Such a tracing on graph paper or on an oscilloscope is called an EKG or electrocardiogram. Physicians use EKGs as diagnostic tools because variations in the normal pattern of electrical activity may indicate cardiac malfunction.

Some cases of malfunction of the heart's impulse conduction system can be corrected by an *artificial pacemaker,* a battery-powered device, that supplies the regular electrical impulses needed to keep the damaged heart pumping. Thousands of persons now depend upon implanted electronic pacemakers to keep their hearts beating at normal rates.

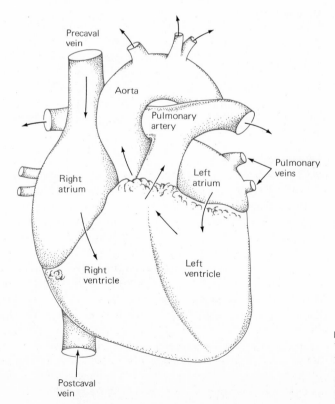

**FIGURE 6-6**
The human heart. Try to follow the path of blood from the veins emptying into the right atrium through the pulmonary circulation and into the aorta.

*Fibrillation* is a disturbance of the heart in which the beats are irregular in both force and rhythm. In extreme cases, the fibers of cardiac muscle contract independently of one another, so that the heart twitches rather than beats. Normal rhythm may be restored by electric shock administered from a special device (cardioverter), or in mild cases by certain drugs such as digitalis.

*Pulse and blood pressure.* When the ventricle contracts, blood is forced out under great pressure into the arteries. Then, as the ventricle relaxes, the pressure decreases. *Pulse* is the rhythmic sequence of high pressure, low pressure that can be felt in an artery. Pulse rate reflects both the speed and regularity of the heartbeat.

The physician uses an instrument called a sphygmomanometer to measure arterial blood pressure. Pressure is reported as a ratio of systolic pressure over diastolic pressure. In a young adult a normal blood pressure would be about 120/80.

By the time blood reaches the veins its pressure is very low. For this reason, veins are equipped with flaplike valves to prevent blood from flowing backwards. Getting blood up through veins in the legs and back to the heart is a constant struggle against gravity. Veins near the surface of the body are poorly supported by connective tissue, and sometimes, when their walls are not strong enough, they stretch, to accommodate blood that accumulates between successive valves. When stretched in this manner, valves no longer meet across the increased diameter of the vein, leading to the pooling of blood within the vein and further stretching the vessel. Such stretched veins are called *varicose veins*. The weak vein walls that predispose one to this condition are thought to be inherited. Varicose veins also occur in the vicinity of the rectum or anus, where they are known as *hemorrhoids*. Hemorrhoids may be painful and may interfere with normal bowel function. When severe, they must be surgically removed.

### Liver and spleen

In Chapter 5 we described the liver as the organ that inventories the nutrient content of the blood and processes whatever surplus may be present. By storing glucose in the form of glycogen and releasing it as needed, the liver regulates the concentration of sugar in the blood. The liver also produces most of the plasma proteins, and along with the spleen, removes worn-out red blood cells from the circulation.

The spleen is a soft, vascular organ in the upper part of the abdominal cavity beneath the diaphragm (Figure 6-7). As the spleen removes abnormal or worn-out red blood cells from the circulation, its cells convert hemoglobin from them into bile pigments, which are then transported by the blood to the liver.

As blood flows through the spleen, it is cleared of bacteria and other harmful matter by macrophages which line the splenic blood vessels. Many lymphocytes and other cells important in immune responses occupy the spleen, and it is an important site of antibody formation.

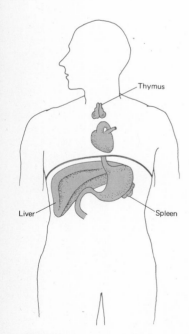

Thymus

Liver

Spleen

**FIGURE 6-7**
Diagram showing position in the body of the liver, spleen, and thymus.

In many animals, the spleen acts as a reservoir of concentrated red blood cells, releasing them to the blood when they are needed, as after heavy bleeding. However, this does not appear to be an important function of the human spleen.

When the spleen has to be removed because of injury, other tissues are able to assume its functions. The liver, however, is indispensable.

## THE PATTERN OF CIRCULATION

Blood is circulated through two different systems: the systemic circulation and the pulmonary circulation (Figure 6-8a). The pulmonary circulation routes blood between the heart and the lungs, while the systemic circulation carries blood between the heart and all other organs of the body, including the functioning cells of the lungs and the very walls of the large blood vessels.

### Pulmonary circulation

The right atrium of the heart receives blood returning from the body cells. This blood is low in oxygen, i.e., *deoxygenated,* having delivered its oxygen content to the cells. It carries carbon dioxide wastes received from the body cells. Deoxygenated blood must be sent into the pulmonary circulation so

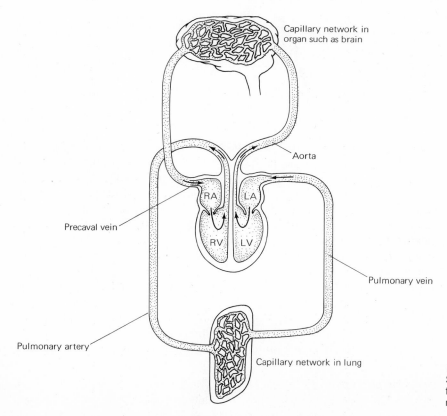

**FIGURE 6-8a**
Simplified diagram of circulation through systemic and pulmonary networks.

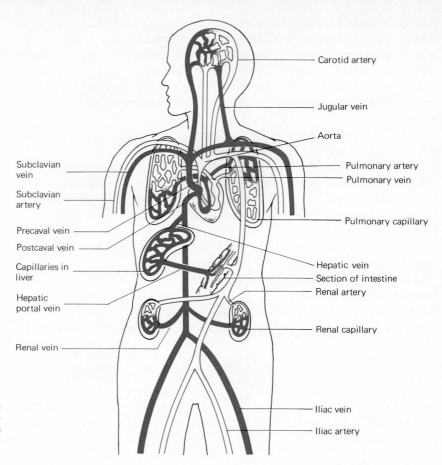

Carotid artery

Jugular vein

Aorta

Pulmonary artery

Pulmonary vein

Pulmonary capillary

Subclavian vein

Subclavian artery

Precaval vein

Postcaval vein

Capillaries in liver

Hepatic portal vein

Renal vein

Hepatic vein

Section of intestine

Renal artery

Renal capillary

Iliac vein

Iliac artery

**FIGURE 6-8b**
Diagram of circulation through systemic and pulmonary networks. This diagram traces the flow of blood to and from several organs.

that it can rid itself of carbon dioxide wastes and pick up a new supply of oxygen. Accordingly, the blood from the right atrium enters the right ventricle, which pumps it into the *pulmonary arteries* leading to the lungs, the only arteries which carry deoxygenated blood. In the lungs the pulmonary arteries branch into an extensive network of pulmonary capillaries. After coursing throughout the lung tissue, pulmonary capillaries merge into pulmonary veins, which carry the blood back to the heart. These veins are the only ones in the body which carry blood rich in oxygen. The pulmonary veins empty into the left atrium.

In summary, blood flows through the vessels of the pulmonary circulation in the following sequence:

Right atrium $\longrightarrow$ right ventricle $\longrightarrow$ pulmonary artery $\longrightarrow$ pulmonary capillaries (in lung) $\longrightarrow$ pulmonary vein $\longrightarrow$ left atrium

### Systemic circulation
Blood returning from the pulmonary circulation enters the left atrium of the heart, then passes into the left ventricle. From the left ventricle blood is

pumped into the largest artery of the body, the *aorta* (Figure 6-8*b*). Arteries going to each organ of the body, including the heart tissue itself, branch off from the aorta. Some of the principal arteries include the *carotid arteries* going to the brain, the *subclavian* arteries to the arms, the *mesenteric arteries* to the intestine, the *renal* arteries to the kidneys, and the *iliac arteries* to the legs. Each of these branches into a capillary network as it enters the organ which it serves. Blood from the brain flows into the *jugular veins,* which carry it back toward the heart. Blood from the shoulders and arms drains into the *subclavian veins*. These veins and others coming from the front part of the body merge to form the large *precaval vein,* which empties into the right atrium. Renal veins from the kidneys, iliac veins from the lower limbs, and other veins returning from organs in the lower regions of the body merge just before they reach the heart to form the large *postcaval vein,* which empties into the right atrium.

As an example, let us trace the path of blood to and from the brain:

Left atrium $\longrightarrow$ left ventricle $\longrightarrow$ aorta $\longrightarrow$ carotid artery $\longrightarrow$
brain capillaries $\longrightarrow$ jugular vein $\longrightarrow$ precaval vein $\longrightarrow$
right atrium $\longrightarrow$ on to pulmonary circulation

*Hepatic portal system.* Generally blood travels from artery to capillary to vein. An exception to this sequence is the *hepatic portal system,* which transports nutrients from the intestine to the liver. Blood reaches the intestines via the mesenteric arteries and enters capillaries in the intestinal villi, where it receives nutrients. Capillaries from the intestinal villi merge to form the hepatic portal vein, which goes not to the heart but to the liver. In the liver the hepatic portal vein branches into a vast capillary network. As the blood courses through these capillaries, the liver cells absorb and process the excess nutrients present. Liver capillaries eventually merge to form the hepatic veins, which in turn empty into the precaval vein. Thus, in the hepatic portal system we find this unusual sequence of blood vessels: capillaries $\rightarrow$ vein $\rightarrow$ capillaries $\rightarrow$ vein (i.e., an extra capillary network).

*Coronary circulation.* The cells of the heart, which require a rich blood supply, are served by two coronary arteries that branch from the aorta just after it leaves the heart. The coronary arteries branch into an extensive system of capillaries within the heart tissue which eventually fuse into coronary veins leading back to the right atrium.

## CARDIOVASCULAR DISEASE

In the United States there are more deaths from circulatory system disorders than from any other single cause. And the primary circulatory disease is *atherosclerosis,* commonly referred to as "hardening of the arteries."

### Atherosclerosis
In this disorder, lipid materials (cholesterol and fatty acids) circulating in the plasma are deposited on the inner walls of arteries. Reacting to the pres-

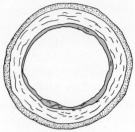

(*a*) Fatty material (in green) is deposited around inner lining of vessel.

(*b*) Fatty material now clogs much of the passageway.

FIGURE 6-9
Diagram showing the progressive clogging of an artery as the atherosclerotic process continues.

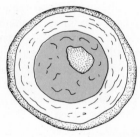

(*c*) A thrombus has formed and vessel is completely blocked.

ence of these materials, the tissue of the arterial walls becomes irritated and inflamed. Eventually calcium combines with the fatty deposits, forming hard, bonelike plaques (thin, flat deposits) (Figure 6-9).

Atherosclerosis most often occurs in the coronary arteries, the aorta, and the arteries supplying the brain. As these arteries become more and more diseased by atherosclerotic build-up, they become increasingly occluded, or clogged. Since less space remains within the vessel for the passage of blood, cells being serviced by such arteries may not receive an adequate blood supply. The stage is set for a variety of serious consequences.

Just what causes the initial deposits of fatty materials is not known, although many contributing factors have been identified. Persons with a high level of cholesterol and/or triglycerides in their blood are much more likely to develop atherosclerosis than those with normal levels. These high blood-fat levels can be traced to both genes and diet. Since lipids are not soluble in blood, they are normally transported while linked to plasma albumin. Recent research has disclosed that some persons lack sufficient lipid-carrying proteins to do the job. They apparently lack the genes required to produce the needed protein. Large quantities of lipid float free in the blood of such persons and may be the source of the initial deposits.

The role of diet in the development of atherosclerosis is controversial. Cholesterol is produced by body cells, especially in the liver, and is essential to normal metabolism. The cholesterol synthesized by the body accounts for about 70 percent of the blood-cholesterol level, and the amount produced remains constant even when no cholesterol is included in the diet. But the remaining 30 percent is apparently influenced by diet. Blood-lipid levels can be lowered by a diet rich in polyunsaturated fatty acids of vegetable origin, as compared to saturated fatty acids and cholesterol of animal foods. Most effective in maintaining low blood-lipid concentrations are the polyunsaturated fatty acids of safflower, corn, peanut, and cottonseed oils. Although the whole story of dietary factors is not yet known, many researchers suggest diets low in cholesterol and saturated fats. Some recent evidence suggests that sugar in the diet may promote atherosclerosis.

Hormones have been shown to be a factor in the development of atherosclerosis. The disorder is much more common in middle-aged men than in women of the same age group. This has been traced to the female hormone estrogen, which somehow seems to inhibit the atherosclerotic process.

Autopsy studies have shown that cigarette smokers have a significantly greater incidence of atherosclerosis of the aorta and coronary arteries than nonsmokers. Recent experiments indicate that this may be because of stimulation by nicotine of fat synthesis within the arterial walls. Habitual tension and other psychologic factors are also thought to promote the development of atherosclerotic lesions.

## Heart disease and heart attacks

As coronary arteries become clogged by atherosclerotic build-up, the blood supply to the heart may become insufficient. Any extra stress placed upon the heart, as by sudden physical exertion or psychologic stress, may result in the discomfort and pain referred to as *angina pectoris*. This condition is often temporarily relieved by nitroglycerin tablets which dilate the coronary arteries, increasing the amount of blood which can reach the heart muscle.

Sometimes platelets adhere to the rough surface of an atherosclerotic plaque and a blood clot, or *thrombus* (Figure 6-10), develops. Because it blocks the flow of blood to the heart muscle a thrombus in a coronary artery might result in a heart attack, or *coronary thrombosis*. The severity of the heart attack depends upon how large a portion of heart muscle is deprived of blood. When the cells of heart muscle are deprived of oxygen for only 30 seconds, they begin to die. Thus, if a thrombus completely blocks an arterial branch which supplies a large part of the heart muscle, death may be immediate. On the other hand, if the thrombus occurs in a small branch of the coronary artery, only the cells serviced by this branch die, and the heart may continue to function. Such localized heart damage is called *myocardial in-*

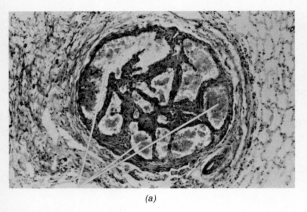

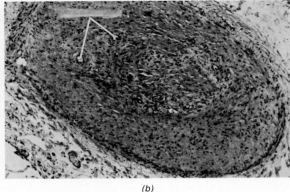

**FIGURE 6-10**

(a) A thrombus completely obliterated the lumen of this artery. Scar tissue has formed, and the artery remains completely blocked. (b) Part of this thrombus has been absorbed, and vascular channels are evident. Some blood is able to pass through this artery. (McGraw-Hill Encyclopedia of Science and Technology, 3d ed.)

*farct.* A heart damaged in this way may be permanently weakened because the dead cardiac muscle of the infarct is replaced with scar tissue which cannot contract. In very severe cases, so much of the heart muscle dies that it cannot contract efficiently or cannot be replaced by scar tissue. In such cases, the victim soon dies.

Sometimes a thrombus, or a part of it, becomes an *embolus* that breaks away from the arterial wall and is carried by the blood until it becomes lodged in another blood vessel, perhaps causing the same types of damage as described above.

Heart attacks are the most common circulatory accidents, because the coronary arteries are the blood vessels most susceptible to atherosclerosis. Heart attacks may also result from a spasm of a coronary blood vessel which closes off the supply of blood to the heart.

On the basis of many studies, authorities generally agree that the risk of dying from a heart attack is significantly increased among those who have high cholesterol or triglyceride levels in their blood, smoke, are overweight, or have high blood pressure. On the other hand regular physical exercise apparently protects a person against heart disease. Men in the 35-to 44-year-old group who have sedentary jobs are seven times more likely to die of heart disease than those who are employed in physical labor.

### Stroke

An interruption of normal blood supply to the brain results in a *stroke,* or cerebral vascular acccident (CVA). Arteries serving the brain may be blocked by a thrombus, by an embolus, or by rupture of a blood vessel. Persons with high blood pressure are more likely to suffer from rupture of a blood vessel than those with normal blood pressure. High blood pressure is itself a common consequence of atherosclerosis. The severity of a stroke depends upon the number of brain cells deprived of oxygen and nutrients. If a vessel serving a very small area is blocked, the brain may be able to function adequately without the damaged tissue. If a more sizable or vital area is damaged, paralysis or death may result.

**External Heart Massage**

WHAT IS IT?
A method for forcing blood through a heart that has stopped beating.

WHEN IS IT USED?
External heart massage is used when a person's heart has stopped beating entirely or is beating inefficiently (too faintly or rapidly). Cardiac arrest sometimes occurs when breathing has stopped and/or the victim is unconscious.

WHY IS IT USED?
Several thousand persons die each year in the United States alone from cardiac arrest. Often an arrested heart can be started again by administering external heart massage.

HOW IS IT DONE?
1 Place the victim on his back on the floor or other firm surface, and kneel at right angles to his chest.
2 Begin mouth-to-mouth resuscitation (Figure 8–7). If no one is there to help, stop heart massage every 30 seconds and give artificial respiration for 10 seconds (4 deep breaths).
3 Place the heel of one hand on the lower third of the breastbone. Place the heel of the other hand at a right angle to and atop the first hand.
4 Apply firm pressure downward so that the breastbone moves about 1 to 2 inches toward the spine. Repeat once each second. The downward pressure must amount to between 70 and 90 pounds with an adult. However, much less pressure should be used on a child.
5 Between downward pressures relax hands to allow the chest to expand.

## THE LYMPH SYSTEM

The lymph system is a second circulatory system which feeds into the blood circulation (Figure 6-11). It returns fluids from the tissues into the main circulatory system, filters out dust and bacteria that enter the body, produces lymphocytes and the cells which produce antibodies, and transports fats absorbed from the intestine.

### Tissue fluid

As blood circulates through the tissues some plasma escapes through the capillaries into the surrounding tissues and it is then called tissue fluid (Figure 6-12). Tissue fluid brings nutrients and oxygen from the capillary directly to individual cells, bathing them in a rich sea of raw materials needed for cellular metabolism. Wastes from the cells can diffuse back into the blood via the tissue fluid.

Most plasma proteins, as well as red blood cells and platelets, are too large to pass out through the capillary walls, and so they remain in the blood. Some of the tissue fluid moves back into the capillaries because of osmotic pressure. Plasma tends to leave the circulatory system at the beginning of a capillary network, and tissue fluid returns to the circulatory system toward the end of the capillary network.

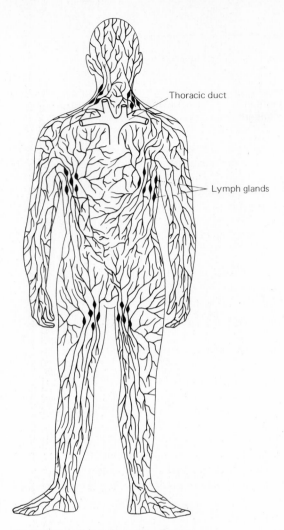

Thoracic duct

Lymph glands

**FIGURE 6-11**
The lymph system. Lymph is emptied into the blood circulation by way of two ducts in the shoulder region.

### Structure of the lymph system

That portion of the tissue fluid not directly absorbed back into the capillaries is returned to the blood by way of the lymph system. The lymph system consists of a network of lymph capillaries that join to form lymph veins. Lymph veins eventually empty into two large blood veins near the heart by way of the thoracic and right lymphatic ducts. In man the lymph system has no heart or other pumping device, although some animals do have lymph "hearts." Lymph arteries are also absent. Several organs, including lymph nodes and the thymus, are associated with the lymph system.

As tissue fluid accumulates, pressure builds up and forces excess fluid through the permeable walls of the lymph capillaries (their ends are closed), where it is known as *lymph*. Wandering leukocytes in the connective tissue also enter in this way, but lymph contains insignificant numbers of erythrocytes and platelets.

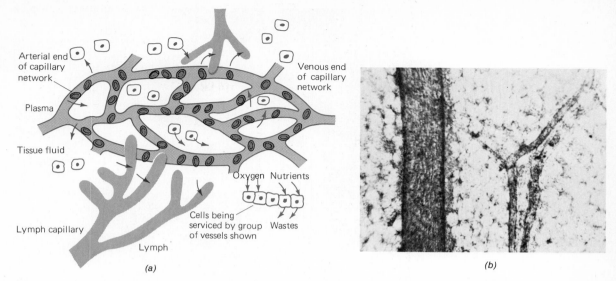

Arterial end
of capillary
network

Plasma

Tissue fluid

Lymph capillary

Lymph

Venous end
of capillary
network

Oxygen  Nutrients

Cells being
serviced by group  Wastes
of vessels shown

*(a)*

*(b)*

**FIGURE 6-12**
(*a*) Exchange of materials among
capillaries, cells of a tissue, and
lymph vessels. Note that tissue
fluid forms when plasma is
forced out of the capillaries at
the beginning of a capillary network. Some of the tissue fluid
reenters capillaries at the venous
end of the capillary network. A
great deal of tissue fluid enters
lymph capillaries and is then referred to as lymph. (*b*) Photomicrograph of lymph vessels.

Lymph veins pass through many *lymph nodes,* small organized masses
of tissue which serve as filters. Vessels passing through lymph nodes are
lined by macrophages which ingest any foreign matter, including bacteria,
which has found its way into the lymph. More than 400 lymph nodes cleanse
the lymph before it is returned to the general blood circulation. Lymph nodes
also function in immune responses and are active sites of lymphocyte
production. During a throat infection or tooth abcess the lymph nodes in the
neck enlarge conspicuously. (You can feel them as hard little knots below
the skin surface.)

The palatine tonsils and adenoids (pharyngeal tonsils) also consist of
lymph tissue and are centers for the proliferation of new lymphocytes. These
centers are situated strategically to meet germs and other foreign matter entering the body through the mouth or nose. By destroying invaders, they protect the respiratory system from infection. Unfortunately, the tonsils and adenoids are sometimes overcome by the invading bacteria, become the site of
frequent infection themselves, and are therefore often considered for
surgical removal.

## Immune responses

The term *immunity,* derived from a Latin word meaning *safe,* refers to the
defense measures taken by the body when invaded by disease organisms or
other foreign matter.

*The body's defense system.* Immune reactions depend upon the combined
action of several types of cells, organs, and specific proteins called *antibodies*. The basis of immunity is the body's ability to recognize its own
cells and molecules and to distinguish them from the cells and molecules of
invading organisms—to discriminate between "self" and "nonself." Such
recognition depends upon subtle chemical differences among the large

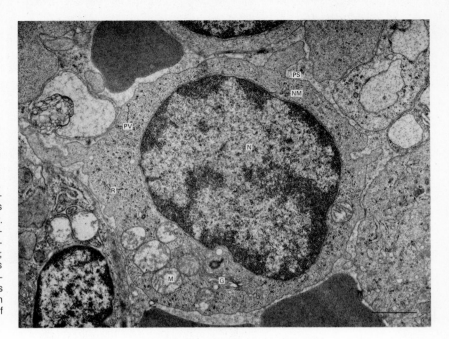

**FIGURE 6-13**
Electron micrograph of a lymphocyte. The black line indicates a length of 1 micrometer. (×24,250) *N*, nucleus; *NM*, nuclear membrane; *M*, mitochondrion; *G*, Golgi apparatus; *PV*, droplet of fluid which has been taken in by cell; *R*, ribosomes; *PS*, pseudopod (cell has flowed into an extended position here, perhaps to engulf a bit of material.) (Dr. Lyle C. Dearden)

molecules such as the proteins and carbohydrates of each organism. A virus or bacterium contains many large molecules which the body recognizes as foreign and, therefore, a threat to its well-being. Any large foreign molecule that enters the body and is capable of producing an immune reaction is termed an *antigen*. Even such common substances as pollen and dust contain antigens. The body defends itself against antigens in two ways. (1) it mobilizes its cellular soldiers, the white blood cells and macrophages which can consume the invaders; (2) it engages in chemical warfare, producing antibodies, specific proteins designed to combine with antigens and render them harmless.

Immune reactions depend upon three different types of cells: phagocytes, lymphocytes, and plasma cells. Phagocytes are any cells which can "eat" foreign substances. They include the granular leukocytes, the monocytes, and the macrophages. The role of the lymphocytes (Figure 6-13) is complex and not entirely understood. Although one lymphocyte looks pretty much like any other, in fact, they may be quite different chemically. There are an estimated 50,000 different types, each designed to react to the presence of a specific type of antigen. When an antigen enters the body, only those lymphocytes chemically competent to respond to it are affected. The affected lymphocytes respond in two ways (Figure 6-14). Some of them undergo mitosis, giving rise to whole companies of identically competent cells. The entire army then migrates to the scene of the invasion and engages in cell-to-cell combat, destroying the foreign cells in a manner which is not completely understood. The second way in which affected lymphocytes are thought to react to antigens has not been conclusively proved. Some immunologists think that some stimulated lymphocytes undergo mitosis and give rise to a population of competent *plasma cells* that produce antibodies. In both responses the mitoses take place within the tissue of the lymph nodes and spleen. Antibody molecules produced by the plasma cells are carried into the blood by way of the lymph circulation.

Once in the blood and in the tissues, antibodies attack antigens in several ways. Some antibodies, referred to as *opsonins,* coat the surfaces of bacteria. Once coated, the bacteria seem to be "tastier" to the phagocytes so that these cells now gobble them up much more rapidly and efficiently! Some antibodies disarm the invader directly. Certain viruses, for example, possess structures which enable them to penetrate and attach themselves to cells of the host. The antibody may cover up this structure so that the virus is unable to do damage. Other antibodies combine with toxins (poisons) produced by bacteria, rendering them inactive.

About 6 days are normally required for the body to mobilize its chemical defenses in sufficient quantity to overcome invading antigens. This may explain why most common colds last for about a week. It also points out the body's need for help in the form of antibiotics and other drugs when a serious illness such as pneumonia attacks. Though the body may be strong enough to withstand the damage being done while it prepares its weapons

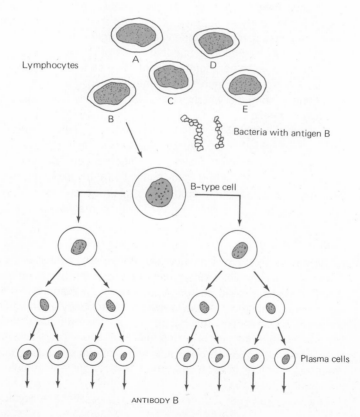

**FIGURE 6-14**
How a lymphocyte responds to an antigen. Each type (A, B, C, etc.) of lymphocyte is slightly different biochemically. When bacteria with antigen type B enters, only lymphocytes capable of making antibody type B are stimulated. A type-B lymphocyte responds by changing into a cell which gives rise to a whole population of type-B plasma cells. Type-B plasma cells then produce the needed antibody B.

(antibodies), it is also possible that the invaders may win the battle and kill the host before it can effectively defend itself.

Once antibodies for a particular antigen have been produced, some of them remain in the body for months or even years. In addition, a small army of plasma cells and its descendants which have been formed for the occasion will live on for a long time. These cells are sometimes referred to as *memory cells* because they are especially equipped to "remember" and deal with specific antigens. These facts explain why we do not usually suffer from the same cold or other disease twice. If the same type of virus should invade the body a second time, it would be dealt with swiftly and efficiently. Certain childhood diseases such as measles and chickenpox are good examples of diseases that people generally have only once. After having the disease, the body is immune from future attacks. Unfortunately, there are so many different types of viruses associated with common colds that each one presents a new antigenic challenge to the body and must be dealt with as a new invader.

A discussion of immune reactions would not be complete without mention of the *thymus,* an organ which plays a key role in the development of competent lymphocytes. This large lymph organ, located in the chest cavity (Figure 6-7), produces a population of lymphocytes known as T cells which are thought to function somewhat differently than other lymphocytes (called B cells). Just how the thymus and the T cells work is the focus of current research. Very early in life, probably during fetal development, the thymus or its lymphocytes somehow "teach" other lymphocytes to be competent—that is, specialized to deal with specific types of antigens. When the thymus is removed from an experimental animal early enough in its development, the animal is unable to launch an immune reaction. The thymus is also thought to secrete a hormone which may affect immunity. After puberty this organ decreases in size.

A protein substance called *interferon* is manufactured by cells in response to infection by viruses, specifically in response to their foreign nucleic acids. Interferon leaves the infected cell and enters neighboring uninfected cells, protecting them against viral invasion. Inferferon may stimulate the uninfected cell to produce a protein which blocks the production of viral nucleic acid or protein. (Viruses will be discussed in greater detail in Chapter 14.)

*Vaccines.* Various types of vaccines are used to confer *artificial immunity* upon the body. Vaccines for typhoid and whooping cough and the Salk polio vaccine consist of dead bacteria or viruses of the respective diseases which still contain enough antigen to stimulate antibody production. When the body meets with these viruses again, it is prepared to deal with them. Other types of vaccine consist of toxins made by specific disease bacteria (e.g., tetanus toxoid) or of live but weakened bacteria or viruses (e.g., the Sabin oral polio vaccine). In each case, the body is artificially immunized against a particular disease.

The so-called "flu vaccines" are not more effective because many different types of viruses cause the symptoms which we refer to as flu or common cold. No one has yet figured out how to gather them all together into one supervaccine. And to complicate the problem further viruses often mutate, or change their genetic structure enough to make themselves antigenically different.

The type of immunity discussed above is referred to as *active immunity* because, whether by natural infection or by planned injection, the body is confronted with specific antigens and actively produces antibodies against them. *Passive immunity* was first used by physicians in 1891 when a young girl dying from diphtheria was injected with serum taken from a sheep that had previously been injected with diphtheria antigen and had actively produced antibodies against the diphtheria—something the child's body was unable to do efficiently in its weakened condition. After receiving the sheep antibodies the girl made a dramatic recovery. Passive immunity may be accomplished by injecting a person with serum from an animal or human being who has actively produced the antibodies needed. However this type of immunity lasts only for a short time. Human gamma globulin contains most of the antibodies normally found in adults and is often used to confer temporary immunity in an emergency situation—for example, to reduce the chances that an already ill child may become infected with mumps or whooping cough during an epidemic.

*Allergy.* About half of us suffer at least mildly from some variety of allergy such as hay fever, asthma, or food sensitivity, while an estimated 10 percent of us have a major allergic disorder. An allergic person makes antibodies to one or more weak antigens referred to as *allergens*. People who are not allergic do not react significantly to these substances. An allergen may enter the body as food or may be breathed in with air and settle upon the lining of a respiratory passageway. When an antibody combines with an allergen, certain cells in the connective tissue react by releasing a substance called *histamine*. Histamine dilates blood capillaries and increases the permeability of the capillary walls so that fluid escapes into the tissues, causing inflammation (swelling). Histamine also causes the muscles of the bronchial tubes to constrict, making breathing difficult. *Antihistamines* are commonly used drugs that block the effects of histamine.

An inherited tendency for allergic disorders explains why some persons are more vulnerable to allergy than others. However even susceptible persons cannot develop an allergy to a particular substance without sufficient exposure. For example, a person cannot become allergic to pollen unless he has inhaled it. When the pollen makes contact with the lining of the respiratory passageways, it may stimulate an allergic response. Food allergies are thought to develop when, through some defect in absorption, an antigen enters the circulation. For example, an entire protein, instead of being properly digested into amino acids, may find its way into the blood.

Young infants may develop allergies to foods which their digestive systems are too immature to tolerate. Breast-fed infants seem less likely to develop food allergies than those given formulas of cow's milk.

*Failures in immune responses.* Allergy is a condition in which the immune system does not function in its owner's best interests. Other disorders, such as arthritis, are also thought to be caused by malfunction of immune mechanisms. Arthritis may be an autoimmune disease in which the immune system turns against "self," causing damage to certain tissues. Some investigators feel that cancer may be linked to a failure in normal immune responses. It has been suggested that cancer cells may arise in each of us by accident every day but that normally they are swiftly destroyed. Why is it that in some persons immune responses suddenly fail to destroy these harbingers of death? So many questions related to immune mechanisms remain unanswered that immunology is one of the most exciting fields of research today.

### Rebuilding man—organ transplants

News stories of heart, kidney, and other organ transplants have become almost commonplace. Most of us find the idea of transplanting a person's organs into the body of another intriguing. And the possibility that such measures might prolong life if one of our organs should suddenly fail is perhaps comforting. At the time of this writing, however, at least one serious problem remains to be solved before successful transplants can be routinely performed.

*Graft rejection.* The chief obstacle to be overcome in achieving successful organ transplants is a form of immune response called *graft rejection*. When an organ is taken from a donor and transplanted into the body of a new host, it brings with it a variety of antigens, molecules foreign to the host. Wandering lymphocytes recognize the transplanted cells as "non-self" and carry the news back to the lymph nodes. In response, the host launches a vigorous attack upon the foreign tissues, often damaging or destroying the transplant. Thus, the same mechanisms that protect the body against disease organisms cause the death of organ transplants. The body seems unable to make the necessary distinction and regards all foreign molecules as "enemies" to be destroyed. Graft rejection is carried out primarily by competent lymphocytes which migrate to the scene of the transplant and carry on cell-to-cell combat with the foreign cells.

Medically, the problem of graft rejection is being dealt with in two ways: careful donor selection and suppression of immune responses. By careful selection of the donor the number of harmful antigens in the graft can be minimized. About 20 genes are involved in making the antigens that stimulate graft rejection. These genes are referred to as histo-(tissue-) compatibility genes. Identical twins are the only people who have exactly the same types of histocompatibility genes and, therefore, identical tissue proteins. This is why the most successful transplants are those performed between identical twins. The body of the host twin regards the donor twin's proteins as "self" rather than foreign, and so does not react to destroy them. Next to identical twins, those most likely to share many of the same types of genes are close members of the same family. A father, mother, or brother makes a good prospect for donating an organ. Among the histocompatibility genes, one referred to as Hu-1 produces a particularly powerful antigen. Even a nonrelated recipient whose Hu-1 gene matches the Hu-1 gene of the donor has a fair chance of successfully accepting an organ. Techniques are being developed for typing histocompatibility genes in the laboratory to find the most suitable donor for a patient in need of a transplant.

The second method for dealing with graft rejection is to attempt to suppress the immune response within the host. Drugs and irradiation are both used to kill lymphocytes and to prevent them from dividing. Unfortunately, these methods are not completely specific to lymphocytes and other cells are injured in the process. A further problem arises because when the immune system is suppressed, the body is unable to respond appropriately to invading disease organisms. The patient is extremely vulnerable to infection and may die of infectious disease. Much research is being carried out in this area, and it is hoped that some means of dealing specifically with graft rejection will be discovered.

*Sources of spare parts.* With the increasing use of organ transplants comes a variety of other problems. Many center about the source of organs to transplant. Although many researchers have been working to devise artificial organs—particularly hearts—we are still far from achieving this goal.[2] Most organs are complex almost beyond imagination, and man cannot even begin to replace them with organs fashioned of nonliving materials. Surprisingly, the heart is one of the least complex organs, since its function is restricted to pumping blood. One obstacle in developing an artificial heart has been the tendency of blood to clot and red blood cells to break, releasing hemoglobin, when they come into contact with anything except a blood-vessel wall. So far researchers have not found an ideal material from which to construct artificial hearts. Still, man-made pumps to replace damaged hearts may be the first successful artificial organs.

At least for the next several years physicians will have to look to nature's handiwork when they are in need of spare parts. Animal donors would be helpful were it not for the immune response, but because animal proteins

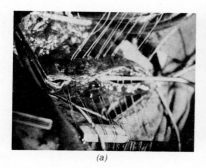

(a)

(b)

**FIGURE 6-15**
Surgical repair of a damaged heart. An aneurysm (weakened, ballooned-out portion of the heart wall which resulted from an infarct) has been removed and the heart wall sutured. (Dr. Dennis Pupello)

---

[2] Artificial arteries, joints, pacemakers, and other structures less complicated than organs are being used successfully.

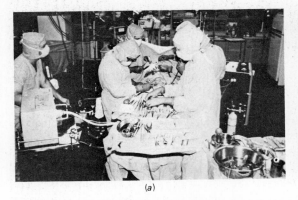

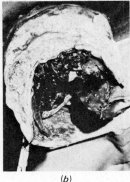

(*a*)  (*b*)

are even less similar to ours than those of an unmatched human donor, organ transplants taken from animals are rejected vigorously by human hosts. Until the problem of graft rejection has been solved, the best source for spare human organs is other human beings.

Taking organs, such as kidneys, from living donors presents a variety of moral and psychologic problems. And of course, some organs such as the heart cannot be spared. Legally, using organs from a recently deceased person, such as an accident victim, has been complicated. Getting permission from relatives before the organs begin to deteriorate, preserving the organ until the new host can be prepared for surgery, and even deciding when the donor is actually dead are a few of the difficulties that have been encountered.

***New parts for old.*** To date, the most successful transplants have been kidney exchanges. For many reasons physicians and researchers have concentrated their efforts on perfecting the techniques involved in replacing diseased kidneys and keeping the recipient alive. A few persons have now received "new" kidneys and have remained alive for several years.

With heart disease a major killer, hundreds of thousands of persons in the United States alone would benefit from exchanging their diseased hearts for more vigorous models. Many have now experienced such surgery, and some have remained alive for a few years. Some surgeons have at least temporarily suspended heart transplants until new solutions to graft-rejection problems can ensure a better prognosis.

Bone marrow transplants may someday provide hope for victims of leukemia and other diseases involving blood cell production. Skin grafts for burn victims have not met with much success yet because of vigorous graft rejection. However, even if eventually rejected, transplanted skin is the best dressing known for burned tissue. It is sometimes desirable to transplant skin from another part of the body to the burned area, especially for cosmetic reasons.

Perhaps the most exciting transplant to envision is that of a brain. In our present state of ignorance, however, one cannot foresee the successful connection of a transplanted brain to the millions of nerves of the host and the securing of an adequate blood supply for it.

1  The circulatory system transports nutrients, oxygen, wastes, and hormones. It helps regulate body temperature, fluid balance, and the level of inorganic substances, and protects against invading microorganisms.

2  Red blood cells transport oxygen. White blood cells protect the body against invading bacteria and other foreign substances. Platelets function in the process of blood clotting.

3  Blood enters the systemic circulation through the aorta, from which arterial branches go to each organ of the body.

4  The right side of the heart receives deoxygenated blood returning from the body cells, and pumps this blood into the pulmonary circulation.

5  Atherosclerosis is the basis of many circulatory diseases, including those commonly referred to as heart attacks.

6  The lymph system returns tissue fluids to the main circulation, filters out dust and bacteria, produces lymphoctytes and cells which produce antibodies, and absorbs fats from the intestine.

7  Immune reactions depend upon the combined action of several types of cells and organs and upon specific proteins called antibodies which are produced by plasma cells.

8  Vaccines are used to confer artificial immunity upon the body.

9  The problem of graft rejection must be solved before successful organ transplants can become a routine procedure.

**FOR FURTHER READING**

Billingham, Rupert, and Willys Silvers: *The Immunology of Transplantation,* Prentice-Hall, Englewood Cliffs, N.J., 1971. An in-depth discussion of the immune mechanisms involved in organ transplants.

Jacob, Stanley W., and Clarice Ashworth Francone: *Structure and Function in Man,* 2d ed., Saunders, Philadelphia, 1970. Chapters 10 and 11 provide a detailed discussion of the human circulatory and lymph systems. Excellent illustrations.

Nossal, G. J. V.: *Antibodies and Immunity,* Basic Books, New York, 1969. An interesting and readable account of immune mechanisms.

Warshofsky, Fred: *The Control of Life. The 21st Century,* Viking, New York, 1969. Chapter 5, "Man-made Man," contains an interesting discussion of organ transplants.

# 7

# DISPOSAL OF METABOLIC WASTES

**Chapter learning objective.** The student must be able to identify the principal human metabolic wastes and describe the organs which excrete them and the mechanisms by which they are eliminated.

*ILO 1*   Define excretion, and identify the principal metabolic wastes in the human body.

*ILO 2*   Describe the role of the liver in processing wastes.

*ILO 3*   Draw and label, identify on a diagram, or describe the organs of the urinary system, giving the function of each.

*ILO 4*   Describe or identify on a diagram the principal parts of the nephron, giving the specific function of each part.

*ILO 5*   Describe the process of urine formation and the composition of urine.

*ILO 6*   Define renal threshold, and describe its relationship to the disorder diabetes mellitus.

*ILO 7*   Describe the regulation of urine volume and give the physiologic basis for diabetes insipidus.

*ILO 8*   Define kidney failure, and compare kidney dialysis to kidney transplant as alternative treatments for chronic kidney failure.

*ILO 9*   Describe the excretory functions of the skin, lungs, and digestive system.

*ILO 10*   Describe how each of the structures discussed in this chapter helps to maintain the steady state of the body.

*ILO 11*   Define each new term introduced in this chapter.

As cells process and utilize nutrients, waste materials are produced. If permitted to accumulate, such waste products would reach toxic concentrations and interfere with vital metabolic activities. *Excretion* is the process of removing metabolic wastes from the body.

Plants do not require sophisticated excretory systems. Their cells are able to use simple nitrogen compounds that animal cells must discard, and they are much more tolerant of excess mineral ions than animal cells are. Any substance no longer needed for metabolic activities can be deposited in water vacuoles or in the form of crystals, simple forms of waste removal.

In most one-celled organisms and very simple animals ordinary diffusion takes care of excretion. Some organisms such as the fresh-water *Amoeba* have specialized contractile vacuoles which pump excess water out of the cell. Larger animals, of course, require more specialized excretory systems. Here we shall consider the human excretory system, which is similar to that of other vertebrates.

Excretion is not the same as elimination (Figure 7-1). Undigested and unabsorbed food materials (substances which never enter body cells) are disposed of by elimination in the form of feces. Excretion, on the other hand, rids the body of metabolic wastes which have been produced as a result of chemical reactions *within* cells.

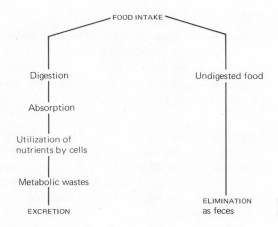

**FIGURE 7-1**
Excretion versus elimination.

The principal metabolic waste products of the human body are water, carbon dioxide, and nitrogenous wastes such as urea and uric acid (Figure 7-2). These wastes are transported by the circulatory system from the cells which produce them to the organs that process and excrete them. The liver is also involved, processing nitrogenous wastes before sending them on to the kidney. Excretion itself is the job of the kidney, the lungs, the skin, and, to a lesser extent, the digestive tube. Not only do these organs rid the body of potentially toxic materials, but by removing excess water, minerals, and other materials they help maintain a steady state.

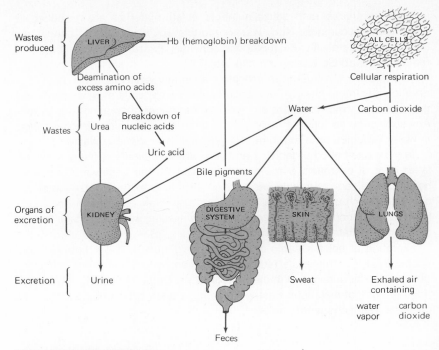

**FIGURE 7-2**
Disposal of metabolic wastes.

## THE URINARY SYSTEM

The kidney excretes nitrogenous wastes, water, and a variety of salts and other substances. Together with the urinary bladder and associated ducts it forms the urinary system. Because people tend to think of urine first when discussing excretion, these structures are sometimes referred to as *the* excretory system.

### Role of the liver

The liver converts excess amino acids into fatty acids by the process of *deamination* (see Chapter 5). Amine groups removed from the amino acids are a nitrogenous waste product, which the circulatory system transports to the kidneys for excretion. The liver also degrades nucleic acids ingested in foods. Nitrogenous units from nucleic acids are converted into uric acid, a second nitrogenous waste excreted by the kidney.

### Organs of the urinary system

The overall structure of the urinary system is illustrated in Figure 7-3. The kidneys resemble a pair of giant lima beans about 4.5 inches long, 2.5 inches wide, and about 1.5 inch thick. Their job is to produce urine from the wastes and excess materials they filter out of the blood. By adjusting the amount of water and salts that they excrete, the kidneys play a vital role in maintaining the internal chemical balance of the body.

Urine passes from the kidneys through one of the paired *ureters,* ducts about 10 inches long that connect the kidneys with the *urinary bladder,* a

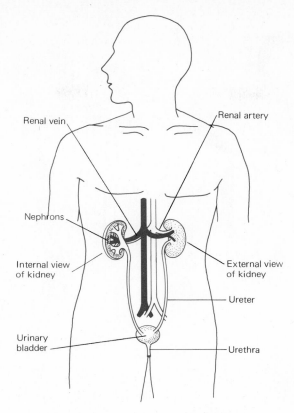

Renal vein

Renal artery

Nephrons

Internal view
of kidney

External view
of kidney

Ureter

Urinary
bladder

Urethra

**FIGURE 7-3**
The urinary system and associated blood vessels.

temporary storage sac for urine. As the bladder fills, its muscular walls stretch to accommodate as much as a pint of urine. The exit of the bladder is normally closed by a ring of involuntary muscle and by bands of voluntary muscle. When the bladder is filled, nerves are stimulated to initiate *urination,* release of urine from the bladder. Smooth muscle in the wall of the bladder contracts, the ring of muscle opens, and urine flows out.

As urine leaves the bladder it flows through the *urethra,* a duct leading to the outside of the body. In the female the urethra is less than 2 inches long and serves only as a passageway for urine. In the male the urethra is about 8 inches long, extends through the penis, and functions to transport semen as well as urine.

## Microscopic structure of the kidney

Each kidney consists of more than a million functional units called *nephrons.* A nephron (Figure 7-4) consists of an *excretory tubule* with an expanded cuplike end called *Bowman's capsule,* which surrounds a cluster of capillaries (called a *glomerulus*) (Figure 7-5). Each human nephron is only about 5 centimeters long, but there are so many of them that if all the tubules of both kidneys could be placed end to end they would extend about 75 miles!

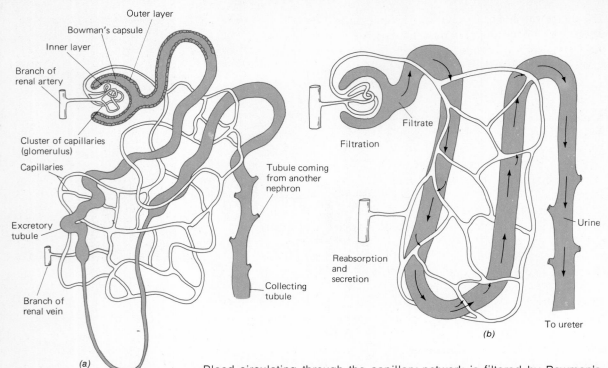

Outer layer

Bowman's capsule

Inner layer

Branch of
renal artery

Cluster of capillaries
(glomerulus)

Capillaries

Excretory
tubule

Branch of
renal vein

Tubule coming
from another
nephron

Collecting
tubule

(a)

Filtrate

Filtration

Reabsorption
and
secretion

Urine

To ureter

(b)

**FIGURE 7-4**
(a) A nephron. (b) Formation of
urine. Highly diagrammatic representation of a nephron to show
where filtration, reabsorption, and
secretion take place.

Blood circulating through the capillary network is filtered by Bowman's
capsule. The filtrate flows into the excretory tubule, where its contents are
inventoried. Useful substances are returned to the blood; waste materials
pass on through the long excretory tubule into a larger collecting tubule
which eventually empties into the ureter.

### Urine formation

Urine formation may be divided into two main phases: *filtration* through
Bowman's capsule and *reabsorption* by the tubules. A third mechanism,
*secretion,* is of minor importance.

*Filtration.* Blood flows to the kidneys through the renal arteries. A tiny arterial branch of the renal artery transports blood to the glomerulus of each
nephron. Blood enters the capillary network under relatively high pressure,
and as a consequence, more than 10 percent of the plasma flowing through
the capillary mass is forced out of the capillaries and into Bowman's capsule. Except for large fat and protein molecules, all the substances normally
dissolved in the plasma pass into the capsule. Once inside the capsule, this
fluid is referred to as filtrate.

About 20 percent of the blood pumped by the heart flows through the
kidneys each minute, so that every 5 minutes the kidneys receive a volume
of blood equal to the total volume of blood in the body. Every 24 hours about
1,700 quarts of blood flow into the glomeruli of the nephrons yielding about

170 quarts of filtrate. Of course a normal person does not produce anywhere near 170 quarts of urine each day. But filtration is not selective and the filtrate contains needed materials such as sugar and salts in addition to water which cannot be allowed to leave the body.

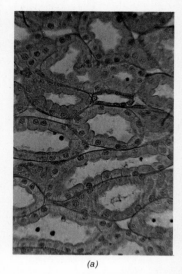

*Reabsorption.* Most of the 170 quarts of filtrate produced each day are reabsorbed into the blood, leaving only about 1.5 quarts of urine. Reabsorption, the function of tubules, reduces the volume of filtrate and returns useful substances to the blood.

Blood leaving the glomerulus flows into a tiny artery, which then divides into a second capillary network surrounding the excretory tubule. (Find these structures on Figure 7-4.) The circulatory pattern here is unusual in that we find the sequence of artery-capillary-artery-capillary. Ordinarily, capillaries flow into a vein. The second set of capillaries joins small veins which empty into the large renal vein draining each kidney. While the first set of capillaries provides the blood to be filtered, the second set receives materials returned to the blood by the tubule.

Reabsorption is *selective.* Wastes and surplus salts and water are retained by the tubule, while glucose, amino acids, and other useful materials are reabsorbed into the blood. This is accomplished by a combination of diffusion, osmosis, and active transport. The large volume of water reabsorbed is thought to be removed mostly by a mechanism called *countercurrent exchange.* In this process salts are actively transported out of the tubule and water then follows osmotically. The tubules will reabsorb more than 40 gallons of water, 2.5 pounds of salt, and 0.5 pound of glucose each day.

(a)

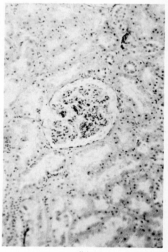

(b)

**FIGURE 7-5**
Photomicrographs of kidney tissue. (a) Kidney tubules cut in various sections. The wall of each tubule consists of epithelial cells. (b) Section through kidney tissue showing part of a glomerulus.

Substances such as glucose which are useful to the body normally are completely reabsorbed from the tubules. However, if a large excess of a particular substance is present in the blood, the tubules may not be able to return all of it. The maximum concentration of a specific substance in the blood at which complete reabsorption can take place is termed the *renal threshold* for that substance. When a substance exceeds its renal threshold, the portion which is not reabsorbed is excreted in the urine. Some substances such as urea have very low thresholds, so that even when present in small concentrations they are not reabsorbed. Other substances such as glucose, amino acids, and hormones have high renal thresholds and are normally completely reabsorbed. Guided by appropriate threshold values, the tubules cleanse the blood of wastes and regulate the internal chemical environment of the body.

An important example of a substance exceeding its renal threshold level is found in the condition *diabetes mellitus.* Because of an insufficiency of the hormone *insulin,* a diabetic suffers from impaired carbohydrate metabolism. Glucose accumulates in the blood instead of being efficiently absorbed and utilized by the cells. The concentration of glucose filtered into the nephron exceeds the renal threshold, and glucose is excreted in the urine. Presence of glucose in the urine is evidence of this disorder.

*Secretion.* A few substances are actively moved from the capillaries surrounding the tubules into the tubules. This mechanism is called *secretion.* Just how important a role secretion plays in human urine formation is not known. Potassium is one substance known to be both reabsorbed and secreted by the tubules. Certain drugs such as penicillin appear to be excreted from the body by tubular secretion.

### Composition of urine

By the time the filtrate reaches the ureter, its composition has been precisely adjusted. All useful materials have been returned to the blood by reabsorption. Wastes and excess materials which entered by filtration or secretion have been retained by the tubules. The adjusted filtrate, called *urine,* is composed of about 96 percent water, 2.5 percent organic wastes (primarily urea), 1.5 percent salts, and traces of other substances.

### Regulation of urine volume

The amount of urine produced depends upon the body's need for retaining or ridding itself of water. When one drinks a great deal of water he produces a correspondingly large amount of urine. Excess water that is absorbed from the digestive tube into the blood is removed by the kidneys, so that the steady volume and composition of blood are maintained.

The kidney receives its information regarding the current *state of the blood* by a rather circuitous route. When fluid intake is low, the body begins to dehydrate. The concentration of salts dissolved in the blood becomes greater, causing an increase in the osmotic pressure of the blood. Specialized receptors in the brain and in large blood vessels are sensitive to such change. The *posterior pituitary gland* in the brain responds to the situation by releasing a hormone called *antidiuretic hormone* (*ADH*), which serves as a chemical messenger carrying information from the brain to the tubules of the nephron, where it causes the walls of the tubules to become more permeable, so that water is more efficiently reabsorbed into the blood. In this way more water is conserved for the body and the blood volume is increased, restoring conditions to normal. A small amount of concentrated urine is produced. A thirst center in the brain also responds to dehydration, stimulating an increase in fluid intake.

When we drink a lot of fluid, the blood becomes diluted and its osmotic pressure falls. Release of ADH by the pituitary gland decreases, lessening the amount of water reabsorbed from the tubules. A large volume of dilute urine is thus produced.

Alcoholic beverages and coffee (because of its caffeine content) stimulate urine production. Such substances, called *diuretics,* are thought to inhibit the secretion of ADH.

Occasionally the pituitary gland malfunctions and does not produce sufficient ADH. The resulting condition is termed *diabetes insipidus* (not to be confused with the more common disorder diabetes mellitus). Water is not efficiently reabsorbed from the tubules, and therefore a large volume of urine is produced. A person with severe diabetes insipidus may excrete up to 25 quarts of urine each day, a serious loss of water to the body. The affected individual becomes dehydrated and is constantly thirsty. He must drink almost continuously to offset fluid loss. Diabetes insipidus can now be controlled by injecting the patient with ADH.

ADH regulates only the rate at which water is excreted by the kidney. Salt excretion is controlled by hormones secreted by the adrenal glands. (These will be discussed in Chapter 11.)

## Kidney disease

Kidney disease ranks fourth among major diseases in the United States. The 8 million persons who suffer from kidney disorders each year bear witness to the multitude of ailments that afflict these complex, vital organs. Kidney function can be impaired by a variety of infections, by poisoning caused by such substances as mercury or carbon tetrachloride, by lesions, tumors, kidney stone formation, shock, or a host of circulatory disorders.

*Kidney failure.* The most serious consequence of kidney disease is kidney failure. In this condition most of the nephrons have ceased to function and the kidneys are unable to cleanse the blood effectively. Instead of being properly excreted, nitrogenous wastes such as urea accumulate in the blood, giving rise to *uremia* (sometimes called uremic poisoning). Metabolic wastes build up in the tissues as well as in the blood, and the patient suffers from fatigue, insomnia, nausea, anemia, and other physical discomforts. In advanced uremia, loss of memory and impaired judgment are common, and untreated cases cause death. Unlike many other chronic diseases which are fatal to older people, chronic kidney disease often kills middle-aged people who are in their most productive years.

*Treatment of kidney failure.* More than 50,000 persons in the United States alone die as a result of kidney disease each year. Many of these deaths are unnecessary, since kidney failure can be treated by *kidney dialysis* or by kidney transplant. At present more than 80 percent of Americans suffering from chronic kidney disease are denied therapy because of lack of funds.

*Kidney dialysis.* Sometimes called an artificial kidney, a kidney dialysis machine is a device which assumes the function of a patient's diseased kidneys. A plastic tube is surgically inserted into both an artery and a vein in the patient's arm or leg. These tubes may then be connected to a circuit of plastic tubing from the kidney machine (Figure 7-6). As blood from the patient's artery flows through the machine, wastes diffuse out of the blood through minute pores in the plastic tubing, and into a cleansing solution surrounding the tubing. The walls of the plastic tubing constitute a semipermeable cellophane-like membrane. After the blood flows through the machine, it reenters the patient's body through the connected vein. Only small amounts of waste are removed each time the blood circulates through the machine, so that the patient must remain hooked up to the machine for several hours (usually about five) while his blood repeatedly recirculates through the dialysis unit.

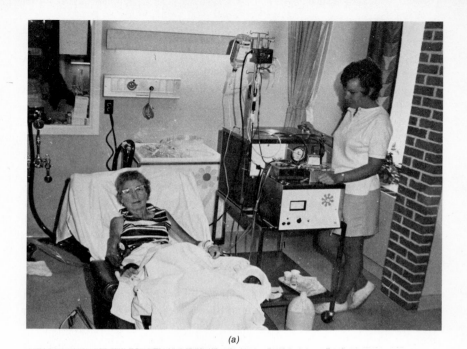

(a)

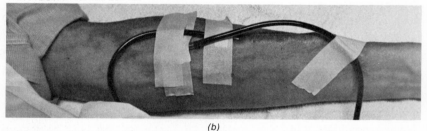

(b)

**FIGURE 7-6**
Kidney dialysis. (a) This patient is hooked up to a kidney dialysis machine. The nurse at right monitors the machine as the patient's blood is dialyzed. (b) Close-up photograph of the arm of a patient being dialyzed to show how the tubes are connected to the patient's blood vessels. In this patient an internal fistula (shunt) has been surgically created so that blood from an artery in the arm flows directly into a vein under increased pressure. (c) Close-up view of disposable-coil dialyzing unit. Blood from the patient flows through the coils of cellulose dialysis tubing. Coils are bathed in dialysis fluid. Tiny holes in the cellulose tubing allow exchange of materials between the blood and the "bath water." Wastes and excess materials leave the blood and pass into the bath water while calcium and other materials from the bath water pass into the blood.

(c)

Kidney dialysis provides a lifesaving treatment for patients with kidney failure. In temporary acute kidney failure dialysis need be used only until the patient's own kidneys regain their normal function. Those suffering from chronic kidney failure with no hope of recovery, however, must depend upon the machine permanently or until a successful transplant can be achieved. Patients with chronic kidney conditions must receive dialysis treatment three times a week. Aside from undergoing this treatment and having a somewhat restricted diet, many of them lead more or less normal lives. More than 80 percent are able to return to their normal occupations.

Kidney dialysis is one of the most controversial medical treatments today because of its enormous cost. Obviously most patients cannot afford the average $24,000-per-year cost of dialysis treatment administered in a hospital. While some have been assisted by federal, state, or private resources, many others have been abandoned to die for lack of funds.

Why is dialysis so expensive? The machine itself is costly, and trained technicians and physicians are required to operate it and to monitor the patient's condition. But the main cost is machine maintenance. Expensive plastic tubing and other components can be used only once, so that each time the machine is used they must be properly replaced.

Many hospitals will not accept a patient for dialysis treatment unless he can guarantee the $20,000 or more annual cost for a period of 3 years. In some places committees have been formed to evaluate applicants for dialysis treatment, and members of such committees have decided who shall live and who shall die on the basis of financial status as well as medical condition. Various plans, including national health insurance programs, have been proposed to help people meet the expense of medical catastrophes such as chronic kidney failure. As a result of social security legislation in 1973 the Federal government is now paying a large share of dialysis costs.

Some patients are now being trained to administer treatment to themselves in their own homes using smaller machines that initially cost about $5,000. Annual cost of home treatment, in addition to the initial expense, is currently about $5,000. To qualify for home treatment, however, a patient must have at least average intelligence, must be judged emotionally stable, and must have a competent, dedicated spouse (or some other person) who can be trained to monitor his treatment.

*Kidney transplant.* Many physicians feel that kidney transplants are a more logical solution to chronic renal disease than dialysis. Existing dialysis machines cannot remove toxic materials as well as a functioning organ can. And with a successful transplant a patient can live a more normal life, since he is not dependent upon a machine. Moreover the expense of transplant surgery is far less than that of years of dialysis therapy.

Unfortunately, until problems of graft rejection (Chapter 6) are solved, kidney transplants remain risky. Furthermore, a more efficient system of obtaining kidneys for transplant purposes is needed. At present there is a shortage of donors and patients must sometimes wait indefinitely for a suitably matched organ to be donated. While they wait, they must rely upon dialysis by machine.

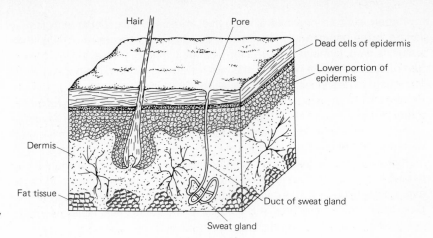

**FIGURE 7-7**
Structure of skin. Note especially the sweat gland.

Labels in figure: Hair, Pore, Dead cells of epidermis, Lower portion of epidermis, Dermis, Fat tissue, Duct of sweat gland, Sweat gland

## OTHER EXCRETORY STRUCTURES

The skin, lungs, and digestive system also function as excretory organs.

### The skin

Skin (Figure 7-7) serves as a protective barrier against disease organisms and mechanical injury, prevents dehydration of body tissues, and helps to regulate body temperature. It contains sensory structures for receiving stimuli of touch, pressure, pain, heat, and cold. Vitamin D is metabolized in skin exposed to ultraviolet rays of sunlight. Here we are concerned with the role of the skin in excretion, a function it performs by the mechanism of sweating.

The sweat glands lie deep in the *dermis* (inner layer of skin) and in the underlying subcutaneous tissue. Each gland consists of a tiny coiled tube connected to a duct which extends through the dermis and epidermis, opening to the outside through a *pore.* An average man has more than 2.5 million sweat glands. Wastes from the blood of nearby capillaries diffuse into the sweat glands and are released from the body through their ducts. Sweat, or perspiration, also cools the body by evaporation and lubricates friction surfaces, such as the palms of the hands and the soles of the feet.

Sweat consists mainly of water with some salts and a trace of urea. Muscular activity, as well as hot weather, increases sweating and, therefore, salt loss. When excessive salt is lost for these reasons, it should be replaced by dietary intake (Figure 7-8).

About a quart of water is excreted daily in sweat. This compares impressively with the quart and a half excreted by the kidney in urine. We are normally unaware of our perspiration. Only when it is produced more quickly than it can evaporate does it accumulate on the skin and become bothersome.

Certain sweat glands found in association with hairs (Figure 7-9) concentrate in specific regions such as the armpits. They probably are not significant as excretory organs. Their secretion is thick and sticky, and at first almost odorless. As this sweat is decomposed by bacteria which inhabit the skin's surface it becomes malodorous, or at least we have been conditioned by television commercials to regard it as such. These sweat glands secrete during states of stress or sexual stimulation. Their odor was most likely important in establishing one's personal scent and in attracting a mate before the age of civilization and certainly before the manufacture of deodorants and perfumes.

## Lungs

The lungs are a complex system of air sacs and blood vessels which exchange gases between the environment and the blood. Details of their structure and respiratory function will be discussed in Chapter 8. As excretory organs the lungs serve to rid the body of excess water and carbon dioxide, both produced as waste products during cellular respiration.

## Digestive system

Bile pigments, which are breakdown products of hemoglobin, are delivered into the intestine by the liver and excreted in the feces. Also, the water content of the feces is regulated by the large intestine, so that excess water from the blood may be added to the feces and excreted in this manner.

## MAINTAINING A STEADY STATE

We have seen how the excretory organs work with the circulatory system to maintain a steady state (a process known as *homeostasis*). The activities of

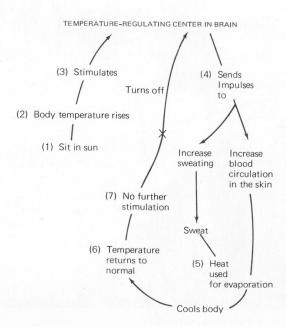

**FIGURE 7-8**
Regulation of sweat excretion. Follow the sequence as indicated by the numbers.

these excretory organs are regulated by the nervous and endocrine systems. Functioning together as a sensitive, self-regulating machine, these organ systems are able to maintain the needed biochemical balance despite constant changes in the outside environment. Whether we travel to the north pole or to the equator, our sweat glands work with our nervous and circulatory systems to maintain a constant body temperature. Whether we drink a gallon of fluid or only a glass on any particular day, our kidneys carefully adjust the volume of water to reflect the steady state of fluid balance required. When any of the mechanisms designed to maintain a steady state fails, the organism suffers disease. If the breakdown is serious enough, the cells are unable to function and death results.

**FIGURE 7-9**
Scanning electron micrograph of human hair. (×5,000) Ends were sliced off to show internal structure. (Emil Bernstein and *Science*)

1 Excretion is the process of removing metabolic wastes from the body.
2 The principal metabolic wastes formed within the cells of the human body are urea, carbon dioxide, and water.
3 Urea is excreted by the kidneys, and, to a lesser extent, by the sweat glands. Water is excreted by the kidneys, sweat glands, lungs, and digestive system. Carbon dioxide is excreted in air exhaled by the lungs.
4 The liver processes excess amino acids and nucleic acids, converting their nitrogen components into nitrogenous wastes, in the form of urea and uric acid.
5 Urine is produced in the kidneys, passed through the ureters to the urinary bladder, and discharged through the urethra.
6 Each nephron consists of Bowman's capsule, an excretory tubule, and an elaborate system of blood vessels.
7 Urine is produced by a process of plasma filtration, reabsorption of useful substances, and secretion.
8 Urine volume is regulated by ADH released at appropriate times by the posterior pituitary gland.
9 Two lifesaving treatments for chronic kidney failure are dialysis and kidney transplant.
10 The skin, lungs, and digestive system perform excretory functions in addition to their other activities.
11 The excretory organs work together with the circulatory system and other organs of the body to maintain a constant internal environment.

Ebra, George: "Rehabilitation Considerations in End Stage Renal Disease," *Journal of Applied Rehabilitation Counseling,* summer 1972, p. 25. Discusses the problems of the dialysis patient and includes some interesting statistics.
Montagna, William: "The Skin," *Scientific American,* February 1965. Discusses the excretory and other functions of skin.

# 8 RESPIRATION

**Chapter learning objective.** The student must be able to trace in sequence the fate of oxygen in the body from its inspiration to its acceptance of hydrogen and conversion to water. The student must also be able to describe the chief chemical reactions and ultimate fate in the body of a fuel substance such as glucose.

*ILO 1* Differentiate between organismic and cellular respiration, and describe their relationship to each other.

*ILO 2* Trace the route traveled by a breath of inhaled air from nostrils to air sacs, and then to recipient cells.

*ILO 3* Describe and give the function of each respiratory structure.

*ILO 4* Describe the mechanical process of breathing and how breathing is controlled. List the steps used to perform mouth-to-mouth resuscitation.

*ILO 5* Compare the composition of inhaled with exhaled air, and describe the exchange of gases between alveoli and blood, and between blood and tissues. Describe how oxygen and $CO_2$ are transported in the blood.

*ILO 6* Write a summary equation for cellular respiration, and give the origin and fate of each substance involved.

*ILO 7* Describe the structure of ATP, and give its functions.

*ILO 8* Summarize the events of glycolysis, giving the key organic compounds formed and the number of carbon atoms in each. Indicate the number of ATP molecules formed and consumed and the transactions in which hydrogen loss occurs.

*ILO 9* Summarize the events of the citric acid cycle, beginning with the conversion of pyruvic acid to acetyl CoA, and describe the fate of carbon-oxygen segments and oxidation by means of the removal of hydrogen.

*ILO 10* Summarize the operation of the electron transport system, indicating the number of ATPs formed by it and the role of molecular oxygen. Describe the effect of oxygen deprivation upon the electron transport system.

*ILO 11* Make a precise comparison in terms of ATP formation between aerobic and anaerobic respiration, and describe two specific examples of anaerobic respiration.

*ILO 12* Describe how cellular respiration is regulated.

*ILO 13* Draw a diagram to show or describe how metabolic activities are interdependent.

*ILO 14* Define each scientific term introduced in this chapter.

Most of us know that we cannot live for more than a few minutes without oxygen, but few can explain why. The "why" and "how" of the body's need for oxygen are the subject of this chapter.

Respiration is the exchange of gases between an organism and its environment. We may divide respiration into two main phases: *organismic respiration* and *cellular respiration.* Organismic respiration is the process of getting oxygen from the air into the body and to each individual cell. Carbon dioxide produced in the cells must then be transported to the lungs and released back into the atmosphere. Cellular respiration refers to the actual consumption of oxygen and the resultant production of carbon dioxide by the cells. During cellular respiration, fuel molecules are degraded and the energy released from their chemical bonds is packaged.

## ORGANISMIC RESPIRATION

In very small organisms no specialized structures are required to ensure an efficient exchange of gases with the environment. Oxygen and carbon dioxide diffuse freely in and out of such organisms and to each of their cells. In plants, air circulating through the intercellular air spaces brings oxygen close to each cell. In large animals the inner cells may be located at too great a distance from the environment to be served efficiently by diffusion alone. These organisms possess highly specialized respiratory and circulatory structures to facilitate the process of gas exchange.

### Respiratory structures

The respiratory system may be thought of as a system of tubes through which air passes on its journey from the nostrils to the air sacs of the lungs, and back. Let us trace the respiratory route followed by a breath of inhaled air (Figures 8-1, 8-2).

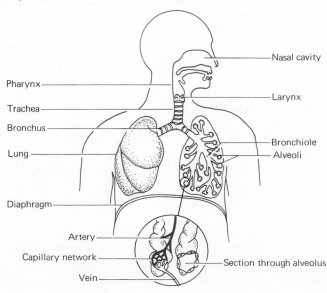

**FIGURE 8-1**
The respiratory system.

**Nose.** The nostrils are the gateway into the body. Each nostril leads into a nasal cavity equipped with large hairs for trapping particles of dirt and other foreign matter. Farther back, the lining of the nasal cavity consists of cells which possess cilia, and mucous cells, which produce about a pint of fluid daily. Dirt particles are trapped in the sticky mucus and pushed along toward the throat by the cilia, which beat about twelve times each second. Foreign particles and mucus are swallowed and so delivered to the digestive system, which is far more capable of disposing of such material than the delicate respiratory system.

Four air *sinuses,* spaces in bones in the back of the nose, in each nasal cavity secrete mucus, which is normally pushed into the nasal cavities by cilia. *Sinusitis,* a condition all too familiar to some of us, occurs when these sinuses become inflamed or infected. It has long been known that the hormone *epinephrine* tends to shrink nasal tissues that have become swollen because of allergy or infection. Drugs, chemically similar to epinephrine, have been developed for temporary relief of a stuffy nose. Many popular nose drops and nasal sprays contain such drugs. As a general rule, however, epinephrine-like drugs tend to irritate the nasal membranes even while they shrink them, so when the effect of the drug wears off, the nose is likely to become even stuffier than before. This *rebound effect* may be very marked when nasal sprays are used to excess. Abuse of nasal sprays can lead to a shrinkage of the blood vessels in the nasal lining, which in turn can lead to the starvation and atrophy of some of the internal structures of the nose.

The lining toward the top of each nasal cavity is specialized as an organ of smell. Another job of the nose is to regulate the temperature of incoming air. Heat conduction from a rich network of blood vessels in the lining of the nose brings inhaled air to body temperature.

**Pharynx.** The back of the nose is continuous with the pharynx, or throat. Whether inhaled through the mouth or the nose, air must pass through the pharynx. Mouth breathing is undesirable as a rule, because the mouth is not designed for filtering or warming incoming air. Air passes through the *glottis,* an opening in the floor of the pharynx which leads into the *larynx.*

**Larynx.** The larynx (sometimes termed the Adam's apple) leads from the pharynx to the *trachea.* Cartilage embedded in its walls prevents the larynx from collapsing and makes it hard to the touch when felt through the neck. Because it contains the vocal cords, the larynx is sometimes called the voice box.

The larynx prevents foreign matter from entering the lungs. When anything but air comes into contact with its sensitive lining, the larynx initiates a cough reflex, designed to expel foreign material quickly from the respiratory system.

Spasms of the larynx are thought to be the cause of death in some drowning victims. As water enters and irritates this organ its muscular wall contracts so that the water cannot move past it and enter the lungs. Unfortunately air also is denied entrance, and the victim dies of asphyxiation (insuf-

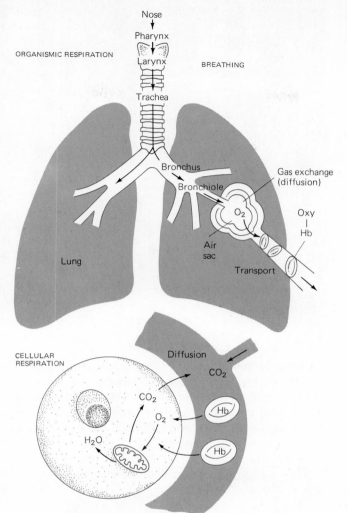

ORGANISMIC RESPIRATION

BREATHING

Nose
↓
Pharynx
↓
Larynx
↓
Trachea

Bronchus
Bronchiole

Gas exchange
(diffusion)

O₂

Oxy
|
Hb

Air
sac

Transport

Lung

CELLULAR
RESPIRATION

Diffusion

CO₂

CO₂

O₂

Hb

H₂O

Hb

**FIGURE 8-2**
Organismic and cellular respiration.

ficient oxygen). This mechanism explains why only very small amounts of water are found in the lungs of some drowned persons.

***Trachea and bronchi.*** From the larynx air passes into the trachea, or windpipe. Like the larynx, the trachea is kept from collapsing by rings of cartilage in its wall. The trachea divides into two branches, the bronchi (singular, bronchus), one going to each lung. Both trachea and bronchi are lined by a mucous membrane containing ciliated cells (Figure 8-3). Medium-sized particles (the smallest commonly found under natural conditions) which have escaped the cleansing mechanisms of nose and larynx are trapped here. Mucous containing these particles is constantly beaten upwards by the cilia to the pharynx, where it is periodically swallowed. This mechanism for keeping foreign material out of the lungs functions as a cilia-propelled mucous elevator.

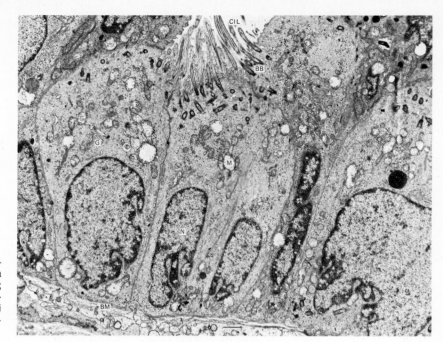

**FIGURE 8-3**
Electron micrograph showing ciliated epithelium lining a bronchus. (×10,000) *CIL,* cilia; *BB,* basal body of cilium; *M,* mitochondria; *N,* nucleus; *G,* Golgi apparatus; *BM,* basement membrane. (Dr. Lyle C. Dearden)

***Bronchioles and air sacs.*** Inside the lungs the bronchi branch into smaller and smaller respiratory tubes, the bronchioles (Figure 8-4). There are more than a million tiny bronchioles in each lung, and each leads into a cluster of tiny air sacs, the *alveoli* (singular, alveolus). These tiny air sacs are lined by walls only one cell thick, through which gases can diffuse freely. Alveoli are

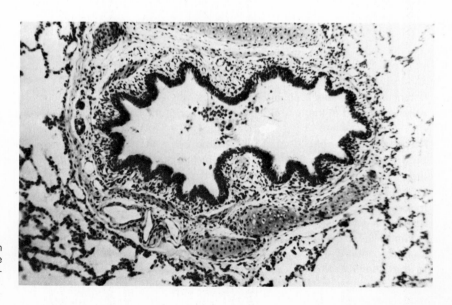

**FIGURE 8-4**
Section through a bronchiole in the lung. Some air sacs are visible at periphery of photomicrograph.

covered by a thin film of lipid-protein which prevents them from collapsing. By greatly increasing the surface area available for gas exchange with the blood, alveoli enhance the efficiency of the lung. Capillaries surround each alveolus (Figures 8-1 and 8-5) so that gases do not have far to travel between air sac and blood.

**Lungs.** The lungs are large organs occupying most of the chest cavity. Each lung consists of a vast system of bronchus, bronchioles, alveoli, and the capillary networks that surround them. These structures are embedded in connective tissue rich in elastic fibers. In addition, nerves and a great deal of lymph tissue are present. The total effect of these structures is to make each lung a spongy, elastic organ with a very large surface area for gas exchange—about forty times the surface area of the entire body. In fact, it has been computed that in the normal adult the surface area of the lung is as large as a tennis court.[1]

Each lung fits into a compartment in the chest cavity which is lined by connective tissue membranes, the *pleura.* A film of fluid between the pleura covering the lung and the pleura lining the chest cavity provides lubrication so that the lung can slide during breathing movements. Inflammation of the pleura is called *pleurisy,* a condition associated with considerable pain during breathing.

### Breathing

Breathing is the mechanical process of moving air from the environment into the lungs, and of expelling air from the lungs back out of the body. A resting adult breathes about 12 times each minute.

**Mechanics of breathing.** Breathing depends upon pressure changes brought about by altering the size of the chest cavity surrounding the lungs. *Inspiration,* or breathing in, takes place when the chest cavity is expanded. This is accomplished principally by the contraction of the *diaphragm,* a large, flat muscle which forms the floor of the chest cavity (Figures 8-1, 8-6). When the diaphragm contracts, it moves downward, increasing the size of the chest cavity. At the same time the rib muscles contract, increasing the circumference of the chest cavity. As the volume of the chest cavity increases, its pressure decreases, and the lungs expand to fill the extra space. When the lungs expand, the air within them spreads out over a larger space, decreasing the pressure. Alveolar pressure becomes less than atmospheric pressure, so that air rushes in from outside the body to equalize the pressure. *Expiration,* or breathing out, occurs when the diaphragm and rib muscles relax, causing the reverse of the events outlined above.

**Control of breathing.** The amount of oxygen required by the body varies from time to time. If you are engaged in strenuous physical activity, for ex-

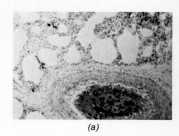

*(a)*

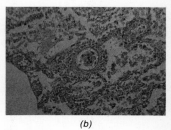

*(b)*

**FIGURE 8-5**
(*a*) Alveolar tissue of lung adjacent to blood vessel. (*b*) Lung tissue from a person with tuberculosis. Note the tubercle in the center.

---

[1] Stanley W. Jacob and Clarice A. Francone. *Structure and Function in Man,* 2d ed., Saunders, Philadelphia, 1970, p. 385.

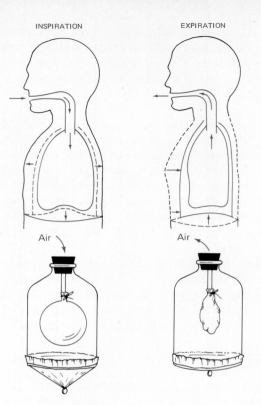

INSPIRATION EXPIRATION

Air     Air

**FIGURE 8-6**
Breathing. The model shown consists of a bell jar with a rubber stopper at the top. A glass tube extends through the stopper and is attached to a balloon. The elastic covering across the bottom of the jar represents a diaphragm. When it is pulled downward, the volume in the jar increases and air rushes into the balloon to fill the vacuum. The balloon represents the lung. To demonstrate expiration the elastic diaphragm is released. The pressure within the jar is thus increased, and air rushes out of the balloon. In the lung the alveoli function as millions of tiny balloons.

ample, more oxygen is required than if you are reading quietly. Breathing is controlled by a respiratory center in the brain which is sensitive to increases in the amount of carbon dioxide in the blood. Nerves from the respiratory center stimulate the contraction of chest muscles. During exercise greater amounts of carbon dioxide are produced. The carbon dioxide stimulates the respiratory center to produce more rapid and more forceful breathing. In this way sufficient oxygen is provided to meet the body's increased need.

At high altitudes air pressure is lower than at sea level. The decreased air pressure causes a reduction in the amount of oxygen that enters the blood, and more rapid breathing results. After a period of time the body adjusts to life at a higher altitude by increasing the number of red blood cells, so that more oxygen is carried by each volume of blood.

Physiologic adaptation to pressure change takes time. Divers who return to the surface too quickly, or pilots who ascend to over 35,000 feet too rapidly, may suffer from *decompression sickness* ("the bends"). While submerged, a diver breathes gases under high pressure, and because of this pressure, nitrogen gas dissolves in his blood and tissues. As he ascends to a lower pressure the dissolved nitrogen is liberated. If he surfaces too rapidly, tiny bubbles form in the blood and tissues, and may block capillaries and cause other damage. These bubbles cause the symptoms of decompression sickness: pain, paralysis, even death.

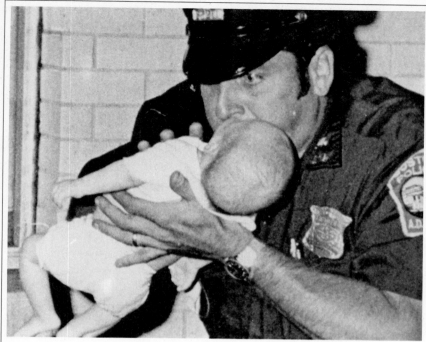

(a)

**Artificial respiration.** *What is it?* Artificial respiration is a method of forcing air in and out of a person's lungs.

*When is it used?* Artificial respiration is used whenever a person has stopped breathing or is breathing in an irregular or shallow manner so that his breathing is not adequate.

*Why is it used?* The purpose of artificial respiration is to provide oxygen to the victim until he can breathe again on his own or until professional medical help is available.

*How is it done?* The most effective method is mouth-to-mouth resuscitation. It is accomplished as follows:

1  Place the person who has stopped breathing on his back.
2  Clear the victim's mouth and throat so that nothing will block the flow of air.
3  Tilt the victim's head back so that his chin points upward, and lift his lower jaw so that it juts out.
4  Pinch the nostrils shut, and blow into the victim's mouth.
5  Remove your mouth and listen for air rushing out of the lungs. If no outflow of air is heard, check for foreign matter blocking the respiratory passageway. Turn the victim on his side and hit him between the shoulder blades to dislodge anything that may be blocking the air tube.
6  Blow vigorously into the victim's mouth about 12 times each minute. Each time remove your mouth and listen for the return of air. Normal breathing may resume after 15 minutes, but could take much longer.

(b)

**FIGURE 8-7**
(a) Mouth-to-mouth resuscitation of an 11-week-old infant. (OMIKRON) (b) Procedure for artificial respiration.

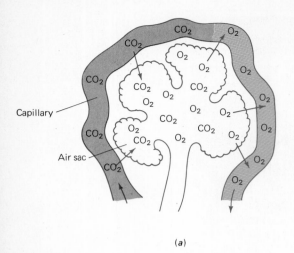
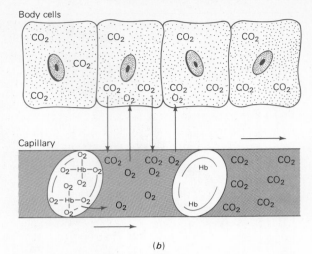

(a) (b)

**FIGURE 8-8**
(a) Exchange of gases between air sacs and capillary. (b) Exchange of gases between capillary and body cells.

## Gas exchange and transport

The respiratory system delivers oxygen to the air sacs, but if oxygen were to *remain* in the lungs, all the other body cells would soon die. The vital link between air sac and body cell is the circulatory system. Each air sac serves as a tiny depot from which oxygen is loaded into blood brought close to the alveolar air by capillaries (Figure 8-8).

Oxygen molecules diffuse from the air sacs into the blood because the air sacs contain a greater concentration of oxygen than does blood entering the pulmonary capillaries. Similarly carbon dioxide moves from the blood, where it is more concentrated, to the air sacs, where it is less concentrated. Each gas diffuses through the cells lining the alveoli and the cells lining the capillaries.

Table 8-1 shows the percentages of oxygen and carbon dioxide present in exhaled air as compared with inhaled air. Because carbon dioxide is produced during cellular respiration, there is more than 100 times as much of this gas entering the alveoli from the blood as there is in air inhaled from the environment.

**Table 8-1. Composition of inhaled air compared with that of exhaled air**

|  | Percent oxygen ($O_2$) | Percent carbon dioxide ($CO_2$) | Percent nitrogen ($N_2$) |
|---|---|---|---|
| Inhaled air* | 20.9 | 0.04 | 79 |
| Exhaled air† | 14.0 | 5.6 | 79 |

Note: As indicated, the body uses up about one-third of the inhaled oxygen. The amount of $CO_2$ increases more than 100-fold because it is produced during cellular respiration.
* The same as atmospheric air.
† Sometimes referred to as alveolar air.

The movement of gases between air sacs and blood is not completely efficient. Not every molecule of inhaled oxygen actually finds its way into the blood, and not every molecule of carbon dioxide is removed from the blood. The amount of exchange that takes place, however, is obviously sufficient to support the metabolic well-being of the body.

When oxygen diffuses into the blood it forms a weak chemical bond with the hemoglobin molecules of the red blood cells. Each hemoglobin molecule can combine with, and thus transport, four molecules of oxygen. Since the chemical bond formed between the oxygen and the hemoglobin is weak, the reaction is readily reversible. In the pulmonary capillaries:

Hemoglobin + oxygen $\longrightarrow$ oxyhemoglobin

When oxyhemoglobin reaches body cells low in oxygen, the reverse reaction occurs:

Oxyhemoglobin $\longrightarrow$ oxygen + hemoglobin

The oxygen released diffuses from the capillaries into the cells.

As blood flows through the capillary networks of an organ, carbon dioxide moves out of the cells where it has accumulated into the blood, where it is less concentrated. Carbon dioxide is transported in the blood in three different ways. About 20 percent is carried on the hemoglobin molecule, 10 percent is transported in the plasma as carbon dioxide itself, and the remainder is dissolved in the plasma after being converted into bicarbonate ions, ($HCO_3^-$).

## CELLULAR RESPIRATION

Cellular respiration is the step-by-step breakdown of fuel molecules, such as glucose, and the packaging of energy released from their chemical bonds. Oxygen is an essential ingredient for this process. A summary equation for the reactions of cellular respiration would be:

$$C_6H_{12}O_6 + 6O_2 \longrightarrow 6CO_2 + 6H_2O + \text{energy}$$

glucose   oxygen   carbon dioxide   water   packaged in ATP

You may recall that this equation is the reverse of the summary equation for photosynthesis. During photosynthesis plants use water, carbon dioxide, and energy from sunlight to manufacture carbohydrates. Oxygen is produced as a by-product. During cellular respiration cells break carbohydrates back down into their components. Only green plant cells can carry on photosynthesis, but every living cell, even green plant cells, carries on cellular respiration. Furthermore, continuous energy input is a prerequisite of life, and so cells must constantly respire, day and night. When respiration ceases, cells die.

Fuel molecules are degraded by the process of biologic *oxidation*. Oxidation occurs when a fuel molecule combines with oxygen, when electrons are removed from it, or when hydrogen is lost. Burning, a form of *non*biologic oxidation, proceeds in one step, so that all the energy from the burning fuel is released in one sudden burst. In contrast, biologic oxidation is carried out in a series of chemical reactions that ensures an efficient step-by-step release of energy. Each reaction is regulated by a specific enzyme. The gradual liberation of energy provides for its efficient use and processing by the cell. Operating at 98.6 degrees Fahrenheit, the body is more efficient than a steam engine which operates at 212 degrees Fahrenheit.

Cellular respiration may be broken down into three processes: glycolysis, the citric acid cycle, and the electron transport system. In each of these phases energy is liberated from the chemical bonds of fuel molecules and is suitably packaged for cellular use.

### Packaging energy

The purpose of cellular respiration is to capture the chemical bond energy of fuel molecules and package them in a form convenient for rapid use by the cell. Most commonly, cellular energy is packaged within the high-energy bonds of *adenosine triphosphate* (ATP) molecules. Adenosine, which is also a component of nucleic acids, consists of a purine base attached to a five-carbon sugar called *ribose.*

When one phosphate group is added to adenosine the resulting compound is *adenosine monophosphate* (*AMP*) (see page 171, top). A great deal of energy is required to add on a second phosphate group to make *adenosine diphosphate* (*ADP*). The chemical bond between the first and second phosphate contains this large amount of energy that was employed to form it, and is referred to as a *high-energy bond.* (It is designated by means of a wavy line.) A great deal of energy is also required to attach a third phosphate group to produce the molecule ATP. Thus, ATP contains two high-energy bonds. The energy contained in each high-energy bond is more than twice the amount found in an ordinary bond.

The important point to remember is that ATP *packages* energy when it is made, then releases the energy for use within the cell when it is broken down.[2] When energy is required for cellular activity, ATP is converted to ADP with the release of the energy stored in the terminal high-energy bond. This is the energy actually used for doing biologic work.

Most of the chemical reactions of cellular respiration are *exergonic* reactions, that is, energy is released during the reaction. On the other hand energy is needed to bond atoms chemically to one another, so that most synthetic reactions are *endergonic* (energy-requiring). ATP serves as a link between these two types of chemical reactions. Energy released from exergonic reactions is packaged in ATP molecules for use in endergonic reactions (Figure 8-9). Large quantities of ATP molecules cannot be stockpiled in

---

[2] Instead of being immediately released, the energy may be transferred with the terminal phosphate to a new substance.

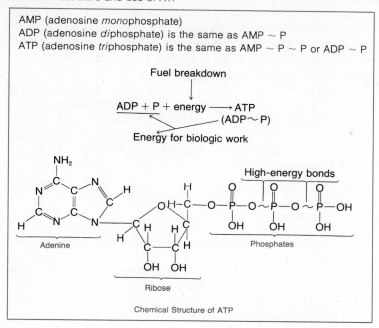

AMP (adenosine *mono*phosphate)
ADP (adenosine *di*phosphate) is the same as AMP ~ P
ATP (adenosine *tri*phosphate) is the same as AMP ~ P ~ P or ADP ~ P

Fuel breakdown

ADP + P + energy ⟶ ATP
(ADP~P)

Energy for biologic work

High-energy bonds

Adenine

Phosphates

Ribose

Chemical Structure of ATP

the cell. They are used almost as quickly as they are produced. Energy from ATP may be used to produce fats or glycogen, molecules stockpiled for long-term energy storage. The cell depends upon the chemical bonds of these organic molecules for its future energy needs.

ATP: Link between exergonic and endergonic reactions

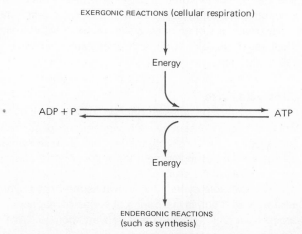

EXERGONIC REACTIONS (cellular respiration)

Energy

ADP + P ⇌ ATP

Energy

ENDERGONIC REACTIONS
(such as synthesis)

**FIGURE 8-9**
ATP: link between exergonic and
endergonic reactions.

## Glycolysis

The biologic oxidation of a molecule of glucose begins with a series of about ten chemical reactions known as *glycolysis.* During this sequence of reactions a molecule of glucose which contains six carbon atoms is chemically broken down into two molecules of *pyruvic acid,* each of which contains three carbon atoms (see Figure 8-10).

For our purposes we will divide glycolysis into two major steps. In the first step the glucose molecule is broken in half to form two molecules of a three-carbon compound called *phosphoglyceraldehyde* (*PGAL*). Since this transformation requires energy, *the cell must invest two molecules of ATP in order to initiate the oxidation of glucose.* We may summarize this step of glycolysis by the following equation:

Glucose + 2ATP →→→ 2PGAL + 2ADP + 2P
(6-carbon compound)      (3-carbon compound)

Several arrows are used to indicate that this equation actually summarizes a sequence of several reactions.

In the second major step of glycolysis two hydrogens are removed from each PGAL and certain other atoms are rearranged, thereby converting each into pyruvic acid. During this transformation enough chemical energy is obtained from the fuel molecule to produce four ATPs.

2PGAL + 4ADP + 4P →→→ 2 pyruvic acid + 4H + 4ATP
(3-carbon compound)               (3-carbon compound)

In the first step of glycolysis two molecules of ATP are used up; in the second step four ATPs are produced. Glycolysis thus yields a *direct net energy profit* of two ATP molecules.

The two hydrogen atoms that are removed from each PGAL immediately combine with the hydrogen carrier molecule, NAD.

$$NAD + 2H \longrightarrow NAD \text{-} H_2$$

The fate of these hydrogen atoms will be discussed in conjunction with the electron transport system and with anaerobic respiration.

The reactions of glycolysis take place in the cytoplasm. Necessary ingredients such as ADP, NAD, and phosphates float freely in the cytoplasm for use as needed.

## Citric acid cycle

Pyruvic acid molecules produced during glycolysis move into the mitochondria, and essentially all subsequent reactions of cellular respiration take place within these organelles. Since these reactions yield most of the energy packaged in the cell, the mitochondria often are referred to as the powerhouses of the cell.

The *citric acid cycle,* also known as the *Krebs cycle* after Sir Hans Krebs who worked it out, is the principal series of reactions that removes hydrogen atoms from fuel molecules and delivers them to NADP for further processing via the electron transport system.

Summary Equation for cellular respiration:

$$C_6H_{12}O_6 + 6O_2 \rightarrow 6CO_2 + 6H_2O + \text{energy packaged in ATP}$$

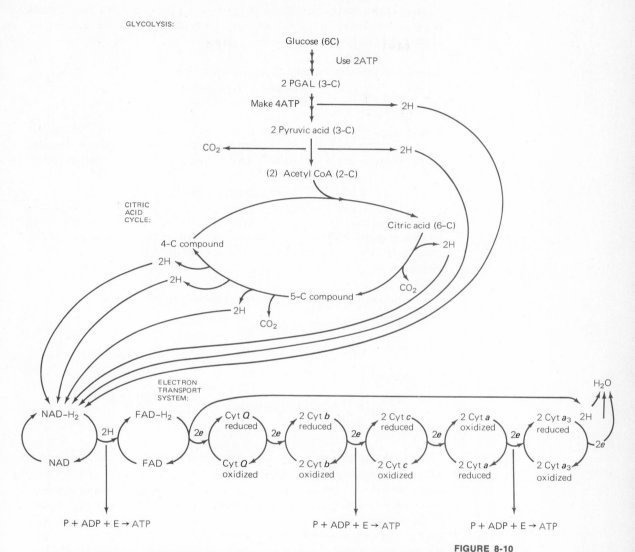

Before the citric acid cycle actually begins, an important preliminary reaction takes place. Each three-carbon molecule of pyruvic acid is degraded by the removal of hydrogen atoms and a carbon-oxygen segment. The carbon-oxygen segment is removed as $CO_2$ which eventually finds its way to the lungs and is exhaled. During a very complex reaction pyruvic acid combines with a substance called *coenzyme A* (CoA) and forms a two-carbon compound, acetyl CoA. Coenzyme A is manufactured in the cell from one of the B vitamins, pantothenic acid.

**FIGURE 8-10**

Summary of the reactions of cellular respiration. (Cyt is an abbreviation for cytochrome, which is discussed on page 175.)

$$\text{2 pyruvic acid} + \text{2CoA} + \text{2NAD} \longrightarrow \text{2 acetyl CoA} + \text{2NAD-H}_2 + \text{2CO}_2$$

The six carbon atoms from the original glucose fuel molecule may now be accounted for as follows. Four of the carbons are located in the two acetyl-CoA molecules, and the other two have been released as carbon dioxide.

Acetyl CoA enters the citric acid cycle by combining with a four-carbon compound (already present in the mitochondrion) to form citric acid, a six-carbon compound.

4-carbon compound + acetyl CoA $\longrightarrow$ citric acid
                        (2-carbon             (6-carbon
                        compound)         compound)

After several chemical reactions citric acid is degraded to a five-carbon compound. The sixth carbon is released (with oxygen) as carbon dioxide. Two hydrogens are also removed and are immediately accepted by NAD.

Citric acid + NAD $\longrightarrow$ 5-C compound + NAD-$H_2$ + $CO_2$

A second sequence of reactions breaks down the five-carbon compound into the original four-carbon compound with which the cycle began. Again carbon dioxide is released, and this time three different pairs of hydrogen atoms are removed.

5-C compound + 3NAD $\longrightarrow$ 4-C compound + 3NAD -$H_2$ + $CO_2$

Note that as long as acetyl CoA is fed into it, the cycle will continue to turn. *Since two acetyl-CoA molecules are produced from the original glucose, two turns of the cycle are necessary to process completely the fuel molecule with which we started.* After these two turns of the citric acid cycle, all that remains of the glucose molecule are carbon dioxide, hydrogens, and energy. The carbon dioxide originates entirely from the food molecule. Even the oxygen it contains comes from the food substance, not from the air. The oxygen of the air combines with the hydrogens that have been removed, as we shall see.

**Electron transport system**
Now we turn our attention to the fate of all the hydrogens removed from the fuel molecule during glycolysis, acetyl-CoA formation, and the functioning of the citric acid cycle. The electron transport system is a chain of several hydrogen-electron carrier molecules. Hydrogen atoms (or in some cases just their electrons) pass along the chain from one carrier molecule to the next in

a specific, orderly sequence. Each carrier molecule is *reduced* (gains electrons) as it picks up the hydrogen, then *oxidized* (loses electrons) as it passes the hydrogen on. Energy is released at three specific points in the chain. As it is released, the energy is captured and packaged in the high-energy phosphate bond of ATP. In this way three ATP molecules are produced for each pair of hydrogens that enter the electron transport system (Figure 8-10).

NAD is the first carrier molecule in the chain. A molecule called FAD is the second. The next five carrier molecules are cytochromes, a group of closely related proteins characterized by a central atom of iron. It is the iron that actually combines with the electrons from the hydrogen atoms. Cytochrome molecules accept only the electron from the hydrogen, rather than the entire atom. When the electrons are separated from the rest of the hydrogen, the remaining ion (a proton) is temporarily set free in the surrounding medium. The last cytochrome in the chain passes on the two electrons (which simultaneously reunite with their hydrogen ions) to molecular oxygen. The chemical union of the hydrogen and oxygen produces water.

Oxygen is the final hydrogen acceptor in the electron transport system. This is the vital role of the oxygen which we breathe. What happens when the body is deprived of oxygen? When no oxygen is available to accept the hydrogen, the last cytochrome in the chain is stuck with it. The preceding carrier molecule then has no acceptor to give its hydrogen electrons to, and so on down the series. The entire system becomes blocked, all the way back to NAD, and no further ATPs can be produced by way of the electron transport system. Most cells of higher organisms cannot live long without oxygen because the amount of energy which they can produce in its absence is insufficient to sustain life.

Lack of oxygen is not the only factor which may interfere with the electron transport system. Some poisons such as cyanide inhibit the normal activity of the cytochrome system. Cyanide combines electrons tightly with one of the cytochrome molecules so that they cannot be passed on to the next member in the chain.

**The fate of glucose**

We have seen how a molecule of glucose is completely degraded during cellular respiration. The carbon-oxygen segments are released as carbon dioxide, while the hydrogens combine with oxygen to form water. In the course of its complete breakdown one molecule of glucose yields sufficient energy to produce 38 molecules of ATP[3] (see Table 8-2).

**Oxidation of other nutrients**

Although glucose is the most common fuel molecule, it is by no means the only one. Excess amino acids and fatty acids, as well as other types of monosaccharides, find their way into the respiratory pathway, sometimes in

---

[3] The maximum number of ATPs which can be produced from one molecule of glucose is 38. Fewer may be produced, in which case more energy is liberated as heat.

the form of pyruvic acid or acetyl CoA. The cell does not know or care about the origin of a fuel molecule. One pyruvic acid is as good as another. These intermediate fuel molecules are processed just as if they had been produced from glucose. If at any time too many molecules of a particular substance accumulate, they may be converted into more needed compounds or may be converted to fatty acids and stored in fat cells.

### Anaerobic respiration

The preceding discussion has centered upon *aerobic respiration,* respiration that depends upon the presence of oxygen (air). *Anaerobic respiration,* or *fermentation,* is fuel utilization without benefit of air, that is, without oxygen (Figure 8-11).

Certain bacteria engage solely in anaerobic respiration. Some of these anaerobic bacteria are associated with polluted water, where they are responsible for putrefaction, the foul smells of which are associated with the anaerobic metabolism of nitrogen and sulfur-containing amino acids. These bacteria displace the normal aerobic inhabitants when waterways become deficient in oxygen as a consequence of pollutional changes. Many anaerobic bacteria inhabit the soil, where they serve as important decomposers. Others cause disease if they get into the body. These include the bacteria causing tetanus, botulism, and gas gangrene.

Some cells which are normally aerobic can shift to fermentation when deprived of oxygen. Anaerobic respiration depends upon the reactions of glycolysis. You will recall that the net profit of two ATPs produced during glycolysis does not require the presence of oxygen. The difference between aerobic glycolysis and anaerobic glycolysis is the fate of the hydrogens removed during the conversion of PGAL to pyruvic acid. Normally, these hydrogens are processed via the electron transport system, but in the absence of oxygen this avenue is blocked, and they must be disposed of in some other way.

When deprived of oxygen, yeast cells use the hydrogens in question to convert pyruvic acid into ethyl alcohol (drinking alcohol). Carbon dioxide is released during this process. Such anaerobic reactions are the basis for the

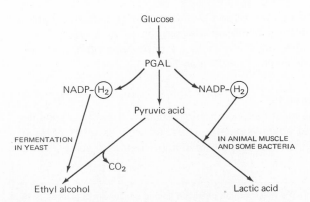

**FIGURE 8-11**
Anaerobic respiration.

**Table 8-2. Energy considerations**

|  | Anaerobic | Aerobic |
|---|---|---|
| From glycolysis | 2 ATP | 2 ATP (oxygen not required) |
| By way of electron transport system: |  |  |
| 4H from glycolysis |  | 6 ATP |
| 4H from pyruvic acid |  | 6 ATP |
| 16H from citric acid cycle |  | 24 ATP |
|  | 2 ATP | 38 ATP |

Note: The H input and output is not balanced because water provides extra H's which enter into the reactions.

production of wine and other alcoholic beverages. Yeast cells are also utilized in the baking industry. The alcohol evaporates from dough, but the carbon dioxide gas accumulates in the dough, causing it to rise.

An alternative method for disposing of the hydrogens removed from glucose during glycolysis is for the cell to give them to the pyruvic acid molecule. When hydrogen atoms are added to pyruvic acid, lactic acid is formed.

Pyruvic acid $+ H_2 \longrightarrow$ lactic acid

This reaction occurs when bacteria sour milk or ferment cabbage to sauerkraut. It also takes place in the muscle cells of man and other animals. During strenuous physical activity the amount of oxygen delivered to the muscle cells may be insufficient to keep pace with the rapid rate of fuel oxidation. Not all the hydrogen atoms accepted by NAD can be processed in the usual manner, since there is a shortage of oxygen. Instead, these hydrogen atoms are given to pyruvic acid to form lactic acid. As lactic acid accumulates in muscle cells it causes *muscle fatigue.* The *oxygen debt* acquired by the muscle cells is repaid by the period of rapid breathing that follows strenuous physical activity. Part of the lactic acid is then oxidized with the help of this extra oxygen. Most of the lactic acid is transported by the blood to the liver cells, where it is converted back to glucose for reuse.

Anaerobic respiration is a very inefficient process, since the fuel is only partially oxidized. Alcohol, the end product of fermentation by yeast cells, can be burned, and could even be used as an automobile fuel. Obviously it contains a great deal of energy that the yeast cells were unable to get out using anaerobic methods. Lactic acid, a three-carbon compound, contains even more energy than alcohol (a two-carbon compound). During aerobic respiration *all* available energy is removed by completely oxidizing fuel molecules. Table 8-2 shows the number of ATPs formed during each phase of cellular respiration, and compares anaerobic to aerobic respiration. A net profit of only two ATPs can be produced anaerobically from one molecule of glucose, as compared to 38 produced when oxygen is available. The two ATPs produced during glycolysis represent only 5 percent of the total energy in a molecule of glucose. Using aerobic methods about 55 percent of the energy in a glucose molecule can be captured. The rest is lost as heat. This compares favorably with the efficiency of the finest man-made machines. In man and other warm-blooded animals some of the heat produced during respiration and other metabolic activities is utilized to maintain a constant body temperature.

The inefficiency of anaerobic respiration necessitates a large supply of fuel. By rapidly degrading many fuel molecules a cell can compensate somewhat for the small amount of energy that can be gained from each. To perform the same amount of work, an anaerobic cell must consume up to twenty times as much glucose or other carbohydrate as an aerobic cell. For this reason skeletal muscle cells, which often respire anaerobically for short periods, store large quantities of glucose in the form of glycogen.

### Regulation of cellular respiration

Cellular respiration requires a steady input of nutrient fuel molecules and oxygen. Under normal conditions these materials are adequately provided and do not affect the rate of respiration. Instead, cellular respiration is regulated by the amount of ADP and phosphate available. As ATP is manufactured, the pool of ADP and phosphate diminishes and respiration slows. Then, as ATP is utilized for cellular activities, ADP and phosphate are released and become available for packaging energy once more. Thus cellular respiration adjusts to a more rapid pace. In this way the rate of cellular respiration is self-regulated and can meet the immediate needs of the cell for energy.

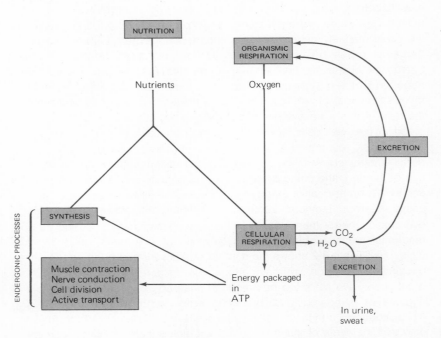

**FIGURE 8-12**
Interaction of metabolic activities. Nutrition and organismic respiration supply raw materials needed for cellular respiration. Cellular respiration provides the ATP required for synthesis and other endergonic processes. Waste products are removed by the process of excretion.

## Metabolism: the big picture

Figure 8-12 illustrates the relationship of cellular respiration to other vital metabolic activities. Nutrition and organismic respiration provide the raw materials for cellular respiration. Waste products generated during the oxidation of fuel molecules are removed by the process of excretion. Energy packaged during cellular respiration is used to perform endergonic biologic activities. Thus, we see that respiration is interdependent with all other metabolic processes. The total interaction of these activities constitutes life itself.

1 Organismic respiration is the process of getting oxygen from the air into the body and to each individual cell, and of delivering carbon dioxide from the cells back out into the environment. Cellular respiration is the actual consumption of oxygen and the resultant production of carbon dioxide, water, and energy by each cell.

2 Inhaled air passes through the nose, pharynx, larynx, trachea, bronchi, bronchioles, and air sacs of the lungs.

3 Breathing is the mechanical process of moving air back and forth between the environment and the air sacs in the lungs.

4 Oxygen diffuses from the air sacs into the blood and is transported to the cells in weak chemical association with hemoglobin.

5 The purpose of cellular respiration is to package energy from fuel molecules in a form convenient for immediate use by the cell.

6 ATP, the molecule most commonly used for short-term energy storage, serves as the link between exergonic and endergonic reactions.

7 During glycolysis a molecule of glucose is degraded into two molecules of pyruvic acid. A net profit of two ATPs is gained, and four hydrogen atoms are removed from the fuel molecule.

8 Breakdown of glucose molecules is completed in the citric acid cycle.

9 Hydrogen atoms removed from fuel molecules are passed along a chain of hydrogen acceptor molecules in the electron transport system. Energy is liberated at three specific points along the chain and is packaged as ATP. The final acceptor is molecular oxygen.

10 When cells are deprived of oxygen, the electron transport system becomes blocked, and energy production is insufficient to support life.

11 Although glucose is the most common fuel molecule utilized by cells, other organic nutrients can be converted into appropriate compounds for use as fuel.

12 Muscle cells can engage in anaerobic respiration for brief periods of time, as when an individual is engaged in strenuous physical activity.

13 Cellular respiration is a self-regulating process. Its rate depends upon the amount of available ADP and phosphate.

Comroe, Julius H., Jr.: "The Lung," *Scientific American* reprint, Freeman, San Francisco, February 1966.

Griffin, Donald R., and Alvin Novick: *Animal Structure and Function,* Holt, New York, 1970. Chapter 5 describes respiration in various animal forms.

Lehninger, Albert L.: *Bioenergetics,* Benjamin, New York, 1965.

Loewy, Ariel G., and Philip Siekevitz: *Cell Structure and Function,* 2d ed., Holt, New York, 1969.

# WHERE HAS ALL THE CLEAN AIR GONE?

***Chapter learning objective.*** The student must be able to describe the major air pollutants, identify their sources, and describe their physiologic effects on living systems; he must also be able to give specific suggestions for improving the quality of the air.

*ILO 1*   Distinguish between unpolluted and polluted air.

*ILO 2*   Describe each of the effects of air pollution: irritation, bronchial constriction, aggravation of existing respiratory disorders, chronic bronchitis, emphysema, lung cancer; give the evidence linking each condition with air pollution, and explain how each can be avoided.

*ILO 3*   Identify and describe each of the major air pollutants given in this chapter, identify the main source of each, and the physiologic disturbances caused by each.

*ILO 4*   Identify and describe the five main sources of air pollution discussed in Table 9-1, and give positive suggestions for abating the pollution emitted by each source.

*ILO 5*   Describe the specific effects of air pollution upon plants and animals other than man.

*ILO 6*   Describe the relationship between weather and air pollution, and describe an inversion layer.

*ILO 7*   Give specific measures for dealing with the problem of air pollution.

*ILO 8*   Describe the evidence that has been amassed supporting the view that cigarette smoking is harmful to human health, and give reasons why cigarette smoking represents a form of air pollution for the smoker and for those in his vicinity.

*ILO 9*   Describe the relationship between cigarette smoking and vascular disease, chronic respiratory disease, cancer, damage to the unborn child, and ulcers.

*ILO 10*   Define all scientific terms introduced in this chapter.

Chemically pure air does not occur in nature. Particles of salt from the sea, dust from erosive processes, and organic compounds from vegetation are a few of the naturally occurring "impurities" in the air we breathe. Such impurities *do not* constitute pollution. The human respiratory system is well adapted to their presence. We may even find natural impurities enjoyable. The fragrance of a pine forest is entirely due to air "impurities" of plant origin, as is the haze of the Great Smoky Mountains, which even the Indians knew.

Contamination of the air by waste products produced by human activities constitutes *air pollution*. The age of air pollution dawned with the discovery of fire. As coal and other fossil fuels became widely used, air quality worsened. Wherever there is fire, there is smoke, and the products of combustion are exhausted directly into the air.

Throughout history man has depended upon natural air currents to carry his aerial garbage away. Until recently the atmosphere cooperated, absorbing huge quantities of wastes. Today, however, smoke from the vast numbers of fires we kindle and the enormous array of complex materials exhausted to the air are overwhelming the atmosphere. Not enough clean air remains to dilute the heavy loads of *pollutants* (materials that pollute) which we are dumping (more than 140 million tons annually by the United States alone) into the atmosphere.

"A 1968 UNESCO conference concluded that man had only about 20 more years before the planet started to become uninhabitable because of air pollution alone."[1] Air-pollution emergencies and even disasters have become familiar events in recent years. Such incidents usually arise not from some abnormal industrial accident (though such disasters have also occurred) but from the normal operations of industry and urban living. In each instance an inversion layer confined the ordinary pollutants of the atmosphere to the immediate air and did not permit them to disperse normally.

In 1970 several major cities throughout the world suffered air-pollution emergencies. In Tokyo 9,000 persons were treated in hospitals as a direct result of the choking smog that hovered over the city. In the United States smog hung over the entire Atlantic Coast for several days, and in New York City unhealthy levels of sulfur oxides were reported over a period of several days. Those who were able fled from the cities to find relief from smarting eyes, sore throats, and aggravated respiratory conditions. Others, especially those with heart conditions or respiratory diseases, remained indoors.

Such air-pollution emergencies, which undoubtedly will become more frequent, dramatically underline the need for change. However, the less dramatic but also serious day-to-day problems of air pollution will also continue to grow worse unless immediate and stringent measures are taken to control the problem.

Although many pollutants remain in the atmosphere for considerable periods of time, they eventually will leave it. Chemically active gases com-

Give me your tired, your poor,
Your huddled masses yearning
to breathe free. . .

—Emma Lazarus

**FIGURE 9-1**
(Frank Peters and *The St. Petersburg Times.*
Redrawn by permission.)

---

[1] Paul R. Ehrlich and Anne H. Ehrlich, *Population Resources Environment; Issues in Human Ecology,* Freeman, San Francisco, 1970, p. 118.

*(a)*

**FIGURE 9-2**
(*a*) Industrial pollution in St. Louis area. (OMIKRON) (*b*) Industrial pollution in the St. Petersburg area. (Selbypik)

*(b)*

bine with soil or water, particulate matter falls out upon all exposed surfaces, and carbon dioxide (in hundreds of years) will be absorbed in part by the ocean. Indeed, the ocean is the ultimate sink for all air pollution, just as it is for water pollution. However, we are rapidly approaching the point where pollutants are being discharged into the atmosphere at a higher rate than they can be removed.

Air pollution cannot be confined. There is no such thing as an "air basin" or "air shed." The wide distribution of air pollution is certain. It is governed by prevailing winds, but localities far from the source may be affected. For this reason air pollution, even more than water pollution, cannot be controlled on a local or even regional basis. National and international controls constitute the only sensible approach.

The problem of air pollution parallels the problem of water pollution, but with a significant difference. Often an individual need not drink polluted water, but even if he must, the dose of pollutants he acquires is not continuous. A fish living in that polluted water, however, is exposed to a continuous dose of toxic substance, all day and every day. Air pollution puts us in the position of the fish. Substances that are not harmful in small occasional doses are likely to be extremely hazardous in small *continuous* doses, and in combination with one another. Furthermore, so great is the absolute quantity of the toxic by-products produced by modern industry that large cities and areas may at any time experience public health disasters resulting from abnormal weather conditions that concentrate them. As the UNESCO conference predicted, in the long run, increasing pollution will certainly produce calamities on a global scale.

And what of the economics of air pollution? Air pollution costs us an estimated 12 billion dollars each year in property damage. For about 3 billion dollars a year we could reduce this pollution by approximately one-third. The economic effects of air pollution are reflected in higher cleaning costs, increased maintenance costs, deterioration of materials, loss of livestock

**FIGURE 9-3**
Burning garbage in a city dump in California. (OMIKRON)

and crops leading to higher food bills, lowered real estate values, and higher medical costs. Air pollution costs each person in the United States an average of $65 per year, money which could be used to purchase a great deal of air-pollution control equipment.

## AIR POLLUTION AND HUMAN HEALTH

Each of us inhales an average of 35 pounds of air every day, about six times as much as the food and drink we consume. And since air is essential, we have no choice but to breathe our polluted air.

The human respiratory system is well equipped to deal with natural impurities. Guarding the nose are hairs which remove large particles from the incoming air (see Chapter 8). The sticky mucous lining of the nasal cavities entraps more particles. Medium-sized particles (the smallest commonly found under natural conditions) are trapped in the mucous coating the trachea, bronchi, and larger bronchioles. These are then carried to the throat by the cilia-propelled mucous elevator, and swallowed. Coughing and sneezing serve to rid the body of some irritating substances and of excessive accumulations of mucus.

The human body is poorly equipped to cope with the multitude of pollutive substances produced by modern civilization. If a pollutant consists of very tiny particles or droplets, or if it is gaseous, it may pass all the body's defenses and lodge in the delicate air sacs of the lungs, sometimes permanently. The lowest bronchioles and the air sacs have no ciliated or mucus cells, and must depend upon phagocytes (janitor cells) to ingest foreign matter. Some phagocytes migrate with their load of debris to the cilia-propelled mucous elevator and are carried to the throat. Many others, however, accumulate in the lungs.

Many occupations and habits inflict permanent damage upon the lungs, so that their condition eventually reflects a person's life style and even habitat. By examining the lungs at an autopsy, a pathologist can determine how

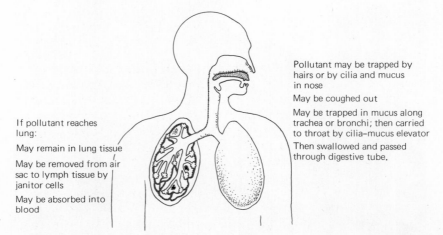

If pollutant reaches lung:

May remain in lung tissue

May be removed from air sac to lymph tissue by janitor cells

May be absorbed into blood

Pollutant may be trapped by hairs or by cilia and mucus in nose

May be coughed out

May be trapped in mucus along trachea or bronchi; then carried to throat by cilia–mucus elevator

Then swallowed and passed through digestive tube.

**FIGURE 9-4**
The body's defenses against inhaled pollutants.

heavily the deceased smoked, if at all, whether he was a miner, whether he worked with asbestos or silicon, whether he lived in a rural or urban area, and many other facts. The clues are embedded within the lung tissue in the form of black particles, scars, lesions, abnormal appearance of the epithelial cells lining the air passageways, and broken air sacs.

### Irritation and bronchial constriction

Epithelial cells lining the nasal passages, throat, and other respiratory pathways react to polluted air by secreting increased amounts of mucus. Cilia are damaged by the pollutants and are unable to clear the mucus and debris from the respiratory tubes. The system resorts to coughing as a mechanism for clearing the airways. Increased irritation results, and the respiratory system becomes susceptible to a host of infective agents. Constant irritation is the basis for many respiratory diseases.

When dirty air is inhaled, the bronchial tubes respond by constricting. The narrower passageway provides a greater opportunity for passing particles to be trapped along the sticky mucus lining the tube. Unfortunately, when the passageway is constricted, less air can pass through to the lungs and a decreased amount of oxygen is available for the body cells. Chain smokers and persons who live in polluted industrial areas experience almost constant bronchial constriction, sometimes amounting to chronic asthma.

Although pollution does not *cause* pneumonia, influenza, or other respiratory infections, it does increase susceptibility to these diseases. Bronchial asthma is a condition in which the body manifests an allergy to certain natural impurities (such as a specific pollen) by narrowing the bronchioles. Difficulty in breathing, wheezing, and coughing result. Bronchial asthma can be worsened by assaulting the already disturbed respiratory system with the noxious particles and gases of polluted air.

### Chronic bronchitis

Chronic bronchitis is an important cause of disability in middle-aged and elderly people. Its increase in recent years has been caused by the rise in air pollution and smoking.

In chronic bronchitis the respiratory passages, especially bronchi and bronchioles, secrete excessive amounts of mucus, and are constricted and inflamed. Respiratory cilia do not function sufficiently to clean the passages of dirt and mucus which partly clog them. The accompanying cough lasts for several months, and the disorder usually returns annually. The condition is diagnosed as chronic bronchitis if a person coughs up mucus on most days during at least a 3-month period and if he does so again the following year.

The respiratory system of a person with chronic bronchitis is quite attractive to bacteria, so that frequent attacks of **acute bronchitis** (a temporary inflammation usually caused by viral or bacterial infection) are common. Eventually chronic bronchitis leads to shortness of breath, permanent damage to the respiratory system, and often emphysema.

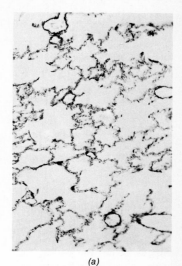

*(a)*

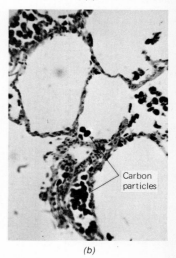

Carbon particles

*(b)*

**FIGURE 9-5**
(a) Normal lung tissue. (*Ward's Natural Science Establishment, Inc..* (b) Lung tissue with accumulated carbon particles.

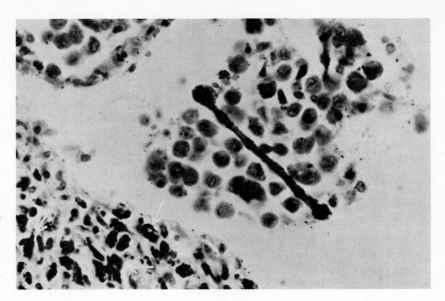

**FIGURE 9-6**
A drumstick-shaped asbestos body is evident in this lung tissue. The alveolar space is dilated and contains many macrophages, some with carbon particles. (*McGraw-Hill Encyclopedia of Science and Technology*, 3d ed.)

Both prevention and treatment of chronic bronchitis depend upon removing major sources of irritation. This means breathing air that is not polluted, and giving up the habit of smoking. For those who work amidst the dust and fumes of certain industries, it means changing occupations.

## Emphysema

Emphysema is now the most common and most crippling chronic respiratory disease. It develops after many years of abuse to the respiratory system. The bronchioles become permanently narrowed, and there is a decrease in the elasticity of the walls of the air sacs (Figure 9-7). These changes make it difficult to exhale air from the lungs, and indeed the first symptom of emphysema is often shortness of breath during physical exertion. As the disease progresses, breathing becomes increasingly difficult until the patient can carry on no physical activity. A large amount of stale air accumulates in the lungs so that the amount of oxygen available to combine with hemoglobin is reduced. The right ventricle of the heart, which works harder to compensate, becomes enlarged. Death often results from heart failure, which is often the cause cited on the death certificate.

Emphysema commonly develops in people with chronic bronchitis. Since the lung tissue is permanently damaged, emphysema cannot be cured, but progress of the disease may be slowed by stopping smoking and reducing other sources of irritation. Life may be prolonged by therapy aimed at dilating respiratory passages and making the patient comfortable. The author of a well-known medical school pathology textbook describes the patient with emphysema as follows: "He suffers with every breath he takes, and he must breathe some 20,000 times in (every) 24 hours."[2]

---

[2] William Boyd, *Textbook of Pathology*, Lea & Febiger, Philadelphia, 1970, p. 701.

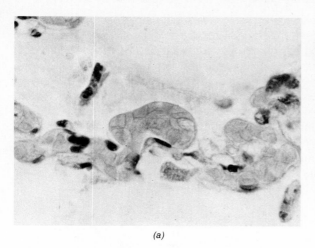

(a)

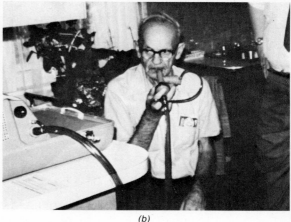

(b)

## Lung cancer

The most common type of lung cancer originates in the epithelial cells lining the bronchi. The precise mechanism by which normal cells are transformed into malignant cells is not yet known; however, certain substances found in tobacco smoke and also in some polluted air have been shown to trigger the change. Whether these *carcinogens* (cancer-producing substances) alter the cells in some way that makes them more vulnerable to a cancer virus or whether some entirely different nonviral mechanism is responsible has not yet been determined. Whatever the mechanism, once cells become cancerous they divide wildly, forming a tumor. The bronchi often become obstructed. Some cells may migrate to invade other parts of the respiratory system or other regions of the body. Without treatment the patient dies in 3 to 5 years, or even sooner, from the time the disease is first detected by x-ray.

**FIGURE 9-7**
(a) Congestion of capillary in the lung of a patient with emphysema. The capillary is dilated and full of red blood cells. (*McGraw-Hill Encyclopedia of Science and Technology*, 3d ed.) (b) This patient, a coal miner most of his life, developed emphysema and pneumoconiosis (black lung disease, caused by inhalation of dust particles in the mine). Photograph shows him learning how to use an inhalation therapy machine which creates a positive pressure, making breathing easier. (Jane L. Baker)

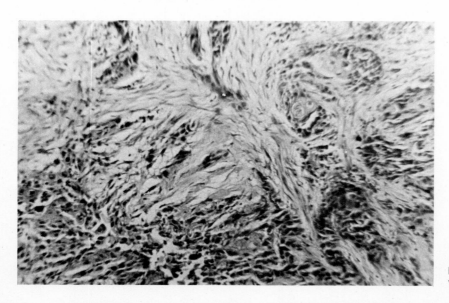

**FIGURE 9-8**
Metastatic lung cancer. An involved lymph node is shown.

As with chronic bronchitis and emphysema, the rapid rise in lung cancer in recent years has been linked with the rise in cigarette smoking and air pollution. People living in urban areas are twice as likely to develop lung cancer as those living in less polluted rural environments. In urban areas the incidence of lung cancer is even greater where certain types of pollutants are highly concentrated in the air. A recent study of residents of Staten Island, New York, strikingly shows the relationship of lung cancer and air pollution. Those who live on the northern part of the island close by a New Jersey industrial area have twice the death rate from respiratory cancer as those who live a few miles south, *on the same island* but farther away from the dangerously polluted air.

## MAJOR POLLUTANTS

The five categories of pollutants that account for most air pollution are carbon monoxide, sulfur oxides, nitrogen oxides, hydrocarbons, and particulates. Their relative contributions to air pollution and their main sources are indicated in Figure 9-9.

### Carbon monoxide

Carbon monoxide (CO), the largest single pollutant, accounts for about 50 percent of all air pollution. It is a colorless, odorless, tasteless gas, produced during incomplete combusion, especially of gasoline in motor vehicles. Cigarettes produce carbon monoxide as they burn, and thus greatly increase the amount that an individual inhales.

**FIGURE 9-9**
Air pollution emissions. (*a*) The major air pollutants (percentages by weight). (*b*) Sources of the major pollutants.

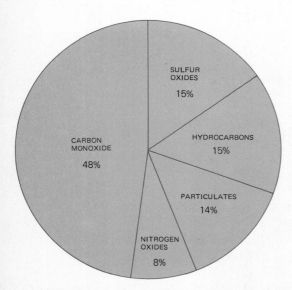

(*a*) The major air pollutants. Percentages by weight

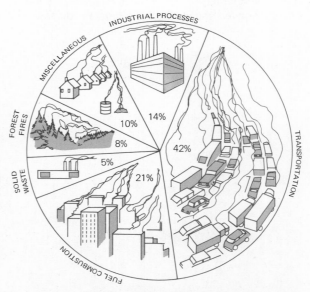

(*b*) Sources of the major pollutants

Carbon monoxide reaches the lungs as part of the inhaled air. It acts upon the blood, rather than on the respiratory system directly. Hemoglobin combines 210 times more readily with carbon monoxide than with oxygen. Carbon monoxide competes for and wins the sites on the hemoglobin molecule which would normally be available for transporting oxygen, thus decreasing the amount of oxygen which can be transported to the cells.

Carbon monoxide forms a strong chemical bond with hemoglobin, and once attached it does not dissociate readily from the hemoglobin molecules. Carbon monoxide-hemoglobin has a half-life of from 2 to 4 hours. That is, about 3 hours are required for half the carbon monoxide to dissociate from hemoglobin and be excreted by the lung. This difficulty in getting the carbon monoxide out of the body is a serious clinical problem in treating acute carbon monoxide poisoning. Even when they are treated speedily, victims may succumb because of the persistence of the carbon monoxide–hemoglobin bond.

Carbon monoxide may reach the lower limits of toxicity on city streets during rush hours, and is often present in ominous amounts in some urban dwellings, especially those built over roadways. Breathing air containing 80 parts per million carbon monoxide (an amount not uncommon in urban areas) for a period of 8 hours reduces the blood's ability to transport oxygen by 15 percent. The physiologic effect is that of losing more than a pint of blood.[3] On the cellular level, lowered oxygen supply may be translated into lowered metabolism. The brain is unable to function efficiently when deprived of normal amounts of oxygen. Recent studies have confirmed that carbon monoxide impairs judgment and alertness. Each of us has a normal, or background, level of about 0.4 percent carbon monoxide-hemoglobin in his blood. In laboratory tests subjects with as little as 2 percent carbon monoxide-hemoglobin above background level were shown to suffer significantly from impaired mental performance. During peak traffic hours the amount of carbon monoxide in the air may cause drowsiness and impaired driving ability. Significantly high levels of carbon monoxide-hemoglobin have been found in some drivers responsible for automobile accidents.

## Sulfur oxides

About 15 percent of air pollution may be attributed to sulfur oxides, mainly sulfur dioxide, a colorless gas that irritates the throat and eyes and causes coughing and choking. Sulfur oxides, produced by burning fossil fuels, originate chiefly with power plants.

In moist air sulfur oxides react with water to form sulfuric acid mist. Inhalation irritates the lining of the respiratory passages and damages cilia. Since other pollutants are generally present in the same polluted air, it is difficult to attribute specific physiologic disorders solely to sulfur oxides. Laboratory experiments on animals have shown that sulfur pollutants interfere

**FIGURE 9-10**
Traffic jam in Rome. Motor vehicles are the main source of carbon monoxide. (Wide World Photos)

---

[3] Paul R. Ehrlich and Anne H. Ehrlich, op. cit., p 119.

with the activity of certain enzymes and decrease the vitamin C content of several organs. Permanent damage to lung tissue and pulmonary hemorrhages have been induced by sulfuric acid mist. There is also some evidence that sulfur dioxide does genetic damage.

Sulfuric acid in the air can be washed out by rain, but in the process the rainwater is acidified. Acid rain damages buildings, outdoor sculpture, and other public works, and in extreme cases contributes to water pollution.

### Unburned hydrocarbons

Hydrocarbon pollutants include a great variety of organic molecules composed of chains of carbon and hydrogen atoms. They are familiar to us as the fossil fuels—petroleum, natural gas, and coal. Gasoline, kerosene, and fuel oil are hydrocarbons made by refining petroleum.

The hydrocarbons that cause trouble are those that enter the air as a result of incomplete combusion or evaporation. Their main source is motor vehicles. Some hydrocarbons such as benzpyrene have been shown to be carcinogenic. Inhalation of other hydrocarbons has been shown to interfere with enzyme activity, carbohydrate metabolism, immune responses, and other vital functions. Some hydrocarbons react with nitrogen oxides to produce photochemical smog (discussed later).

### Particulates

Particulates are solid particles in the air, such as soot (black particles resulting from burning wood, oil, or coal), dust, or fly ash (unburnable particles from fossil fuels). Power plants and industries are primarily responsible for these pollutants, which account for an estimated 13 percent of air pollution.

**FIGURE 9-11**
Power plant emissions can be substantially reduced by use of pollution control devices. (*a*) Stack with precipitator operating normally. (*b*) Same stack with control turned off. (Tampa Electric Company)

(a)

(b)

When released into the air some particulates settle to the ground or make surfaces such as window sills and automobiles dirty and dusty. Others remain suspended in the air and enter the body in inhaled air. Particles that enter the body irritate the respiratory passages; those that reach the alveoli interfere with normal gas exchange. Asbestos, lead, and fluorides are among the most hazardous and widespread particulates.

Asbestos is sprayed upon the steel skeletons of buildings to fireproof them. But about half the spray used misses its mark and becomes suspended in surrounding air. Asbestos also enters the air as brake linings and clutch facings of vehicles wear down. Most of us undoubtedly have asbestos particles in our lungs. Unfortunately the lungs have great difficulty in disposing of these particles and react to their presence by forming little growths, called *asbestos bodies,* around them. There is strong evidence that after a period of years these growths often become cancerous.

Lead fumes and dust are also dangerous particulates. Some historians attribute the decline of the Roman Empire to gradual lead poisoning of the upper classes. It seems that the pots used in preparation of their wine were lined with lead. The resulting lead poisoning is held to be responsible for the high death rate and low birth rate of the Roman aristocracy. The lower classes apparently drank less wine. This theory has recently been supported by the finding of high concentrations of lead in the skeletons of upper-class Romans.

Lead enters the body by way of the digestive system, the respiratory system, or directly through the skin. It is stored in bone tissue and accumulates as more and more finds its way into the body. One of its effects on metabolism is to inhibit the production of hemoglobin, resulting in anemia. Lead also damages the nervous system and the kidneys.

Chronic lead poisoning is found among industrial workers who come into constant contact with lead fumes and dust. Recently lead has been recognized as an important new ingredient of urban air. Its most important source is leaded gasoline. Chronic lead poisoning is difficult to diagnose. Since Americans are carrying about half the estimated toxic level, some authorities believe that some of us may already be suffering from mild forms of this condition. Use of high-octane unleaded gasolines would greatly reduce the amount of lead pollution.

Fluorides are discharged into the air as a waste product from the manufacture of phosphates, steel, aluminum, and many other industrial products. They settle upon vegetation and are ingested by domestic animals and people. Fluoride poisoning has resulted in sizable losses to agricultural and ranching interests.

**Nitrogen oxides**

Any high-temperature combustion process tends to unite chemically the oxygen and nitrogen of the air to produce nitrogen oxides. Their primary sources are industrial smokestacks, power plants, and motor vehicle tail pipes. Nitrogen oxides are responsible for the smell and color of the brown haze that now hangs over many of our cities. These pollutants are an important component of photochemical smog.

**Table 9-1. Sources of air pollution**

| Source | The problem | Outlook |
|---|---|---|
| Transportation | Ejects carbon monoxide, unburned hydrocarbon fuel, particulates, nitrogen oxides, and other compounds into the atmosphere. High-powered engines, especially when run at low speeds or idled in city traffic, use fuel less efficiently and cause more pollution than low-powered engines. | (1) Congress has passed legislation requiring a 90 percent decrease in CO and hydrocarbon emissions by 1975 (as compared with 1970 model cars) and a decrease in nitrogen oxide emissions by 1976. Major auto manufacturers are currently seeking a postponement, claiming that they cannot meet these deadlines. (2) Japanese manufacturers have developed an engine in which gasoline components are separated before entering the cylinder, thus promoting more efficient burning and less pollution. (3) The Wankel, or rotary, engine employs low combustion temperatures, thus exhausting less nitrogen oxide. (4) The engine of the future may be a modernized steam engine, as this may be designed almost pollution-free. (5) Increased public transportation would help solve this pollution problem. |
| Electric power generation | Most electricity today is generated by thermoelectric plants (plants that burn coal, oil, or gas). Immense amounts of fuel are burned to create steam fed to turbines which drive generators. Sulfur oxides, nitrogen oxides, and smoke and other particulates are produced. Efficient combustion which reduces smoke leads to high temperatures, which produce nitrogen oxides. An average power plant burning coal with 3% sulfur content produces 68,000 tons of sulfur dioxide annually. Use of very tall stacks reduces the pollution concentration in the immediate area, but the pollution eventually returns to earth, sometimes in the form of acid rain. | (1) Particulates can be controlled by *scrubbers* (effluent gases are washed with large amounts of water, but dirty water then usually causes water pollution), *electrostatic precipitators* (they produce a gritty substance that has some industrial use), and other devices. (2) No proved, inexpensive way of eliminating nitrogen oxides now exists. (3) New techniques are being investigated for chemically removing sulfur from coal, but at present no satisfactory way of reducing amount of sulfur oxides is known. |
| Industrial sources | Pollutants so diverse as almost to defy description. Examples: sulfur oxides and particulate heavy metals from smelting; dense smoke from iron foundries and steelmaking; fluorides from phosphate fertilizer production. | (1) Specific technologies available to control many types of industrial pollution. (2) Legislation to control industrial pollution has been passed but is often not enforced. (3) Some changes in economic philosophy with respect to planned obsolescence, high-profit–oriented production, and recycling needed. |
| Incineration | Produces complex mixtures of air pollutants. | (1) Particulates can be controlled as in power generation. (2) Regulation of plastics and other packaging materials and recycling materials which could be used again would reduce this source. Ultimate solution is the recycling of solid wastes. |
| Space heating | Difficult to regulate since each household has its own furnace. | In the future all homes may be heated by electricity or hydrogen gas so that pollution could be controlled at a central generating plant. |

Nitrogen oxides irritate the eyes, nose, and throat. Recent studies indicate that they may interfere with the ability of phagocytic cells to remove dirt from the lungs. At high concentrations nitrogen oxides may be lethal because they combine with hemoglobin 300,000 times more readily than oxygen, leading to death by asphyxiation. A recent study showed that families living in an area with a high nitrogen oxide level suffered 18.8 percent more respiratory illness than control families living in less polluted areas. Laboratory investigations confirm that nitrogen oxide lowers resistance to both viral and bacterial infections. Children living in the area of high nitrogen oxide level also had significantly lower breathing capacity, suggesting that chronic respiratory difficulties were developing.

## Other pollutants

Carbon dioxide ($CO_2$) is not usually considered a pollutant, since it is a natural, necessary component of the atmosphere. However, by burning fossil fuels we have released great quantities of this gas into the atmosphere. Since 1880 the amount of carbon dioxide in the atmosphere has increased by 12 percent. We do not know what the consequences of such interference with the normal balance of atmospheric gases may be.

Commonly used aerosol sprays have also been cited as pollutants. Inhalation of the Freon propellant utilized in most aerosol sprays, and perhaps in some instances the aerosol product itself, has recently been cited as hazardous. Certain deodorant sprays, hair sprays, and other aerosols have been linked with respiratory disease, heart damage, and the presence of premalignant cells in the respiratory system.

Radionuclides constitute still another group of atmospheric pollutants. They are emitted by nuclear reactors, and later find their way into plants and higher members of food chains. The radiation emitted by these pollutants causes *mutations*, permanent changes in the genes of living cells. As nuclear power plants spring up to meet the demands of a growing population, radionuclides will become an increasing threat to human health and safety.

## Smog

Pollutants in the air react with one another to form new, sometimes dangerous, compounds. Breathing these compounds may be much more hazardous than breathing the individual ingredients separately.

Air pollution has been divided into two main types: sulfur oxide–particulate (*SOP*) smog, and *photochemical smog*. SOP smog is common in the eastern United States as well as in London and Japan. In moist air sulfur dioxide may form sulfuric acid which clings to particles such as fly ash in the air. The particulate–sulfur compound combination is then inhaled.

Photochemical smog is the Los Angeles variety of air pollution, but is also found in other areas of the United States. This type of smog forms when nitrogen oxides react with oxygen under conditions of bright sunlight to form ozone, and then with hydrocarbons to produce odd organic oxidizing compounds such as PAN (peroxyacyl nitrate). PAN irritates the eyes and dam-

(a)                                                    (b)

**FIGURE 9-12**
(a) New York smog. (b) Chicago smog. (OMIKRON)

ages foliage. Other chemical reactions may produce various other hydrocarbons. Laboratory experiments suggest that ozone and nitrogen oxide may adversely affect human fertility, since they chemically kill sperm cells. Ozone is also thought to suppress the body's ability to fight infection.

Intermediate types of smog have also been observed. The synergistic (combined) effects of these combinations have not yet been studied extensively but will no doubt prove more damaging than either SOP or photochemical smog alone.


## ECOLOGIC EFFECTS OF AIR POLLUTION

Air pollution affects plants and animals other than man. Indirectly, of course, we are affected in terms of decreased plant life, damage to crops, and harm to domestic animals.


### Effects on plants
Air pollution at times reduces the amount of sunlight that reaches Chicago by about 40 percent, and the amount reaching New York by 25 percent. Photosynthesis, as you know, depends upon the energy of sunlight.

Air pollutants are responsible for the loss of hundreds of millions of dollars annually in United States crops. In Florida and elsewhere thousands of acres of citrus groves have been damaged by fluorides. Cut-flower industries have been wiped out by air pollution in many areas.

Trees and other plants absorb great quantities of pollutants directly from the air. For the most part damage occurs in the photosynthetic tissue (mesophyll) of the leaf. The surfaces of mesophyll cells, which are moist to facilitate gas exchange in photosynthesis, are vulnerable to attack by toxic substances in the air. Fluoride, photochemical smog, sulfur dioxide, and even soot can kill vegetation in this way. At a concentration of less than 1 part per

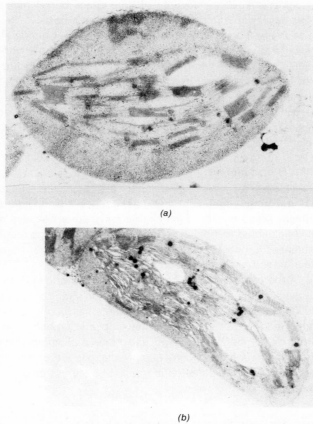

(a)

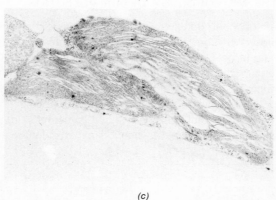

(b)

(c)

**FIGURE 9-13**
Air pollution damages chloroplasts. (a) Normal control. (b) Chloroplast structure has been damaged after the plant was subjected to 1 part per million sulfur dioxide for 1 hour. (c) Chloroplast damage after plant was subjected to 3 parts per million nitrogen dioxide for 1 hour.
(A. R. Wellburn, O. Majernik, and Florence A. M. Wellburn and *Environmental Pollution*)

million nitrogen dioxide reduces the growth of tomato plants by 30 percent. Some species of plants are more susceptible to pollution than others. Certain species are especially sensitive to one pollutant, while others are more sensitive to a different one. These differences complicate the picture.

In Los Angeles County and areas up to 60 miles away, forests are being damaged by photochemical smog. Photosynthesis is reduced 66 percent by smog concentrations of 0.25 part per million. This decrease in photosynthesis slows the flow of resins under the bark, which in turn renders the trees susceptible to plant disease and insect pests.

With the increase in air pollution that present trends are producing, vast damage to vegetation may occur over the face of the North American continent, and even the world. Marine explorer Jacques Picard recently reported that 1.8 million tons of hydrocarbons from automobile emissions precipitate into the ocean each year (about 20 percent of the total amount of hydrocarbon pollutants thrust upon the sea annually). In this, and other ways, air pollution is contributing to the decline of sea life, which has decreased an estimated 30 to 50 percent during the past 20 years.

### Effects on animals

Animals are exposed to air pollution both by inhaling pollutants directly and by ingesting vegetation that has been contaminated by air pollutants. "Smoky areas become undesirable for grazing because of the poor quality of vegetation and inhalation of smoke."[4] The acreage required to raise cattle increases in polluted areas, and illness and death in sheep and cattle have been traced to specific pollutants.

"Airborne fluorides (F) have caused more worldwide damage to domestic animals than any other air pollutant."[5] In Florida and elsewhere fluoride pollution from phosphate fertilizer plants has contaminated grass on which cattle feed. Exposed livestock develop a condition known as fluorosis, in which abnormal bone growth and other pathologic changes render them practically worthless. Honey bee populations are also decreased by fluoride pollution. Whether pollution is good for business or not obviously depends upon one's point of view.

## THE METEOROLOGY OF AIR POLLUTION

Atmospheric pollution is not equally severe everywhere or at all times. Almost everyone has noticed that smog and soot are likely to be increased near large cities or industrial operations, and that air pollution is more intense on some days. In large cities incidence of respiratory diseases and number of deaths are increased (*excess deaths*) on many days by intensified pollution. How do these incidents come about?

The first point to be understood is that most large cities have overloaded their atmosphere with wastes to such an extent that the polluted air does not

---

[4] *Air Pollutants Affecting the Performance of Domestic Animals: A Literature Review,* Agricultural Handbook No. 380, Agriculture Research Service, U.S. Department of Agriculture, 1970, p. 9.
[5] Ibid., p. 41.

dissipate (that is, it does not diffuse quickly or effectively) unless abnormally windy, clear weather prevails. Thus, cities are in a perpetually precarious position; the least atmospheric change can turn a situation which often borders on the serious into a real emergency.

Then, too, on calm days the interior of the city is likely to be considerably warmer than the surrounding countryside. Warm air, heated by contact with buildings and pavement, rises, carrying dirt and gases upward with it. As it rises it cools and eventually begins to sink. As it sinks, however, it tends to be drawn into the city again, where it repeats the cycle. Whatever noxious contents are in the air tend to become progressively concentrated throughout the day, for they cannot leave the urban area.

The most serious hazard, however, is the development of an *inversion layer*. To understand it, we must first consider the normal differences of temperature that occur in conjunction with altitude. During the day the sun-warmed earth tends to heat the air that is in contact with it. This hot air expands, becomes lighter, and rises, almost as if it were a balloon without walls. If the area of upwelling is not too localized, the warm air may continue to rise to a great height. As it rises the pressure of the surrounding atmosphere on it lessens and it expands further. Such expansion cools it greatly.

Normally air near the ground tends to be warmer than higher air, but under some conditions that relationship is reversed. At night the ground normally radiates its heat to outer space so rapidly that the lower air is cooled by contact with it. Often the upper air actually remains warmer than the lower. This is especially likely to occur on still, clear nights. The lower air may not become really warm again until about 10 o'clock the following morning. The warm upper air acts as a kind of lid, sealing pollutants below it and keeping them from being dispersed. In almost every city, daily records of air pollution show a peak around 9 o'clock in the morning. Exhausts from commuter traffic, increased power demands associated with arising and breakfast, and the smoky starting of many fires combine with the inversion layer to produce a temporary condition that would swiftly become dangerous if the inversion layer did not break up with the warming ground temperatures.

Inversion layers may have other origins. In Los Angeles County cool air from the sea comes to lie in the bowl-shaped topography formed by the surrounding mountains. At the same time, a layer of warmer air moves over those mountains from the deserts to the east to form an inversion layer. From time to time any city will experience days in which no vertical dispersal of its effluents takes place. Such events should be of as much concern as an epidemic.

In time air pollution may permanently alter the climate of our planet. Some scientists fear that if we continue to inject great quantities of particulates into the atmosphere we may bring on another ice age. Particulate matter could serve as nuclei for the condensation of high clouds which would reflect sunlight away from the earth. One scientist has calculated that the addition of only 50 million tons of pollutant particles to the atmosphere could reduce the average surface temperature of our planet from its present 60 degrees Fahrenheit to about 40 degrees. Most forms of plant life could not survive in such a cold climate.

Lid of warm air seals in pollutants

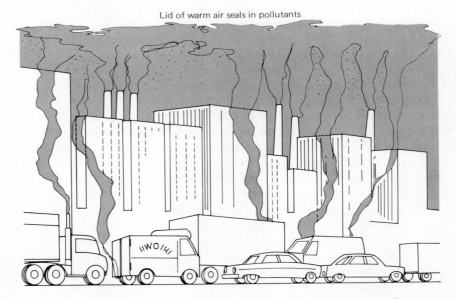

**FIGURE 9-14**
A thermal inversion.

## DEALING WITH THE PROBLEM

The Clean Air Act of 1967 allowed the states to set their own standards for air pollution control. With the growing realization that air pollution cannot be confined within state boundaries, this legislation was amended in 1970 to give the federal government the responsibility for setting standards for the

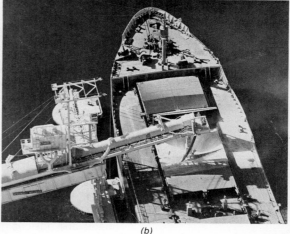

(a)                                                    (b)

FIGURE 9-15
(a) Unloading phosphate—without control devices. (b) Unloading phosphate—using covering device. (Selbypik)

entire country. Standards have been set for carbon monoxide, hydrocarbons, nitrogen oxides, sulfur oxides, particulates, and photochemical oxidants. Primary standards—those necessary to ensure public health—must be met by 1977. In 1972 some 170 million Americans were being exposed to air pollution violating the primary standards. No national date has been set for meeting secondary standards, those which would protect public welfare (preservation of vegetation, wildlife, buildings, etc.), but some local areas have set goals for meeting them.

Technology can reduce some of the effects of air pollution, but it simply is not practical to remove *all* air pollutants by *any* technology. *Complete* removal of pollutants is usually accompanied by costs of operation and investment so high as to be out of the question. To illustrate this fact, consider the example of Los Angeles County.

Los Angeles County may be considered a world leader in air pollution control technology and legislation. Despite this, in more than 20 years of operation, their pollution control measures have achieved *no* reduction in air pollution. The reason is that the population has increased. So many more persons have moved into the county, and purchased so many automobiles, that even with the best safeguards available, they pollute the atmosphere as much as the smaller population of 1948 did with no controls.

In the long run, then, no solution to this problem of environmental deterioration is feasible without a population reduction and some limitation upon the consumption of power and manufactured goods.

Short-term measures which may alleviate air pollution include:

1  Development of pollution control technology. Research will be stimulated by setting standards of air pollution *higher* than present-day technology can attain.
2  The development of less pollutive modes of transportation.
3  The encouragement of mass transit and the discouragement of road building.
4  Institution of a national or international pollution alert system with the authority to

shut down industrial operations in localities threatened with acute episodes or disasters.

5 Greatly increased epidemiologic and other research on long-term effects of air pollution.

6 Most important of all, the enforcement of national standards of air purity and pollution control, providing both civil *and criminal* penalties for polluters.

7 Commitment by each of us to learn what we can do on an individual basis to help solve the problem of air pollution.

## SMOKING: A SERIOUS FORM OF AIR POLLUTION

In this age of environmental awareness many of us are genuinely concerned by the sight of smoke billowing from industrial smokestacks or the stench of automobile fumes on a crowded city street. Ironically, we tend to neglect an even more deadly form of air pollution—the highly concentrated particulates and noxious gases of burning cigarettes.

Nobel Prize–winning chemist Linus Pauling has computed that for every cigarette a person smokes he shortens his life by about 14.4 minutes. In 1967 the Surgeon General of the United States, Dr. William H. Stewart, stated that more than 300,000 Americans die each year as a consequence of smoking. Major killers among smokers are heart and vascular diseases, cancer, emphysema, and chronic bronchitis. The damaging effects of smoking on lung function can be measured less than 1 year after people begin to smoke.

Smoking is one of the most serious public health problems in the United States today. An estimated 42 percent of American adults (50 million persons) smoke cigarettes at the present time. About 36 percent of young men and 22 percent of women begin smoking cigarettes by age 18, and per capita cigarette smoking is increasing steadily.

### A matter of personal choice?

Even if every smoker were aware of the damage he inflicts upon his own tissues and was willing to accept the consequences of poor health and shortened life expectancy, the decision to smoke still affects others. First, the cloud of smoke produced envelops smoker and nonsmoker alike, greatly increasing the concentration and variety of pollutants inhaled by all. Cigarette smoke accumulates in the air recirculated in office buildings, and damaging effects have been found in the lungs of nonsmokers. Second, the smoker represents a burden to the nonsmoker. Workers who smoke a pack of cigarettes or more per day are absent from their jobs because of illness 33 percent more often than nonsmokers. This amounts to 77 million days lost from work each year in the United States alone.

Another consideration is the chronic long-term illnesses associated with cigarette smoking. When adults become disabled by emphysema or heart disease, the care and support of their families must be assumed by others. Furthermore, about 1,800 persons die each year from fires caused by smoking, and such fires kill smokers and nonsmokers indiscriminately. Obviously, smoking is a problem to society as a whole.

## Why is cigarette smoking harmful?

Smoking produces highly polluted air. Although the total volume of polluted air produced by smokers does not begin to compete with the amount spewed forth by industry, the health consequences are more severe, for the air polluted by the smoker is the air closest to him—the very air he must breathe. Industrial pollution is generally somewhat diluted by the time it wafts our way.

Tobacco smoke consists of about 60 percent gases and 40 percent particulates. The gaseous portion consists of about 50 different substances, including carbon monoxide, nitrogen oxides, formaldehyde, and hydrogen cyanide. The concentration of carbon monoxide may reach as high as 42,000 parts per million, an amount that would certainly be lethal if the smoker did not also breathe less-polluted air. Hydrogen cyanide is a respiratory poison which is considered dangerous at levels of about 10 parts per million. In pure cigarette smoke it is found at 1,600 parts per million.

Tars and nicotine are among the particulates of tobacco smoke. Tobacco tar contains more than 200 different chemical compounds. At least 10 of these are hydrocarbons that have been shown to cause cancer in animals. Others promote the growth of certain cancers, while still others are toxic in different ways.

Concentration of particulates in inhaled cigarette smoke may be up to 100 times greater than that found in polluted urban air. As the pollutants are drawn through the burning cigarette they are heated. The burning zone of the cigarette may reach a temperature of more than 1600 degrees Fahrenheit. As they are heated, the pollutants become more reactive, and some even vaporize. The smoker's lungs serve as a test tube in which the many different inhaled chemicals interact, producing a Pandora's box of chemical compounds, many of them harmful to living cells.

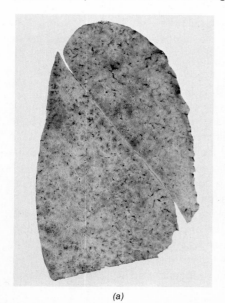

(a)

(b)

**FIGURE 9-16**
(a) Whole-mount lung section from normal lung. No emphysema. (b) Whole-mount lung section, advanced emphysema. (Dr. Oscar Auerbach and *The New England Journal of Medicine*)

## What about pipes and cigars?

Cigar and pipe smokers do not suffer as much physiologic damage as do cigarette smokers. The heavy smoke from pipes and cigars is not as readily inhaled as the lighter, more pleasant smoke produced by cigarette smoking. This difference in inhalation is thought to be the chief reason why pipe and cigar smokers have lower overall death rates than cigarette smokers. However, for men who smoke only cigars the death rate is still 22 percent higher than for nonsmokers (45 to 64 age group). For pipe smokers the corresponding death rate is 11 percent higher than for nonsmokers.

## The evidence

No rational person would suggest today that cigarette smoking is good for human health. Evidence to the contrary is overwhelming. Thousands of studies have been carried out covering virtually every aspect of smoking and its effects upon the body.

Many experiments of the most direct cause-and-effect type have been carried out on animals, because human beings do not ordinarily volunteer to live in laboratories under rigorously controlled conditions for years at a time. As a fortunate alternative, dogs can be taught to smoke, and rodents and other animals can be subjected to carcinogenic (cancer-producing) agents in smoke. Enough is known about the physiologic similarities between these animals and man to permit scientific adjustment of the data, allowing meaningful applications to human beings. Results of such experiments have been confirmed by clinical studies of human beings by autopsy studies of human tissues, and by epidemiologic studies involving thousands of persons.

In one type of epidemiologic study, smoking habits of hundreds of thousands of persons have been correlated with their physical health, or lack of same, and this information has been statistically analyzed. For specific information on experiments, we encourage you to refer to the periodic report to the Surgeon General, *The Health Consequences of Smoking,* prepared by the Public Health Service.

## Effects of smoking on human health

Cigarette smokers suffer a significantly higher death rate *from all causes* than nonsmokers (Figure 9-17). They are more likely to die from cancer, heart and vascular disease, chronic respiratory disease, and ulcers. Even unborn babies suffer a great mortality risk if their mothers smoke. The average life expectancy of a 25-year-old man who smokes less than a pack a day is 5 years less than that of his nonsmoking counterpart. If he smokes two or more packs each day, the smoker reduces his life expectancy by 8 years.

The death rate is higher for smokers who inhale and for smokers who begin smoking at an early age. *The risk of death* increases with the number of cigarettes smoked each day. For men who smoke fewer than 10 cigarettes per day the *death rate is 40 percent higher* than for nonsmokers. For those who smoke 40 or more cigarettes daily the death rate is *120 percent higher.*

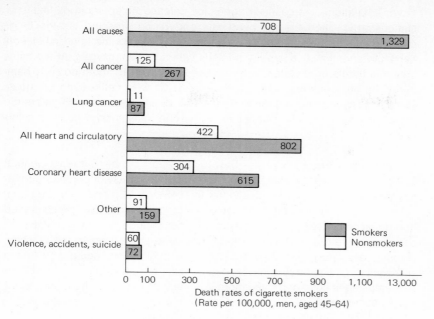

| | |
|---|---|
| All causes | 708 |
| | 1,329 |
| All cancer | 125 |
| | 267 |
| Lung cancer | 11 |
| | 87 |
| All heart and circulatory | 422 |
| | 802 |
| Coronary heart disease | 304 |
| | 615 |
| Other | 91 |
| | 159 |
| Violence, accidents, suicide | 60 |
| | 72 |

Smokers
Nonsmokers

0    100    300    500    700    900    1,100   13,000

Death rates of cigarette smokers
(Rate per 100,000, men, aged 45–64)

**FIGURE 9-17**
Smokers have higher death rates from all causes than nonsmokers. Note that the heart disease death rate is twice as high for smokers as for nonsmokers, and the lung cancer rate nearly eight times higher. (The U.S. Department of Health, Education and Welfare booklet, "Facts: Smoking and Health.")

Women smokers appear to have lower death rates than men smokers, probably because most women inhale less deeply than men. Women also tend to smoke fewer cigarettes per day and to smoke filtered and low-"tar" cigarettes. In addition, women smokers formerly began their smoking careers at a later age than men. Still, death rates for women who do smoke are significantly higher than for nonsmoking women.

**Heart and vascular disease.** In terms of the number of persons affected, coronary artery disease is the most important consequence of cigarette smoking. Among 45- to 54-year-old men the death rate from coronary artery disease is 300 percent greater than among nonsmokers of the same age.

Autopsy studies on the hearts of people who have been killed in war or accidents show a much greater degree of atherosclerosis in smokers than in nonsmokers. Recent studies have confirmed that nicotine speeds atherosclerosis by stimulating the synthesis of fat within walls of major arteries. Carbon monoxide from cigarette smoke may also contribute to the development of atherosclerosis. Cholesterol and lipoproteins in the blood are at higher levels in smokers, especially in the 20 to 29 age group. There is some evidence that smoking causes an increase in the adhesiveness of platelets, which could contribute to thrombus (clot) formation.

Smoking places an increased stress upon the heart. As arteries become clogged by fatty deposits the heart must work harder to maintain a sufficient blood supply to the cells. Then, because of the carbon monoxide in the smoke, the blood of smokers cannot transport oxygen as readily as the blood of nonsmokers. Again, the heart must work harder to compensate. Furthermore, inhalation of nicotine triggers the release of epinephrine-like substances which increase the heart rate and blood pressure. As the lungs become less and less efficient as a result of smoking, the heart must work harder in an attempt to supply the cells with sufficient oxygen. At the same time that the heart is called upon to work harder, its ability to do so is hampered by the coronary stress that deprives its own cells of an adequate supply of oxygen and nutrients. No wonder 16 smokers under 50 years old die suddenly from heart attacks for every one nonsmoker who suffers this fate!

*Chronic respiratory disease.* The 1971 Report of the Surgeon General states that cigarette smoking is the most important cause of chronic respiratory disease in the United States today. In 1945 only 2,300 persons died from chronic bronchitis and emphysema; in 1967, more than 25,000 persons died from these diseases. These statistics do not include the large number of persons whose direct cause of death was some other disease but in whom chronic respiratory disease contributed to death. The increase in these diseases parallels the rise in both cigarette smoking and air pollution. Of the two, smoking is considered the more important cause, since it represents the more concentrated source of pollution. More than 75 percent of patients with chronic bronchitis have a history of heavy cigarette smoking.

Both the particulates and gases produced by smoking interfere with the normal beating of respiratory cilia. As a result mucus accumulates along the respiratory lining, providing a suitable habitat for invading bacteria. Without the cilia-propelled mucous elevator, foreign matter accumulates and irritates the lining. Mucus production increases. The system resorts to coughing as a means of getting rid of both the excess mucus and the accumulated debris. In response to the constant irritation, cells lining the air tubes multiply abnormally (a process somewhat like callus formation on irritated skin). When the lining is thickened in this way, the small bronchial tubes become narrowed and are easily clogged by the excessive mucus secretion. Air becomes trapped in the air sacs. When the smoker coughs the delicate walls of these air sacs may rupture. As more and more air sacs rupture, the lungs become less and less efficient and the smoker suffers from shortness of breath. Particulates also accumulate on the lining of the alveoli and interfere directly with the ability of the surface film to absorb oxygen. Smoker's cough is a symptom of chronic bronchitis.

Chronic bronchitis and emphysema are not normally found in canine populations. Yet dogs trained to smoke cigarettes develop both these diseases.

Early warning signs of emphysema have been found in 97 out of 100 cigarette-smoking men. Among ex-smokers similar symptoms were found in only 6 out of 10, indicating improvement in the lung function of at least some ex-smokers. A control group of nonsmokers who were also not exposed to cigarette smoke at work showed no warning signs. However, nonsmokers who were exposed to cigarette smoke inflicted upon them by smoking coworkers showed some symptoms.

Risk of death from chronic bronchitis and emphysema is up to twenty

**Table 9-2. Smoking and emphysema**

| Degree of emphysema | Subjects who never smoked regularly, % | Pipe or cigar smokers, % | Cigarette smokers, % | |
|---|---|---|---|---|
| | | | <1* | >1† |
| None | 90.0 | 46.5 | 13.1 | 0.3 |
| Minimal or slight | 7.1 | 46.0 | 50.1 | 47.8 |
| Moderate | 2.9 | 6.3 | 25.1 | 32.7 |
| Advanced | 0 | 1.2 | 11.7 | 19.2 |

* Less than one pack of cigarettes per day.
† More than one pack of cigarettes per day.
SOURCE: Data from Oscar Auerbach, E. Cuyler Hammond, Lawrence Garfinkel, and Carmine Benante: "Relation of Smoking and Age to Emphysema," *The New England Journal of Medicine,* Apr. 20, 1972.

times greater among smokers. The degree of risk depends upon the number of years the person has smoked and upon how many cigarettes he has smoked per day.

Respiratory infections are also more common and more severe among smokers than among nonsmokers. In one study male cigarette smokers suffered 22 percent more cases of influenza than males who had never smoked. Female smokers had 74 percent more acute bronchitis than did nonsmoking females.

*Cancer.* Cigarette smoking is the main cause of lung cancer. One authority has estimated that at least 90 percent of deaths from lung cancer would not occur if people did not smoke cigarettes. Carcinogenic compounds accumulate in the lining of the respiratory passages in smokers and are thought to disturb the normal metabolic balance of these cells. Studies of tissue lining the air tubes have identified premalignant changes among smokers. For example, large numbers of abnormal cells containing atypical nuclei are prevalent.

Dogs subjected to tobacco smoke inhalation over a period of more than 2 years develop lung cancer. The disease can also be experimentally induced in mice. The cancer that develops is similar to lung cancer in man. The incidence of human lung cancer increases with the amount of smoking, the number of years the person has smoked, and the degree of smoke inhalation. The cure rate for lung cancer is less than 10 percent.

Other forms of cancer more common among smokers than nonsmokers are cancer of the mouth, throat, larynx, esophagus, and bladder. Pipe smoking often leads to cancer of the lip.

*Other effects.* A woman who smokes during pregnancy decreases the amount of oxygen available for the developing child. The consequence is retarded fetal growth. Babies born to smoking mothers weigh less than babies of nonsmoking mothers. Also, the rate of unsuccessful pregnancies is significantly greater among women who smoke. A study of 2,000 pregnant women who smoked five or more cigarettes per day showed a higher rate of spontaneous abortion (miscarriage), stillbirth, and infant death up to the age of 4 months. A British follow-up study of 7-year-old children born to mothers who smoked heavily during pregnancy showed significantly decreased height, retarded reading ability, and lower ratings on "social adjustment" than a comparison group of children born to nonsmoking mothers.

Smoking by pregnant women appears to deprive offspring of sufficient oxygen during the critical developmental period. The amount of carbon monoxide–hemoglobin in umbilical-vein-cord blood has been measured at 7.3 percent for babies of smoking mothers as compared with 0.7 percent for those of nonsmoking mothers. Nicotine and carcinogens apparently also accumulate in the placenta and are passed on to the fetus. A great deal of research remains to be done before we will know the details of how the unborn child is affected by the smoking habits of its mother. Meanwhile, it would seem prudent to help the unborn get off to a good start by not subjecting them to the effects of tobacco smoke.

Smokers have a higher rate of peptic ulcers and are more likely to die from this condition than nonsmokers. Gum disease, tooth loss, and other oral problems are also more prevalent among smokers.

### What about quitting?

After years of abuse one cannot expect his respiratory system to return magically to health with the cessation of smoking. Some of the damage is permanent. Yet there is a significant amount of recovery, as reflected by the decreased risk of death from several diseases in those who have quit smoking. For example, ex-smokers have lower death rates from chronic respiratory disease, heart and vascular disease, and lung cancer than smokers who continue to indulge. The precise risk figures vary with the number of years a person smoked before he quit, the number of cigarettes he smoked, the age at which he began to smoke, and the number of years since he quit.

Some individuals have a very difficult time breaking the smoking habit, and a few seem entirely unable to quit. It has been suggested that these people may be physiologically addicted to nicotine.

**CHAPTER SUMMARY**

1  Clean air contains natural impurities, but these should not be confused with the man-made pollutants which characterize polluted air.
2  The increase in air pollution and cigarette smoking in recent years has been linked to a dramatic increase in chronic bronchitis, emphysema, and lung cancer.
3  Carbon monoxide, which is emitted primarily by motor vehicles, combines with hemoglobin, thereby depriving body cells of a normal oxygen supply.
4  Other major pollutants are sulfur oxides, unburned hydrocarbons, particulates, and nitrogen oxides; each class has been associated with physiologic damage. Certain pollutants in the air react with one another to form smog.
5  Air pollution damages plant life and causes physiologic disorders and death in domestic animals and wildlife.
6  The level of air pollution in urban areas is affected by weather conditions; the most serious hazard is the development of an inversion layer which prevents pollutants from being normally dispersed.

7 Tobacco smoking is a serious form of air pollution that affects nonsmokers as well as smokers.

8 Cigarette smoking has been linked to heart and vascular disease, chronic respiratory disease, and cancer.

Ehrlich, Paul R., and Anne H. Ehrlich: *Population Resources Environment; Issues in Human Ecology*, Freeman, San Francisco, 1970. Detailed account of the relationship between population and the environment.

Esposito, John C.: *Vanishing Air*, Center for Responsible Law, Grossman, New York, 1970. Ralph Nader's study group report on air pollution.

*The Health Consequences of Smoking, A Report to the Surgeon General;* 1971, U.S. Dept. of Health, Education and Welfare, 1971. Essential reading for anyone who would like to quit smoking.

*The Health Consequences of Smoking,* U.S. Dept. of Health, Education and Welfare, 1973.

Wagner, Richard H.: *Environment and Man*, Norton, New York, 1971. An easy-to-read and highly interesting account of man's interaction with the environment; includes an interesting chapter on air pollution.

**FOR FURTHER READING**

# 10 RESPONSIVENESS: NEURAL CONTROL

***Chapter learning objective.*** Using appropriate terminology, the student must be able to describe the anatomic and biochemical basis for neural control with respect to reception, conduction, integration, and effection. He must also be able to discuss the problem of drug abuse, identifying and describing the effects of each class of drugs (as given in the chapter) of current social concern.

*ILO 1* Define responsiveness, give its structural basis in simple animals and plants, and describe the level of response possible in each case.

*ILO 2* Identify and describe the principal divisions of the human nervous system.

*ILO 3* Draw, or label on a diagram, a neuron. Identify each part; give its function.

*ILO 4* Give four functions of glial cells, and describe the structural deterioration which accompanies multiple sclerosis.

*ILO 5* Draw, or label on a diagram, a reflex arc consisting of three neurons. Label each structure, and indicate the direction of information flow.

*ILO 6* Identify different types of receptors found in the human body, and describe the process of reception.

*ILO 7* Describe the mechanisms by which impulses are transmitted along neurons, between neurons, and between neuron and effector.

*ILO 8* Give the basis for directional limitation of impulse conduction, and identify two factors which influence speed of conduction.

*ILO 9* Describe how a neuron "decides" whether or not to conduct an impulse.

*ILO 10* Describe the structure and functions of the spinal cord.

*ILO 11* Give the functions for each division of the brainstem: medulla, pons, midbrain, thalamus, and hypothalamus.

*ILO 12* Give the functions of the cerebellum, and describe the structure and functions of the cerebrum.

*ILO 13* Give the functions of the reticular activating and limbic systems.

*ILO 14* Identify and give the functions of four types of transmitter substances released by neurons.

*ILO 15* Describe the physiologic state known as sleep.

*ILO 16* Describe current theories of memory and learning.

*ILO 17* Describe the effects of the environment upon behavior and brain structure, and describe new techniques being used to control or modify behavior.

*ILO 18* Define effection, and describe the structure of skeletal muscle and the biochemical basis for contraction.

*ILO 19* Discuss each of the following drugs in terms of their biologic actions and the social effects resulting from their abuse: alcohol, stimulants, depressants, opiates, hallucinogens, and marijuana.

*ILO 20* Define all scientific terms introduced in this chapter.

The ability of an organism to maintain its steady state depends largely upon the effectiveness with which it can respond to changes in its external or internal environment. An ameba in need of nutrients responds positively to food in its aqueous surroundings by flowing toward it and engulfing it. A plant can orient itself so as to receive maximum amounts of available sunlight. Vibrations from footsteps provoke an earthworm to retreat quickly into its burrow. During the course of a working day a man processes and responds to hundreds of stimuli that bombard him from every direction—the traffic light that turns red as he approaches, the restaurant menu from which he selects his lunch, the shortage of trained personnel which requires him to make a new executive decision. All these situations require him to interact with conditions in his surroundings. His survival and his success depend upon the effectiveness with which he responds.

The ability of an organism to react to changes in its internal or external environment is termed *responsiveness.* In order for an organism to respond it must receive a stimulus (that is, perceive a change), transmit a message to other parts of its system, and then effect an appropriate reaction. Cytoplasm itself is *irritable,* that is, sensitive to changes in its environment. In a one-celled organism, such as an ameba, the entire cellular surface may be sensitive to stimuli such as light, heat, touch, and certain chemicals. Messages are conducted throughout the cytoplasm, and the entire ameba responds appropriately, either by withdrawing or by moving toward the stimulus, depending on its nature (Figure 10-1). In some unicellular organisms, structures specialized for carrying on the function of responsiveness are evident.

The simple multicellular animal *Hydra* has a primitive nervous system known as a nerve net (Figure 10-2). Certain cells are specialized to receive stimuli while others conduct news of the stimuli to cells specialized to react by contracting. No central control organ and no definite neural pathways are present. Neural messages are simply conducted throughout the nerve net. As a consequence of its primitive nervous system, the hydra's behavior is

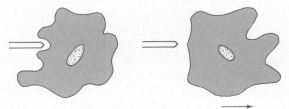

(a) Ameba responds to a negative stimulus by withdrawing

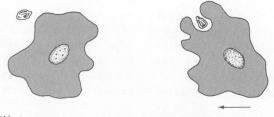

(b) Ameba responds positively to the presence of food

**FIGURE 10-1**
Responsiveness in ameba. (a) Ameba responds to a negative stimulus by withdrawing. (b) Ameba responds positively to the presence of food.

**FIGURE 10-2**
Nerve net in Hydra.

rather simple. When hungry, a hydra may sting a small animal, but it does not actually eat until stimulated by a chemical, glutathione, which is released by the tissues of the injured prey. When glutathione alone is placed in the water near a hungry hydra the animal responds just as though food were present, opening its mouth and moving its tentacles in feeding movements. Its behavior is programmed, almost mechanical, and quite restricted.

Somewhat more complex than the *Hydra,* the flatworm *Planaria* is the simplest animal to have a brain (Figure 10-3). This uncomplicated organ consists of a concentration of nerve cells that serves as a relay center. Two nerve cords connected to the brain extend the length of the body and at regular intervals are connected to one another by cross nerves arranged like the rungs of a ladder.

More complex animals have highly developed nervous systems for receiving stimuli and coordinating responses. Elaborate systems of muscles and glands carry out appropriate responses. Hormones, which are chemical messengers secreted by endocrine glands, function with the nervous system to regulate physiologic activity.

Plants possess neither nerves nor muscles and consequently are unable to uproot and run when danger threatens. They do, however, respond to changes in their environment. Plant responses depend primarily upon hormones which regulate their pattern of growth appropriately and which react to changes in light, temperature, water, and minerals.

## ORGANIZATION OF THE NERVOUS SYSTEM

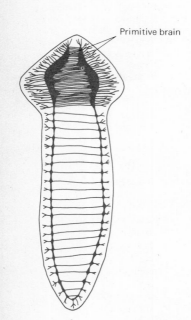

Primitive brain

**FIGURE 10-3**
Nervous system in the flatworm.

The tissues and organs of the nervous system are made up of specific arrangements of nerve cells, or *neurons.* These nerve cells are supported by nonnerve glial cells.

### Major divisions of the nervous system

The two main divisions of the nervous system are the *central nervous system* (CNS) and the *peripheral nervous system* (PNS) (Table 10-1). The CNS consists of the brain and the spinal cord. Serving as central control, these organs integrate all incoming information and determine the response to be made. The PNS is made up of 12 pairs of cranial nerves and 31 pairs of spinal nerves which serve as the communication lines linking central control with all its outposts. These nerves continually inform central control of changing conditions and also transmit orders from central control to appropriate muscles and glands.

That portion of the PNS which controls the organs of the body such as the heart, lungs, digestive system, and kidneys is the *autonomic nervous system.* Ordinarily we do not consciously control internal processes such as the beating of the heart or the movements of the intestine. These activities normally are regulated automatically by the autonomic system. Recent experiments have shown that people can learn conscious control of certain autonomic activities. Some individuals have been taught to alter their heart-

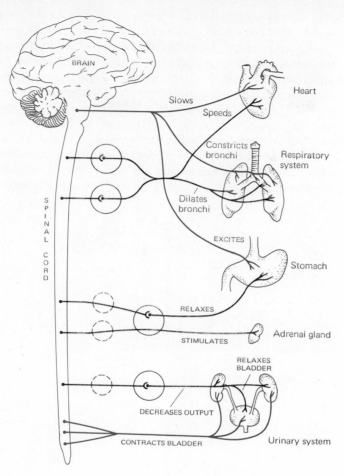

BRAIN

SPINAL CORD

Slows

Speeds

Heart

Constricts bronchi

Respiratory system

Dilates bronchi

EXCITES

Stomach

RELAXES

STIMULATES

Adrenal gland

RELAXES BLADDER

DECREASES OUTPUT

CONTRACTS BLADDER

Urinary system

**FIGURE 10-4**
The CNS and the autonomic portion of the PNS. Green, parasympathetic nerves; black, sympathetic nerves. Note that most of the organs shown are controlled by a nerve coming from each system. Circles represent ganglia (masses of nerve-cell bodies).

beat and blood pressure, and it is hoped that certain psychosomatic disorders may be cured in this way.

The autonomic nervous system consists of *sympathetic* and *parasympathetic* nerves. Each organ is regulated by nerves coming from each system, and these nerves work antagonistically. If one inhibits activity, the other nerve stimulates activity. For example, the heart rate is increased by the action of its sympathetic nerve and decreased by orders from its parasympathetic nerve. Working in a seesaw relationship, these nerves maintain an appropriate balance of activity. We may compare this type of control mechanism to the way in which we might control the speed of a car by balancing the pressure on accelerator and brake.

## Neurons and glial cells

The neuron is the basic unit of the nervous system. In the vertebrate body the most common type is the *multipolar neuron* (Figure 10-5). The bulk of the

**Table 10-1. Divisions of the nervous system**

I. Central nervous system
  A. Brain
  B. Spinal cord
II. Peripheral nervous system (cranial and spinal nerves)
  A. Nonautonomic portion
  B. Autonomic
    1. Sympathetic
    2. Parasympathetic

cytoplasm of a neuron, as well as its nucleus, is located in the *cell body.* From the cell body extend two types of cytoplasmic fibers, the *dendrites* and a long, single *axon.* Dendrites are usually short, highly branched fibers which receive neural messages, or impulses, and transmit them to the cell body. The cell body also can receive impulses directly, and it integrates incoming signals. Although microscopic in diameter, an axon may be 3 feet or more in length. The axon conducts neural messages from the cell body to another neuron, or to a muscle or gland. At its end the axon branches and terminates in tiny structures called *boutons.*

Several types of supporting cells known as glial cells give mechanical support, protect the neurons, and are thought also to serve a nutritive function. Some glial cells appear to receive nutrients (at least amino acids)

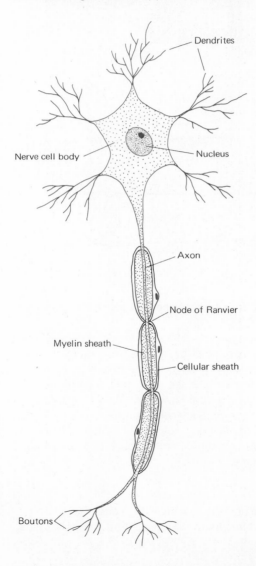

Dendrites

Nerve cell body

Nucleus

Axon

Node of Ranvier

Myelin sheath

Cellular sheath

Boutons

**FIGURE 10-5**
A neuron.

delivered by the blood and pass them on to the neurons, perhaps also functioning to filter out harmful substances which may be present. Other glial cells surround fibers of peripheral neurons, forming a *cellular sheath* about them. Many fibers are also encased in a fatty covering, the *myelin sheath* (Figures 10-5, 10-6). The myelin sheath of a peripheral neuron fiber is composed of cell membranes of cellular sheath cells wrapped about the fiber in spiral fashion. Indentations, called *nodes of Ranvier,* occur at intervals between successive glial cells. At these points the fiber is not insulated by myelin.

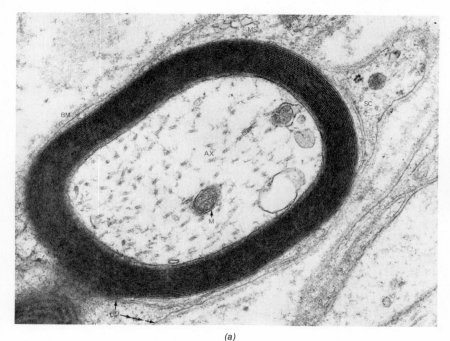

*(a)*

**FIGURE 10-6**
(*a*) Electron micrograph of a cross section through a single myelinated axon. (×82,000) *AX,* axon; *CM,* cell membrane; *M,* mitochondrion; *MS,* myelin sheath; *SC,* glial cell of cellular sheath. (Dr. Lyle C. Dearden) (*b*) Nerve-cell body.

One of the most common diseases of the nervous system is *multiple sclerosis,* in which patches of myelin deteriorate at irregular intervals along the length of the neurons and are eventually replaced by scar tissue. This damage interferes with conduction of neural messages, so that the victim is unable to coordinate voluntary movements, and experiences vision impairment, tremor, and other symptoms. The cause of multiple sclerosis is not yet known, though recently a virus has been implicated.

The cellular sheath is important in the regeneration of injured neurons. When an axon is cut, the portion separated from the cell body deteriorates and is digested by surrounding glial cells, but the cellular sheath remains intact. The cut end of the axon regrows slowly through the empty cellular sheath, and after a long period of time at least partial neural function may be restored. Cells of the central nervous system lack cellular sheaths and apparently are unable to regenerate when injured.

### Nerves

What we ordinarily think of as a nerve is a complex cord consisting of hundreds or even thousands of neuron fibers. We may compare a nerve to a telephone cable. The individual fibers correspond to the wires that run through the cable. These fibers are bound together by glial cells and by connective tissue into an organic cable.

## HOW THE NERVOUS SYSTEM WORKS

Neural response depends upon four processes: *reception, conduction, integration,* and *effection.* Reception is the process of detecting or receiving a stimulus; it is the job of specialized sense organs as well as of neurons themselves. Conduction is the transmission of information (such as a sensory stimulus) along the neurons, from one neuron to another, or from a neuron to a muscle or gland. Integration, the process of sorting and integrating incoming information and determining the appropriate mode of response, is primarily the function of the central nervous system. Effection, the actual reaction, is carried out by *effectors,* the muscles and glands.

### Reflex action

The simplest example of responsiveness in man is the *reflex action,* a relatively fixed response pattern to a simple stimulus. The response is predictable and automatic, not requiring conscious thought. Most of the activities that we engage in daily—walking, regulating activities of the body, even many of the movements associated with eating and drinking—are reflex actions.

Although most reflex actions are much more complex, let us consider one in which a sequence of only three neurons is needed to carry out a response to a stimulus (Figure 10-7). Suppose you touch a hot stove. Almost instantly, and before you are consciously aware of the situation, you jerk your hand away from this unpleasant stimulus. But before you actually withdraw your hand a message is carried from the skin to the spinal cord by a *sensory neuron.* In the tissue of the spinal cord the message is transmitted from the sensory neuron to a *connector neuron,* then to an appropriate motor neuron, which conducts the message to groups of muscles which respond by contracting and pulling the hand from the stove. Actually, many neurons located in sensory, connector, and motor nerves participate in such a reaction, and complicated switching is involved. One moves his hand *up* from a hot stove, but *down* from a hot light bulb. Generally, we are not even consciously aware that all these responding muscles exist.

Quite probably at the same time that the connector neuron sends a message out along a motor neuron it also sends one up the spinal cord to the conscious areas of the brain. As you withdraw your hand from the hot stove you become conscious of what has happened, and aware of the pain. This awareness, however, is a feature apart from the reflex response. That the brain is not necessary to a reflex activity has been demonstrated by experi-

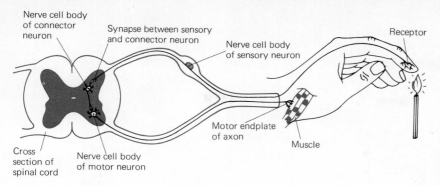

Nerve cell body of connector neuron

Synapse between sensory and connector neuron

Nerve cell body of sensory neuron

Receptor

Motor endplate of axon

Muscle

Cross section of spinal cord

Nerve cell body of motor neuron

**FIGURE 10-7**
A simple reflex action. Receptors in the skin are stimulated and transmit a message over a sensory neuron to a connector neuron in the spinal cord. The connector neuron in turn triggers the appropriate motor neuron, which causes the muscle to contract, pulling the hand away from the stimulus.

ments on laboratory animals. For example, a frog is decapitated, so that it has no brain, and is suspended from a stand. Its leg is then dipped into an acid solution. By means of a reflex activity the frog immediately lifts its leg and withdraws it from the noxious solution. Some reflex actions such as the pupil reflex of the eyes do involve parts of the brain, but these are areas that are functionally similar to the spinal cord and have nothing to do with conscious thought. Sometimes, though, reflex activities are subject to conscious inhibition.

## Reception

Receptors are cells or organs specialized to initiate nerve impulses in response to specific changes in their environment. Perhaps the most familiar receptors are the complex sense organs located in the head—the eyes, ears, nose, and taste buds (Figure 10-8). The sensory cells of the eyes are specialized to react to light waves while cells in the ears are sensitive to transmitted sound waves. Dissolved chemicals stimulate the taste buds, and information concerning odors is received by olfactory receptors in the nose.

Thousands of tiny touch receptors are located in the skin (Figure 10-11). Some are especially sensitive to pressure, others to light touch. One type of skin receptor is specialized to receive information regarding cold, another type receives information on warmth, while bare endings of sensory neurons react to pain. Specialized receptors located in muscle tissue are sensitive to changes in movement, tension, and position. Their continuous reports to the central nervous system help ensure that muscle movement will be properly coordinated. Still other types of receptors are located deep within the body and even within the brain itself. These receptors send messages to the CNS concerning changes in the internal environment.

Each type of receptor is specialized to react to one type of stimulus. Receptors in the ear will not be stimulated by visual stimuli, nor will touch receptors in the skin be triggered by heat. Once stimulated, however, each type of receptor sends the same type of message. Our central control "knows" what the message says not by reading it, but by knowing which neuron delivered it.

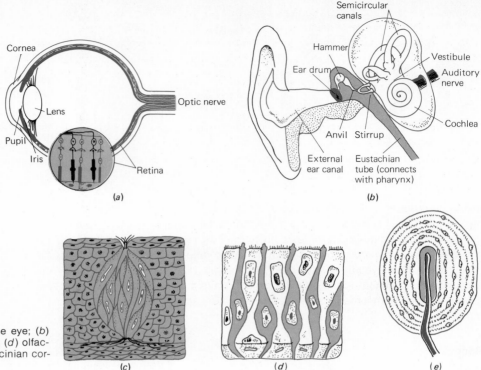

**FIGURE 10-8**
Sensory organs: (*a*) the eye; (*b*) the ear; (*c*) taste buds; (*d*) olfactory epithelium; (*e*) pacinian corpuscles.

(*a*)     (*b*)

(*c*)     (*d*)     (*e*)

***Sensory organs.*** *The eye.* Sight is man's dominant sense. More than 80 percent of the incoming information about his environment is received by the eyes. The eye forms images as a camera does. The eyelids act as a shutter. Like a transparent window covering the eye, the *cornea* keeps dust and other foreign matter out. The colored portion of the eye, the *iris,* acts as a diaphragm. In bright light it contracts, narrowing its opening, the *pupil.* In weak light it expands, allowing more light to enter. Light passes through the *lens* and is received by light receptors in the *retina,* which functions as a light-sensitive film.

Two types of light receptors in the retina are the *rod cells* and *cone cells.* Rods are sensitive to weak light but not to color. They contain a chemical, rhodopsin (or visual purple), which is necessary for the conversion of radiant energy of light into nerve impulses. Vitamin A is required for synthesizing rhodopsin. In the cones a chemical called iodopsin functions similarly to rhodopsin. Three types of cone cells are present, each sensitive to one of the primary colors—red, blue, or yellow. People who are colorblind lack cones. Sensory information is conducted from the retina to the cerebral cortex of the brain by the optic nerve (one of the cranial nerves). In the cerebral cortex the information is perceived as an image.

*The ear.* Sound waves travel through the external ear canal and strike the *eardrum,* causing it to vibrate. Three little bones—the hammer, anvil, and stirrup—in the middle ear transmit vibrations from the eardrum to the *cochlea* of the inner ear. Various types of cells within the cochlea are sensitive to specific frequencies of sound. The brain can distinguish between sounds of different frequency in accordance with the particular cells of the cochlea which have been stimulated, and consequently with the nerve cells which relay the information to the brain. The semicircular canals and vestibule in the inner ear are concerned with the sense of balance.

*Taste buds.* Reception of taste stimuli is the job of little clusters of cells called *taste buds.* Different cell types are sensitive to each of the four kinds of tastes—sweet, sour, salty, and bitter. Taste buds are located in the tongue and in other parts of the mouth. Figure 10-9*c* shows a single taste bud embedded in epithelial cells of the tongue.

*Olfactory epithelium.* Olfactory neurons themselves serve as receptors, as well as conductors, of incoming information regarding odor. These neurons, which end in the epithelium lining the upper part of the nose, are stimulated by odorous molecules in the air. A good bit of what we usually think of as taste is actually the sense of smell. When we suffer from a head cold food does not "taste" nearly as good.

*Pacinian corpuscles.* *Pacinian corpuscles* are pressure receptors. They are distributed throughout the connective tissue underlying the dermis and in many organs of the body. Each is composed of layers of connective tissue surrounding layers of epithelial cells, giving the layers the appearance of an onion. Sensory neurons enter into the corpuscles. It is thought that when a corpuscle is squeezed by pressure the neuron fiber inside it becomes stretched and, as a result, is depolarized. There are several types of "touch" receptors besides pacinian corpuscles.

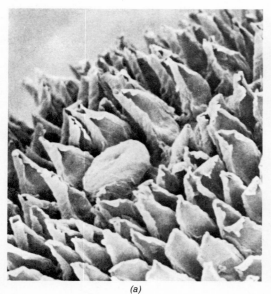

(a)

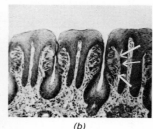

(b)

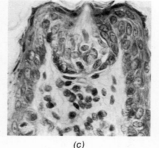

(c)

**FIGURE 10-9**
(*a*) A scanning electron micrograph of the surface of the rabbit tongue. (*b*) Tongue surface. Note positions of taste buds. (*c*) Taste bud on the surface of tongue.
(Parts *a* and *c* from the McGraw-Hill Encyclopedia of Science & Technology, 3d ed.)

## Conduction

Once a receptor has been stimulated, it conducts the information to central control, and then the message is conducted back to the effectors. The message must be transmitted along neurons and also from one neuron to the next.

***Conduction along a neuron.*** A resting neuron, that is one which is not conducting a neural message, is electrically charged. The charge is a consequence of the uneven distribution of ions between the inside of the cell and the immediate external environment. Sodium ions ($Na^+$) are continuously pumped out of the nerve cell and are replaced by potassium ions ($K^+$). However the cell membrane is far more permeable to potassium ions than to sodium, and some of the potassium ions tend to diffuse back out. As a result

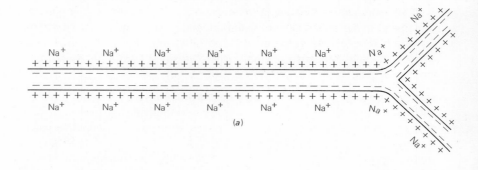

(a)

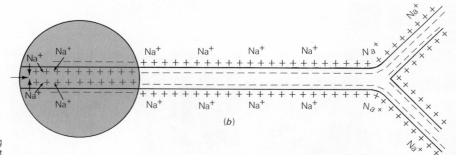

(b)

**FIGURE 10-10**
Conduction of an impulse along an axon. (a) This axon segment is shown "at rest," i.e., no impulse is being conducted. The sodium pump continuously pumps sodium ions out of the cell, and the inside of the cell is negatively charged as compared to the outside. (b) An impulse is being conducted as a wave of depolarization, a momentary breakdown of polarity. (c) The impulse has been transmitted farther along the axon. The portion of the axon along which the impulse has already passed has returned to normal polarized condition.

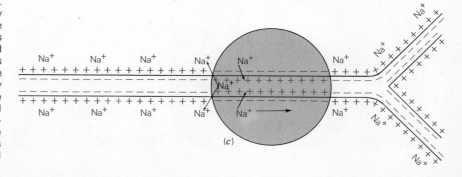

(c)

positively charged ions are slightly more concentrated on the outside of the cell, so that the outside is positively charged as compared to the inside. Negatively charged protein molecules which are too large to pass out of the cell also contribute to the relative negative charge along the inside of the cell membrane. The resting nerve cell is said to be electrically polarized, or to have a *resting potential.*

A neural message, or impulse, is actually a disturbance of the normal polarity of the neuron. The impulse is conducted as a wave of depolarization. When stimulated, the cell membrane of the neuron becomes permeable to sodium for a brief instant; that is, the sodium pump momentarily ceases to function. As sodium rushes in, the neuron becomes depolarized at that point, and the wave of depolarization sweeps along the axon in a self-propagating manner from one node of Ranvier to the next. After the impulse has passed a given part of the axon the resting polarity is restored. DDT and other chlorinated hydrocarbon insecticides apparently act by interfering with the action of the sodium pump (perhaps by poisoning a certain enzyme) so that impulses cannot be transmitted. Nervous systems of insects and some other animals are more sensitive to these poisons than those of mammals.

***Conduction between neurons.*** Neurons are associated with one another to form circuits or pathways, as demonstrated by the reflex circuit which has been discussed. A message must be transmitted from one neuron to the next in the sequence, but the wave of depolarization stops when it reaches the end of an axon because the two neurons are separated from each other by a tiny space, the *synapse.* Instead, information is relayed from one neuron to the next by a chemical messenger that diffuses across the synapse.

When the impulse reaches the boutons of the axon it stimulates the re-

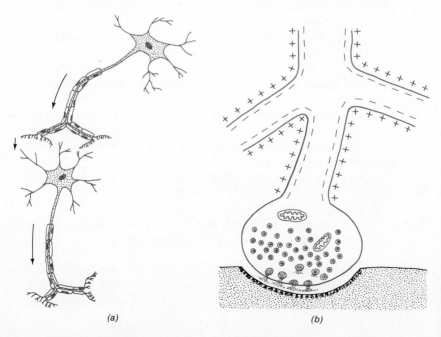

*(a)*          *(b)*

**FIGURE 10-11**
Conduction of an impulse between neurons. (*a*) The synapse between two neurons presents a problem because the wave of depolarization cannot jump across the gap. (*b*) The problem is solved by the secretion of transmitter substance, a chemical which diffuses across the synapse and triggers an impulse in the postsynaptic neuron. Transmitter substances are being released from one bouton.

lease of a chemical called a *transmitter substance,* or neurohumor. The transmitter substance diffuses quickly across the gap of the synapse (less than a millionth of an inch) and depolarizes the dendrites of adjoining neurons so that a new impulse is initiated.

Soon after the transmitter substance is released, any excess is broken down by an enzyme. In this way the transmitter substance is prevented from triggering more than one impulse in the next neuron. Without this enzyme neurons would continue to be stimulated and too many impulses would be transmitted.

*Conduction between neuron and effector.* A synapse also separates the ends of a motor neuron from the muscle cell it innervates. When an impulse reaches the end of a motor neuron, a transmitter called *acetylcholine* is released, diffuses to the muscle cell, and stimulates it to begin the complex business of contraction. Excess acetylcholine is immediately inactivated by the enzyme *cholinesterase* so that the muscle will not continue to contract indefinitely. Calcium ions are necessary for the release of acetylcholine, so that if calcium concentration is lowered, release of acetylcholine is blocked and the muscle will not contract properly. Magnesium ions, on the other hand, inhibit acetylcholine release.

The most poisonous substance known is the toxin from the bacterium *Clostridium botulinum,* which causes botulism. It acts by preventing the release of acetylcholine from neuron endings. About ½ pound of botulism toxin would be enough to poison all the people in the world. Curare, the substance used by South American Indians to poison their arrow tips, blocks the normal action of acetylcholine by itself combining with receptor sites in the muscle. Nerve gases and the related organophosphate pesticides such as Malathion and Parathion poison the vertebrate body by inactivating cholinesterase. In the presence of these poisons acetylcholine continues to stimulate muscle cells so that they go into a state of sustained contraction called *tetany.*

*Direction and speed of conduction.* Neurons are one-way streets in the body. Impulses are always conducted from dendrite to cell body to axon. This limitation is the result of the localization of transmitter substance at the end of the axon. Only when the axon is depolarized can transmitter substance be released. Dendrites have no transmitter substance, and therefore cannot transmit an impulse across a synapse.

Compared to the speed of an electric current or to the speed of light, a nerve impulse travels rather slowly. In the human body impulses are conducted at speeds varying from 5 to 120 meters per second. Myelinated axons conduct impulses faster than unmyelinated fibers. Also, the greater the diameter of the fiber, the faster it conducts an impulse.

*Firing neurons.* In physiology "firing" a neuron means stimulating it so that it transmits an impulse. Each neuron has a specific threshold or minimum level of stimulation that will produce depolarization. A stimulus which is weaker than the threshold level will not fire the neuron. On the other hand, a stimulus just strong enough to fire the nerve cell or one far stronger results in the *same* impulse conduction. The neuron is either stimulated to transmit an impulse, or it is not. There is no variation in the *strength* of a single impulse. This is referred to as the *all-or-none* law.

From our own experience we all know that there are different levels of intensity of sensations. We know the difference between a little pain and a big pain, for example. This is not a contradiction to the all-or-none law, but can be explained in terms of the **number** of neurons which are stimulated. The more neurons which are stimulated, the more intense the feeling of pain. Continued stimulation of the receptor would also result in repetition of the message. Thus, differences in stimulus intensity are reported partly by differences in the frequency of firing of individual neurons, and partly by the number of neurons recruited in response to the sensory stimulus.

Many neurons may interconnect at a single synapse, and several may work together to trigger a postsynaptic neuron. The first neuron in a circuit may deliver a subthreshold message to the second neuron. By itself the first could never fire the second neuron with transmitter substance from a single impulse. However, the transmitter released by two or more impulses coming from either the same neuron or from a team of neurons would be sufficient to fire the second neuron in the circuit. The amounts of transmitter substance released into a synapse are thus added together to fire the postsynaptic neuron.

Some neurons act to inhibit the activity of other neurons. Inhibitory transmitter substance is released and acts to raise the threshold level of affected neurons. Both excitatory and inhibitory transmitter substances act upon the permeability of the cell membrane, but in different ways.

## Neural integration

Neural integration is the process of sorting and interpreting incoming information and determining an appropriate course of action. More than 90 percent of the body's neurons are located in the central nervous system, so it is here that most neural integration takes place. The brain and spinal cord are responsible for making the "decisions."

**FIGURE 10-12**
(*a*) The human brain. (OMIKRON) (*b*) Surface view of the brain. Note how the brain merges into the spinal cord. (*c*) View of brain sliced down the middle so that structures internal to the cerebrum may be studied.

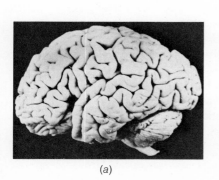

(*a*)

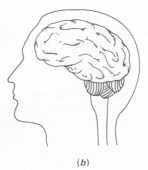

(*b*)

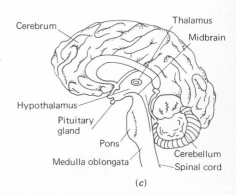

Cerebrum
Thalamus
Midbrain
Hypothalamus
Pituitary gland
Pons
Medulla oblongata
Cerebellum
Spinal cord

(*c*)

By a process of adding and subtracting, the membrane of the cell body integrates all the messages converging upon it. This process of integration is quite mechanical and is carried out on the molecular level. Some transmitter substances raise the threshold, others lower it. Thus, one transmitter may cancel out the effect of another transmitter substance. After these molecular tabulations are carried out, the type and amount of transmitter substance which predominates determines the result. If sufficient excitatory transmitter is present, the neuron will be stimulated and a message will be transmitted.

**The spinal cord.** The spinal cord is a tubular structure protected by the bony vertebral column which surrounds it. Besides its role in reflex activity, the spinal cord functions to conduct messages to and from the brain.

When examined in cross section the spinal cord is seen to consist of an inner portion of *gray matter* surrounded by *white matter.* The gray matter consists primarily of the cell bodies of neurons present within the cord. White matter is composed of myelinated fibers that run up and down the spinal cord.

Neurons which carry messages to the brain are arranged into bundles called *ascending tracts,* while neurons which conduct messages from the brain towards the effectors are referred to as *descending tracts.* As they pass through the spinal cord some nerve fibers cross over from one side to the other. For this reason the right side of the brain controls the left side of the body and the left side of the brain controls the right side of the body.

**The brain.** A soft, wrinkled mass of tissue weighing about 3 pounds, the human brain is the most complex mechanism known. Man's most intricate computer does not begin to rival the complexity of his own brain. A large portion of the genetic information of the human organism is required to direct the brain's construction.

Each of the brain's 10 billion neurons is functionally connected with as many as 60,000 others. Man may never unravel all the tangled neural circuits that govern his behavior, but many mysteries of brain function are being solved.

In accordance with its importance, the brain is the most protected organ in the body. Encased within the bones of the skull, the brain is further shielded by three layers of connective tissue, the *meninges.* A special tissue fluid, *cerebrospinal fluid,* also cushions it against mechanical injury. More than a fifth of the store of circulating oxygen and about two-thirds of the circulating glucose are delivered to the brain cells continuously by a vast network of capillaries.

The brain is an anterior (toward the front) and expanded extension of the spinal cord. Both structures develop from the same tube in the embryo (as will be discussed in Chapter 13). Functions of the three main divisions of the brain—the *cerebrum,* the *cerebellum,* and the *brainstem*—are given in Table 10-2.

**Table 10-2. The brain**

| Structure | Description | Function |
|---|---|---|
| **Brainstem** | | |
|   Medulla | Continuous with spinal cord; primarily made up of nerves passing from spinal cord to rest of brain | Contains centers (clusters of neuron cell bodies) which control heartbeat, respiration, swallowing |
|   Pons | Consists of fibers connecting various parts of the brain with one another | Connects medulla with other parts of brain |
|   Midbrain | Just above pons. Largest part of brain in lower vertebrates; in man most of its functions are assumed by the cerebrum | Center for visual and auditory reflexes (e.g., pupil reflex, blinking, adjusting ear to volume of sound) |
|   Thalamus | At top of brainstem | Main relay center for conducting information between spinal cord and cerebrum. Neurons in thalamus sort and interpret incoming messages before relaying them to appropriate neurons in cerebrum. |
|   Hypothalamus | Just below thalamus. Pituitary gland is connected to hypothalamus by a stalk of neural tissue | Contains centers for control of body temperature, appetite, fat metabolism, and certain emotions; regulates pituitary gland |
| **Cerebellum** | Second largest division of brain | Reflex center for muscular coordination and refinement of movements. When it is injured, performance of voluntary movements are uncoordinated and clumsy. |
| **Cerebrum** | Largest, most prominent part of human brain. More than 70% of brain's cells located here. Groove divides cerebrum into right and left hemispheres. Each hemisphere divided by shallow furrows into 5 lobes: frontal, parietal, temporal, insular, occipital. | Center of intellect, memory, consciousness, and language. Also controls sensation and motor functions. |
|   Cerebral cortex (outer gray matter) | Arranged into convolutions (folds) which increase surface area. White matter within cortex consists of myelinated axons of the neurons that connect various regions of brain. These axons are arranged into bundles (tracts). Functionally, cerebral cortex is divided into: | |
| |   Motor cortex | Controls movement of voluntary muscles |
| |   Sensory cortex | Integrates incoming information from eyes, ears, pressure and touch receptors, etc. |
| |   Association cortex | Site of intellect, memory, language, and emotion |

*Action systems of the brain.* In order to understand how the brain actually works it is necessary to study specific neural pathways which mediate behavior. Since there are 50 trillion or so synapses in the human brain, it is not surprising that researchers have not been able to trace very many meaningful circuits from amidst the tangled web of neurons. A few action systems have been traced, and two will be mentioned here—the *reticular activating system* (*RAS*) and the *limbic system.*

The RAS is a diffuse network of neurons in the brainstem which regulates consciousness. All incoming messages en route to the cerebrum are funneled through the RAS, where they are sorted and integrated. Important mes-

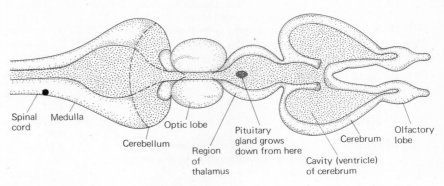
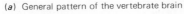

Spinal cord　Medulla

Optic lobe

Cerebellum

Region of thalamus

Pituitary gland grows down from here

Cavity (ventricle) of cerebrum

Cerebrum

Olfactory lobe

(*a*) General pattern of the vertebrate brain

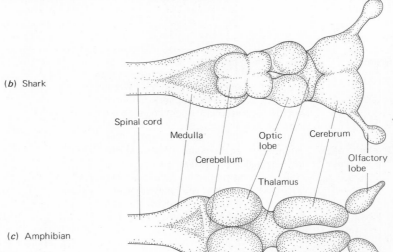

(*b*) Shark

Spinal cord

Medulla

Cerebellum

Optic lobe

Thalamus

Cerebrum

Olfactory lobe

(*c*) Amphibian

**FIGURE 10-13**
Comparison of brain structure in different vertebrates. (*a*) A section through a generalized vertebrate brain to show the basic arrangement of structures. The brain has been cut horizontally to show the ventricles (cavities within the brain). The brain ventricles are continuous with the spinal canal. In the series *b* through *d*, brains of increasingly complex vertebrates are shown. The same basic structures are present in each brain, but the cerebrum becomes progressively larger and more complex. (*b*) Shark brain. (*c*) Amphibian brain. (*d*) Mammal brain.

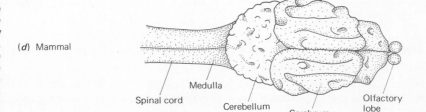

(*d*) Mammal

Spinal cord

Medulla

Cerebellum

Cerebrum

Olfactory lobe

sages requiring the attention of the higher faculties are relayed into the cerebrum; those which can be handled elsewhere are dispatched to appropriate areas. In this way the RAS protects the cerebrum from being overwhelmed by information. For example, it is not normally necessary for the cerebrum to evaluate the internal working of the kidney, or the moment-to-moment status of sweat glands or blood vessels. These and other life-support systems are regulated by automatic reflex mechanisms.

A second action system is the limbic system, which apparently generates emotions which influence behavior. This complex system is intimately interconnected with the RAS. By influencing hormonal activity, the limbic system helps prepare the body for action. It has been found that when electrodes are implanted in certain regions of the limbic area of rat brains, the experimental animals prefer to stimulate their brains above all other activities. A hungry rat would rather press a self-stimulation lever than eat. Apparently such areas of the limbic system are neurally connected to the pleasure center (an area not yet identified) of the brain.

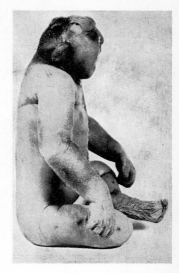

FIGURE 10-14
A human infant born without a cerebrum (anencephalic). (Nicholson J. Eastman and L. M. Hellman, *Williams Obstetrics*, and Appleton-Century-Crofts)

*Transmitter substances in the brain.* Four different transmitter substances thus far studied are acetylcholine, norepinephrine, dopamine, and serotonin. A perplexing feature of these chemicals is that they do not always induce the same reaction; a transmitter may stimulate some types of neurons and inhibit other types.

Acetylcholine, the transmitter that stimulates muscle fiber, is also found within the brain where it is thought to reduce the resting potential of brain cells so that they may be more easily fired. Norepinephrine is associated with alertness and arousal. Dopamine seems to be concentrated in areas of the brain associated with motor activity. Chemically related to norepinephrine, dopamine seems to enhance norepinephrine effects. A deficiency in dopamine has been linked to Parkinson's disease, a condition characterized by rigid tremors of the head and limbs. Serotonin appears to have an inhibitory effect and to be necessary in order for an individual to fall asleep. Neurons which secrete acetylcholine or serotonin are said to be *cholinergic,* while those which secrete norepinephrine are *adrenergic.*

*Sleep and dreaming.* When the RAS slows the process of relaying incoming information to the cerebrum, we lose awareness of stimuli in our surroundings, and go to sleep. During sleep electrical activity of the cerebrum decreases and the body relaxes. Every hour and a half or so during the night this relaxed sleep pattern shifts to what is called **REM** (*rapid-eye-movement*) sleep. In REM sleep the individual dreams, electrical activity greatly increases, and beneath the closed eyelids the eyes move about rapidly. Everyone dreams, though some persons do not recall their nocturnal adventures.

When experimental subjects are deprived of the normal experiences of REM sleep for several nights by awakening them when these rapid-eye-movements begin, they become disoriented. There is evidence that continued deprivation in some instances may result in psychosis (severe mental illness).

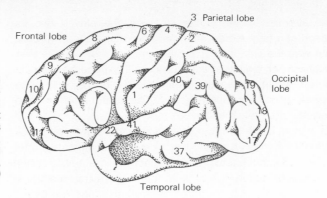

*Memory and learning.* Even very simple animals can be taught certain behavioral responses by the process of conditioning. By carefully rewarding or reinforcing desired behavior, one can teach animals to repeat it. In 1927 the Russian physiologist Pavlov performed a classical conditioning experiment. By ringing a bell each time he presented food to the animal, he taught a dog to salivate in response to the bell. Eventually the dog would salivate at the sound of the bell alone, even without the food. The bell had become a stimulus which resulted in a conditioned response, salivation. With the use of this classical conditioning technique and a host of refinements, animals have been taught to run mazes, react to light or dark, dance, walk tightropes, and perform many other types of behavior. The knowledgeable parent knows how to reinforce desirable behavior on the part of a child with the reward of praise, while disregarding much undesirable behavior so that through lack of reward it is extinguished. Psychologists have used deconditioning techniques to cure phobias in some patients.

The mechanisms by which the brain stores information and retrieves it on demand is the subject of much research, but little understanding has as yet been achieved. A great deal of the information that continuously bombards us is thought to pass into, and quickly out of, a short-term memory storage system. Thus, when you look up a phone number and remember it just long enough to place your call, the digit sequence briefly has been a part of your short-term memory. After the call is made the number is quickly forgotten. Everyday decision making and problem solving depend upon the working of the short-term memory.

An event that makes a strong impression upon us or information which we especially want to remember is further processed and deposited within a long-term memory bank. When information is deemed sufficiently important to remember, it is thought to be fed into the RAS by the limbic system. Then the RAS arouses the cerebral cortex with a message that says, "Store this information." Present evidence points to a mechanism involving transmitter substances. It is thought that as learning takes place specific neurons become more sensitive to transmitter substance. The increased sensitivity in the neural circuit involved may be the memory itself. With time, sensitivity declines and we may forget stored information. RNA and protein are thought by some researchers to play a central role in this complex biochemical mechanism.

The limbic system has been shown to be important in processing stored information. When parts of this system are injured or removed, an individual is unable to retain and store new information. He forgets all recent events within a few minutes but is able to recall memories which had been stored prior to the time of brain damage.

There appears to be no specific area of the cerebrum where long-term memories are stored. When cerebral tissue is damaged or destroyed in experimental animals, information is lost proportional to the amount of tissue destroyed but not according to the specific area which is damaged.

Retrieval of information stored in the long-term memory bank is of considerable interest—especially to students! Some researchers believe that once information is deposited in long-term storage it remains in the brain forever. The only problem is finding the information when we need it! When we seem to forget a particular bit of information it is because we have not efficiently searched for it.

*Controlling behavior.* In 1925 the behaviorist psychologist John Watson wrote:

> Give me a dozen healthy infants, well-formed, and my own specified world to bring them up in, and I'll guarantee to take any one at random and train him to become any type of specialist I might select—doctor, lawyer, merchant, chief, and yes, even beggarman and thief, regardless of his talents, penchants, tendencies, abilities, vocations, and the race of his ancestors.[1]

Watson's conception of the impact of the environment was considered extreme at that time, but since then his view of the role of the environment in shaping the development and intellectual potential of infants and children has gained wide acceptance. An expanding array of multicolored and multishaped crib mobiles, educational toys for babies, and emphasis on early childhood education are producing more intelligent children and yielding greater opportunity for future academic achievement. Recent studies indicate that the way in which a mother relates to her child during the first 18 months of life is critical in establishing personality and academic potential. Mothers who interact warmly with their young children and who stimulate their intellectual curiosity from the earliest months of life help to ensure future social and academic success. Brain structure and biochemistry itself may be changed by such environmental enrichment.

A rigorous series of studies spanning several years has shown that experimental animals provided with a stimulating environment show changes in brain structure as compared with animals kept in less stimulating environments. When exposed to enriched environments rats exhibit an increase in the size of their cell bodies and the nuclei of their brain neurons, they

---

[1] J. B. Watson, *Behaviorism*, Peoples Institute, New York, 1925, p. 82.

**FIGURE 10-16**
Electroencepalographs (EEGs). (*a*) Normal *alpha* activity is present posteriorly over both hemispheres at 9 to 10 waves per second. The arrows signify eye opening and closure. Note the striking reduction in alpha activity while the eyes are open. (*b*) Hyperventilation activates a generalized 3-per-second *spike-and-wave* discharge in this patient with petit mal epilepsy. Note the abrupt onset of the abnormal activity, which arises from a normal background. (The abbreviations refer to areas of the skull, for example, LF stands for left frontal, RC, right central, etc.)

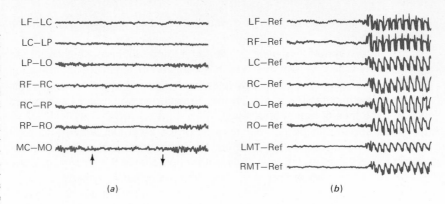

(*a*)                                    (*b*)

develop more glial cells, and show increased concentration of synaptic contact. Characteristic biochemical changes have also been identified.

Modifying or controlling behavior has long been the goal of psychologists. To their inventory of conditioning and deconditioning techniques researchers have added ever more sophisticated electrical and chemical tools for manipulating the mind. Psychologist James McConnell believes that with techniques already at their disposal scientists could change the behavior of anyone with normal intelligence "from whatever it is now into whatever you want it to be. . . ."

By placing electrodes in appropriate places on the skull, one may record the electrical activity of the brain. A tracing of brain waves obtained in this way is called an *EEG* (*electroencephalogram*). Brain wave patterns change characteristically with a subject's mental state. When alert, for example, we have a different EEG than when we are asleep or dreaming. EEG patterns are used as diagnostic tools in determining certain disease states such as epilepsy.

Alpha wave patterns associated with relaxed activity have been the focus of recent experiments. Subjects have been taught to control the activity of their brains by voluntarily producing alpha waves. By seeing and listening to their brain waves on a machine called an alpha pacer, subjects receive continuous feedback to reinforce their efforts. Researchers hope that those with psychosomatic disorders could be taught in this way to relax without the use of drugs. And can you imagine the advantage of being able to induce the relaxed alpha state while taking an examination?

Recent breakthroughs in understanding and manipulating behavior have provided inroads toward curing mental illness. Patients suffering from mental disorders account for as many hospital admissions as patients suffering from all other illnesses combined. Sigmund Freud, the father of psychoanalysis, predicted that scientists would eventually discover a biochemical basis for mental disease. Among the approaches to the treatment of psychopathologies, use of drugs has provided significant relief. Though scientists have not learned how to cure mental illness by means of drugs, they now use them routinely in treatment. Patients who spent years in the back wards of mental institutions have been thereby returned to society. Most of the drugs used act by blocking or enhancing the effect of specific transmitters within the brain. Antidepressant drugs, for example, enhance the activity of norepinephrine in the RAS, changing the patient's mood for the better.

Recently, a chemical related to LSD has been identified in the brain. When produced in too large a quantity, this substance is associated with, and thought by some to be the cause of, the serious mental illness schizophrenia.

## Effection

We have discussed how stimuli are received, how they are conducted, and how they are integrated within the CNS. We have described how appropriate neural messages are sent along motor neurons to appropriate glands or muscles. Now we turn our attention to the response itself, the contraction of the muscle.

*The structure of skeletal muscle.* The three types of muscle tissue in the body are *skeletal, smooth,* and *cardiac.* Smooth muscle is the involuntary muscle which moves our internal organs such as the intestines, while cardiac muscle is the functional tissue of the heart. Our discussion here will be confined to skeletal muscle, the voluntary muscle tissue of the body.

The structure which we usually think of as a muscle, the biceps in your arm, for example, consists of thousands of muscle cells wrapped in connective tissue. Because muscle cells are elongated in shape, they are generally referred to as fibers. Each muscle fiber is so large that it contains more than one nucleus.

Each fiber contains bundles of tiny parallel threads, called *myofibrils,* which extend the length of the cell. Cross striations in the myofibrils give skeletal muscle its striped, or striated, appearance. In an electron micrograph it is possible to see that the myofibrils are in turn made up of two types of *myofilaments,* coarse and fine filaments (Figure 10-18). The dark stripe seen with a light microscope is produced by the combination of interdigitating (overlapping) myofilaments. Muscle contraction occurs when the two types of myofilaments slide between one another, thereby shortening the cell.

*The biochemistry of muscle action.* The coarse myofilaments are composed of the protein *myosin,* the fine filaments of the protein *actin.* When acetylcholine is released by associated motor neurons, the cell membrane of the muscle fiber is depolarized. This depolarization extends into the cell along an elaborate system of tubules which are inward extensions of the cell membrane. These tubules come into contact with the ER (called *sarcoplasmic reticulum* in muscle). As a result of the depolarization, calcium is released from specific sites along the sarcoplasmic reticulum. The release of calcium, in a way not yet completely understood, causes myosin to catalyze the conversion of ATP to ADP. The energy released by this reaction is used to form temporary chemical bonds between the actin and myosin fil-

**FIGURE 10-17**
Man's muscles according to the eighteenth century Dutch anatomist Albinus. The rhinoceros was included in Albinus's engraving because of its rarity. (Albert Bettex, Simon & Schuster, and *Droemersche Verlagsanstalt.*)

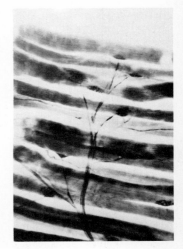

**FIGURE 10-18**
Motor end plates of axon extend into muscle tissue. When they release acetylcholine the muscle is stimulated to begin the complex business of contraction.

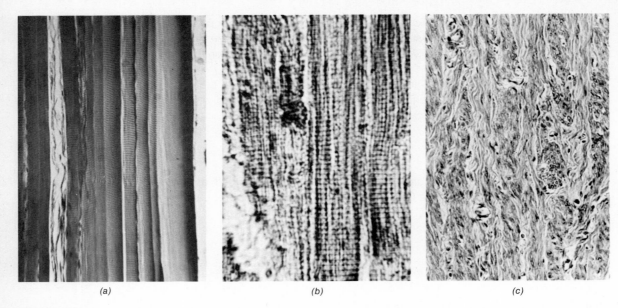

(a)                    (b)                    (c)

**FIGURE 10-19**

(a) Skeletal muscle is striated. (b) Cardiac muscle, though also striated, is involuntary. (c) Smooth muscle is the type found in the internal organs. (Carolina Biological Supply Company)

**FIGURE 10-20**

Electron micrograph of skeletal muscle. Black line indicates length of 1 micrometer. (×48,000) GLY, glycogen; MY, myosin filaments; ACT, actin filaments; M, mitochondria; TS, transverse tubular system; TR, triad; A, H, I, and Z, zones in the muscle tissue. (Dr. Lyle C. Dearden)

aments along their length. As they combine, they slide past one another, with an apparent tendency to combine along as great a length as possible. Relaxation of the fibers occurs when the sarcoplasmic reticulum binds the calcium once again.

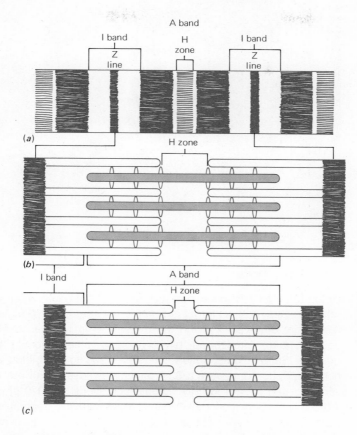

(a)

(b)

(c)

**FIGURE 10-21**
(a) Diagram shows the repeating units in a muscle fiber. The I band consists of actin filaments, the A band of myosin filaments. (b) Diagram of one unit of muscle fiber to show how the filaments interdigitate with one another. (c) The same unit as shown in (b) but now in a contracted state. (H. E. Huxley)

As you might imagine, muscle cells require an enormous amount of energy. Fuel is stored in them in the form of muscle glycogen, large molecules consisting of hundreds of glucose molecules. Glycogen is broken down into glucose as it is needed, and the glucose is utilized as fuel in cellular respiration. Since the ATP produced cannot be stored in appreciable quantity, muscle cells have a special kind of molecule in which energy can be stored. This is creatine phosphate, composed of creatine and the high-energy phosphate from ATP. As needed, the high-energy phosphate is used to charge the ADP associated with actomyosin.

Muscle contraction is not very efficient. Only about 30 percent of the energy is actually converted to mechanical work. The remaining energy is lost as heat, produced mainly as a result of friction. This is why we get hot when we work hard physically. It is also why we shiver when we are cold. The muscle contraction involved in shivering is one way of producing heat to warm the body.

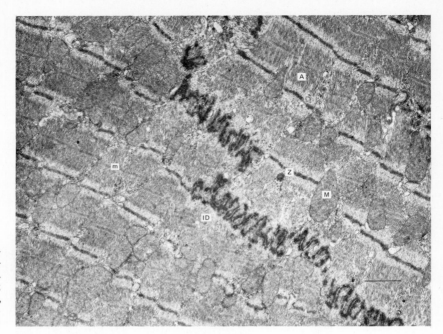

**FIGURE 10-22**
Electron micrograph of cardiac muscle. Black line indicates length of 1 micrometer. (×20,550) *A*, A band; *m*, m line; *Z*, Z line; *M*, mitochondrion; *ID*, intercalated disk (junction between two cells.) (Dr. Lyle C. Dearden)

## THE DRUG SCENE

Drugs to help us fall asleep, drugs to help us stay awake, drugs to help us relax, drugs to help us forget our problems, drugs to help us feel like part of the group, drugs to "expand the mind"! Some purchased freely over the counter, some prescribed by physicians, others bought furtively from pushers. Sale of prescription drugs amounts to $4 billion annually, and nonprescription drugs are a $2.4 billion business. Of all prescribed drugs, some 23 percent are for psychologic conditions. In 1970 physicians wrote more than 225 million prescriptions for sedatives, tranquilizers, and pep pills. More and more people are attempting to control their moods with an ever-growing variety of stronger and stronger drugs. We are a pill-popping society and may someday be remembered as "the drugged generation."

Why is this emphasis on drugs undesirable? One reason is that no drug is completely harmless. The body is a delicately balanced machine, and even a mild drug is a chemical trespasser. Even aspirin has recently been shown to cause noteworthy side effects, such as gastric bleeding. But besides the individual matter of doing harm to one's own body, the use and abuse of drugs have become an important social issue. Drugs affect not only the user but also his family, his employer, and even strangers with whom he shares the highway or from whom he must steal to support his habit.

Some major classes of drugs now commonly abused are alcohol, stimulants, depressants, opiates, hallucinogens, and marijuana. All these are capable of causing *psychologic dependence,* by which we mean that they are emotionally or psychologically habit forming, and that the pursuit of the drug may become so important to the user that he is likely to forego other, more acceptable goals. With continued use most of these drugs induce *tolerance,* which means that to produce the desired effect the user must continually increase the dose. Tolerance often leads to *physical addiction,* because with larger and continued doses the body becomes physically dependent upon the drug. If an addict is denied his drug he suffers from a characteristic set of physiologic withdrawal symptoms, depending on the specific drug involved.

### Alcohol

Alcohol is the most abused drug in the United States today. An estimated 95 million Americans drink alcoholic beverages, and more than 9 million have developed serious drinking problems. Those with drinking problems represent an annual $15 billion loss to the economy in lost work time and property damage. Alcohol is involved in half (28,000) of our annual traffic deaths and in one-third of all homicides.

Contrary to popular belief, alcohol is a depressant rather than a stimulant. After briefly depressing certain neurons which inhibit activity of the cerebral cortex (thus in effect stimulating activity), the depressant effect of alcohol spreads to neurons in the cerebral cortex itself and to other cerebral and spinal neurons. Its major effect is overall depression. Performance of all motor tasks is impaired by alcohol. Movement becomes slower and less accurate, so that even simple activities such as walking or speaking are adversely affected. Mental processes are impaired as well. Studies show that the ability to learn is slowed and the ability to reason and make rational judgments is diminished.

Ethyl alcohol ($C_2H_6O$), the intoxicating ingredient in alcoholic beverages, can be measured in the blood within 5 minutes after ingestion. Maximum concentration is reached in from 30 minutes to 2 hours. From the blood, alcohol enters all the body tissues. The body gets rid of alcohol almost exclusively by oxidation, and this process is begun in the liver. Because alcohol can be utilized as fuel, many alcoholics manage to get along without eating properly, and consequently suffer from nutritional deficiencies as well.

A person who drinks alcoholic beverages habitually develops a tolerance which impels him to drink larger quantities in order to achieve the same intoxicating effects. A typical alcoholic drinks about eleven times as much as the "average" normal drinker. It is thought that upon repeated ingestion the body "learns" how to metabolize alcohol more quickly. In the alcoholic the nervous system seems to adjust biochemically to the continual presence of this drug. As a result the alcoholic can function more effectively when inebriated than a nonalcoholic can. The price the alcoholic pays is psychologic and physical addiction, impairment of the digestive system, and physical damage to the nervous system. Alcoholics have an exceptionally high incidence of peptic ulcers, diseases of the pancreas, and cirrhosis of the liver. Physical damage to brain cells results in eventual loss of control over motor activities. Recent studies indicate that alcohol also damages chromosomes. An alcoholic's life expectancy is reduced by about 10 to 12 years.

## Stimulants

Amphetamines such as Dexedrine, Benzedrine, and Methedrine are widely used as stimulants. They are commonly referred to as pep pills, uppers, or speed. Amphetamines stimulate the release of norepinephrine from neurons in the RAS so that conduction of impulses through the RAS is enhanced. Thus, the cerebrum is kept in a state of superarousal. Heart action and body metabolism are also accelerated. When amphetamines are used over long periods, or in large doses, the user suffers mental fatigue and depression. A user who abuses amphetamines to the point of going for several days without sleep may develop drug-induced mental illness almost identical to schizophrenia.

The mental confusion characteristic of amphetamine abusers is illustrated by a true story of a truck driver who used amphetamines to stay awake and drive for more than 2 days. While driving at 60 miles per hour the truck driver decided to take a nap. When removed from the total wreckage of his truck, he explained that he thought his copilot would take over the driving. The problem was that the "copilot" was an illusion induced by the amphetamines, and the driver was entirely alone in the truck!

Of the 12 billion amphetamine tablets produced in 1971, an estimated 50 percent found its way into the black market. Evidence indicates that amphetamines have been overprescribed by physicians as well, for many patients have become hooked on amphetamines after first using them for help in dieting (see Chapter 5). As many as one out of every five adults may be a long-term or habitual amphetamine user, and abuse of these drugs has been reported even among elementary school children. Although these stimulants are normally ingested in pill form, some abusers dissolve them and inject them directly into a vein (the practice of "speeding").

Beginning in 1972 the Justice Department imposed severe restrictions upon the amount of amphetamines which can be legally produced. Whether this will reduce the problem remains to be seen.

Many users develop a tolerance for amphetamines within 3 to 4 weeks, probably because their liver cells "learn" how to break down the drug rapidly. Some people then make the mistake of increasing dosage in order to produce the desired effect. Some physicians feel that amphetamines are even more addictive than heroin. Amphetamine withdrawal produces symptoms of depression, sleepiness, and extreme fatigue, as well as muscle cramps and severe gastric intestinal disturbance. In one study, the EEG pattern of amphetamine addicts indicated a doubling of REM sleep during withdrawal from the drug, and the EEG pattern did not return to normal for 3 to 8 weeks. There is some evidence that severe amphetamine abuse causes permanent or long-term brain damage.

One common stimulant which is not an amphetamine is caffeine. Although it presents no serious problem of abuse, caffeine does stimulate the nervous system as well as the heart. When drowsy, many people drink coffee, which contains 100 to 150 milligrams of caffeine per cup, in order to stay awake. Excessive caffeine can cause insomnia, mild delirium, abnormal heart rhythms, and other symptoms. There is some recent evidence that caffeine may damage chromosomes.

## Depressants

Barbiturates such as Nembutal and Seconal, as well as milder tranquilizers, relax the central nervous system and are commonly prescribed as sedatives and sleeping pills. When taken in overdose they can be dangerous. Barbiturates account for an estimated 20 percent of acute poisonings admitted to hospitals, 6 percent of suicides, and 18 percent of accidental deaths. Often the user becomes confused, forgets that he has already taken his pills, and ingests additional amounts. Ingestion of 3 grams or more by an adult may be fatal because the drug depresses vital metabolic activities such as respiration. Alcohol and barbiturates have an additive effect, so the combination of the two drugs is especially dangerous and is a common cause of toxicity.

Barbiturates decrease the excitability of nerve cells, and apparently increase the serotonin content in the brain. The RAS is especially susceptible to their action. Barbiturate addiction often begins with a doctor's prescription designed to calm a patient who has insomnia or other symptoms of anxiety. As the patient's body becomes tolerant to the drug, the dosage is gradually increased until he becomes physically addicted as well as psychologically dependent. While a usual dosage of barbiturate might be less than 0.3 gram daily, an average addict ingests about 1.5 grams. At 0.4 gram an individual who is not addicted shows mental dulling, and at 0.8 gram almost any task is performed with decreased efficiency. Lack of coordination due to effects on the cerebellum is characteristic of chronic users. The EEG pattern is abnormal in chronic barbiturate intoxication and in subjects experiencing withdrawal. When deprived of the drug, addicts experience nervousness, weakness, tremor, generalized seizures (often with loss of consciousness), and sometimes hallucinations and delusions. In the medical treatment of a barbiturate addict, the drug is withdrawn slowly over a period of several days.

## Opiates

Opiates—opium, heroin, codeine, and morphine—are drugs made from the opium poppy. Along with certain synthetic drugs they are classified as *narcotics,* drugs which relieve pain and produce a state of stupor. Heroin, the most dangerous, frequently abused opiate, claims 250,000 addicted Americans.

Generally heroin is dissolved in a liquid and injected into a vein. It depresses certain areas of the brain so that incoming neural messages are dulled. The effect is one of relaxation, reduction of tension, and euphoria, and is often referred to as a "high." A user quickly becomes psychologically dependent upon the drug, and as tolerance develops, he requires a larger and larger dose to achieve the same effect. Specific neurons in the brain become dependent upon the drug and this dependency constitutes the basis for addiction.

The life expectancy of a heroin addict is shortened an average 15 to 20 years. Sudden death from overdose is a constant threat, since the illegal heroin traffic does not guarantee uniform doses, and an unusually pure or large dose can be fatal. In New York City alone, 900 persons, 200 of them

**FIGURE 10-23**
Shooting drugs. (Armed Forces Insitute of Pathology)

teenagers, died of heroin overdose in 1969. Heroin is expensive, and the addict requires an ever-increasing dose. His desperate need for funds (often $100 or more a day) to support his habit has led to a rise in the crime rate. In 1970 addicts stole an estimated $10 billion in New York City to pay for their drugs.

Breaking the habit is extremely difficult. When deprived of heroin, the addict suffers withdrawal symptoms within 18 hours. He sweats, shakes, and suffers from chills, nausea, vomiting, diarrhea, severe aches, and pains. These symptoms may last for several days. Even after rehabilitation many addicts yield to the temptation of available drugs and quickly become addicted again. Unless psychiatric or other help is available and successful, whatever personality disorder made the addict vulnerable to drug abuse in the first place may cause him to accept heroin's offer of escape from reality once more.

One way in which heroin addiction is being dealt with medically is with methadone treatment. Methadone is a drug which eliminates the addict's craving for heroin. Although methadone is itself addicting, its effects are less incapacitating, and it is far less expensive (costing about 2 cents per day). Unlike a heroin addict, a person on methadone maintenance is able to function productively in society. Withdrawal from methadone is also less painful than withdrawal from heroin.

### Hallucinogens

Hallucinogens such as LSD (lysergic acid diethylamide), mescaline, and psilocybin cause the user to see, imagine, and hear things in an exaggerated and abnormal manner; thus they are often described as mind-expanding drugs. Mescaline is made from the peyote cactus; psilocybin is derived from a Mexican mushroom. LSD is a synthetic chemical originally developed for use in treatment of mental illness. Its clinical value has not been established, and it is now produced and sold illegally. LSD is a very potent drug—so strong that a single ounce can provide about 300,000 doses.

LSD and other hallucinogens are thought to enhance the activity of norepinephrine and dopamine while inhibiting the depressing effects of serotonin. Information coming through the RAS is not properly filtered, so that the cerebral cortex is bombarded with impulses. Familiar objects appear to take on new features. Shapes change, colors become brighter, vivid hallucinations envelop the user, and he may even have mystic experiences. Users take LSD in order to experience this "trip" into the realm of perceptual distortions. Sometimes the user's personality seems to split so that he feels two opposite emotions, such as happiness and sadness, simultaneously.

An average dose of LSD causes dizziness, nausea, drowsiness, and blurring of vision. One hazard is the "bad trip," in which beauty gives way to horror and the user experiences frightening nightmares and feelings of pain and panic. Suicides have resulted from perceptual distortions which convince the user that he can fly out of a high window or perform some equally impossible feat. In some cases use of LSD has precipitated mental illness. Another hazard is that weeks or even months after taking LSD a user may suddenly experience a recurrence of a former trip. Although there have been studies linking LSD with chromosome damage and birth defects, these have not been confirmed.

### Marijuana

No other drug in recent years has excited the controversy of marijuana. By January 1972 an estimated 24 million Americans had tried this drug, and more than 8 million were still using it. A Gallup poll in 1972 reported that 51 percent of college students had used marijuana. Though new to middle-class America, marijuana, along with alcohol, is one of the two oldest known intoxicants. It is made of chopped leaves, flowers, and stems of the hemp plant, *Cannabis sativa,* and contains only a small amount of the active ingredient THC (tetrahydrocannabinol). Hashish, made from the resin secreted by the flowering tops of female hemp plants, is a much stronger drug, containing up to ten times as much THC.

**FIGURE 10-24**
Marijuana and its derivatives.
(Carolina Biological Supply Company)

Marijuana appears to be a relatively mild intoxicant which acts upon the RAS, resulting in exaggeration of incoming sensory stimuli. It is thought to act upon serotonin as well as norepinephrine pathways. In low doses marijuana produces feelings similar to those of mild alcohol intoxication. In larger doses it interferes with performance of tasks involving short-term memory, and there is some evidence that it impairs driving ability. With large doses of marijuana, users experience hallucinations. Effects of long-term marijuana use are still being studied. Though it may turn out to be a relatively harmless drug, additional research is needed before an accurate evaluation can be made.

### The future?

The problem of drug abuse now extends into every sector of our society. A 1972 report by a New York State commission on education stated that nearly half the high school students in New York use psychoactive drugs. Heroin addiction afflicts children 7 years old, and younger. Upper- and middle-class groups abuse amphetamines and barbiturates to a greater extent than lower socioeconomic groups.

Some observers predict that widespread use of drugs to control mood will eventually be replaced by the more sophisticated technique of direct electrical stimulation of the brain (*ESB*). When depressed or in need of a lift, one would simply press a button in a small control unit which could be carried in his pocket. By remote control specific areas of the brain such as the pleasure center could be stimulated. This has already been done experimentally on a few severely depressed patients. At present it requires surgical implantation of electrodes into the brain, but researchers are looking for a more refined technique.

If mind control should become a popular reality, will each individual retain control of his ESB system? Or shall we find ourselves in George Orwell's *1984*, with a master controller pushing our buttons? ESB and drugs share a common feature. They both provide a means of escaping from the real world. Perhaps we should ask ourselves why so many of us want to escape? Could we perhaps channel our energies into the task of building meaningful lives and producing a world from which we and our children would not want to run?

**Table 10-3. Some commonly abused drugs and their effects**

| Name of drug | Effect on mood | Action on the body | Dangers associated with abuse |
|---|---|---|---|
| Alcohol | Euphoria,* relaxation, release of inhibitions | Depresses CNS, impairs judgment and memory | Psychologic dependence,† tolerance,‡ physical addiction,§ damage to pancreas, cirrhosis of liver, brain damage, eventual loss of motor control, possible chromosome damage |
| Marijuana (grass, pot, Mary Jane) | Euphoria | Impairs coordination, inflames eyes | Psychologic dependence |
| Amphetamines (e.g., Dexedrine) (pep pills, uppers) | Euphoria, excitation, hyperactivity | Enhance conduction of impulses in RAS (by stimulating release of nor-epinephrine), increase heart rate, raise blood pressure, dilate pupils | Psychologic dependence, tolerance, addiction, psychosis, and death from overdose |
| Barbiturates (e.g., Nembutal, Seconal) (downers, goof balls) | Euphoria, reduce anxiety, induce sleep | Inhibit impulse conduction in RAS, depress CNS, impair coordination, depress skeletal muscle and heart, lower blood pressure and breathing rate | Psychologic dependence, tolerance, addiction, death may result from overdose or withdrawal, in combination with alcohol may be fatal |
| LSD (acid) | Euphoria, excitation, hyperactivity, hallucinations | Alters level of transmitters in brain, increases heart rate, raises blood pressure, distorts perception, dilates pupils, causes irregular breathing | Psychologic dependence, tolerance |
| Heroin (H, junk, horse) | Euphoria, relaxation, feeling of self-confidence | Depresses certain areas of brain, depresses reflexes, impairs coordination, constricts pupils | Psychologic dependence, tolerance, addiction, convulsions, hepatitis from dirty needles, death from overdose |
| Cocaine (snow, coke, C) | Euphoria, excitation, hyperactivity | Dilates pupils, causes tremors, hyperactive reflexes, insomnia | Psychologic dependence, psychosis, convulsions, hepatitis, death from overdose |
| Methaqualone (sopro, soap, quaalud) | Euphoria, reduces anxiety | Sedative hypnotic, mechanism of action unknown | Psychologic dependence, tolerance, convulsions, possible liver and kidney damage |

\* Feeling of well-being.
† The user becomes emotionally dependent upon the drug and suffers psychologic distress when deprived of it.
‡ An increasingly larger quantity is required to obtain the desired effect.
§ Habitual use of the drug causes physiologic changes. When the drug is withheld the addict suffers physical illness and characteristic withdrawal symptoms.

1  All living systems possess the ability to react adaptively to their environment.
2  Tissues and organs of the nervous system consist of neurons and the glial cells which support them.
3  A simple reflex pathway is a neural circuit consisting of sensory, connector, and motor neurons.
4  Receptors are specialized cells which are sensitive to some specific type of stimulus in the external or internal environment.
5  Conduction of nerve impulses is an electrochemical mechanism depending upon depolarization of the cell membrane and upon transmitter substances which carry neural messages across synapses.
6  Integration, the sorting and interpretation of incoming information, is primarily performed within the central nervous system.
7  Involuntary activities are controlled by parts of the brainstem and by the cerebellum.
8  The cerebrum is the center for memory, consciousness, language, and intellectual activity.
9  Consciousness and alertness are regulated by the reticular activating system; the limbic system is concerned with emotional behavior.
10  Information deposited in the long-term memory storage system of the brain is probably never lost; the only problem is in retrieving it when needed.
11  Environmental stimuli have been shown to have great impact upon development of behavior and intellect, and perhaps even on the brain tissue itself.
12  ATP provides the immediate energy for muscle contraction, which is brought about mechanically by the interdigitation of actin and myosin filaments.
13  Drug use and abuse have become an important social issue as an increasing number of persons experiment with drugs in an attempt to modify their own feelings and moods.

**FOR FURTHER READING**

Brecher, Edward M., and the editors of *Consumer Reports: Licit and Illicit Drugs,* Little, Brown, Boston, 1972.
McCashland, Benjamin W.: *Animal Coordinating Mechanisms,* William C. Brown, Iowa, 1968.
"The Mind," a four-part series on neural control and drugs, *Life Magazine,* September–November, 1972.
Rosenzweig, Mark R., Edward L. Bennett, and Marian Diamond: "Brain Changes in Response to Experience," *Scientific American,* February 1972, p. 22.
Warshofsky, Fred: *The Control of Life: The 21st Century,* Viking, New York, 1969.

# 11 RESPONSIVENESS: ENDOCRINE CONTROL

***Chapter learning objective.*** The student must be able to describe the interaction of endocrine glands in regulating metabolic activities in the human body, and to describe the consequences of specific endocrine malfunction.

*ILO 1* Identify the human endocrine glands, and locate them anatomically in the body.

*ILO 2* Identify the hormones secreted by each endocrine gland, and describe the functions of each.

*ILO 3* Describe the physiologic consequences of malfunction of each endocrine gland, giving the results of an insufficiency and of hypersecretion discussed in the text.

*ILO 4* Describe the mechanisms of hormone action, including the role of second messengers such as cyclic AMP.

*ILO 5* Describe the role of the hypothalamus and its releasing factors in regulating endocrine activity.

*ILO 6* Diagram or describe the regulation of each of the endocrine glands studied, paying special attention to feedback mechanisms.

*ILO 7* Describe the role of the adrenal glands in helping the body to adapt to stress, and identify noise as a stressor which should be limited.

*ILO 8* Define each new scientific term in this chapter.

The endocrine system works closely with the nervous system to maintain a steady state. Endocrine control depends upon chemical messengers called *hormones* produced by one type of cell or tissue, and dispatched to influence the metabolic activities of other cell or tissue types. The term "hormone" is from a Greek word meaning "I arouse to activity."

The study of endocrine activity, functions, and malfunctions is called *endocrinology*. For many years the field of endocrinology was limited to the study of hormones secreted by specialized organs known as *endocrine glands*. In man and other vertebrates about ten different endocrine glands located throughout the body (Figure 11-1) produce and secrete specific hormones. Endocrine glands secrete their hormones into surrounding tissue fluid and capillaries. Hormones then reach their destination via diffusion or by circulatory system transport. Thus, endocrine glands differ from *exocrine glands*, such as sweat glands and salivary glands, which pour their secretions into ducts through which they are delivered to the body surface or into body cavities.

With recent discoveries the field of endocrinology has been broadened to include the study of chemicals such as prostaglandins, which are produced

ENDOCRINE SYSTEM

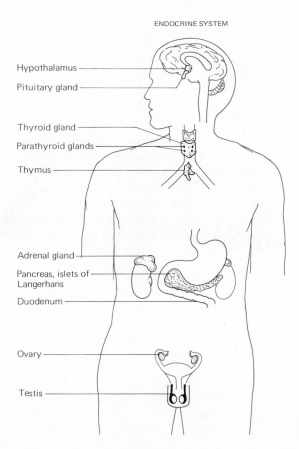

Hypothalamus

Pituitary gland

Thyroid gland

Parathyroid glands

Thymus

Adrenal gland

Pancreas, islets of Langerhans

Duodenum

Ovary

Testis

**FIGURE 11-1**
Location of the major endocrine glands.

by cells that are widely distributed throughout the body rather than by a single discrete organ.

In plants, where behavior depends largely upon hormones, there are no discrete endocrine glands. Instead hormones are secreted by specific types of tissues.

Chemically, hormones do not fall into any single category. Though they are all organic compounds, their structures vary widely. Some hormones are derivatives of amino acids, some are polypeptides, and others (like insulin) are proteins. Still others are steroids. Only a very small amount of any hormone is required to bring about the appropriate response.

## HOW HORMONES WORK

Such complex and varied biochemical interactions are involved in hormone activity that scientists are just beginning to work out the details. At any moment traces of 30 to 40 hormones may be present in the blood. Generally, a hormone is secreted by a particular endocrine gland into the blood and transported to its target tissues. Though the hormone passes through many tissues of the body seemingly "unnoticed," the target tissues respond to its presence by binding the hormone to its cells, a kind of decoding process. Various target tissues may respond differently to the presence of the same hormone. For example the male hormone testosterone stimulates the growth of the beard but may also cause hair on the head to fall out, resulting in baldness.

### Mechanisms of hormone action

Once a hormone is taken up by a particular tissue, how does it influence metabolic activity? In many cases, the hormone activates a second messenger to relay the message. One second messenger, called *cyclic AMP* (adenosine monophosphate), has been extensively studied. When a hormone reaches the target tissue it alters the level of cyclic AMP within the cell. Cyclic AMP (Figure 11-2) is formed directly from ATP (adenosine triphosphate) in a reaction catalyzed by the enzyme adenyl cyclase. Adenyl cyclase is attached to the cell membrane of almost all cells in the body. Once it has served its purpose, cyclic AMP may be inactivated by another enzyme.

$$\text{ATP} \xrightarrow{\text{adenyl cyclase}} \text{cyclic AMP}$$

Cyclic AMP relays a hormonal message by activating a specific enzyme which produces the desired response either directly or by way of a long chain of biochemical events. The particular action initiated by cyclic AMP depends upon the type of enzyme systems present in the cell. Thus the same hormone can promote two different responses in two different types of cells if their enzyme systems are distinct. In some cases cyclic AMP may affect the activity of specific genes. Cyclic AMP is also important in regulating the *release* of several hormones. In some cases it is essential both for hormone

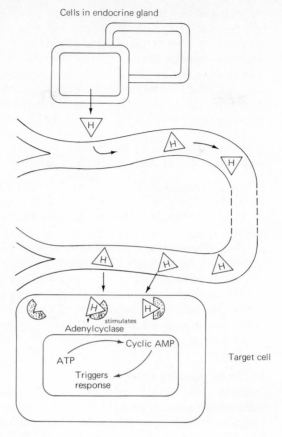

Cells in endocrine gland

stimulates
Adenylcyclase

Cyclic AMP

ATP

Triggers
response

Target cell

**FIGURE 11-2**
The action of a hormone is often mediated by a second messenger, such as cyclic AMP.

secretion by the endocrine gland and for the action of the hormone on the target tissue.

Many hormones appear to stimulate the release of a group of chemicals called *prostaglandins*, which are produced from fatty acids in the cell membranes of many cells. Because prostaglandins appear to influence cyclic AMP levels, they are suspected of playing a part in the response of tissues to hormones. Their precise mode of action is not yet known.

## Regulation of hormone secretion

Hormone secretion generally is self-regulated by a mechanism known as *feedback control.* Information regarding the hormone level or its effects is *fed back* to the gland, which then responds in a homeostatic manner, that is, in such a way as to maintain approximately constant conditions within the organism. Feedback mechanisms are commonly employed in mechanical devices. For instance a constant temperature is maintained in a house by signals between the thermostat and furnace. When the temperature falls below the thermostat setting, the furnace is automatically turned on. Then as the temperature rises above the desired level the thermostat signals the furnace to shut off. In the body a low level of calcium in the blood serves as a signal for the parathyroid gland to secrete more parathyroid hormone. This is *positive feedback*. On the other hand when calcium concentration rises to a certain level, the parathyroid responds negatively by slowing its output of hormone. This is an example of *negative feedback control*. Many endocrine secretions are regulated by a system of feedback control involving the pituitary gland and the hypothalamus.

## HYPOTHALAMUS AND PITUITARY GLAND

Nervous and endocrine systems are linked by the hypothalamus, which regulates the activity of the pituitary gland. Because it controls the activity of several other endocrine glands, the *pituitary gland* has been dubbed the master gland of the body. Truly a biologic marvel, the pituitary gland is only the size of a large pea, yet secretes at least nine distinct hormones which exert far-reaching influences over body activities. Connected to the hypothalamus by a stalk of neural tissue, the pituitary gland consists of two main lobes, referred to as the anterior and posterior pituitary glands. In the embryo the anterior pituitary develops from tissue which forms the roof of the mouth, while the posterior pituitary forms from an outpocketing of the developing hypothalamus.

### Posterior pituitary gland

Two hormones, *oxytocin* and *antidiuretic hormone* (ADH), are secreted by the posterior pituitary gland. These hormones are actually produced by specialized nerve cells in the hypothalamus, and reach the posterior pituitary by flowing through axons which connect the hypothalamus with the pituitary. When they reach the posterior pituitary, these hormones are stored until nerve impulses from the hypothalamus stimulate their release. (Figure 11-3) Oxytocin and ADH are peptides.

*Oxytocin.* When an infant sucks at his mother's breast, sensory neurons report to the posterior pituitary, thereby stimulating the release of oxytocin. This hormone stimulates the actual release of milk from the breast, and also stimulates the contraction of the uterus during and after childbirth. Physicians sometimes administer oxytocin to induce or speed labor. Men have about the same quantity of oxytocin circulating in their blood as women, but its function in the male is unknown.

*Antidiuretic hormone.* ADH acts primarily to increase water resorption from the tubules of the kidney (see Chapter 7). It works by increasing the amount of cyclic AMP in the kidney cells, which in turn increases the permeability of the cell membranes to water and sodium. Apparently, prostaglandins can inhibit this action by blocking the formation of cyclic AMP. The hormone norepinephrine also inhibits the action of ADH on the kidney tubules.

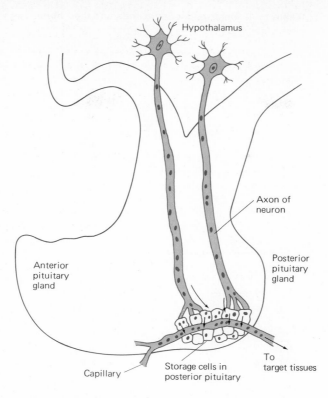

Hypothalamus

Axon of
neuron

Posterior
pituitary
gland

Anterior
pituitary
gland

To
target tissues

Capillary

Storage cells in
posterior pituitary

**FIGURE 11-3**
Hormones secreted by the posterior pituitary gland are actually manufactured in cells of the hypothalamus. They reach the pituitary gland by flowing through the axons of certain neurons. These hormones are then stored in certain cells of the posterior pituitary gland and released into the blood as needed.

Secretion of ADH is regulated by the amount of water in the blood. Receptors in the hypothalamus respond to changes in the concentration of dissolved substances in the blood plasma. These solutes become more highly concentrated when the plasma water content decreases; when their concentration is too high, the hypothalamus signals the pituitary to secrete more ADH. The kidney tubules resorb more water, and this tends to minimize further water loss. On the other hand when the blood becomes too dilute, receptors in the hypothalamus are distended and send the appropriate messages to decrease ADH secretion.

ADH works with another hormone, aldosterone (from the adrenal cortex), to regulate water balance throughout the body. When present in large quantities, ADH increases blood pressure by constricting small arteries.

**Anterior pituitary gland**

Seven distinct hormones are secreted by the anterior pituitary. These include growth hormone, lactogenic hormone, and several tropic hormones which stimulate other endocrine glands. Each of the anterior pituitary hormones is in some way regulated by a separate releasing factor, and sometimes also by an inhibiting factor produced in the hypothalamus. At appropriate times these *neurohormones* (hormones secreted by neural tissue) are released by the hypothalamus and diffuse into a special system of capillaries through

**Table 11-1. Major endocrine glands and their hormones**

| Gland | Hormone* | Function |
|-------|----------|----------|
| Anterior pituitary | Growth hormone | Stimulates body growth; promotes protein synthesis, affects carbohydrate and lipid metabolism |
| | Thyroid-stimulating hormone (TSH) | Stimulates thyroid gland to make and secrete thyroid hormones |
| | Lactogenic hormone | Stimulates cells of mammary glands to secrete milk |
| | Adrenocorticotropic hormone (ACTH) | Stimulates adrenal cortex to secrete glucocorticoids |
| Posterior pituitary | Antidiuretic hormone (ADH) | Stimulates reabsorption of water by kidney tubules |
| | Oxytocin | Stimulates contraction of uterus; stimulates milk release from breast |
| Thyroid | Thyroxine | Stimulates rate of metabolism |
| | Calcitonin | Lowers blood-calcium level |
| Parathyroids | Parathyroid hormone | Regulates calcium and phosphorous metabolism |
| Islets of pancreas | Insulin | Stimulates uptake of glucose by muscle and other cells; lowers blood glucose |
| | Glucagon | Stimulates conversion of liver glycogen to glucose; stimulates breakdown of protein and fat; increases blood glucose |
| Adrenal medulla | Epinephrine | Stimulates conversion of liver glycogen to glucose; stimulates conversion of muscle glycogen to lactic acid; prepares body to deal with stress situations |
| | Norepinephrine | Constricts small blood vessels; raises blood pressure |
| Adrenal cortex | Cortisol | Stimulates protein breakdown; stimulates synthesis of glucose; helps body deal with stress |
| | Aldosterone | Regulates salt metabolism |
| Duodenum | Secretin | Stimulates pancreatic secretion of fluids |
| | Pancreozymin | Stimulates pancreas to secrete enzymes |
| | Cholecystokinin | Stimulates gallbladder to contract |
| Stomach | Gastrin | Stimulates secretion of gastric juice |

* Reproductive hormones will be discussed in Chapter 12.

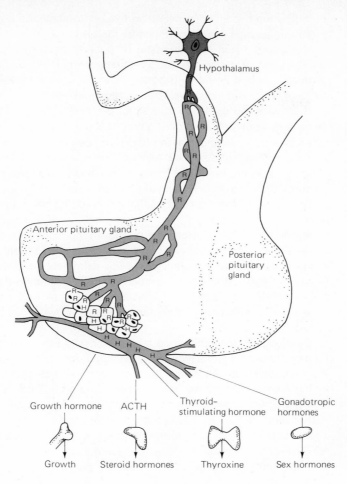

FIGURE 11-4
The hypothalamus secretes specific releasing factors which reach the anterior pituitary gland by way of capillaries. Each releasing factor stimulates the synthesis of a particular hormone by the cells of the anterior pituitary gland.

which they are carried to the anterior pituitary. When we speak of the pituitary as being stimulated or inhibited, it should be understood that certain receptors in the hypothalamus are first affected. They in turn control the pituitary (Figure 11-4).

***Growth hormone.*** Growth hormone stimulates body growth, especially of the long bones, by promoting protein synthesis. Growth hormone also affects the metabolism of several salts, of fat, and of glucose. Fasting and lowered blood-sugar levels stimulate the secretion of growth hormone in adults as well as children. Under conditions of insufficient carbohydrate fuel, growth hormone mobilizes fatty acids which can be used as an alternate fuel for energy production. Several other hormones work with growth hormone in regulating metabolic activities.

Secretion of growth hormone is regulated by both a growth hormone–releasing factor (GRF) and a growth hormone–inhibiting factor (GIF) se-

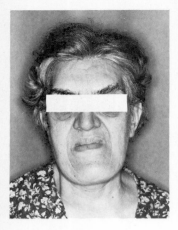

**FIGURE 11-5**
The prognathous jaw shown in the photo is characteristic of acromegaly. (Armed Forces Institute of Pathology)

creted by the hypothalamus. GRF is thought to stimulate both the synthesis and the release of growth hormone by the pituitary.

The growth spurt characteristic of puberty occurs only if the male or female hormone is also present. Eventually sex hormones promote the fusion of growth centers in the long bones, so that increase in height, even in the presence of growth hormone, is no longer possible.

There is some evidence that sleep promotes growth, since growth hormone is normally released during non-REM sleep (see Chapter 10). Recently researchers have found that in children whose physical needs are met but who are deprived of emotional attention, growth may be retarded. Apparently cuddling, playing, and other forms of nurture are essential to normal development. Many emotionally stunted children also exhibit abnormal patterns of sleep, and this may be responsible for inhibiting secretion of growth hormone.

Hyper- (too much) secretion of growth hormone during childhood is responsible for the condition known as gigantism in abnormally tall people. At the other extreme are pituitary dwarfs, extremely small persons who suffered from hypo- (too little) secretion of growth hormone during childhood. These individuals, familiar as circus midgets, are well proportioned and have normal intelligence and normal sexual development. They are just miniature.

By injecting growth hormone it is now possible to stimulate growth in patients suffering from this type of pituitary malfunction if the growth centers in the long bones are still open. Other mechanisms, including certain releasing factors from the hypothalamus, also bear on growth. Thus, some people may have sufficient growth hormone, yet still attain only short stature. In these individuals injection of growth hormone will not stimulate growth.

If pituitary malfunction leads to hypersecretion of growth hormone during adulthood, the individual develops acromegaly (large extremities). Long bones cannot grow in length after maturity, but bones in the hands, feet, and face are stimulated to enlarge.

*Lactogenic hormone.* Lactogenic hormone, also called prolactin, stimulates the cells of the mammary glands to secrete milk. When male animals, particularly birds, are injected with this hormone they become maternal and demonstrate feminine interest in caring for the young. For this reason lactogenic hormone is suspected of stimulating maternal behavior characteristic of normal females. Lactogenic hormone also plays a role in establishing hormone balance during pregnancy. Although normally present in both sexes, its function in males is unknown.

*Tropic hormones.* Several tropic hormones are secreted by the pituitary, and each stimulates the activity of another endocrine gland. Their interaction with the target gland is best described in conjunction with a discussion of that gland. Accordingly, *thyroid-stimulating hormone* will be discussed in association with the thyroid gland, *ACTH* with the adrenal gland, and the gonadotropic hormones, FSH and LH, will be studied with reproduction in Chapter 12.

## THYROID GLAND

The thyroid gland is located in the throat region, in front of the trachea and below the larynx (see Figure 11-1). Its principal hormones are thyroxine and calcitonin. The latter will be discussed in connection with the parathyroid gland.

### Thyroxine

Thyroxine is synthesized from the amino acid tyrosine. Its most distinctive chemical feature is the four iodine atoms attached to each molecule. When iodine is lacking in the diet, thyroxine cannot be synthesized. People with limited diets who might otherwise be deficient in iodine intake are assured proper amounts by the use of iodized salt.

It is likely that thyroxine stimulates the rate of metabolism by stimulating the electron transport system (as evidenced by increased oxygen consumption), and by increasing protein synthesis. Perhaps as a result of these actions, many body activities are speeded by thyroxine. For example, it enhances the activity of growth hormone. Thyroxine is also necessary for cellular differentiation. In tadpoles, development into adult frogs cannot take place without thyroxine, as the hormone appears to regulate selectively the synthesis of needed proteins.

### Regulation of thyroxine secretion

Secretion of thyroxine is regulated by the hypothalamus and anterior pituitary gland. A releasing factor secreted by the hypothalamus stimulates the pituitary to secrete thyroid-stimulating hormone (TSH). TSH acts on the thyroid gland by stimulating increased uptake of iodine, synthesis and secretion of thyroxine, and increased size of the thyroid gland. TSH action upon the thyroid gland is mediated by cyclic AMP (Figure 11-6).

When the concentration of thyroxine in the blood exceeds a certain level, both the hypothalamus and the pituitary are signaled to reduce their secretion. As the amount of circulating thyroid hormone decreases, the hypothalamus is stimulated to secrete TSH-releasing factor. This hormonal interaction among hypothalamus, pituitary, and thyroid gland is a striking example of regulation by means of feedback mechanisms.

Secretion of thyroxine may also be influenced by stress. Both emotional factors and physical factors such as cold result in increased thyroxine secretion. During cold weather an increase in thyroxine secretion reduces the efficiency of the citric acid cycle, thus increasing body heat.

### Disorders of the thyroid gland

A child who suffers from underactivity of the thyroid gland, known as *hypothyroidism,* has a very low metabolic rate. If untreated, his physical and men-

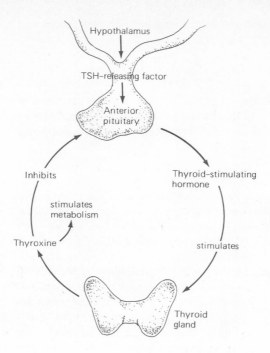

Hypothalamus

TSH–releasing factor

Anterior pituitary

Inhibits

Thyroid-stimulating hormone

stimulates metabolism

Thyroxine

stimulates

Thyroid gland

**FIGURE 11-6**
Feedback control between the pituitary gland and the thyroid gland.

tal growth are retarded and he may become a *cretin* dwarf. (Such an individual should be distinguished from a pituitary dwarf.)

Sometimes an adult becomes hypothyroid. His slow, lethargic behavior correlates with a reduction in metabolic rate. In adults the condition is termed *myxedema*. Hypothyroidism is easily treated with thyroid pills, usually made from beef thyroid extracts.

Overactivity of the thyroid gland, hyperthyroidism, results in a superenergetic, often irritable individual. Formerly treated by removing part of the thyroid gland, the condition can now be controlled with appropriate drugs.

An abnormal enlargement of the thyroid gland is termed *goiter*, often a result of iodine deficiency. Without iodine the thyroid cannot synthesize thyroxine, and the level of this hormone in the blood is reduced. In compensation, the pituitary continues to secrete TSH in even larger amounts. In response to the TSH, the thyroid gland enlarges, often to monstrous size, though it still cannot manufacture thyroxine. Thanks to iodized salt, goiter is no longer common in the United States. Nevertheless, hundreds of thousands of people throughout the world still suffer from this easily preventable disease.

## PARATHYROID GLANDS

Four tiny parathyroid glands are embedded in the connective tissue which surrounds the thyroid gland. Their secretion, parathyroid hormone, helps regulate the amount of calcium and phosphorus in the blood. As the calcium

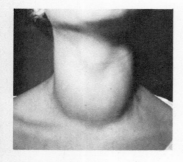

**FIGURE 11-7**
Malfunction of the thyroid gland sometimes results in goiter. (Dr. Gilbert Echelman)

level in the blood decreases, there is a corresponding increase in phosphorus. An appropriate concentration of calcium and phosphorous salts is essential for nerve and muscle function. Formation and maintenance of bone also depend upon a normal supply of these salts.

Parathyroid hormone works to maintain an appropriate level of calcium in the blood by stimulating reabsorption of calcium from the kidney tubules. When additional calcium is needed, parathyroid hormone stimulates release of calcium from the bones. Vitamin D promotes the absorption of ingested calcium from the intestine.

### Regulation of parathyroid secretion

When the level of calcium in the blood falls even slightly, parathyroid secretion increases. On the other hand when the calcium level increases above normal, parathyroid secretion is turned off. *Calcitonin*, a hormone secreted by the thyroid gland, also helps prevent calcium from reaching excessive levels. Calcitonin inhibits the release of calcium from bone, and in general antagonizes the action of parathyroid hormone (Figure 11-8).

### Disorders of calcium metabolism

Parathyroid deficiency is one cause of a low level of calcium in the blood. This condition causes muscles and nerves to become highly irritable so that

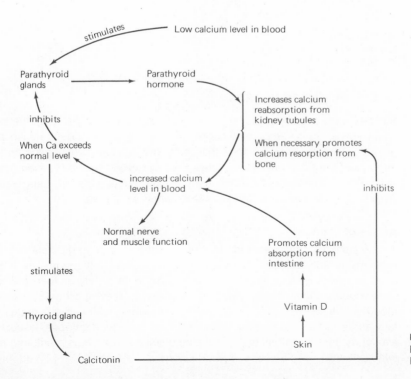

**FIGURE 11-8**
Regulation of calcium metabolism.

stimuli which normally produce little or no response cause them to discharge. Discomfort, tremor, and tetany occur. Such symptoms can be relieved by the injection of either calcium solution or parathyroid hormone. On the other hand, hyperactivity of the parathyroid gland results in the removal of so much calcium from the bones that they become brittle and are easily fractured. The level of calcium in the blood may become so high that calcium precipitates (comes out of solution) in the kidney, and may aggregate to form kidney stones.

During pregnancy large quantities of calcium are required for the production of bone in the developing fetus. If sufficient calcium is not provided by the mother's diet, parathyroid secretion increases and causes calcium to be metabolized from the mother's bones.

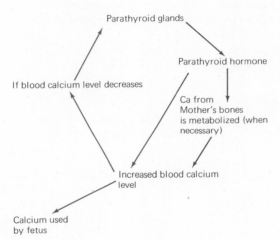

**FIGURE 11-9**
Calcium for the developing fetus.

## ISLETS OF THE PANCREAS

Besides secreting digestive enzymes (as described in Chapter 5), the pancreas serves as an important endocrine gland. Its hormones, insulin and glucagon, are secreted by cells that form little clusters, or islets, dispersed throughout the pancreas. These islets, first described by the German histologist Paul Langerhans, are sometimes called the islets of Langerhans. About a million islets are present in the human pancreas.

### Action of insulin

Insulin stimulates cells, especially muscle cells, to take up glucose from the body by enabling them to transport it across the cell membrane. Once glucose enters muscle cells, it is either used immediately as fuel or stored as glycogen. Insulin activity results in lowering the glucose level in the blood, and also influences fat and protein metabolism. It reduces the use of fatty acids as fuel and instead stimulates their storage in adipose tissue. In a similar manner, it inhibits the use of amino acids as fuel, thus promoting protein synthesis.

## Action of glucagon

The principal action of glucagon is to raise the level of glucose in the blood. In the liver, glucagon stimulates the conversion of glycogen into glucose by a process involving cyclic AMP. Glucose thus formed moves into the blood, elevating the blood-sugar level.

Glucagon may also promote tissues to take up fatty acids from the blood and to use them as fuel. Recent work has shown that the small intestine contains a glucagon-like hormone, GI glucagon, that may stimulate the secretion of insulin when sugar is eaten.

## Regulation of insulin and glucagon secretion

In healthy people, the insulin level of the blood is self-regulating. After eating, when the blood-sugar level rises above a critical level (80 to 100 milligrams per 100 milliliters) the glucose itself stimulates the secretion of insulin. The insulin stimulates the uptake of glucose by the cells. Then, when the blood-sugar level falls to a normal level, insulin secretion is halted.

When one has not eaten for several hours, the glucose supply in the blood decreases. This fall in blood-sugar level stimulates glucagon secretion, and the blood-sugar level is restored by mobilizing the glycogen supply in the liver. As we shall see presently, another hormone, epinephrine, also plays a role in the regulation of blood sugar.

## Diabetes mellitus

Diabetes mellitus is the principal disorder associated with the pancreatic hormones. Sometimes referred to as *sugar diabetes*, this condition afflicts an estimated 4 million Americans, nearly 2 percent of the total population. Almost half these diabetics have only minor symptoms, and many are not even aware that they have the disorder. Yet it is important that they be identified, because early treatment of diabetes increases the life expectancy of diabetics. Those most likely to have diabetes are obese, have diabetic relatives and are over 45. Some forms of diabetes appear to have a genetic basis, but the exact mechanism of inheritance is not known.

The diabetic is characterized by an abnormally high level of glucose in his blood, resulting from an insulin deficiency. Recent evidence suggests that in addition target cells do not respond efficiently to whatever insulin is present. Instead of the normal blood-sugar level of about 80 milligrams per 100 milliliters (after fasting), the diabetic may have 300 milligrams per 100 milliliters or even more. The amount of glucose that filters into the kidney nephrons exceeds the renal threshold level so that not all of the glucose can be reabsorbed into the blood. Therefore, sugar is excreted in the urine; this is one of the symptoms of the disorder. Because of the glucose molecules in the filtrate, more water than normal remains in the tubules, and an abnormally large volume of urine is produced. This tends to dehydrate the body, making the diabetic constantly thirsty. Because glucose is not readily available as fuel for cellular respiration, cells metabolize proteins and fats; the untreated diabetic therefore becomes thin and emaciated. Furthermore, a serious complication arises as a consequence of the cells' using large quantities of fat as fuel. *Ketone bodies*, toxic compounds formed during the breakdown of fats, accumulate in the blood and urine and interfere with the normal alkalinity of the blood. The resulting acidosis has a widespread and harmful influence on many body functions and may lead to coma. Diabetes may also result in vascular damage, including increased likelihood of atherosclerosis, though the precise relationship between excessive blood sugar and vascular damage is not yet understood.

Diabetes is treated with injections of insulin generally prepared from beef or pork pancreas. Since it is a protein, insulin cannot be taken orally because it would be broken down by enzymes in the intestine. Mild cases can be treated with tolbutamide and related drugs taken orally, which serve to stimulate the islets to produce insulin. Unfortunately, these drugs may have harmful side effects when used over long periods of time.

If a diabetic injects himself with too much insulin, or if the islets secrete too large a quantity because of some malfunction, muscle cells are stimulated to rapidly take up too much glucose from the blood. The blood-sugar level falls drastically, and brain cells, which cannot store glucose, are deprived of their constantly needed supply. Insulin shock may result. The victim appears to be drunk, may suffer convulsions, become unconscious, or even die. To ensure that he will receive appropriate treatment in the event of shock or coma, a diabetic should wear an identifying bracelet.

Treatment of diabetes includes a carefully managed, balanced diet, designed to prevent excessive intake of carbohydrates and to maintain a desirable body weight. Carbohydrate balance is achieved more easily when the diabetic eats five or six small meals each day rather than three large ones.

## ADRENAL GLANDS

The paired adrenal glands are located just above the kidneys. Each gland consists of a central portion, the *adrenal medulla*, and an outer section, the *adrenal cortex*. Although they are wedded anatomically, the adrenal medulla and cortex have very different origins in the embryo and function as distinct endocrine glands. Both help to regulate normal metabolic activities, and both play important roles in preparing the body to deal effectively with stress.

### Adrenal medulla

In the embryo the adrenal medulla develops from nerve tissue. Its cells are regarded as modified neurons, and its activity is regulated directly by the nervous system. Some physiologists look upon the adrenal medulla as a special part of the sympathetic nervous system.

Two hormones secreted by the adrenal medulla are *epinephrine* (trademark Adrenalin) and *norepinephrine* (noradrenaline). These hormones are often referred to as *catecholamines*. You should recall that norepinephrine is also secreted as a transmitter substance by sympathetic nerve endings and by neurons in the central nervous system.

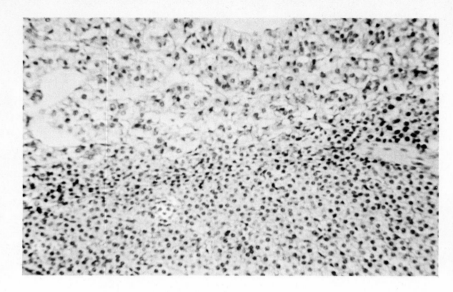

**FIGURE 11-10**
Adrenal gland. This pho-
tomicrograph shows the junction
between the adrenal medulla
(above) and the adrenal cortex
(below). Note that the tissue is
visibly different in each part.

The adrenal medulla is often referred to as the emergency gland of the body, because its hormones prepare us physiologically to cope with sudden stress. If a monster were suddenly to appear before you, secretions from the adrenal medulla would enable you to run faster and think more quickly!

Both epinephrine and norepinephrine are continuously secreted in small amounts. Both increase the heart rate, raise blood pressure, and increase the strength of muscle contraction. Norepinephrine, which raises blood pressure by constricting small blood vessels, is secreted in amounts needed to maintain normal blood pressure.

Epinephrine acts antagonistically to insulin in maintaining a normal blood-sugar level. It raises blood glucose by increasing the amount of glucose absorbed from the intestine, stimulating the conversion of glycogen into glucose by the liver, and stimulating the conversion of muscle glycogen to lactic acid, much of which is reconverted into glycogen in the liver.

When the body is under intense stress, an increased secretion of epinephrine constricts blood vessels going to the skin and kidneys, while dilating those of muscle and brain. Such a redistribution of blood in favor of those organs required for emergency action enables the body to respond quickly and effectively. Mental alertness is increased, and the ability for fight or flight is enhanced. At the same time the appropriate blood-sugar level is maintained.

Secretion of adrenal medulla hormones is controlled by the nervous system. Taking biology tests, anxiety, pain, fright, noise, and other stressors stimulate the dispatch of neural messages from the brain. These messages are carried by sympathetic nerves to the adrenal medulla, where the axons of these neurons secrete acetylcholine, which in turn stimulates the secretion of the hormones.

Tissues that respond to epinephrine and norepinephrine have specialized receptors. Cyclic AMP acts as a second messenger in mediating the actual reaction within the target tissue. Thyroxine and a hormone secreted by the adrenal cortex enhance the effects of the adrenal medulla hormones.

Norepinephrine is used clinically to increase the blood pressure of patients who are in shock. It is also used in some cases to stimulate a heart which has stopped beating, by injection directly into the heart muscle.

### Adrenal cortex

About fifty different steroids have been identified in the adrenal cortex. (Steroids, you will recall, are a chemical group classified as lipids; cholesterol is a well-known example.) Most of the adrenal steroids are intermediate substances from which the actual hormones are synthesized. Several different types of hormones are associated with the adrenal cortex. One group consists of sex hormones; small quantities of both male and female hormones are produced in the adrenal cortex of both sexes. (These are intermediates in the production of other hormones, and under normal conditions may not always be secreted.) A second type of adrenal cortical hormone, collectively referred to as mineralocorticoids, functions to regulate the salt balance of the body. The principal one is aldosterone. A third group, the *glucocorticoids,* helps to regulate carbohydrate metabolism. Cortisol and cortisone belong to this group.

*Aldosterone* is important in regulating the salt and water balance of the

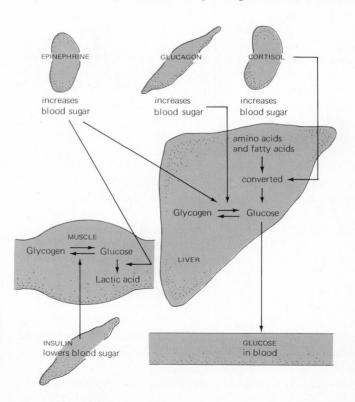

**FIGURE 11-11**
Regulation of blood-sugar level is the job of several organs in the body. Hormones from the islets of the pancreas and the adrenal glands act in a seesaw fashion to keep the blood-sugar level constant.

body. Its principal action is to promote reabsorption of sodium from the kidney tubules, and also from the sweat glands. Aldosterone is thought to work with ADH in maintaining appropriate concentrations of salt and water in brain, liver, and muscle tissue. Intracellular levels of potassium and magnesium may also be influenced by aldosterone.

Aldosterone secretion is regulated by blood volume. When blood volume decreases, aldosterone secretion increases. A greater quantity of sodium is reabsorbed so that a larger volume of water is retained. When blood volume returns to normal, aldosterone secretion decreases.

*Cortisol* is the principal glucocorticoid. When necessary, cortisol acts to increase the blood-sugar level by increasing the quantity of amino acids sent to the liver by other tissues, and then by stimulating their conversion to glucose. Glycogen is also converted to glucose, and the blood-sugar level is thus maintained. At the same time that cortisol increases the glucose supply of the liver, it inhibits the uptake of glucose in the blood by muscle and fat tissue.

Cortisol works with epinephrine and growth hormone in mobilizing fatty acids from fat tissue. It also works with aldosterone in promoting sodium reabsorption. The role of cortisol in helping the body to deal with stress will be discussed shortly.

Cortisol secretion is regulated by the hypothalamus and pituitary gland. A releasing factor, CRF (corticotropin-releasing factor), is secreted by the hypothalamus. CRF stimulates the anterior pituitary gland to secrete ACTH (adrenocorticotropic hormone), which in turn stimulates the adrenal cortex to secrete cortisol. When the amount of cortisol in the blood exceeds the appropriate level, the activity of the hypothalamus and pituitary is inhibited. This normal control mechanism can be overridden in stressful situations, when the central nervous system takes over and orders the secretion of unusually large amounts of ACTH.

Insufficient secretion of hormones of the adrenal cortex results in *Addison's disease*. A patient with this condition suffers from low blood pressure, abnormally low levels of sodium in his blood, general weakness, low blood-sugar level, and a characteristic bronze skin color. Addison's disease is now treated by administering the missing hormones. Hyperactivity of the adrenal cortex leads to *Cushing's disease*, a condition associated with high blood pressure, a variety of metabolic disturbances, obesity, and a characteristic moon-shaped face.

Because they reduce swelling, glucocorticoids are used clinically to relieve the symptoms of rheumatoid arthritis and other conditions characterized by inflammation. Cortisol prevents swelling by stabilizing lysosomes so that they do not release their digestive enzymes. They must be used with caution, however, because they depress normal immune mechanisms, and may have undesirable delayed side effects.

## Adrenal glands and stress

The adrenal medulla and cortex work together to help the body deal with stress. We have seen how the medulla functions to enable an individual to cope with a sudden emergency. It is thought that epinephrine secreted by the medulla also stimulates the secretion of ACTH by the pituitary. In this roundabout way the medulla stimulates the adrenal cortex. The cortex then provides a stabilizing backup system. For example, the medulla promotes the rapid oxidation of glucose so that the muscles and brain can act efficiently. The cortex promotes the conversion of amino acids to glucose so that the fuel supply will not be depleted. The adrenal cortex is thought to function also in adapting the body to nonspecific stress. An example of nonspecific stress would be general anxiety and tension. The body's response to such stress is referred to as the *general adaptation syndrome*.

When the adrenal glands are surgically removed from experimental animals, their ability to cope with stress is greatly reduced. This ability can be temporarily restored by injecting them with adrenal cortical hormones.

There is some evidence that a certain amount of stress in infancy is needed in order for an organism to develop appropriate adaptive behavior later in life. Recent studies also have revealed that in adult life some stress is necessary for effective learning and behavioral response. Just what the optimum level is, has not been determined.

Physiologist Hans Selye has suggested that continual stress is harmful because of the deleterious side effects of the adrenal cortical hormones. While cortisol is helpful in reducing inflammation, for example, it also interferes with normal immune processes so that infection can spread. By injecting experimental animals with large amounts of adrenal cortical hormones researchers have induced disease states. Among them are arthritis, high blood pressure, and ulcers.

One stressor which has been the object of much current debate is noise. Even preschoolers know—from watching Sesame Street—that noise is unhealthy. Yet environmental noise is becoming louder every year. Though some of us may voluntarily subject ourselves to the deafening sounds of a rock band or blaring stereo, most noise assaults us without our personal consent. The noises of motor cycles and trucks, construction machinery, and jet aircraft are among those that bombard us daily.

Sounds as low as 70 decibels affect us physiologically. Whether or not we happen to be enjoying the noise, adrenal hormones are secreted, muscles tense, skin pales, blood vessels constrict, and an array of biochemical changes take place.

The stress which results from noise is not alleviated by ordinary tranquilizers. According to recent legislation a worker cannot be exposed to noise levels of 90 decibels or more for longer than 8 hours. Longer exposure may permanently impair hearing. Even brief exposure to sounds of 140 decibels or higher can rupture eardrums and cause permanent loss of hearing. Even when persons are able to adapt to stress such as loud noise, adverse effects become evident after the subject is removed from the stress situation. Recent laboratory experiments on human beings suggest that this delayed reaction is directly related to exposure to the stressor.

Human beings are not the only organisms affected by noise. Mice go into

**Table 11-2. Noise intensity**

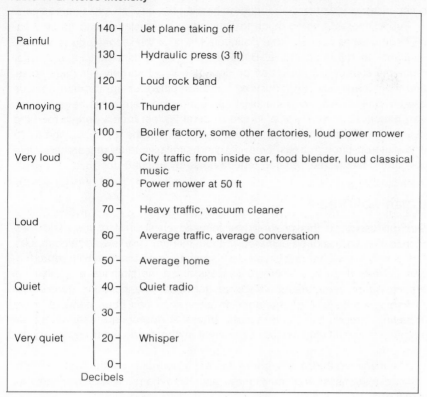

| | Decibels | |
|---|---|---|
| Painful | 140 | Jet plane taking off |
| | 130 | Hydraulic press (3 ft) |
| | 120 | Loud rock band |
| Annoying | 110 | Thunder |
| | 100 | Boiler factory, some other factories, loud power mower |
| Very loud | 90 | City traffic from inside car, food blender, loud classical music |
| | 80 | Power mower at 50 ft |
| | 70 | Heavy traffic, vacuum cleaner |
| Loud | 60 | Average traffic, average conversation |
| | 50 | Average home |
| Quiet | 40 | Quiet radio |
| | 30 | |
| Very quiet | 20 | Whisper |
| | 0 | |

convulsions when subjected to loud sounds. In recent laboratory experiments 90 percent of mice tested suffered convulsions in response to the noise of an electric drill. On the other hand noise may have a beneficial aspect. There is now evidence that some plant seeds sprout faster when subjected to loud sounds. Further research is needed before farmers begin to invest in loudspeakers!

Noise has been an important issue in the debates over the development of the SST (supersonic transport plane). When these planes take off they make twice as much noise as conventional jets, and in flight they generate thunderous sonic booms. (Sonic booms are pressure waves swept back by the jet's movement which travel toward the ground at the speed of sound.) Sonic booms startle children and adults alike and may cause fetal damage. They also cause damage to buildings and have caused landslides. By 1971 the Air Force had paid about $1 million in sonic boom damage. If overland flights of SSTs are permitted we will all be subjected to the startling, stress-engendering booms. Future technology may reduce this noise problem, but at present it remains a source of serious concern.

Objectionable noise in urban areas doubles annually even without the

SST, and millions of dollars are now awarded annually by industries as compensation to workers who suffered hearing losses. Can anything be done?

Public interest can do much to stimulate the legislation and its enforcement needed to control noise pollution. Much of the technology is already available to decrease noise. Buildings can be constructed to absorb or block out sounds. Vehicles can be designed which largely eliminate noise, and machinery used by industries can be muffled, or appropriate devices used to shield workers. Recently it has been discovered that belts of trees very effectively absorb noise. In one study at 25 feet from a certain road the decibel level was recorded at 82 without trees, but the level dropped to 73 when filtered through trees. From this and similar studies researchers have concluded that a tree belt can reduce noise by about 8 decibels.

## OTHER HORMONES

Our discussion in this chapter has focused on several of the major hormones, but would not be complete without mention of a few additional ones. Sex hormones will be discussed in Chapter 12 in connection with reproduction. Several hormones, secreted by cells lining the stomach and intestine, are important in regulating digestive activities. The hormone *gastrin,* for example, is released by the stomach when food has been delivered to it. Traveling through the blood, gastrin arrives at appropriate cells lining the stomach and stimulates them to secrete hydrochloric acid necessary for digestion.

The thymus gland is thought to secrete a hormone which plays a role in immune responses, and the kidneys secrete a hormone which influences blood pressure.

Some researchers regard prostaglandins as a group of local hormones, since they appear to be produced by many different tissues in the body. Besides their probable role in regulating actions of other hormones, prostaglandins directly influence a wide variety of metabolic activities. They relax the smooth muscle of the bronchial tubes, and apparently help regulate blood pressure. Because of their stimulating effect on uterine contraction, they are being used experimentally to induce labor, as well as abortions. Prostaglandins synthesized in the temperature-regulating center of the hypothalamus cause fever, and it is now thought that the mode of action of aspirin, long a mystery, involves a specific antagonism of this prostaglandin activity. Prostaglandins are being studied for possible use in the treatment of such diverse conditions as asthma, ulcers, high blood pressure, inflammation, shock, and stroke.

In this chapter we have considered how hormones work together to regulate blood-sugar level, salt and water balance, and other aspects of metabolism. Their action plays a major role in the ability of the body to maintain a steady state. We are just beginning to understand the endocrine system and its complex control mechanisms. No doubt new hormones and new mechanisms remain to be discovered. In many ways endocrinology is the new frontier of biology and medicine.

**1** Hormones are chemical messengers which regulate metabolic activities in the body. Their secretion is regulated by feedback control systems.

**2** Many hormones affect the target tissue by activating a second messenger such as cyclic AMP.

**3** Oxytocin and ADH are produced by the hypothalamus and are stored and secreted by the posterior pituitary gland.

**4** The anterior pituitary gland secretes at least seven distinct hormones, including growth hormone, lactogenic hormone, and several tropic hormones.

**5** Thyroxine, secreted by the thyroid gland, regulates the rate of metabolism, and is necessary for cellular differentiation.

**6** Parathyroid hormone regulates the amount of calcium and phosphorus in the blood.

**7** Insulin and glucagon work antagonistically to maintain the appropriate level of glucose in the blood. Glucose level is also affected by epinephrine and cortisol.

**8** The principal disorder of pancreatic endocrine function is diabetes mellitus, a condition characterized by abnormally high levels of blood glucose.

**9** The adrenal medulla prepares the body physiologically to deal with emergency situations.

**10** Hormones of the adrenal cortex regulate salt and water balance and help to maintain blood-sugar level. The adrenal cortex is thought to help the body to adapt to nonspecific stress.

**11** Noise is becoming increasingly important as a stressor, and should be limited.

**12** Prostaglandins, a group of local hormones produced by many types of cells, exert a variety of influences upon metabolic activities.

**FOR FURTHER READING**

Clegg, P. Catherine, and Arthur G. Clegg: *Hormones, Cells, and Organisms,* Stanford University Press, Stanford, Calif. 1969. After the student has mastered the basic facts of endocrinology he will enjoy reading this book, which unifies the facts and applies them to a study of various metabolic processes.

Gardner, Lytt I.: "Deprivation Dwarfism," *Scientific American*, July 1972, p. 76. Discusses how environmental factors may influence growth-hormone secretion.

Guillemin, Roger, and Roger Burgus: "The Hormones of the Hypothalamus," *Scientific American,* November 1972, p. 24. Summarizes current knowledge about hypothalamus-pituitary relationships.

Sawin, Clark T.: *The Hormones; Endocrine Physiology*, Little, Brown, Boston, 1969. An interesting basic account of human endocrinology. Easy to read and highly recommended for students who wish to delve more deeply into the subject of hormones.

# 12

# HUMAN REPRODUCTION

**Chapter learning objective.** Using appropriate terminology the student must be able to discuss human reproduction, including gamete formation and meiosis, hormonal control, conception, contraception, sex determination, population control, and venereal disease.

*ILO 1* Compare asexual with sexual reproduction, giving three specific examples of asexual reproduction.

*ILO 2* List the biologic advantages of sexual reproduction.

*ILO 3* Describe the process of meiosis, describe its function, and compare and contrast its main features with those of mitosis.

*ILO 4* Trace the development of sperm cells and their passage through the male reproductive system, describing (or labeling on a diagram) each male structure and giving its function; describe the role played by male hormones.

*ILO 5* Describe and give the function of, or label on a diagram, each structure of the female reproductive system.

*ILO 6* Trace the development of the ovum in time and space (i.e., through each anatomic structure) until it is either fertilized or deteriorates.

*ILO 7* Describe the hormonal control of the menstrual cycle, and identify the timing of important events of the cycle such as ovulation and menstruation.

*ILO 8* Describe the process of fertilization, and list its functions; identify the time of the menstrual cycle at which sexual intercourse is most likely to result in pregnancy.

*ILO 9* Describe how the sex of the offspring is determined, explaining its chromosomal basis, and give the biologic basis for sex planning.

*ILO 10* Describe the mode of action and give the advantages and disadvantages of each of the methods of contraception discussed: the pill, IUD, spermicides, condom, diaphragm, rhythm method, douche, withdrawal method, and sterilization.

*ILO 11* Discuss abortion, differentiating among the three types of abortion, and give several arguments for and against liberalized abortion legislation.

*ILO 12* Distinguish between birth control and population control, and describe five different procedures which would serve to reduce population.

*ILO 13* Define venereal disease, and describe the symptoms, effects, and treatment for gonorrhea and syphilis.

*ILO 14* Define all scientific terms introduced in this chapter.

Perpetuation of a species depends upon the ability of its members to produce offspring. Most very simple organisms reproduce asexually (without sex), but among many of them sexual reproduction is an often employed alternative. All complex plants and animals reproduce sexually.

## ASEXUAL REPRODUCTION

Three common forms of asexual reproduction are *binary fission, budding,* and *spore formation* (Figure 12-1). Bacteria and unicellular forms such as the ameba can reproduce by binary fission: An individual simply undergoes mitosis, dividing to form two smaller, but identical, copies of itself. Each new cell ingests food, metabolizes, and grows until it attains adult size, then repeats the process.

Sponges, hydras, and yeasts reproduce by budding, in which each new individual grows as an extension of the parent's body. It may remain attached, becoming part of a colony of closely associated but independent organisms, or may separate completely from the parent.

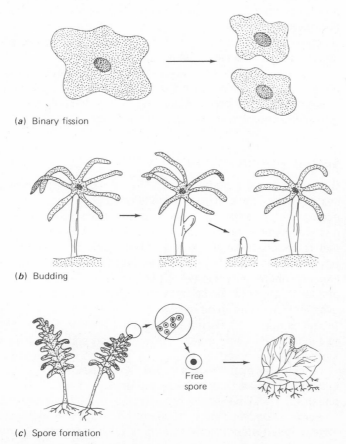

(*a*) Binary fission

(*b*) Budding

Free
spore

(*c*) Spore formation

**FIGURE 12-1**
Methods of asexual reproduction. (*a*) Most unicellular organisms can reproduce asexually by binary fission. Chromosomes are duplicated and distributed by mitosis, and the mother cell divides to form two daughter cells. (*b*) Hydra and many other organisms reproduce asexually by budding. A part of the parent body separates and develops into a new individual. The portion of the parent body which buds is not specialized exclusively for performing a reproductive function. (*c*) Spore formation is a method of asexual reproduction in plants. A spore is a cell specialized as a reproductive unit. It develops directly into an adult.

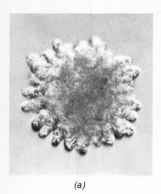

(a)

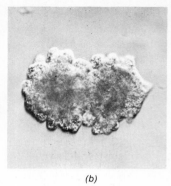

(b)

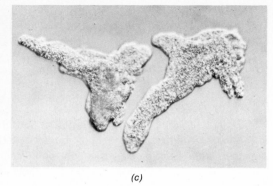

(c)

**FIGURE 12-2**
Actual photographs of binary fission in an ameba. (Carolina Biological Supply Company) (*a*) Ameba begins the process of binary fission. Though not visible in these photographs mitosis is taking place. (*b*) After about 8 minutes the cell begins to divide. (*c*) After about 21 minutes the cell has completely divided and the two new daughter cells have separated. Each is now capable of leading an independent existence.

Many plants produce offspring at some stage of their life cycle by forming *spores.* A spore is a reproductive cell that can develop directly into a new individual without going through the process of gamete formation or fertilization. Spores are often equipped with protective coatings which can resist the adverse effects of drying and are ideally suited for dispersal by wind or water. They seem to compensate for the plant's inability to travel about. Still another method for reproducing and dispersing plants is accomplished artificially by plant *cuttings.* A part of a plant, usually a stem, is cut from the parent plant and rooted in soil. This method propagates plants asexually and is useful to the farmer as well as the home gardener.

In asexual reproduction the parent endows the offspring with a set of genes exactly identical to his own so that the new individual becomes a carbon copy of the parent. Mutation provides the only means of varying the genetic composition of members of the species.

## SEXUAL REPRODUCTION

Sexual reproduction requires the fusion of two different types of *gametes,* or sex cells, to form a *zygote,* or fertilized egg. The genetic composition of each gamete is different, so the zygote contains genetic information which is a unique combination of the genes contributed by each. Usually, in species that reproduce sexually there are two types of individuals—males which produce actively motile sperm cells, and females which produce egg cells. However, there are some types of animals (certain parasitic worms, for example) in which the same individual produces both types of gametes and even fertilizes itself. In other species, such as the earthworm, each organism produces eggs and sperm, but does not fertilize its own gametes. Instead two individuals come together in copulation and exchange sperm cells, a kind of double reproductive act.

All complex plants have an alternation of sexual and asexual generations in their life cycles. During sexual reproduction the male gamete produces a pollen tube that serves to bring it into contact with the female gamete.

*Fertilization,* the fusion of sperm and egg, may take place inside the female body (*internal fertilization*), or outside (*external fertilization*). Most aquatic animals, as well as some insects, practice external fertilization. Mating partners simultaneously release eggs and sperm into the water. Dispersion frequently results in a loss of many gametes, but so many are released that a sufficient number of sperm and egg cells meet and unite to ensure propagation of the species.

In internal fertilization matters are left less to chance. The male delivers sperm cells directly into the body of the female. Her moist tissues provide the watery medium required by the sperm cells. Most terrestrial animals, as well as a few fish and other aquatic animals, practice the internal method of fertilization.

## Advantages of sexual reproduction

Sex has certain biologic advantages besides those which might immediately come to mind. Species that reproduce sexually exhibit a great deal of *variety* among their individual members, and variety has been described as nature's grand tactic of survival. In sexual reproduction the offspring is the product of the genes contributed by both parents, rather than a genetic copy of a single parent. What is more, the genes are thoroughly shuffled during the production of sex cells so that no two offspring even of the same parents are likely to be very similar.

Sexual reproduction provides for the rapid spread of advantageous genes through a population. Because there is variety among the individual members of a population, some members may by chance possess more favorable genes than their fellows. "Favorable" genes are those which enhance an individual's ability to survive and therefore reproduce in his environment. The term "favorable" is relative, since environmental conditions change. For example, the heavy coats of polar bears are an adaptation to the cold climate in which they live. If for some reason the climate should change and become warmer, genes for heavy coats would no longer be advantageous. Instead those bears which by chance possess genes for skimpy fur coats would have an advantage. Sexual reproduction would enable these

**FIGURE 12-3**
Sexual reproduction. (*a*) In frogs and other amphibians, fertilization is external. The male clasps the female, and as she discharges her eggs into the water, he releases sperm over them. (Carolina Biological Supply Company) (*b*) In grasshoppers, fertilization is internal.

(a)

(b)

new favorable genes to spread through the bear population at a far more rapid rate than would asexual reproduction.

In sexual reproduction only the tiny gamete of the parent lives on in the new generation. The rest of the parental body ages and eventually dies. On the other hand each individual enjoys the advantages of specialization, not possible in asexual forms. Only a small part of the organism is occupied with the reproductive function, while other cells are free to specialize in other tasks, such as thinking and locomotion.

## Meiosis

Meiosis is the special form of cell division by which gametes are produced. In Chapter 3 we examined mitosis, the ordinary form of cell multiplication in which the chromosomes of the mother cell are duplicated and a complete set is distributed to each new cell (see Chapter 3 for review). In meiosis the daughter cells receive only *half* the number of chromosomes present in the mother cell. These two processes are contrasted in Figure 12-4.

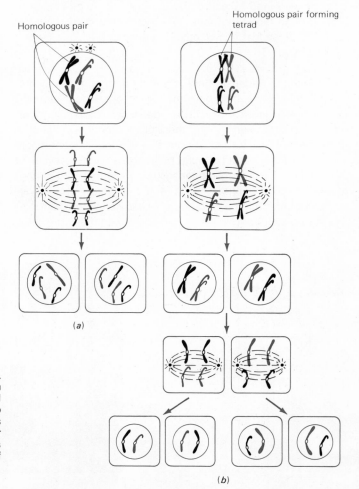

**FIGURE 12-4**
Meiosis as compared to mitosis. (a) Mitosis. Note that each daughter cell has an identical set of four chromosomes (two pairs). (b) Meiosis. Two divisions take place, so that four daughter cells are produced. Each has only two chromosomes (one of each pair).

Each species has a characteristic chromosome number, known as the *diploid* number. In man the diploid chromosome number is 46; accordingly a complete set of 46 chromosomes is present in every cell of the human body except the gametes. Gametes cannot have 46 chromosomes because when sperm and egg fuse in fertilization they become one cell, the zygote. If each gamete contributed 46 chromosomes, the zygote would have 92, and each succeeding generation of offspring would be endowed with double the chromosome complement of its parents. The results, were all this possible, would be a genetic nightmare. Meiosis neatly solves the problem by producing gametes which bear only half the normal chromosome number, the *haploid* number. When two haploid gametes unite in fertilization the diploid number is restored. Then, by mitotic divisions the zygote and its progeny distribute identical sets of diploid chromosomes to each of the cells which develop into the adult organism. In this way the chromosome number of the species remains constant.

The 46 human chromosomes in each cell actually consist of 23 pairs. One set of 23 is inherited from the father (the *paternal* chromosomes), while the other set is contributed by the mother (the *maternal* chromosomes). Each pair of chromosomes is identical in size and shape and bears a set of genes governing *the same traits*. Such a pair of chromosomes is referred to as a *homologous pair*. Let us consider eye color as a specific example. On one of your maternal chromosomes you have a gene with instructions for eye color which you inherited from your mother. In a corresponding location on the homologous paternal chromosome there is a gene with instructions for eye color, also. Both genes may specify blue eyes, both may specify brown, or one may specify blue and the other brown. Together they constitute your genetic endowment with respect to eye color.[1]

Meiosis differs from mitosis in several important ways. In meiosis the mother cell gives rise to four cells, each of which has only a haploid set of chromosomes. During the prophase of the first division the homologous pairs of chromosomes are attracted to each other. Each pair comes together in a process called *synapsis*. Since each chromosome has already duplicated itself to form two chromatids, a synapsed pair actually consists of four chromatids and is referred to as a *tetrad*. Synapsis is important because it provides the opportunity for homologous pairs to exchange genes, a process referred to as *crossing-over*. By this mechanism genetic information can be exchanged between homologous chromosomes, and the prospects for variety among the offspring of sexual partners are greatly enhanced.

During the first meiotic division the homologous pairs separate and *one of each pair* moves to each new cell. The chromatids do not separate from each other until the second division. Each of the first two cells produced contains only half the total genetic information that was present in the original (mother) cell. For example a maternal chromosome with a gene for blue eye color would be in one cell, while the corresponding paternal chromo-

---

[1] Actually inheritance of eye color is more complicated and involves several gene pairs, but the example given illustrates the point.

some—perhaps bearing a gene for brown eye color—would be in the other cell.

Distribution of the homologous chromosomes into the two cells is random. Though it is *possible* that one gamete could receive a complete set of 23 maternal chromosomes and the other gamete a complete set of 23 paternal chromosomes, the chromosomes are not likely to be distributed in this manner. It is much more *probable* that each gamete will receive some chromosomes donated from each parent.[2] In this way each gamete becomes a genetic combination of both parents (and of all four grandparents). The possible combinations for shuffling the 23 pairs of differing chromosomes is so great that it is unlikely that two gametes would ever be identical. Crossing-over further ensures this unlikelihood.

No further chromosome duplication occurs between the two meiotic divisions. The second division simply permits the chromatids to separate. Each cell divides into two new cells, so that a total of four gametes are produced, each containing 23 chromosomes.

Meiosis serves two important functions: (1) it ensures variety through the random distribution of homologous chromosomes and by crossing-over; (2) it provides for chromosome constancy among the members of a species.

## THE MALE

In reproduction the male produces sperm cells and delivers them into the female reproductive tract. The sperm cell which combines with the egg contributes half the genetic endowment of the offspring and determines its sex.

### Male reproductive system

The male reproductive structures are illustrated in Figure 12-5. They include the *testes,* the *conducting tubes* which lead from the testes to the penis, *accessory glands* that produce semen, and the *penis.*

*Testes.* The paired testes are the male *gonads* (organs that produce gametes). They produce sperm cells and also the principal male sex hormone, *testosterone.* Each testis is an oval organ about 1.5 inches in length. Packed into each testis is a system of tiny coiled tubules (seminiferous tubules), which if uncoiled and placed end to end would span a length of almost half a mile. Within these tubules sperm cells are produced. Clumps of endocrine cells (*interstitial cells*) located between the tubules secrete testosterone.

In the embryo the testes develop in the abdominal cavity. About 2 months before birth they descend into the *scrotum,* a bag of skin suspended between the thighs of the male. Sperm cells cannot survive at normal body

---

[2] It should be understood that although chromosomes contributed by the mother are called maternal chromosomes, they were not always so. Half the *mother's* chromosomes were contributed by *her* father and half by *her* mother.

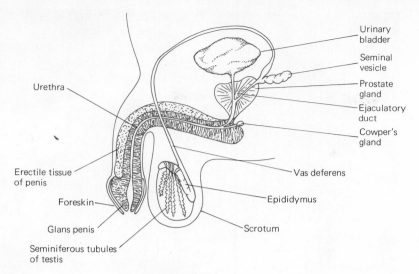

Urinary bladder

Seminal vesicle

Prostate gland

Ejaculatory duct

Cowper's gland

Urethra

Vas deferens

Erectile tissue of penis

Epididymis

Foreskin

Glans penis

Scrotum

Seminiferous tubules of testis

**FIGURE 12-5**
Reproductive system of the human male.

temperature and the scrotum serves to maintain the testes at an optimal temperature of about 3 degrees below body temperature. Involuntary muscle fibers in the wall of the scrotum relax in hot weather, allowing the testes to be positioned away from the body. An abundant supply of sweat glands also contributes to the constant cooler temperature. In cold weather the muscles in the scrotum contract by an involuntary reflex, and draw the testes up close to the abdominal wall, where they are kept warm.

Occasionally the testes do not descend into the scrotum by the time of birth, but remain in the abdominal cavity. Viable sperm are not produced by abdominal testes, and the male is sterile, that is, unable to father offspring. A person with this condition is fully male, because masculinity depends on the presence of male sex hormone, the secretion of which is not affected by the high temperature of the abdominal cavity. Fortunately, this condition is easily corrected by surgery in childhood so that future fertility is restored.

The scrotal cavity is connected with the abdominal cavity by the remnant of the *inguinal canal,* through which the testes descended into the scrotum. This constitutes a weak place in the abdominal wall and is a common site of hernia in the male. An *inguinal hernia* is the bulging of a loop of intestine through a tear in the inguinal canal. The tear is often caused by straining the abdominal muscles by lifting a heavy object in an incorrect manner.

*Conducting tubes.* Sperm cells leave the testis through a series of ducts emptying into a larger tube, the *epididymis.* Each epididymis is a tortuous tubule positioned on the posterior side of each testis. Within the epididymis the sperm cells complete their maturation. The epididymis continues as a straight tube called the *vas deferens,* which leaves the scrotum through the inguinal canal and passes into the abdominal cavity. (Bear in mind that each structure mentioned above is actually paired; that is, there are two testes, two epididymides, and two vasa deferentia.)

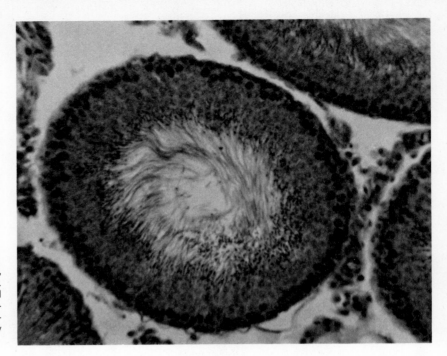

**FIGURE 12-6**

Section through tubules in testis, showing developing sperm. Least mature cells are found around periphery of tubule. Mature cells are around lumen (cavity of tubule), and their tails may be seen projecting into lumen.

On their journey out of the body sperm travel through the vas deferens into the paired *ejaculatory ducts,* then through the single urethra as the latter leaves the urinary bladder. The urethra is about 8 inches long and extends through the penis, opening to the outside. Thus, sperm must pass through an epididymis, vas deferens, ejaculatory duct, and urethra on their journey from the tubules in the testes.

***Accessory glands.*** Three different accessory glands, the *seminal vesicles,* the *prostate,* and *Cowper's glands,* secrete fluid into the conducting tubes. The mixture of sperm cells suspended in these secretions constitutes *semen,* the whitish fluid ejaculated during orgasm.

The paired seminal vesicles empty into the ejaculatory ducts. Their alkaline secretion neutralizes the acidity of the semen coming from the testes, activating the sperm and permitting them to swim. Seminal fluid also contains sugar, mostly fructose, which the sperm cells may utilize as an energy source.

The unpaired prostate gland contributes another alkaline fluid to the semen. It empties its contents into the urethra. In older men the prostate sometimes becomes inflamed and infected and may block the normal passage of urine through the urethra. Cancer of the prostate is a common affliction of this gland, but fortunately such cancer is usually slow growing and tends to be confined within the gland. Just prior to orgasm Cowper's glands secrete a drop or two of an alkaline mucous secretion into the urethra at the base of the penis.

*Penis.* The copulatory organ, or penis, is usually flaccid, or limp. It consists primarily of three columns of erectile tissue. One of these surrounds the portion of the urethra that courses through the penis. When the male is sexually stimulated, nerve impulses cause the arteries that supply the spongy tissue to dilate and the cavities of the spongy erectile tissue fill with blood. At the same time, veins in the penis are compressed, so that for a while not as much blood can leave as enters. The penis becomes stiff, enlarged, and erect.

The *glans penis,* which forms a kind of cap over the end of the penis, contains a rich supply of nerve endings. The loose fold of skin, the foreskin, which covers the glans penis is often removed at birth by a surgical procedure referred to as *circumcision.* Removal of the foreskin prevents accumulation of secretion, called smegma, under it. If not regularly washed from the penis in uncircumcised men, smegma provides potential medium for infection.

## Sperm cells

Millions of sperm cells are produced constantly in the tubules of the testes from undifferentiated cells in the walls of the tubules. These divide by mitosis to produce a large population of pre-sperm cells. After puberty some of the pre-sperm cells begin to enlarge and then undergo meiosis to form im-

FIGURE 12-7
Human sperm cell. (a) *AC,* acrosome (part that makes contact with ovum); *N,* nucleus; *MP,* middle piece; *F,* flagellum. (b) *M,* mitochondrion. (Dr. Lyle C. Dearden)

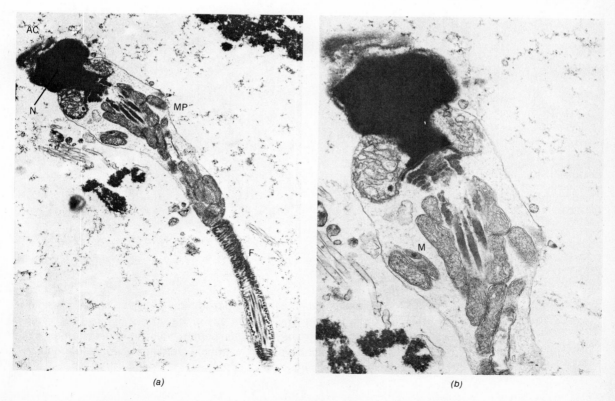

*(a)*

*(b)*

mature sperm cells, which subsequently differentiate to become mature sperm cells.

The mature sperm cell, one of the smallest cells in the body, has very little cytoplasm. The head of the sperm consists mainly of the nucleus. At the tip there is a mass of Golgi bodies, which are thought to help the sperm penetrate the membrane around the egg. Mitochondria are packed into a middle section, where they carry on aerobic metabolism to provide energy needed for locomotion. The sperm propels itself by means of its long "tail," or flagellum.

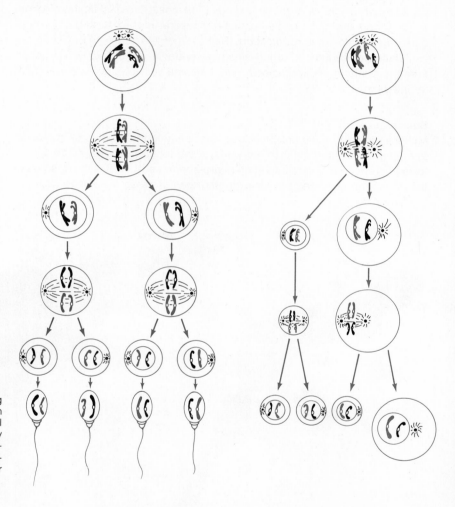

**FIGURE 12-8**
Comparison of sperm and egg production. Though meiosis in the male results in four functional sperm cells, in the female only one functional egg cell is produced; the other three cells produced are nonfunctional polar bodies.

A high concentration of carbon dioxide in the testes serves to maintain a low pH. In this acidic environment sperm are unable to move their flagella and are essentially paralyzed. This permits them to conserve their stored food so that they will have sufficient energy to swim when the opportunity to fertilize an egg presents itself.

## Male hormones

With the onset of puberty, at about age 12, the male reproductive system begins to mature. At that time some change takes place in the hypothalamus which stimulates the anterior pituitary gland to secrete larger quantities of gonadotropic hormones. One of these, *FSH* (follicle-stimulating hormone), stimulates the development of the tubules in the testes and promotes the process of sperm production. The other gonadotropic hormone. *LH* (luteinizing hormone), regulates the activity of the endocrine cells in the testes that secrete the male hormone, *testosterone.*

As large quantities of testosterone begin to circulate through the body, changes in the secondary sexual characteristics associated with puberty take place. The beard and pubic hair begin to grow, the voice begins to deepen, muscles and fat distribution become characteristically masculine, and the sex organs begin to enlarge.

The adrenal cortex also secretes male hormone which is capable of producing a certain amount of masculinity. A male who has suffered *castration* (removal of the testes) before puberty does not develop secondary masculine characteristics. However, if castration occurs after puberty, secretions from the adrenal cortex are able to maintain a certain amount of the individual's masculinity. Males normally possess some female sex hormone, while females normally have a small amount of male hormone secreted by the adrenal gland. Sexuality is determined not so much by whether male or female hormones are present or absent as by the balance that exists between them.

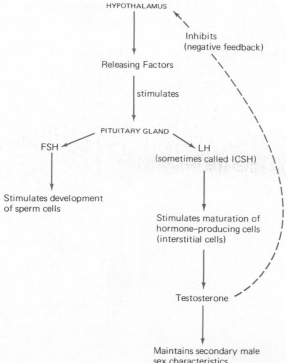

**FIGURE 12-9**
Male reproductive hormones.

## THE FEMALE

The role of the female in reproduction is more complex than that of the male. Not only does she produce egg cells and receive the sperm, but she houses and nourishes the resulting embryo during the 9 months of life before birth.

### Female reproductive system

The female reproductive system includes two *ovaries,* two *oviducts,* a *uterus,* a *vagina, external genital structures,* and the *breasts* containing the *mammary glands.*

*Ovaries.* The ovaries are the female gonads. Like the testes they serve the dual function of gamete production and sex hormone secretion. Each ovary is an almond-shaped body about 1.5 inches in length. Its internal structure consists primarily of connective tissue (stroma), through which are scattered eggs, or ova, in various stages of maturation.

*Oviducts.* When an egg is released from one of the ovaries, it is ejected into the abdominal cavity. This process is known as *ovulation.* Although the oviducts (also called fallopian tubes) are not actually attached to the ovaries, they are in a strategic position to receive the eggs. The funnel-shaped opening of each oviduct is lined with cilia which create a current that helps propel the egg into the oviduct. The muscular walls of the oviduct contract, moving the ovum toward the uterus. Ciliated cells in the folded lining of the oviduct probably assist in moving the egg along its 4.5-inch length. Fertilization takes place in the upper portion of the oviduct.

*Uterus.* Each oviduct joins the hollow, muscular uterus, or womb. Normally about the size of a fist, the uterus is a pear-shaped organ that occupies a central position in the pelvic cavity. The thick mucous lining (*endometrium*)

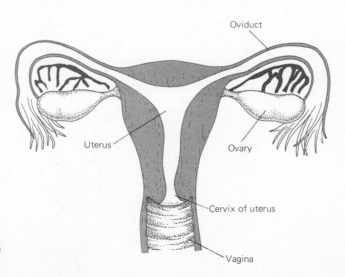

**FIGURE 12-10**
Reproductive system of human female.

of the uterus becomes thicker during the course of each month in preparation for a possible pregnancy. If an egg is fertilized, the tiny embryo finds its way into the uterus and implants itself in the thick lining. Here it grows and develops, sustained by the nutrients and oxygen delivered by surrounding maternal blood vessels. It is one of nature's wonders that the uterus, ordinarily only about 3 inches in length, can expand to accommodate a fully developed fetus. At the time of birth, contractions of the thick muscular walls of the uterus help to expel the baby. If fertilization does not occur during the monthly cycle, the mucous lining sloughs off and is discharged in the process of *menstruation.*

The lower portion of the uterus, called the *cervix,* projects slightly into the vagina. The canal within the cervix is continuous with the vaginal canal.

*Vagina.* The vagina is an elastic, muscular tube, about 3 inches long, which extends from the cervix to the external genital structures. During sexual intercourse the vagina receives the penis, and when a baby is born the vagina functions as the birth canal. During menstruation the discarded uterine lining is discharged through the vagina.

*External genital structures.* The external female genital structures are referred to collectively as the *vulva.* Two pairs of liplike folds surround the opening into the vagina. The delicate inner folds of the skin are the *labia minora.* External to these are the heavier, hairy, *labia majora.* In the embryo the labia majora develop from the same structures that form the scrotum in the male.

Just above the labia minora is the *clitoris,* a very small erectile structure comparable to the male glans penis. Crowded with nerve endings, the clitoris serves as a center of sexual sensation in the female. During sexual excitement it becomes enlarged and stiff.

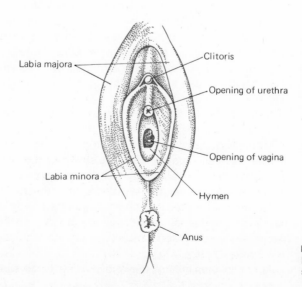

**FIGURE 12-11**
External female reproductive structures.

The *hymen,* a thin ring of tissue that may partially block the entrance to the vagina, has through the ages been considered the symbol of virginity. All the attention which has been focused upon the hymen is not merited, since its presence or absence is not a reliable indicator of the previous occurrence or nonoccurrence of sexual relations. Some women are born without a hymen. In others it is destroyed by strenuous physical exercise during childhood, or by use of tampons inserted to absorb the menstrual flow. In a few women the hymen persists after years of sexual relations or even after childbirth.

Just anterior to the vaginal opening is the opening of the urethra. In the female, the urinary system is entirely separated from the reproductive system.

**Breasts.** The breasts (containing the mammary glands) are composed of fatty and glandular tissue. Their primary purpose is to provide nourishment for the young. After childbirth, milk is produced and secreted by the many separate mammary glands and is conducted to the nipples by ducts. Muscle tissue in the nipples contracts to release the milk from the ducts.

### Production of ova

At the time of birth the ovaries of a female infant contain an estimated 400,000 immature egg cells, or pre-ova. Each is surrounded by a ring of epithelial cells. Together the pre-ovum and its surrounding epithelial cells constitute a *follicle.* During fetal development some of the pre-ova develop into larger cells (*primary oocytes*) and begin the prophase of the first meiotic division, but further development is arrested until after puberty.

At puberty the menstrual cycle begins; generally only one ovum is released each month throughout a woman's reproductive life. Thus, about 400 eggs are ovulated during the 30 or so active reproductive years. The remaining thousands of ova eventually degenerate.

At the beginning of each menstrual cycle a few follicles begin to develop. The follicle cells proliferate by mitosis so that the ovum is surrounded by a growing cluster of cells. A fluid secreted by the follicle cells forms a pool in the center of the follicle and contributes substantially to its large size. Surrounding connective tissue of the ovary becomes organized about the follicle, forming a thick membrane around it. The cells of this membrane, known as *theca cells,* are thought to be the main source of the female hormone estrogen.

As the follicle grows, the developing ovum completes its first meiotic division. The resulting two cells are strikingly disproportionate in size. While the large cell develops further to become the ovum, the tiny cell, called the *first polar body,* eventually disintegrates. The second meiotic division is not completed until after fertilization. In contrast to gamete production in the male, which yields four functional sperm cells, meiosis in the female gives rise to only one functional ovum and three apparently useless polar bodies (see Figure 12-8).

As the follicles mature, they move closer to the surface of the ovary. At the

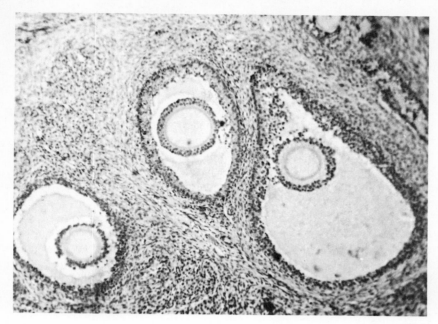

**FIGURE 12-12**
Section through a small part of
ovary to show follicles in various
stages of maturity. The largest is
the most mature.

time of ovulation the largest follicle ruptures, and the egg, surrounded by
only one layer of follicle cells, is ejected through the wall of the ovary into
the body cavity. That portion of the follicle which remains behind in the ovary
develops into an important endocrine structure, the *corpus luteum.*

## Female hormones and the menstrual cycle

When a female reaches puberty, the gonadotropic hormones FSH and LH
secreted by the anterior pituitary stimulate the ovaries to begin their reproductive function. Interaction of these hormones with the hormones secreted
by the ovary regulate the menstrual cycle, which runs its course about every
28 days from puberty until *menopause,* the end of a woman's reproductive
life. The purposes of the menstrual cycle are to stimulate the production of
an ovum each month and to prepare the uterus to receive an embryo.

The onset of menstruation marks day 1 of each menstrual cycle. For the
following 2 weeks follicles develop in the ovary. About 14 days prior to the
beginning of the next menstrual period the most mature follicle extrudes its
ovum from the ovary, the process of ovulation. The ovum finds its way into
the oviduct, where, if sperm are present, it may be fertilized. If it is fertilized,
development begins and the tiny embryo implants itself in the thick uterine
lining which has developed specially for this purpose. In the event that fertilization does not occur, the ovum begins to degenerate after about 36 hours.

The occurrence of ovulation is sometimes marked by mild abdominal discomfort, referred to as middle pain (*mittelschmerz*). Body temperature also
changes shortly after ovulation. The temperature first dips, then rises about
0.5 to 0.75 degree Fahrenheit above average body temperature and remains

elevated for several days. Women who practice the rhythm method of contraception, or those attempting to conceive, should keep a daily record of their temperature and plot a monthly temperature curve in order to determine when ovulation occurs. A newer method for pinpointing the time of ovulation is based on cyclical changes in the cervical mucus. At the time of ovulation the cervical mucus is rich in glucose, perhaps a source of nourishment for entering sperm cells. This monthly increase in the concentration of glucose can be measured by means of a color change on special tape available in most drugstores. Recent studies also indicate that the level of an enzyme called alkaline phosphatase rises in the saliva a few days before ovulation occurs. A tape sensitive to this enzyme has been developed, and a woman need only test her saliva each day to determine whether ovulation is imminent.

*Pituitary gonadotropic hormones.* During the first several days of the menstrual cycle FSH (follicle-stimulating hormone) is the dominant hormone secreted by the pituitary. It stimulates growth of the follicles within the ovary. LH (luteinizing hormone), acting in the presence of FSH, stimulates the

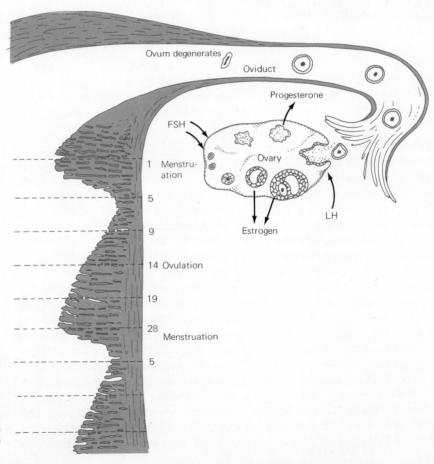

**FIGURE 12-13a**

(a) The menstrual cycle. This diagram illustrates changes that take place in the lining of the uterus and in the ovary during the menstrual cycle. When pregnancy does not occur, menstruation takes place and the cycle begins anew about every 28 days.

most developed follicle to ovulate. Then LH acts upon the remains of the follicle, stimulating its transformation into a corpus luteum.

**Ovarian hormones.** As the follicle matures, some of its cells secrete *estrogen,* the principal female sex hormone. Estrogen is responsible for secondary female sex characteristics: breast development, growth of pubic hair, characteristic female distribution of fat, and increase in size of sex organs. During the menstrual cycle estrogen stimulates the lining of the uterine wall to thicken. Under its influence both the blood vessels and glands of the uterine lining increase in number and in size. Estrogen probably also stimulates the pituitary gland to secrete LH, while inhibiting FSH secretion.

When the corpus luteum develops following ovulation, it secretes a second ovarian hormone, *progesterone,* which stimulates the uterine lining to complete its preparation for reception of an embryo. Progesterone also stimulates the glands of the uterine wall to secrete glycogen, which serves as nourishment for the early embryo.

If the ovum is fertilized, resulting in pregnancy, tissues of the embryo secrete a hormone, *chorionic gonadotropin, CG,* which signals the corpus

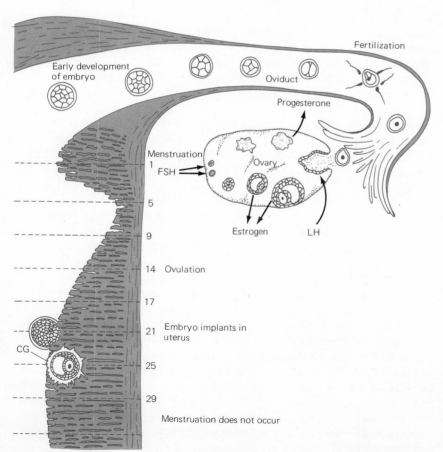

**FIGURE 12-13b**
(*b*) The menstrual cycle when pregnancy occurs. The corpus luteum does not degenerate, and menstruation does not take place. The wall of the uterus remains thickened, so that the embryo can develop within it.

**Table 12-1. The female sex hormones and their functions**

| Name of hormone | Secreted by | Function |
|---|---|---|
| FSH | Anterior pituitary | Stimulates development of follicles |
| Estrogen | Cells of developing follicles in ovary | Maintains secondary female sex characteristics; stimulates the uterine lining to thicken; stimulates growth of mammary gland ducts; stimulates pituitary gland to secrete LH |
| LH | Anterior pituitary | Stimulates ovulation; stimulates follicle to develop into corpus luteum |
| Progesterone | Corpus luteum | Stimulates growth of uterine and mammary glands; inhibits secretion of both FSH and LH |
| CG | Membranes of developing embryo | Maintains corpus luteum |

luteum to continue to function much in the fashion as the pituitary hormones. Continued secretion of progesterone maintains the uterus in the thickened condition essential to the maintenance of the embryo. If fertilization does not occur, the corpus luteum degenerates by the twenty-eighth day. As the concentration of circulating progesterone falls, the lining of the uterus begins to slough off. The monthly discharge of the unneeded thickened uterine lining is menstruation.

*Interaction of hormones.* The menstrual cycle is regulated by the interaction of releasing factors from the hypothalamus, along with pituitary and ovarian hormones. These hormones signal the organs involved by a system of feedback control similar to the thyroid-pituitary relationship discussed in Chapter 11.

The hypothalamus is keenly sensitive to the concentration of ovarian hormones in the blood. When the progesterone level falls following breakdown of the corpus luteum, the hypothalamus secretes an *FSH-releasing factor.* Reaching the anterior pituitary by way of blood vessels connecting the two organs, FSH-releasing factor stimulates the anterior pituitary to secrete FSH. This normally occurs on *day 1* of each menstrual cycle.

As the follicles develop they secrete estrogen, which affects the hypothalamus, causing it to release LH-releasing factor. The latter stimulates the pituitary to secrete LH, which in turn triggers ovulation and stimulates development of the corpus luteum. The production of FSH is inhibited first by the high concentration of estrogen in the blood near the time of ovulation, then by the progesterone secreted by the corpus luteum. Inhibition is thought to come about both by an effect upon the hypothalamus and by a direct effect on the pituitary. Exact details are not yet known, but it is clear that the system is self-regulating. Once an ovum has matured and been discharged from the ovary, no further follicles are needed during that cycle. Thus, there is no current need for FSH.

The amount of progesterone secreted by the corpus luteum rises continuously until a critical concentration of progesterone in the blood is reached. At that time the production of LH needed for maintenance of the corpus luteum is inhibited. In this way the corpus luteum seemingly arranges for its own demise. If pregnancy has not been initiated, the corpus luteum is not needed and deteriorates. The progesterone level falls, and a new cycle begins. If pregnancy has been established, the corpus luteum is maintained for about 3 months by the CG hormone secreted by membranes (the *placenta*) surrounding the developing embryo. After that time the placenta begins secreting steroid sex hormones.

The precise timing of the events of the menstrual cycle depends upon this complex series of chemical messages. Each month the female reproductive system prepares for pregnancy so that if the ovum should be fertilized the uterus will be ready to support its development.

*Menopause.* At about age 50 menopause begins, marking the end of a woman's reproductive years. The ovary begins to degenerate, and the amount of estrogen and progesterone secreted diminishes. In response to these changes the pituitary secretes larger quantities of FSH and LH, and the entire system of feedback control breaks down. Hot flashes sometimes occur, apparently because of the effect of decreased estrogen upon the temperature-regulating center in the hypothalamus. Estrogen deficiency also may contribute to headaches and feelings of depression experienced by some women. The first symptom of menopause is irregularity of menstrual periods, and eventually menstruation ceases altogether. After menopause begins, the ovaries become increasingly shrunken, the lining of the vagina becomes very thin, and the external genital structures and breasts begin to atrophy. Despite these physical changes, menopause does not affect a woman's interest in sex or her sexual performance.

Many of the symptoms of menopause can be alleviated by the judicious replacement of the missing hormones. Although medical opinion on the use of hormone therapy varies, many physicians feel that such therapy slows the aging process in skin, in the vagina, and perhaps also in the circulatory system.

## SEXUAL INTERCOURSE

Sexual intercourse, or *coitus,* is the sex act during which the erect penis releases semen into the vagina of the sexual partner. Biologically, coitus is the act which ensures internal fertilization; its function is reproduction. Man has succeeded in fulfilling this biologic purpose only too well. In our overpopulated world, reproduction has ceased to be the prime purpose of sexual intercourse. More often, sex is considered the ultimate means of expressing love and of deriving physical pleasure.

In order to engage in sexual intercourse the penis must be erect. Sexual foreplay provides the needed stimulation and promotes movement of the sperm into the vas deferens. Other tissues sensitive to sexual excitement—scrotum in the male, and vulva, vagina, and breasts in the female—become engorged with blood (tumescent). Muscle tension increases, and in the female the clitoris becomes erect.

Coitus actually commences when the penis is inserted into the vagina. The penis is moved back and forth in the vagina by movements called pelvic thrusts. Erectile tissue in the walls of the vagina itself narrows the passageway somewhat, increasing the friction between vagina and penis. Physical and psychic sensations resulting from this friction and from the entire intimate interaction between two individuals lead to *orgasm,* the climax of sexual excitement.

Orgasm is marked by a series of muscle contractions and by the release of sexual tension. In the male, contraction of the vas deferens propels sperm through the urethra, while the accessory glands contract, adding their secretions and culminating in the ejaculation of semen from the penis. In the female, the clitoris, vagina, and uterus contract. Although some fluid is secreted during sexual stimulation, orgasm in the female is not marked by fluid ejaculation. Women often require prolonged stimulation to achieve orgasm.

The semen ejaculated amounts to about a teaspoonful of whitish fluid, consisting of sperm cells dispersed in the secretions of the accessory glands. As many as 500 million sperm cells may be present in a single ejaculation. Sperm are released directly adjacent to the cervix of the uterus, an advantageous starting point for their journey up the female reproductive tract in search of an ovum.

A major reason for male sterility is an inadequate number of sperm cells in the semen. Sperm counts are undertaken in clinical laboratories and hospitals when a couple's attempts to produce a child are repeatedly unsuccessful. Sometimes large numbers of abnormal sperm cells are produced, and occasionally for unknown reasons sperm may not be produced at all. The fever of an infection or a localized infection of the testis (without fever) may temporarily reduce the sperm count. Sterility should not be confused with *impotence,* the inability to sustain an erection. Impotence is experienced from time to time by almost all men, but prolonged impotence generally is associated with psychologic problems, while sterility has a more obvious physiologic basis. Both, however, may result in failure to produce children.


## CONCEPTION

The initiation of a new life is termed conception. Life begins with fertilization, but its progress is ensured only if the body of the female is prepared to nurture it.

As soon as sperm are released into the vagina they begin their journey toward the ovum. During most of the menstrual cycle the rather highly acidic vagina presents a hostile environment to the sperm cells, and a thick plug of mucus blocks their way through the cervix. Near the time of ovulation, however, the vagina becomes slightly alkaline, and the mucous plug thins, allowing sperm cells to pass into the uterus. The thin film of fluid within the reproductive tract provides an aqueous medium through which sperm can propel themselves.

Although many sperm lose their way, large numbers succeed in traversing the uterus and reaching the oviducts. If an ovum is present within the oviduct, fertilization can take place. The ovum can be fertilized only for about 36 hours after it leaves the ovary. After that time it begins to degenerate and can no longer produce a new life. Sperm cells remain alive inside the female reproductive system for approximately 48 hours, but their ability to penetrate the ovum declines rapidly after 12 to 24 hours. Thus, there is only about a 3-day period during the middle of the menstrual cycle, perhaps days 13 to 16, when conception is likely to occur. The time of ovulation is the critical factor, but unfortunately this frequently is not a predictable event.

Fertilization normally occurs in the upper third of the oviduct. Only one sperm actually fertilizes an egg, but large numbers are apparently required to traverse the barrier created by the layer of follicle cells (corona radiata) which surround the ovum. These follicle cells seem to be connected to one another by a kind of protein cement. Each sperm is thought to secrete a small amount of enzyme that dissolves the cement, but a large number of sperm are required to provide sufficient enzyme to clear a path for sperm penetration.

As soon as one sperm penetrates an ovum, a change takes place in the surface of the ovum which prohibits the entrance of additional sperm cells. Sperm entry stimulates the ovum to complete the second meiotic division. Then, sperm and egg nuclei fuse, forming a single nucleus containing the chromosomes of both.

Fertilization serves several functions: (1) the diploid number of chromosomes is restored, as the sperm contributes a new set of chromosomes bearing a unique complement of genes; (2) the sex of the offspring is determined by the sperm cell; (3) fertilization initiates the developmental events leading to the birth of a new individual.

## SEX OF THE OFFSPRING

Throughout history kings have disposed of wives who could not produce male heirs, and lesser men have blamed their spouses for producing daughters when sons were needed to work in the fields. Then science came along and showed that in human beings it is actually the male who determines the sex of the offspring. More recently techniques have been proposed which may permit planning conception so as to favor producing a baby of a desired sex.

### Determination of sex

Among the 23 pairs of human chromosomes, one pair constitutes the sex chromosomes. They are the only chromosome pair not completely homologous. One, the X chromosome, is longer and consists of many more genes than the other, the Y chromosome. The Y chromosome contains the genes that determine maleness. Every normal human female has two X chromo-

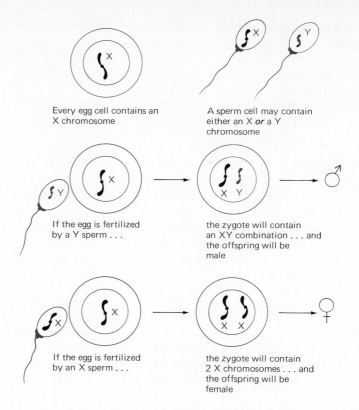

Every egg cell contains an
X chromosome

A sperm cell may contain
either an X *or* a Y
chromosome

If the egg is fertilized
by a Y sperm . . .

the zygote will contain
an XY combination . . . and
the offspring will be
male

If the egg is fertilized
by an X sperm . . .

the zygote will contain
2 X chromosomes . . . and
the offspring will be
female

**FIGURE** 12-14
Determination of sex.

somes in each of her cells, while every normal male has an X and a Y chromosome.

In sperm production, when homologous chromosomes separate during meiosis the X chromosome goes into one cell and the Y chromosome into the other. Thus, two distinct types of sperm cells are formed with respect to the sex chromosomes, X sperm cells and Y sperm cells. On the other hand, since the female has only X chromosomes, every egg cell contains an X chromosome.

When an X-bearing sperm fertilizes an ovum, the resulting zygote contains two X chromosomes and develops into a female. When a Y-bearing sperm fertilizes an egg, the zygote contains the XY combination and develops into a male. In this way the sperm determines the sex of the offspring.

The way in which the Y chromosome determines maleness is not completely clear. In the early embryo the developing sexual structures are identical. The genes of the Y chromosome apparently direct the development of the ambiguous gonads into testes. In response to male-determining genes on the Y chromosome, cells of the developing testes give off a brief burst of male sex hormone that permanently establishes masculinity as the direction of development. The hypothalamus itself becomes sexually differentiated by the male hormone and thereafter directs the masculine pattern of gonadotropic hormone secretion. In the absence of the Y chromosome no burst of male hormone is secreted in the embryo, and as a result the undifferentiated structures develop into feminine structures.

## Planning sex of offspring

An obstetrician-researcher, Dr. Landrum Shettles, recently suggested that a couple could plan the sex of their offspring with some success by timing intercourse in relation to ovulation. There is some very tentative evidence that X sperm cells are slightly larger than Y sperm cells, and that under favorable conditions of pH, Y sperm cells can move faster. Dr. Shettles has suggested that a couple wishing to produce a boy should have intercourse at the time of, or immediately after, ovulation. The alkaline nature of the cervical mucus at that time is conducive to the survival of the Y chromosome. Under these favorable conditions the Y sperm may win the race to the egg and produce a male child. Engaging in intercourse about 2 days before ovulation may shift the odds in favor of a girl. The acidic environment of the cervix and vagina at that time may ensure that all the Y sperm cells die before the egg presents itself for fertilization. One of the hardier X sperm cells may remain to fertilize the ovum. Just how effective this procedure may be has not yet been established.

Devices are under investigation that could be inserted prior to intercourse and would serve to differentially inactivate one type of sperm, thus allowing the other type to fertilize the egg. It is quite possible that in the future couples will routinely select the sex of their offspring. This ability to choose may help somewhat to alleviate the population crisis. In the past couples who might have stopped with two children, had they been lucky enough to have one of each sex, have tried over and over when chance did not play in their favor.

## CONTRACEPTION

Contraception means prevention of conception. Any method for separating sexual intercourse from production of offspring may be considered contraception. Married women of childbearing age who use no form of contraception have an average pregnancy rate of 70 percent a year. In underdeveloped countries only about 10 percent of the people make use of modern contraceptive methods. Most couples agree that it is best to have babies by choice, not by chance, but unfortunately the majority of couples who engage in sexual intercourse have only vague notions of how to prevent conception. In fact an estimated half of all pregnancies in the United States are accidental. A recent study indicates that among unwanted children antisocial behavior, crime, psychiatric illness, and public dependence are twice as common as among wanted children. And the steadily increasing number of abused and battered children is now estimated at 50,000 each year in the United States.

Recent Gallup polls have reflected the changing attitudes toward large families as awareness of the population crisis has grown. In 1945, 49 percent of Americans felt that four or more children was the ideal number for a family. In 1967 the number of Americans who felt that way had declined to 40 percent, and by 1973 it had declined further to 20 percent.

Many persons have become aware of the financial responsibility that comes with each new baby. Recently the Institute of Life Insurance estimated that raising one child to age 18 costs a bare minimum of $30,000. Their estimate includes just basic necessities and of course does not include the expense of a college education—an item which can easily cost an additional $20,000 over a 4-year period.

Sex education is important in solving the problem of unwanted babies. In one survey of unmarried pregnant teenagers, 25 percent of the girls had not even considered the possibility of becoming pregnant as a result of having sexual intercourse, and 70 percent did not use any form of contraception on a regular basis. A rising number of pregnancies among 12- and 13-year-olds is reflected in current public health statistics. Such pregnancies are a social problem, and in addition are often physically damaging to the health of both the immature mother and her offspring. Youngsters old enough to produce babies should know where they come from and how to prevent conception. Because parents, for a host of reasons, often neglect to provide such information, many feel that sex education should be included in the public school curriculum.

Since ancient times man has searched for effective contraceptive methods. Modern science has developed a variety of contraceptives with a high percentage of reliability, but the ideal method of birth control has not yet been devised. Some of the more common methods of contraception are described below.

### The pill

Oral contraceptives are the most popular form of birth control in the United States. (About 9 million women were taking "the pill" in 1973.) When used properly, oral contraceptives are almost 100 percent effective in preventing pregnancy.

Oral contraceptives consist of hormones that alter the normal sequence of the menstrual cycle in such a way as to prevent ovulation. There are many different types and brands of oral contraceptives on the market, each slightly different from the others.

There are times when the body itself prevents pregnancy by inhibiting ovulation. For example the hormones of pregnancy inhibit the pituitary gland from secreting FSH and LH. No further follicles develop, and ovulation does not occur. In some women ovulation does not resume after a pregnancy for as long as they continue to breast-feed their babies, because a hormone (lactogenic hormone) associated with breast feeding apparently inhibits the secretion of FSH and LH. Attempting to use this as a means of population control, some societies encourage women to nurse their offspring until they are 3 or 4 years old. Unfortunately this method is not very reliable, and some lactating women do become pregnant.

Most existing oral contraceptives consist of estrogen and progesterone. These hormones deceive the pituitary, causing it to react as though the ovaries themselves were secreting these hormones. In the concentrations used, these hormones prevent the pituitary system from secreting both LH and FSH, and ovulation does not occur.

Two main groups of oral contraceptives are the combination pills and the sequential pills. Each combination pill contains both estrogen and synthetic progesterone (called progestogen). Starting on the fifth day of the menstrual cycle a woman takes one each day for 20 days. About 3 days after the last pill in the series, menstruation begins. With the sequential pills a woman takes a series of 15 pills containing only estrogen, then a combined estrogen-progesterone pill for the last 5 days. This regimen is designed to provide a more normal hormonal pattern. The problem with the sequential pills is that if a woman forgets even one pill, ovulation might occur. Thus, there is a slightly higher risk of pregnancy with these than with the combination pills.

A new "mini-pill," containing only small amounts of progestogen, appears to make the cervical mucus unreceptive to sperm cells so that they are unable to pass into the uterus. Although this pill appears to have fewer side effects than the older types, it is only about 97 percent effective.

*Advantages.* Short of sterilization, oral contraceptives are the most effective contraceptive method known. They afford sexual freedom to a couple, since no special preparations for birth control must be made prior to each coitus. As the pill precisely controls the timing of hormonal events, it consequently regulates the menstrual cycle.

*Disadvantages.* Some women suffer minor discomforts when they first begin taking the pill. Recent studies have linked use of oral contraceptives with vitamin C and perhaps other vitamin deficiencies. There is also some evidence that the pill can lead to serious side effects in a few women. A condition called thromboembolism, in which blood clots develop in abnormal places, has been linked with use of oral contraceptives, but the statistical risk is quite small. Only about 3 deaths per 100,000 in women who use the pill have been associated with its use. In comparison, pregnancy itself is responsible for about 25 deaths per 100,000 women each year.

Use of oral contraceptives requires a certain amount of sophistication, as one must be able to keep track of the days of the cycle and have some basic knowledge of the reproductive process. An episode that occurred in one of the underdeveloped countries striving to reduce population is a case in point. The technician who had distributed birth control pills went back to check on one of the women who was taking them. Her husband informed the aide that the woman was visiting her mother for a few days, but explained that there was no need for concern because he was taking the pills himself in his wife's absence.

Some drug companies package full sets of 28 pills so that one is taken each day with no periods in between to monitor. The last seven pills of the 28 are actually vitamin pills and are a different color than the hormone pills. This should prove more successful among the less educated—if they will remember to take the pills in the correct sequence.

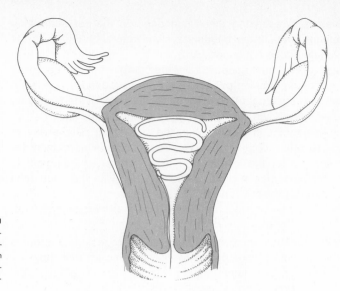

**FIGURE 12-15**
Intrauterine device (IUD) in place within uterus. (This figure, Figures 12-16 to 12-18, and acknowledged figures in Chapter 13 redrawn by permission of Robert Demarest and John Sciarra, *Conception, Birth, and Contraception,* McGraw-Hill, 1969)

### The IUD

The IUD, or intrauterine device, is a small plastic loop or coil which is inserted into the uterus by a physician. It may be left inside the uterus indefinitely or until a woman desires to become pregnant.

The mode of action of the IUD is not known. One explanation is that the IUD makes the uterus more irritable than normal and that resulting uterine contractions expel the fertilized egg from the body before it can implant itself. Another suggestion is that the IUD causes a minor inflammation. White blood cells are mobilized and destroy the developing embryo before it can implant. Still a third idea is that the IUD changes the cervical mucus so that it is *spermicidal* (kills sperm cells). A recent development has been the addition of copper to the IUD, an innovation that appears to make the IUD almost as effective as the pill but also has raised problems of possible harmful effects on the uterine lining.

*Advantages.* Some of the newer IUDs have been shown to be more than 99 percent effective. The IUD is probably the most convenient of all contraceptive devices, for once inserted, it requires little further attention. Perhaps for this reason the IUD has been more popular in underdeveloped countries than the pill. An estimated 6 million are currently in use, mainly in Asian countries. IUDs will probably become increasingly popular in the United States as women learn of the newer designs, which are both more comfortable and more effective.

*Disadvantages.* In a few women, primarily young women who have not borne children, the IUD is expelled spontaneously. In others, side effects such as discomfort and mild bleeding call for their removal. Occasionally women have become pregnant with the IUD still in place. In these instances the fetus does not appear to be harmed by the presence of the IUD, and both are expelled at the time of birth.

## Spermicides

Contraceptive foams and jellies contain chemicals that kill sperm cells. The appropriate dosage is simply inserted with an applicator into the vagina just prior to intercourse. These preparations are available without prescription. Disadvantages of spermicides are that they are only about 80 percent effective and that they tend to be costly. Contraceptive jellies are often used in combination with a diaphragm.

## Mechanical barriers

The condom and the diaphragm function by presenting a mechanical barrier between sperm and ovum.

*Condom.* The condom is a thin latex sac that fits over the erect penis during sexual intercourse. Semen is trapped within the sealed end of the sac so that sperm cells are never free within the vagina. After intercourse the condom is removed and discarded with the semen still inside. Condoms are available without prescription and are considered to be free of harmful side effects. They afford some protection against venereal disease because direct contact between the sex organs of male and female is avoided.

One disadvantage of the condom is that sexual foreplay is interrupted while the condom is fitted onto the erect penis. Another is that this contraceptive device is considered only about 90 percent effective because the condom sometimes loosens or breaks, allowing sperm to spill into the vagina.

*Diaphragm.* The contraceptive diaphragm is a thin rubber cup with a spring in the rim which makes it flexible. It is coated with spermicidal cream or jelly and inserted into the vagina before intercourse. When properly positioned, the diaphragm blocks entrance of the sperm into the cervix and is not felt by either partner during coitus (Figure 12-16).

When used with a spermicide the diaphragm is more than 85 percent effective in preventing conception. Failure generally results from improper insertion or from neglecting to add a spermicide. To ensure proper fit, the size of a diaphragm must be determined by a physician.

## Rhythm method

The rhythm method is keyed to the sequence of fertile and infertile periods of the menstrual cycle. Those who practice this method of birth control abstain from sexual intercourse during the woman's most probable fertile period, about 4 days before expected ovulation and 3 or 4 days after ovulation. The method is good in theory, but only about 75 percent reliable in practice. Its high rate of failure may be attributed to the irregularity of the menstrual cycle in so many women.

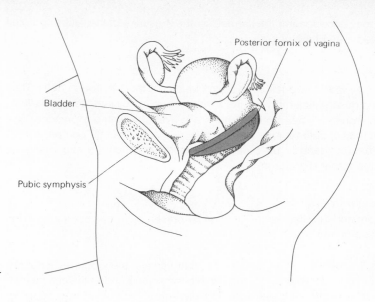

Posterior fornix of vagina

Bladder

Pubic symphysis

**FIGURE 12-16**
Contraceptive diaphragm in position in vagina.

In a recent interview Dr. Leo J. Latz, the man who popularized the rhythm method about 40 years ago, stated, "If there is a great deal of variation between menstrual cycles . . . rhythm won't work. And it won't work if the woman is under emotional or physical strain, if she is subjected to a change in climate or altitude, if she is taking certain medication or if there is intercourse during her fertile period."[3]

One wonders how many young women in today's jet-paced society are not under emotional or physical strain. Still, the rhythm method is the only form of birth control not prohibited by certain religious groups. Those who choose to practice it can increase the chances for its success by taking the utmost care in determining the time of ovulation each month. Once a woman has established that ovulation has occurred, the "safe" days for the second half of the cycle can be reliably computed.

### Unsatisfactory methods
*Douching* (flushing the vagina with some type of solution) is not an effective means of birth control. The rationale of the method is to wash the sperm out of the vagina. Douching does not work because immediately after ejaculation thousands of sperm have reached the cervical canal and are beyond reach of a douche.

In the withdrawal method (*coitus interruptus*) the male is supposed to withdraw his penis from the vagina immediately prior to ejaculation so that sperm are not released into the vagina. Sperm must also not be released onto the external genital structures as some might find their way into the

---

[3] Quoted from an Associated Press interview, 1971.

vagina and to the ovum. This method does not consider the few drops of semen containing thousands of sperm cells that are often released prior to ejaculation. Experience has shown that these are sometimes more than enough to make the couple parents. Besides, withdrawal just prior to ejaculation is contrary to powerful drives present as orgasm is reached.

## Sterilization

The only virtually foolproof method of contraception is sterilization. This procedure is becoming increasingly popular, particularly with middle-aged couples who have already produced the desired number of children.

***Male sterilization.*** About 75 percent of those currently volunteering for surgical sterilization are men. About one million men in the United States undergo such surgery each year. ***Vasectomy,*** the surgical procedure used to render the male sterile, is simple, relatively inexpensive, takes only about 30 minutes, and can be done in a doctor's office. Each vas deferens is severed through tiny incisions made in the top of the scrotum. The cut ends of the vasa deferentia are cauterized, or tied, so that sperm cannot travel through the vasa deferentia to become part of the semen. In a newer and faster procedure each vas deferens is clamped shut with a vanadium clip.

A vasectomy does not interfere in any way with a man's sexual performance. In fact many couples report a positive effect on their sex life when the fear of pregnancy has been erased. Since most of the semen is produced by the accessory glands, which are still intact and functioning, there is no noticeable change in the amount of semen ejaculated. Only the tiny sperm cells are absent. Sperm cells which are produced in the testes are simply reabsorbed by the body, so that no unusual build-up takes place. Male hormone, secreted by the cells in the testes, is transported as usual by the

**Table 12-2. Effectiveness of contraceptive methods**

| Method | Approximate failure rate* |
|---|---|
| Oral contraceptives | 0.5–1 |
| IUD (newer types) | 0.5–1 |
| Spermicides | 20 |
| Condom | 10 |
| Diaphragm plus jelly | 15 |
| Rhythm method | 25 |
| Douche | 40 |
| Withdrawal method | 30 |
| Sterilization | 0 |
| No. of women, per 100, who become pregnant using no contraception | 75–100 |

\* Number of women who become pregnant each year per 100 women using the method.

(*a*) Vas being ligated and divided

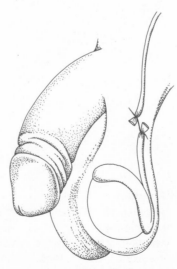

(*b*) Final appearance

**FIGURE 12-17**
Vasectomy.

blood, so that there is no effect on the hormonal balance. By surgically reuniting the ends of the vasa deferentia, vasectomies have been reversed and the male has been made fertile again in about 30 percent of attempts made. A new surgical procedure is being investigated in which a tiny piece of plastic tubing with a plug in the middle is inserted into the vas deferens. The plug could be removed if a man should decide that he wanted to father a child.

An alternative to reversible vasectomy is the storage of frozen sperm in sperm banks. If the male should decide to father another child after he has been sterilized, he simply "withdraws" his sperm for use in artificially inseminating his wife. Sperm banks are currently being established throughout the United States.

*Female sterilization.* *Tubal ligation* (referred to as "tying the tubes") is the female counterpart of vasectomy. In this procedure the oviducts are severed and the cut ends tied so that ova can no longer reach the uterus. Like a vasectomy, a tubal ligation does not in any way interfere with sexuality.

Surgical procedures in current use require opening the abdominal cavity and therefore are considered major surgery. Since there is some risk, female sterilization is generally undertaken only when pregnancy would jeopardize the woman's health. However new methods of surgery are being developed which may make female sterilization as safe and as popular as vasectomy.

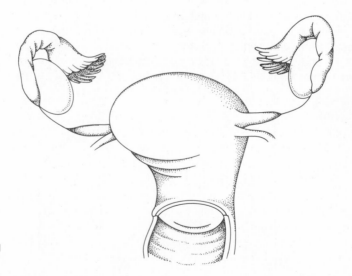

**FIGURE 12-18**
Sterilization in the female: tubal ligation.

### Abortion

Abortion is not really a form of contraception, since obviously conception has already occurred. However it is an effective means of birth control. In fact, abortion is probably the most widely used single method of birth control in the world. All societies appear to employ abortion for this purpose. An estimated 25 to 40 million abortions are performed each year.

There are three types of abortion. *Spontaneous abortions,* popularly known as miscarriages, are initiated by natural processes, often as a way of destroying an abnormal embryo. *Therapeutic abortions* are performed by physicians when the health of the mother is jeopardized by the pregnancy, or sometimes when there is good reason to suspect that the embryo will develop in a grossly abnormal manner. The third type of abortion is performed to terminate an unwanted pregnancy, that is, *abortion solely as a means of birth control.*

Although the medical risks connected with abortion naturally depend upon the circumstances under which it is performed, an abortion performed by a competent physician in a modern hospital or well-equipped clinic is among the safest of surgical procedures (3 deaths per 100,000 abortions). In fact, a properly performed early abortion is markedly safer than the average pregnancy (25 maternal deaths per 100,000). The earlier in pregnancy an abortion is performed, the safer the procedure is.

At the present time most abortions are performed using a suction method. The cervix is dilated and a suction aspirator is inserted and used to remove the thickened lining of the uterus. This lining, the same tissue expelled monthly during menstruation, contains the tiny embryo. Abortions performed later in pregnancy involve injection of a salt solution into the uterus. Severe contractions result, and the fetus is expelled. In the future abortions may be induced chemically by administration of prostaglandins (Chapter 11). This method appears to be an extremely safe, nonsurgical means for terminating pregnancy.

In January 1973 the U.S. Supreme Court ruled that during the first 3 months of pregnancy the decision to have an abortion rests entirely with the pregnant woman and her physician and that the state cannot interfere. After the first 3 months the state may regulate abortion procedure in order to protect maternal health. A state can prohibit abortion during the last 10 weeks of pregnancy, when the fetus can survive outside the uterus. The Supreme Court's decision reflects changing attitudes[4] toward abortion and came after several states had liberalized restrictive abortion laws. In 1971 about 500,000 legal abortions were performed in the United States.

Historically, abortion legislation has discriminated against the poor. Abortions have always been available to those who could afford the time and money to travel to places where they are legal. The poor woman took her chances with the unqualified, often unsanitary, but less expensive, practitioner. Thousands of women have died as a result of such illegal abortions.

One argument that has been used against legalized abortion is that women might suspend other means of birth control in favor of routine, frequent abortions. This has not proved to be the case in nations with extremely liberal abortion laws. An abortion is an unpleasant, somewhat painful experience, not one that a woman would rush to repeat.

The principal moral issue surrounding the matter of abortion relates to

---

[4] In January 1973 a Gallup poll reported that 46 percent of Americans favor legal abortion.

the theologic question of when the embryo acquires a soul. St. Thomas Aquinas felt that there was no way to tell at what time in its development the embryo becomes possessed with a soul, and consequently he saw nothing morally wrong with abortion. Many Protestants and most liberal Jewish theologians seem inclined to agree with St. Thomas's viewpoint. Their feeling is that an early embryo is only a *potential* human being, and that there is nothing morally wrong with removing it. Some theologians believe that ensoulment takes place at birth when the baby draws his first breath. Most Catholic theologians, as well as some Protestant ones, feel that since no one knows when an embryo becomes possessed with a soul it is morally unsafe to terminate the life of an embryo.[5] But others take the viewpoint that the embryo is the *blueprint* of the human being, not the finished product until the brain has developed to a stage where consciousness is possible. When performed early in pregnancy, abortions appear preferable to grossly malformed or unwanted babies.

### The future of contraception

Research for the ideal contraceptive is continuing at a rapid pace. Researchers are seeking a method that will be 100 percent effective, will be completely safe, and will require little effort to use. Several interesting possibilities seem just around the corner.

For instance, studies are being carried out on prostaglandin tampons, which a woman could insert to bring on menstruation each month. At that early stage of possible pregnancy the woman would not even know whether she had or had not been pregnant.

Long-term contraception is another promising possibility. Tiny capsules containing a form of progesterone might be injected under the skin of the hip or arm with a hypodermic needle. The hormone would be released slowly into the body over a long period of time, perhaps as long as a year. Hormone-containing silastic rings that could be worn high in the vagina are also being tested.

Other research is centered upon the releasing factors synthesized by the hypothalamus. If synthetic antireleasing factors could be developed, the pituitary gland might not release the hormones that trigger ovulation.

"Morning-after" pills which could be taken within a few days after intercourse and which hormonally would prevent the embryo from implanting in the uterus are also being developed. Still other research is focused upon new methods of male contraception. To date hormonal intervention in the male has been unsuccessful because though sperm production can be prevented, secretion of male hormone is reduced at the same time.

### POPULATION CONTROL

An increasing number of authorities feel that the population explosion cannot be dealt with successfully by purely voluntary measures of birth con-

---

[5] The Catholic Church did not specifically forbid abortion until 1869.

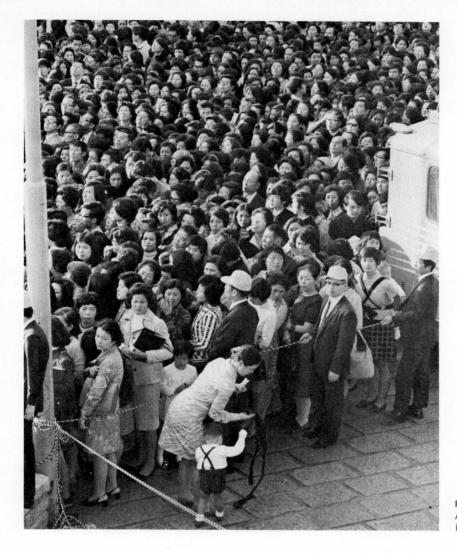

**FIGURE 12-19**
A crowd awaiting a discount sale
in Tokyo. (OMIKRON)

trol. For one thing, people often wait until they have three or four children before seeking effective means of contraception. Their desire for contraception at that point will do little to stabilize population growth. A more effective alternative is population control, the planned regulation of population size by society. In its extreme and most unpleasant form population control would involve involuntary fertility control of the entire population.

No responsible person desires any such drastic intervention in the private lives of citizens, and yet a *failure* to take any steps to encourage population control is an immoral course that can only lead to social disaster. In order to avoid coercive measures in the future, a number of steps can be undertaken in the present.

Education is an obvious first step. The relationship between overpopulation and environmental deterioration of several types should be made clear to everyone, not only via formal educational channels but also in magazines, television, and other universal media. (We will be dealing with this in later chapters.) The basic facts of overpopulation should be included in the public school curriculum at all levels. Debate and dialogue in legislatures and in the press may unearth potential solutions to the population crisis which have not been carefully considered heretofore.

Certain legislative measures would be immediately helpful—legislation, that is, which does not impose restrictions but which lifts restrictions imposed in former times when our ecologic crisis was not acute. For example, there is no reason why contraceptive methods should be unavailable to poor families merely because of inability to pay for them. Contraceptive devices should be distributed without cost through local health departments, hospital outpatient clinics, and similar channels, including perhaps social welfare agencies. On the other hand, opportunities for birth control should be available to all groups and classes of citizens, and should not be aimed preferentially at any one socioeconomic group.

At the present time it is not easy for those wishing to be surgically sterilized to obtain such operations in many locales. Public health programs should be initiated to make both vasectomies and tubal ligation widely available and inexpensive. Indeed, such operations could be provided free of charge or be fully covered by low-cost medical insurance. Finally, from the viewpoint of population control, abortion is an important alternative when birth control methods fail.

## VENEREAL DISEASE

Any disease transmitted by sexual contact is considered a venereal disease (VD). Among the most common venereal diseases are syphilis and gonorrhea. During the past few years the incidence of VD has increased dramatically. The incidence of VD among teen-agers in the United States is one case every 2 minutes. Some public health officials blame the increase in VD on increased sexual promiscuity. The pill is a contributing factor, since it has significantly replaced the condom which provided some protection against VD. In addition the pill appears to increase the moisture and alkalinity of the female reproductive tract, conditions which promote the growth of gonorrhea bacteria. A woman taking oral contraceptives stands an almost 100 percent chance of contracting gonorrhea if she engages in coitus with an infected partner, as compared to about a 40 percent risk for a woman not taking the pill. Another reason cited for the rise in VD is insufficient funds for tracing contacts, that is, tracking down people with whom an infected person has had sexual contact. The faster these people are found and treated, the fewer people will contract the disease from them. Still another reason is inadequate educational programs concerning venereal disease.

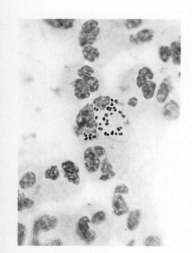

**FIGURE 12-20**
Smear showing bacteria which cause gonorrhea.(Ward's Natural Science Establishment)

## Gonorrhea

The incidence of gonorrhea has reached epidemic proportions in the United States and in many other parts of the world. More than 2 million cases are treated each year in the United States, making it the second most common communicable disease, second only to the common cold. A great number of additional cases are not reported. An estimated 80 percent of women with gonorrhea lack symptoms and remain untreated. The disease occurs most frequently among 15- to 19-year-olds.

Gonorrhea is caused by a bacterium. From 3 to 9 days after infection most men experience a severe burning upon urination and a discharge of yellow pus from the penis. Generally these symptoms prompt a male to seek medical treatment, though some men have a form of gonorrhea that is symptomless. Most infected women experience no symptoms, but some notice a vaginal discharge.

If untreated, the infection spreads within the body, with serious consequences. Sterility, chronic invalidism, and arthritis are common complications. Gonorrhea is treated with penicillin, but unfortunately, resistant strains of the infective bacteria are becoming common. The dosage of penicillin required to cure the disease has been continuously increasing, and in some cases penicillin does not effect a cure. Streptomycin and other antibiotics are used when penicillin is ineffective.

### Syphilis

An estimated 500,000 Americans are unaware that they have syphilis and are in urgent need of medical treatment. Syphilis occurs most frequently in the 20-to-24 age group, followed closely by the 25-to-29 age group. Although less common than gonorrhea, syphilis is a more serious disease.

Syphilis is caused by a corkscrew-shaped spirochete bacterium called *Treponema pallidum* (Figure 12-21). Because this spirochete cannot survive outside the body for more than a few seconds, one cannot become infected from a dirty toilet seat or contaminated towel. The disease is contracted almost exclusively by *direct sexual contact*.

The first sign of infection is the appearance of a sore, the *primary chancre,* that usually appears near the site of infection. Generally, the primary chancre appears about 3 weeks after infection (though the time may vary from 10 days to 10 weeks.) If untreated, the sore usually heals by itself within a month. The primary chancre is highly infectious, so that engaging in sexual relations (even kissing if the chancre is orally located) at this stage of the disease will most certainly transmit the infection to an uninfected partner. Syphilis is also transmissible by homosexual contact.

About 6 weeks after the primary chancre has disappeared, a rash may appear, marking the second state of the disease. Then syphilis goes into a long latent period, which may last up to 20 years or even longer. During this time the spirochetes invade organs of the body—often the brain, the heart, or the liver. The infection can be detected at this time only by blood tests. When the disease becomes obvious again, lesions have formed in the infected organs. The resulting damage causes insanity, impairment of the heart function, or even death.

Syphilis can be cured with massive doses of penicillin. However, penicillin-resistant strains are beginning to appear. When syphilis is diagnosed,

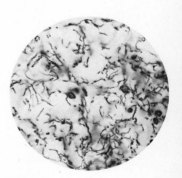

**FIGURE 12-21**
The bacteria (*Treponema pallidum*) which cause syphilis. (TURTOX)

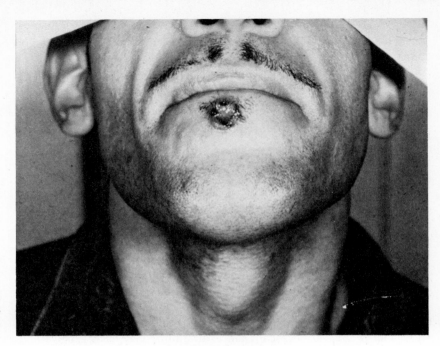

**FIGURE 12-22**
Primary syphilitic chancre. (Center for Disease Control, Dept. of Health, Education and Welfare)

usually by blood tests, public health officials attempt to track down all of the infected person's contacts and treat them. Physicians in private practice are frequently reluctant to report cases to public health officials for fear of embarassing their patients. For this reason the incidence of early syphilis is estimated to be nearly four times the number of cases actually reported.

There is urgent need for vaccines that could be used against syphilis and gonorrhea. At present there is a constant struggle to outwit the adaptive mechanisms of the disease organisms involved. Medical science must always have one more antibiotic ready in case resistant strains to the prevailing choice of treatment should appear. Vaccines would shift the emphasis to prevention rather than treatment, but to date efforts to develop vaccines have been unsuccessful.

### Other forms of VD

*Trichomonas vaginalis* is a flagellate protozoan which causes a common, though not serious, venereal disease. *This* infection *can* be transmitted by dirty toilet seats, as well as by sexual contact. An estimated 20 to 30 percent of women have this infection. The protozoa live in the vagina or, less commonly, in the male urethra. Trichomonas is treated with a variety of drugs.

Pubic lice and crab lice that inhabit the pubic area may be considered a form of VD, since they may be transmitted during sexual intercourse. Pubic lice feed on human blood. Infection can be cured by application of a specially medicated shampoo.

1 In asexual reproduction a single parent endows its offspring with a set of genes identical to its own; in sexual reproduction each of two parents contributes half of the offspring's genetic endowment.

2 Sexual reproduction, by providing for a thorough shuffling of the genes between generations, ensures variety among the members of a species and also provides for the rapid spread of advantageous genes throughout a population.

3. Meiosis results in haploid gametes and thus ensures that the chromosome number of the species will remain constant.

4 Sperm production takes place in the tubules of the testes. To reach the outside of the body, sperm travel through a series of conducting ducts. Three different accessory glands contribute fluid to the semen.

5 Ova are formed in the ovaries. At ovulation an ovum is ejected from the ovary and finds its way into the oviduct, where it may be fertilized.

6 The first day of menstruation constitutes day 1 of the menstrual cycle. In a regular cycle ovulation occurs on about day 14. A woman's most fertile period occurs during the 2 days immediately prior to ovulation and during a 24- to 28-hour period after ovulation.

7 The menstrual cycle is regulated by the interaction of pituitary and ovarian hormones.

8 Fertilization restores the diploid chromosome number to the zygote, determines the sex of the offspring, and triggers development.

9 The sex of the offspring is determined by whether an X sperm or a Y sperm fertilizes the egg.

10 The pill is the most effective means of contraception currently available, with the exception of sterilization. The IUD is also highly effective.

11 Abortion is probably the most widely used method of birth control in the world today.

12 The venereal disease gonorrhea has reached epidemic proportions. About 80 percent of women with the disease have no symptoms and consequently remain untreated.

13 An estimated 500,000 Americans are unaware that they have syphilis and are in urgent need of treatment. In its later stages syphilis can cause serious physiologic damage and even death.

**FOR FURTHER READING**

Demarest, Robert J., and John J. Sciarra: *Conception, Birth, and Contraception,* McGraw-Hill, New York, 1969. We highly recommend this visual presentation of human reproduction as one of the finest books of its type on the market.

Rorvik, David M., with Landrum B. Shettles: *Your Baby's Sex: Now You Can Choose,* Dodd, Mead (and Bantam), New York, 1970. An interesting, thought-provoking guide to a concept of sex planning.

Rugh, Roberts, and Landrum B. Shettles: *From Conception to Birth,* Harper & Row, New York, 1971. An excellent account of the reproductive process and of development.

Vleck, David B.: *How and Why Not to Have That Baby,* Optimum Population, Vermont, 1971. This short booklet discusses human reproduction, contraception, and related problems.

Warshofsky, Fred: *The Control of Life: The 21st Century,* Viking, New York, 1969. Discusses possible future methods of contraception.

# 13

## THE ORIGIN OF THE ORGANISM

**Chapter learning objective.** The student must be able to describe the pattern and regulation of prenatal development and to describe the various environmental factors that may affect development.

*ILO 1* Describe the preformation theory and the theory of epigenesis, and relate these theories to current concepts of development.

*ILO 2* Define growth, morphogenesis, and cellular differentiation, and describe the role of each process in the development of an organism.

*ILO 3* Describe the principal events and characteristics of each of the early stages of development: zygote, cleavage, blastocyst, gastrula, and nervous system development.

*ILO 4* Describe the implantation of the zygote and the development of the placenta; give the functions of the placenta and of the amnion.

*ILO 5* Describe the general course of the later development of the human being, from 1 month after conception until the time of birth.

*ILO 6* Describe the birth process, distinguishing among the three stages of labor.

*ILO 7* List four advantages and one disadvantage of breast feeding an infant.

*ILO 8* Identify and describe the genetic and nongenetic factors that interact to regulate the process of development; be able to describe the experiments discussed and to apply them to the developmental process.

*ILO 9* Describe specific steps that the pregnant woman can take to promote the well-being of her developing child.

*ILO 10* Describe how nutrients and drugs may influence the development of the embryo. (Recount the thalidomide tragedy.)

*ILO 11* Describe the effects of oxygen deprivation on the embryo and relate them to cigarette smoking.

*ILO 12* Describe the effects of infection by German measles and venereal disease upon the developing embryo.

*ILO 13* Define each scientific term introduced in this chapter.

Many scientists of the seventeenth century thought that the egg cell contained a completely formed, though miniature, human being. They believed that all the parts were *already* present, so that the embryo had only to grow in size. This concept is known as the *preformation theory*.

By the end of the seventeenth century two competing groups of preformationists emerged. One group, the ovists, thought that the preformed organism resided within the egg; the opposing group, the spermists, were certain that the "little man" was housed in the sperm. Using their crude microscopes some investigators even imagined that they could see a completely formed tiny human being within the head of the sperm (Figure 13-1).

Some scientists carried the theory to an extreme form, arguing that every woman contained within her body a miniature of every individual who would ever descend from her. Her children, grandchildren, great grandchildren, and so on were thought to be preformed, each within the reproductive cells of the other. Some men of the time even computed mathematically how many generations could fit, one within the other's gametes. They concluded that when all these generations had lived and died, the human species would end. Jan Swammerdam, a renowned preformationist, felt that this concept explained original sin. He wrote, "In nature there is no generation, but only propagation, the growth of parts. Thus, original sin is explained, for all men were contained in the organs of Adam and Eve. When their stock of eggs is finished the human race will cease to be."

The idea of preformation was not restricted in its application to human beings alone. All plant and animal species were included. For almost 200 years this theory was seriously debated by scientists and philosophers.

An opposing view, the *theory of epigenesis,* gained experimental support as better techniques for investigation were developed. This theory held that the embryo develops from an undifferentiated zygote and that the structures of the body emerge in an orderly sequence, developing their characteristic forms only as they emerge.

Today we know that development is largely epigenetic. No invisible "little man" waits preformed in either gamete. Development proceeds from one cell to billions, from a formless mass of cells to an intricate, highly specialized, and organized organism. However, a spark of truth can be found in the preformationist view. Although the "little man" himself is not to be found within the zygote, his blueprint *is* there, precisely encoded in the form of chemical specifications within the DNA of the genes.

**FIGURE 13-1**
The preformed little man within the sperm as visualized by seventeenth-century scientists. (Reproduction of Hartseeker's drawing from "Essay de Dioptrique," Paris, 1694)

## THE PATTERN OF DEVELOPMENT

How does the microscopic, unspecialized zygote give rise to the blood, bones, brain, and all the other complex structures of the completed organism? As we shall see, development is a balanced combination of three processes: *growth, morphogenesis,* and *cellular differentiation.*

Growth of the embryo includes both cellular growth and mitosis. An orderly pattern of growth provides the cellular building blocks of the organism. But growth alone would produce only a formless heap of cells. Cells must ar-

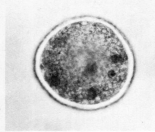

**FIGURE 13-2**
Human zygote. This single cell contains the genetic instructions for producing a complete human being. (Carnegie Institution of Washington)

range themselves into specific structures and appropriate body forms. The precise and complicated cellular movements designed to bring about the form of a multicellular organism with its intricate pattern of tissues and organs constitute morphogenesis.

Not only must cells be arranged into specific structures, but they must also be able to perform varied and specialized functions. Cells must be *specialized,* as well as organized. In order to function in different fashions, body structures must be composed of different components. During early development cells begin to differentiate from one another, specializing biochemically and structurally to perform specific tasks. More than 200 distinct types of cells can be found in the human body. The process by which cells become specialized is known as cellular differentiation.

As you read the following section on development, bear in mind that growth, morphogenesis, and cellular differentiation are intimately interrelated. Although our discussion will center upon human development, the pattern of early development is basically similar for all vertebrates.

### The zygote

Although it appears to be a relatively simple cell, the zygote, or fertilized egg, has the potential to give rise to all the diverse cell types of the complete individual. Its single nucleus contains the chromosomes contributed by both parents. Since the sperm cell is quite tiny in comparison to the egg, the bulk of the zygote cytoplasm comes from the ovum. However, both sperm and egg make an equal genetic contribution to the developing organism.

In most vertebrate zygotes the cytoplasm contains yolk bodies which serve as food for the developing embryo. The amount and distribution of yolk vary among different animal groups. Whereas bird eggs contain great quantities of yolk, those of mammals have only small amounts. Yolk is absent from the human zygote.

**FIGURE 13-3**
Cleavage in the human embryo. (a) Two-cell stage. (b) Four-cell stage. (c) Morula. (Carnegie Institution of Washington)

### Cleavage—from one cell to many

Shortly after fertilization the zygote undergoes a series of rapid mitoses, collectively referred to as the *cleavage stage*. By about 24 hours after fertilization the human zygote has completed the first mitotic division and reached

*(a)*

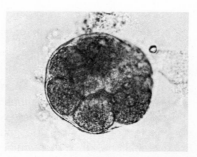

*(b)*  *(c)*

the two-cell stage. Each of the cells of the two-cell-stage *embryo* undergoes mitosis, bringing the number of cells to four. Repeated divisions continue to increase the number of cells making up the embryo. At about the 16-cell stage the embryo consists of a tiny cluster of cells called the *morula*. As cleavage takes place the embryo is pushed along the oviduct by ciliary action and muscular contraction. By the time the embryo reaches the uterus on about the fifth day of development it is in the morula stage.

During cleavage each interphase period is so brief that the cells do not have time to grow. For this reason the mass of cells produced is not larger than the original zygote. The principal effect of early cleavage, then, is to partition the very large zygote into many small cells. These serve as basic building units which determine the early form of the organism. Their small size allows the cells to move about with relative ease, arranging themselves into the patterns necessary for further development. Each cell is able to flow along, ameba-style, probably guided by the proteins of its own cell coat and those of other cell surfaces. Specific properties conferred upon cell membranes by these surface proteins are important in helping cells to "recognize" one another, and so, in determining which ones will adhere to form tissues.

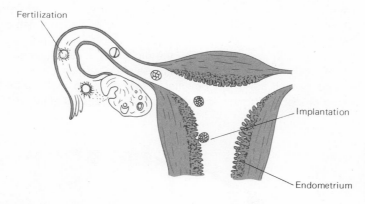

**FIGURE 13-4**
Cleavage takes place as the embryo moves along the oviduct to the uterus. (Redrawn from Demarest and Sciarra)

## Early development in the uterus

When the embryo enters the uterus, the membrane (the *zona pellucida*) which had surrounded the embryo, is dissolved. Now the embryo is bathed in a nutritive fluid secreted by the glands of the uterus. Nourished in this manner, the embryo continues its development for 2 or 3 days, floating free in the uterine cavity. During this period its cells arrange themselves into the form of a hollow ball called a *blastocyst* (or blastula in many other animal groups) (Figure 13-5). The outer layer of cells, the *trophoblast,* eventually forms the protective and nutritive membranes (the *chorion* and *placenta*) which surround the embryo. A little cluster of cells, the *inner cell mass,* which projects into the cavity of the blastocyst, gives rise to the embryo itself.

Occasionally the inner cell mass subdivides to form two independent groups of cells, and each develops into a complete organism. Since each cell has an identical set of genes, the individuals formed are exactly alike — identical twins. Very rarely, the two inner cell masses are not completely separated and give rise to conjoined (Siamese) twins. Fraternal twins develop when a woman ovulates two eggs and each is fertilized by a separate sperm. Each zygote has its own distinctive genetic endowment, and so the individuals produced are not identical. Triplets (and other multiple births) may similarly be either identical or fraternal. In the United States twins are born once in about 88 births, triplets once in 88 squared, and quadruplets once in 88 cubed.

*Implantation.* On about the seventh day of development the embryo begins to implant itself in the wall of the uterus. The trophoblast cells in contact with the uterine lining secrete enzymes which erode an area of the uterine wall just large enough to accommodate the tiny embryo. Slowly the embryo works its way down into the underlying connective and vascular tissues. The opening through which the blastocyst enters the wall of the uterus is closed, first by a blood clot, and eventually by the overgrowth of regenerated epithelial cells. All further development of the embryo takes place *within* the wall of the uterus.

During implantation enzymes destroy some tiny maternal capillaries in the wall of the uterus. Blood from these capillaries comes into direct contact with the trophoblast of the embryo, temporarily providing a rich source of nutrition.

Implantation is completed by the ninth day of development. This would correspond to about the twenty-third day of a woman's menstrual cycle, so that despite all the developmental activities taking place within, a woman would probably not even know that she was pregnant at this time.

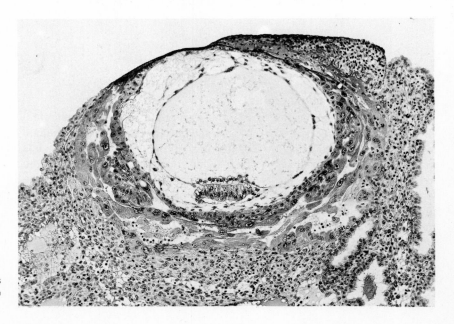

**FIGURE 13-5**
Implanted blastocyst at 12 days after fertilization. (Carnegie Institution of Washington)

Once the embryo has embedded itself within the uterine wall the tropho-blast begins to secrete the hormone CG (chorionic gonadotropin). CG acts upon the corpus luteum, preventing it from degenerating, as it otherwise nor-mally would at this time of the menstrual cycle. In this way CG serves as a messenger, informing the body that pregnancy has taken place. Some tests for early pregnancy are based upon the presence of CG in the maternal urine. If for any reason CG is *not* secreted before the corpus luteum has begun to degenerate, the embryo is aborted, sloughed off with the uterine lining, and disposed of with the menstrual flow. Under such circumstances a woman would not know that she had been briefly pregnant.

***Development of the placenta.*** The placenta is the organ of exchange between the embryo and its mother. It provides for the nutrition and respira-tion of the embryo. The placenta is also an excretory organ, providing a means for the wastes produced by the embryo to be delivered to the blood of the mother so that her body can dispose of them. Finally, the placenta serves as an important endocrine organ, producing CG and later estrogen and progesterone. Thus, the placenta functions as a combination lung, kidney, digestive system, and endocrine gland for the developing embryo.

The placenta is composed of both embryonic and maternal tissues.

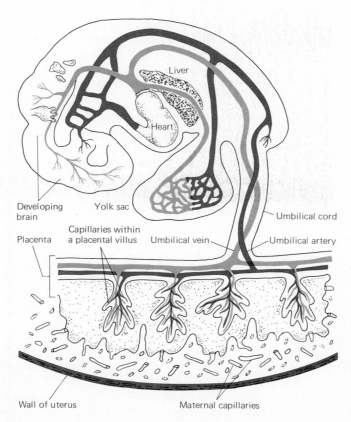

Developing brain

Yolk sac

Placenta

Capillaries within a placental villus

Umbilical vein

Umbilical cord

Umbilical artery

Wall of uterus

Maternal capillaries

**FIGURE 13-6**
Relationship of an embryo to its placenta. Blood from the embryo and maternal blood both circu-late through the placenta, where exchange of materials takes place. Maternal and embryonic blood do not mix. Although the human embryo does not depend upon yolk for nourishment, a yolk sac develops and its walls serve as a temporary site of blood-cell formation.

After implantation has been completed, the trophoblast begins to grow rapidly, forming finger-like projections called *villi,* which penetrate deep into the uterine wall. While the villi are forming, the circulatory system of the embryo is also developing. An *umbilical cord* forms, connecting the embryo with the placental portion of the trophoblast. Blood vessels from the embryo grow through the umbilical cord and connect with a vast network of capillaries developing within the placental villi. The placenta consists of the portion of the trophoblast which has formed villi, together with the uterine tissue between the villi. Embryonic blood vessels extend into the villi, and on the other side of the trophoblast tissue, maternal capillaries and small pools of maternal blood traverse the tissue between the villi. Thus, maternal blood comes into close proximity to the embryonic blood circulating through the capillaries in the villi (Figure 13-6). Nutrients and oxygen from the mother's blood diffuse across the placenta into the embryo's blood, while carbon dioxide and nitrogenous wastes from the embryo diffuse into the maternal circulation. The blood of the embryo does not actually mix with the maternal blood. The two circulatory systems are completely separate from one another, but the placenta brings them close together so that materials may be exchanged by diffusion. By about the twentieth day of development the placenta has begun to function.

The placenta constitutes a differentially permeable barrier separating the mother from her developing child. Not long ago it was thought that the placental barrier prevented harmful substances in the mother from reaching the embryo. Now we know that some viruses infecting the mother can cross the placenta and also infect the embryo. Many types of drugs ingested by the mother, vitamins, hormones, and occasionally even cells of both fetus and mother also manage to traverse the placental barrier.

*The amnion.* The amnion is a membrane which develops between the embryo and the placenta, and surrounds the entire embryo. It is penetrated only by the umbilical blood vessels. The cavity between the embryo and the amnion is filled with a clear liquid called *amniotic fluid,* which bathes the embryo and in effect provides it with its own private swimming pool. Embryos of most egg-laying land vertebrates are surrounded by an amnion, which ensures that they will not dry out. Amphibians, such as frogs and toads, which do not possess this membrane or an egg shell, must pass their early life as aquatic larvas and, for the most part, cannot stray far from the water. The embryo moves easily within the amniotic fluid, which also serves to cushion it against mechanical injury. The amnion and the membranes of the placenta are referred to as *fetal membranes.* They are discarded at the time of birth.

*Development of the embryo proper.* While fetal membranes are developing, important changes are also taking place within the inner cell mass. After implantation the blastocyst begins the process of *gastrulation* whereby it becomes a three-layered embryo. Cells of the blastocyst arrange themselves to form the three basic embryonic tissues, the *ectoderm* (outer skin), *mesoderm* (middle skin), and *endoderm* (inner skin).

Until the time of gastrulation, development appears to depend upon growth and morphogenesis. In the human embryo the first *visible* sign of differentiation occurs during gastrulation when one group of cells of the inner cell mass becomes flattened and the remaining cells become elongated (columnar). The flattened cells develop into the endoderm tissue, while the columnar cells give rise to the ectoderm. The mesoderm develops between these two layers. During gastrulation the embryo is referred to as a *gastrula*.

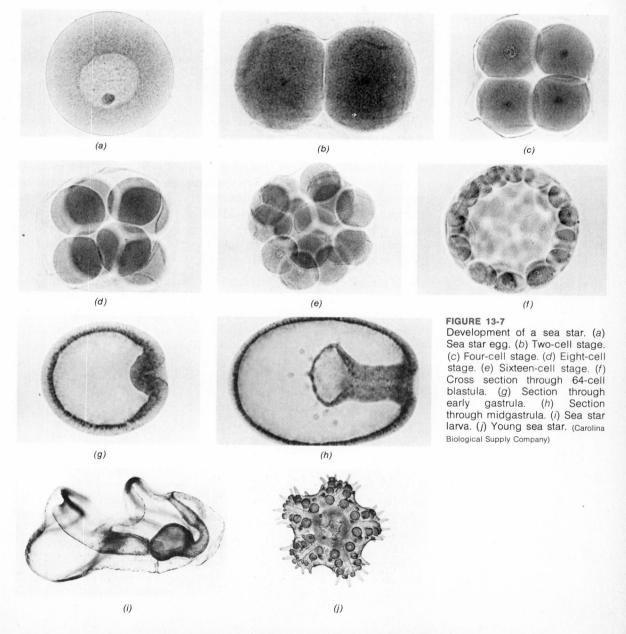

(a)

(b)

(c)

(d)

(e)

(f)

(g)

(h)

**FIGURE 13-7**
Development of a sea star. (*a*) Sea star egg. (*b*) Two-cell stage. (*c*) Four-cell stage. (*d*) Eight-cell stage. (*e*) Sixteen-cell stage. (*f*) Cross section through 64-cell blastula. (*g*) Section through early gastrula. (*h*) Section through midgastrula. (*i*) Sea star larva. (*j*) Young sea star. (Carolina Biological Supply Company)

(i)

(j)

**Fate of the germ layers**

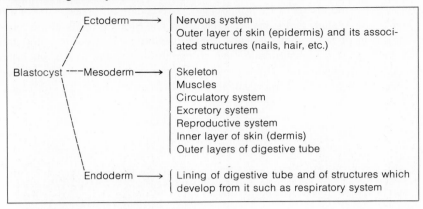

| | | |
|---|---|---|
| | Ectoderm ⟶ | Nervous system<br>Outer layer of skin (epidermis) and its associated structures (nails, hair, etc.) |
| Blastocyst - - -Mesoderm ⟶ | | Skeleton<br>Muscles<br>Circulatory system<br>Excretory system<br>Reproductive system<br>Inner layer of skin (dermis)<br>Outer layers of digestive tube |
| | Endoderm ⟶ | Lining of digestive tube and of structures which develop from it such as respiratory system |

*Fate of the germ layers.* Each embryonic tissue layer gives rise to uniquely specific structures. The development of these structures follows a predictable pattern which is similar for all vertebrate embryos. Ectoderm covers the embryo, specializing to form the outer layer of skin. Ectoderm also gives rise to the nervous system. Endoderm becomes the lining of the digestive and respiratory systems, and also forms the lining of organs such as the digestive glands. Almost all the other structures of the body—muscle, bone, circulatory, reproductive, and excretory structures—are derived from mesoderm. In fact, tissue of mesodermal origin accounts for more than 90 percent of our body weight. (See the accompanying diagram.)

*Development of the nervous system.* The brain and spinal cord are among the first organs to develop in the early embryo. Late in the gastrula stage the ectoderm thickens in the midline of the embryo to form the *neural plate.* The neural plate develops just above a rod of tissue called the *notochord,* a structure which serves as a flexible skeletal axis in all chordate embryos. (In vertebrate embryos it is eventually replaced by the vertebral column.) The notochord actually stimulates the tissue above it to differentiate into nervous tissue, a process which will be discussed presently.

Central cells of the neural plate move downward, forming a depression, the *neural groove,* while the cells flanking the groove on each side form *neural folds.* Continued proliferation of their cells brings the folds closer together until they meet to form a tube, the *neural tube.* The neural tube sinks slowly down into the tissue of the embryo, while growth of the outer ectoderm serves to cover the outer surface. Later, the outer ectoderm differentiates to form the outer layer of skin. The front (anterior) portion of the neural tube grows and differentiates to form the brain, while the remainder of the tube develops into the spinal cord. A thin crest of cells which flanks the neural tube on each side develops into cranial and spinal nerves.

This brief description of the origin of the nervous system serves to show how well-defined structures emerge as a result of precise movements and rearrangements of cells. While the nervous system is forming, the heart,

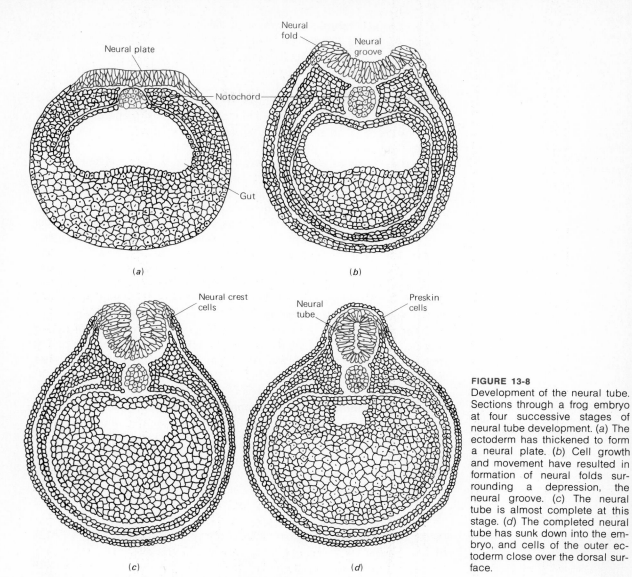

Neural plate

Notochord

Gut

(a)

Neural
fold

Neural
groove

(b)

Neural crest
cells

(c)

Neural
tube

Preskin
cells

(d)

**FIGURE 13-8**
Development of the neural tube.
Sections through a frog embryo
at four successive stages of
neural tube development. (a) The
ectoderm has thickened to form
a neural plate. (b) Cell growth
and movement have resulted in
formation of neural folds surrounding a depression, the
neural groove. (c) The neural
tube is almost complete at this
stage. (d) The completed neural
tube has sunk down into the embryo, and cells of the outer ectoderm close over the dorsal surface.

blood vessels, digestive tube, and many other structures also begin their
development.

## Later development

Development proceeds as an orderly, predictable sequence of events. Two
embryos of the same species and age are nearly identical with respect to
the stage of their development. Conversely, one can determine the age of an
embryo by studying the state of its development. Normally, one can predict
with startling precision which structures will begin to develop, or begin to
function, on a particular day or even at a specific hour of development.

***The embryo at 1 month.*** At 4 weeks of development the human embryo is 0.2 inch long and weighs about 0.02 gram. (There are 454 grams in 1 pound.) Although tiny, it is 10,000 times heavier than the zygote from which it emerged. It consists of about 94 percent water. (In comparison, an adult consists of about 65 percent water.)

The 4-week-old embryo has already developed the rudiments of many of its organs. A simple but functional circulatory system is in operation. At this stage the heart is an S-shaped tube which beats about sixty times each minute. The neural tube has formed, and parts of the brain are beginning to differentiate. Because the head end of the embryo develops faster than the posterior end, the embryo appears to have a huge head in proportion to the rest of its body. Limb buds have just begun to develop and will eventually give rise to arms and legs. The digestive tube is developing and is giving rise to structures which will form the respiratory system, digestive glands, and thyroid gland. Gill arches have developed in the region of the pharynx; they are separated from one another by rudimentary pharyngeal gill slits. These structures develop similarly to the gill structures found in fish and some amphibians, but in these animals gills develop along the arches and water circulates through the gill slits. In man and other terrestrial vertebrates the gill structures are modified to form organs more appropriate for terrestrial life.[1] For example, the first gill slit and associated pharyngeal pouch develop into the ear canal, eardrum, and eustachian tube. Another temporary structure is the thin tail which is prominent in the embryo but which later regresses.

***The second month.*** During the second month all the organs continue to develop. The face forms, with a prominent nose, upper and lower lips,

---

[1] In man and other mammals the gill slits are more accurately called branchial grooves, because they are separated from the pharynx by a thin plate of tissue; i.e., the slits normally do not open completely.

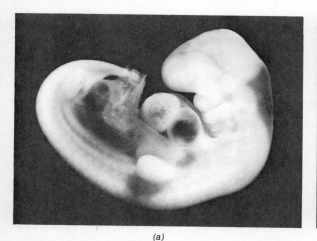

(a)

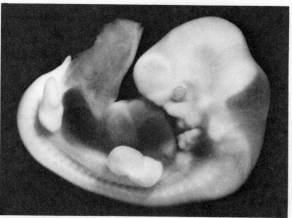

(b)

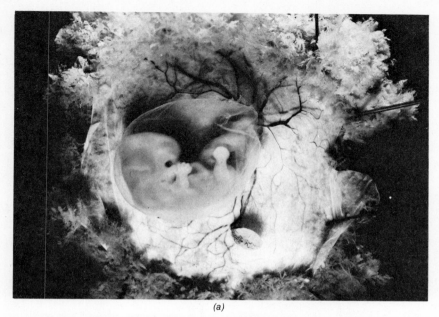

(a)

**FIGURE 13-11**
During the second month growth, differentiation, and morphogenesis continue rapidly. The embryo begins to look recognizably human. (a) Sixth week of development. (b) Seventh week. (c) Human embryo at 7 weeks. (Carnegie Institution of Washington)

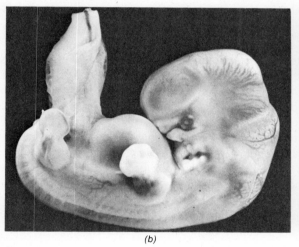

(b)

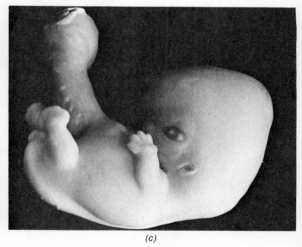

(c)

eyelids, cheeks, and ears. The ears, however, are located in the neck region and do not assume their proper place until later in development. Limb buds give rise to arms and legs, and fingers and toes form. The tail, prominent during the fifth week, becomes inconspicuous by the end of the second month. Rudiments of external genital structures are visible, but sex cannot yet be distinguished. The liver has become quite large, and the small intestine is already coiling. Major blood vessels assume their final positions. Muscles are forming, and the embryo is capable of some movement. The brain begins to send impulses to regulate the functions of other organs, and simple nerve reflexes occur. At 2 months the embryo measures 1.2 inches and weighs 1 gram (about one-thirtieth of an ounce).

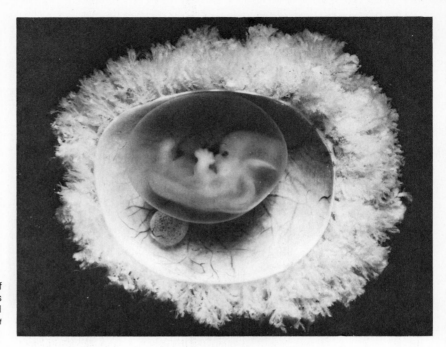

**FIGURE 13-12**
Embryo in its seventh week of
development, surrounded by its
intact amnion and other fetal
membranes. (Carnegie Institution of
Washington)

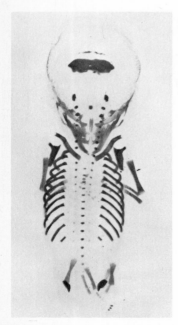

**FIGURE 13-13**
Dorsal (back) view of embryo at
about 9 weeks, showing development of skeleton. (Carolina Biological Supply Company)

**The third month.** After the second month, the embryo is often referred to as
a fetus. By this time the basic forms of all the organs have been laid down.
During the third month the fetus becomes recognizable as human. The
remaining months of development are a period of rapid growth and of more
final differentiation. During the third month the external genital structures differentiate, indicating the sex of the fetus. The ears and eyes approach their
final position. The fetus performs breathing movements, pumping amniotic
fluid in and out of its lungs, and can also carry on sucking movements. By
the end of the third month the fetus is almost 3 inches long.

After 3 months of pregnancy the uterus begins to increase significantly in
size in order to accommodate the rapidly growing fetus. The mother's
breasts enlarge, and her abdomen begins to become stretched.

**The fetus at 5 months.** By 5 months of development the fetus measures
about 10 inches, half the length it will attain by the time of birth. It weighs
only slightly more than a pound. Most of its body is covered by a downy fetal
hair referred to as *lanugo*. The fetus can move freely within the amniotic cavity, and during the fifth month the mother becomes aware of fetal movements
(quickening).

**Six months to birth.** At 6 months of development the skin of the fetus has a
wrinkled appearance, probably because it is growing much faster than the
underlying connective tissue. If born prematurely at this age the fetus will
move, cry, and try to breathe, but almost always dies. Its brain is not yet suf-

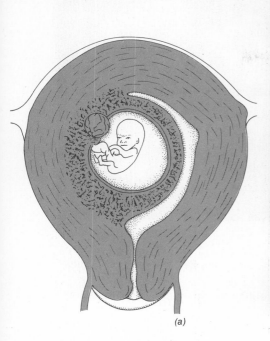

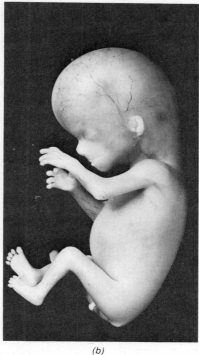

(a)          (b)

**FIGURE 13-14**
(a) Diagram of 10-week embryo to show position in uterus. (Redrawn from Demarest and Sciarra) (b) 10-week embryo. (Carnegie Institution of Washington)

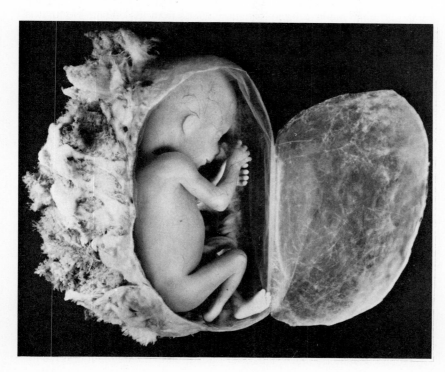

**FIGURE 13-15**
At 16 weeks the human fetus measures a little more than 5 inches in length. (Carnegie Institution of Washington)

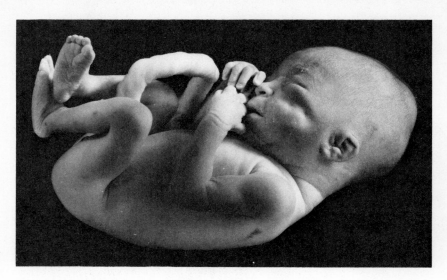

ficiently developed to sustain vital functions such as breathing and regulation of body temperature.

During the seventh month the cerebrum grows rapidly, and the cortex becomes convoluted. The sucking reflex, the grasp reflex, and several other reflex responses are present. In the last 3 months or so of prenatal life the fetus practices sucking movements and may suck its thumb. During the eighth and ninth months the fetus hears sounds from the outside world. It is capable of simple learning and can be taught to respond to certain stimuli.

During the last months of prenatal life the fetus gains weight rapidly. Fat is deposited in the subcutaneous tissue so that the skin no longer appears wrinkled. Hair may begin to grow on the head, and during the last few weeks the lanugo disappears. The skin becomes covered by a protective creamlike substance, the *vernix*. At birth, the full-term baby weighs about 7 pounds (3,000 grams) and measures about 20 inches in length.

**Birth**

From the time of conception about 266 days are required for the baby to complete its prenatal development. Several days before birth the fetus usually assumes an upside-down position so that its head will enter the birth canal first. Late in pregnancy the placenta begins to degenerate.

The term *labor* refers to the process by which the baby is expelled from the uterus. As the fetus begins to outgrow its home, the uterus is distended to its fullest capacity. The tissue of the uterus may become very irritable and may tend to contract in order to rid itself of its burden. Hormonal events are also known to be important in initiating labor.

Labor may be divided into three stages. During the first stage the cervix dilates in order to allow passage of the fetus. This stage begins when the mother starts to feel uterine contractions. The sensation accompanying each contraction usually begins in the lower-back area and extends around to the front of the abdomen. As labor proceeds, the contractions become more intense, more frequent, and more rhythmic, so that they occur at regular and decreasing intervals. The first stage of labor ends when the cervix has dilated sufficiently to allow passage of the fetus. This stage is the longest period of labor, lasting as long as 12 hours or more in a woman having her first child.

During the second stage of labor the baby is delivered. The second stage begins when the baby's head passes into the vagina which has become greatly distended to accommodate it. By contracting her abdominal muscles the mother can help push the baby along through the vagina. The amnion often ruptures during the second stage of labor, releasing the amniotic fluid. This event, popularly referred to as the "breaking of the bag of waters," sometimes occurs earlier in labor or may even precede labor.

As the baby moves through the vagina, its head rotates in order to facilitate passage. Just before birth the physician often makes a surgical incision, called an *episiotomy,* extending from the vagina toward the anus. This procedure facilitates expulsion of the baby and prevents tearing of maternal tissue which might otherwise occur. After delivery the cut is neatly sutured and heals within a few weeks.

The infant's head emerges slowly from the vagina. Once the head is born two or three additional contractions serve to eject the entire baby into the waiting hands of the physician. At this time the infant is still connected to the placenta by the umbilical cord. At present it is common practice to clamp and cut the cord immediately after the infant has been delivered. Usually, the newborn takes his first breath of air within a few seconds of birth and cries within half a minute.

During the third stage of labor the placenta separates from the uterus and is expelled from the mother's body. Generally this is accomplished within 10 minutes after the birth of the baby. Now referred to as the afterbirth, the placenta has no further function and is discarded.

At birth the infant begins life as an independent organism. For the first time he must breathe and eat on his own. Important changes take place within the circulatory system so that blood is routed separately through pulmonary and systemic circulations. Since the lungs cannot function during prenatal life, an extensive circulation through them would be pointless. Accordingly, blood is routed from the pulmonary artery through a special blood vessel (ductus arteriosus) into the aorta. At birth, muscle in the wall of this vessel contracts, constricting the flow of blood through it, and in effect rerouting blood into the lungs. By about 8 weeks after birth tissue grows into the vessel, permanently closing it and turning it into a solid cord of tissue. Another change involves the flow of blood through the heart itself. During development the wall between the right and left atria of the heart is not complete and blood entering the two sides of the heart mixes. At birth, when the infant begins to breathe, a flap of tissue is pressed into place closing the hole. Eventually this tissue flap grows firmly into place. When the umbilical cord is cut the flow of blood through the umbilical blood vessels is immediately interrupted. Slowly they become filled with connective tissue, and eventually they, too, are transformed into cords of fibrous tissue.

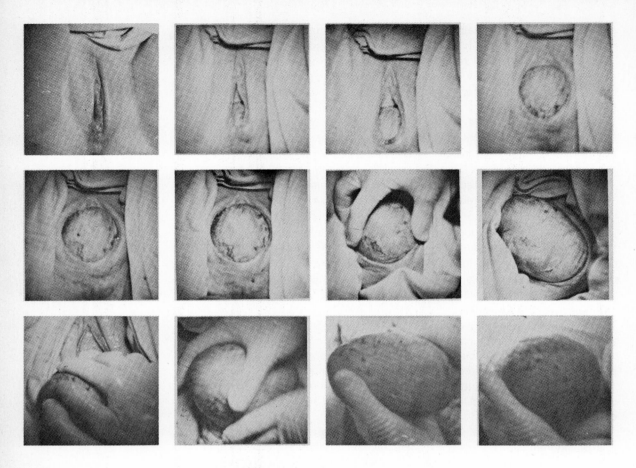

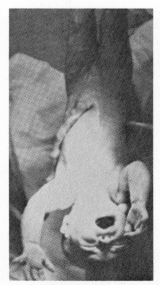

**FIGURE 13-17**
Normal delivery of a baby. (Dr. Roberts Rugh and Dr. Landrum B. Shettles, from Rugh and Shettles, *From Conception to Birth: The Drama of Life's Beginnings,* Harper & Row, 1971)

The termination of pregnancy initiates important changes within the maternal body also. Immediately after the placenta is expelled the uterus contracts, forming a hard structure, and within a few weeks it shrinks to one-twentieth its pregnant size. The thick lining of the uterus is shed slowly during a supermenstruation (technically known as *lochia*) which may last for several weeks following pregnancy.

## Nursing the infant

The large quantity of progesterone secreted during pregnancy stimulates glandular development of the breasts. Both the milk glands and the system of ducts develop in preparation for lactation (milk secretion).

Immediately after giving birth the mother's anterior pituitary begins to secrete sufficient quantities of *lactogenic hormone* to stimulate the prepared mammary glands to produce milk. For the first 2 or 3 days after birth a fluid called *colostrum* is secreted. Colostrum contains a higher concentration of protein than milk, but less fat. This fluid is nutritional and also serves to clean out the mucus, bile, and other substances that have accumulated in the newborn's digestive tract. By the third day after delivery, milk itself is produced by the mammary glands.

While lactogenic hormone is responsible for stimulating milk secretion, another hormone, oxytocin, secreted by the posterior pituitary gland, actually causes the milk to be released from the breast. The nipple contains erectile tissue, which becomes firm when stimulated by the baby's sucking. At this time a nerve reflex is initiated which results in the release of oxytocin by the pituitary. Oxytocin causes the smooth muscles around the milk ducts in the breast to contract and push the milk out.

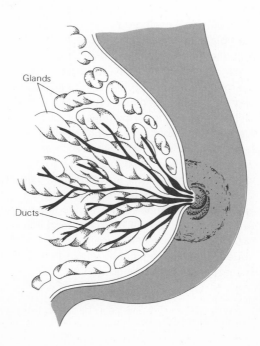

Glands

Ducts

**FIGURE 13-18**
Lactating breast. (Redrawn from Demarest and Sciarra)

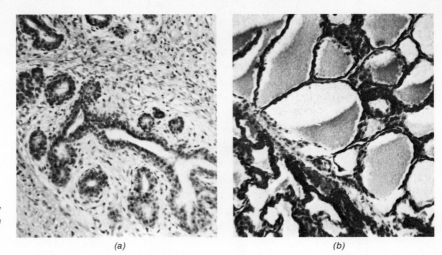

**FIGURE 13-19**
Tissue from inactive mammary gland as compared with tissue from lactating mammary gland.

*(a)*                                                                                     *(b)*

***Advantages of breast feeding.*** Breast feeding promotes the rapid recovery of the uterus, because as the infant nurses, oxytocin is released and stimulates the uterus to shrink back to its normal size. Oxytocin also helps to control uterine bleeding.

Another important advantage of breast feeding is the composition of human milk. Just as cow's milk is tailored for the nutritional needs of the calf, human milk is especially tailored for the nutritional needs of the human infant. Cow's milk contains more butter fat than human milk, and must be modified in order to make it nutritionally acceptable to the human infant. Thus, formulas composed of milk and other ingredients must be used, rather than plain homogenized milk. "Feeding difficulties such as 'spitting,' colic and allergic reactions are fewer and less severe in breast-fed infants."[2] Infants fed only human milk during the first weeks of life do not develop an allergy to cow's milk. It has also been reported that breast-fed infants have a slightly lower incidence of respiratory infections during the second 6 months of life. Mother's milk contains various viral and bacterial antibodies, but there is some controversy as to whether or not these are digested by the infant before they can be absorbed. In any event, such antibodies must bathe the oral and pharyngeal membranes of the infant, affording some protection to the upper respiratory and digestive gateways to the body.

Other advantages of breast feeding are that the milk is always available, is at the correct temperature, and is free from bacteria that can cause digestive upsets. Successful breast feeding affords a natural close physical relationship that is psychologically satisfying to both mother and child. As an added benefit, women who nurse their babies have a lower incidence of breast cancer than women who do not.

---

[2] W. E. Nelson, V. Vaughan, and R. J. McKay (eds.), *Textbook of Pediatrics,* W. B. Saunders, Philadelphia, 1969, p. 144.

**Disadvantages of breast feeding.**  Some mothers feel that breast feeding is inconvenient. In our society the breast has become so much a sex symbol that it is not considered socially acceptable to nurse babies in public. Mothers who do not wish to be confined to a schedule revolving about the privacy of the home or ladies' room may prefer the public convenience of the bottle. Many women have solved this problem by utilizing a combination of breast and bottle feeding. When it is inconvenient to nurse the infant, a formula is substituted. Another alternative is to extract milk by means of a breast pump. The milk can be frozen and given by bottle in the mother's absence.

Recently another disadvantage of breast feeding has developed. Human breast milk has been found to contain more DDT than cow's milk. This is principally because human beings are more carnivorous than cows; consequently they occupy a higher level on the food chain and ingest greater quantities of DDT. At its present concentration there is no evidence that the DDT level in human milk is harmful to the infant. Most authorities feel that the advantages of breast feeding far outweigh these disadvantages.

Many women who attempt to breast-feed their infants give up within a few days or weeks because of misinformation, lack of encouragement, and ignorance of what to expect. Even physicians are often not helpful. A woman who wants to nurse her baby should seek to inform herself thoroughly. Those interested in additional information on breast feeding should contact the La Leche League, an organization dedicated to helping mothers successfully nurse their infants.

## Novel origins

About 10,000 children born each year are products of *artificial insemination.* Usually this procedure is sought when the male partner of a couple desiring a child is sterile or carries a genetic defect. Although the sperm donor remains anonymous to the couple involved, his genetic qualifications are carefully screened by physicians.

*Artificial inovulation* is a technique by which an ovum is removed from a female's ovary, fertilized in a test tube, then reimplanted in her uterus. Still in its experimental stages, this procedure could be useful in some cases of infertility and would also provide the opportunity for checking the early embryo for certain defects.

Another new development is *host mothering*. In this procedure a tiny embryo is removed from its natural mother and implanted into a female substitute. The foster mother can support the developing embryo either until birth or temporarily until it is implanted into the original mother or another host. This technique has already proved useful to animal breeders. For example, embryos from a prize sheep can be temporarily implanted into rabbits for easy shipping by air, then reimplanted into a foster mother sheep, perhaps of inferior quality. Host mothering also has the advantage of allowing an animal of superior quality to produce more offspring than would be naturally possible. In one recent series of experiments mouse embryos were frozen for up to 8 days, then successfully transplanted into host mothers. Host mothering may someday be popular with women who can produce embryos but are unable to carry them to term. Similarly, professional women who prefer not being bothered with pregnancy might employ host mothers to carry their offspring for them. An entirely new occupation may thereby emerge!

## REGULATION OF DEVELOPMENT

From our earlier discussion it should be clear that development proceeds in a predictable, orderly manner. When we consider the complexity of the process, we marvel at the precision with which a single-celled zygote gives rise to the intricate structures and organ systems of a functioning organism. How does each cell "know" where to go and in what specific way to differentiate? Such questions have challenged investigators for many years. Thanks to the perseverance of many biologic detectives searching for clues in their laboratories, we now have at least a basic understanding of developmental processes.

### Role of the genes

The genes contained within the nucleus of the zygote are an elaborate set of instructions for making a specific organism. A human zygote always develops into a human baby, never into a calf or puppy, because the information in its genes is specific for making a human organism. Furthermore, one human may be tall with blue eyes, while another is short with brown eyes. These, as well as thousands of other characteristics, are determined by the types of genes present in the zygote.

When the zygote divides by mitosis to form two cells, a complete set of genes is distributed to each new cell. As cleavage and later development proceed, more and more cells are formed by mitosis, and every one contains the exact number of genes and type of genetic information present in the zygote. Now if all the cells of the organism have an identical set of genes with the same instructions, how can they become different from one another?

Genes control the activities of the cell by providing the instructions for making specific types of proteins. (Details of this concept are discussed in Chapter 14). Apparently, then, different genes function in different types of cells. In a muscle cell the genes for synthesizing muscle protein are active, while in a nerve cell genes specifying muscle protein do *not* operate. Instead, genes with instructions for making protein specific to nerve cells are active. The next obvious question is, what mechanisms operate to activate certain genes while repressing others?

### Cytoplasmic factors

Perhaps the initial influence upon differentiation is the distribution of materials in the cytoplasm of the zygote itself. If the zygote cytoplasm is not homogeneous, then during cell division the cytoplasmic substances portioned out to each new cell might also vary. The two new cells would differ from one

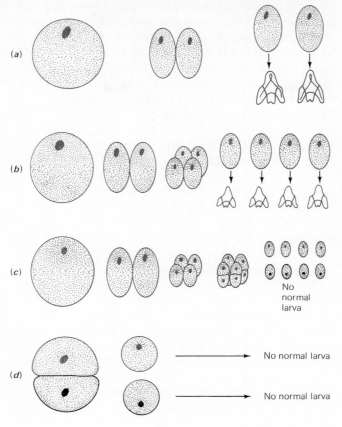

**FIGURE 13-20**
Experiments that illustrate the influence of cytoplasmic factors upon development. (*a*) When the first two cells of a sea urchin embryo are separated from one another, each cell is able to give rise to a normal larva. (*b*) If each of the first four cells of the sea urchin embryo are separated from one another, four normal sea urchin larvae develop. (*c*) When the cells of the eight-cell stage are separated from one another, not even one is able to develop into a normal larva. (*d*) If an experimenter causes the plane of cleavage of the first division to extend horizontally, instead of from pole to pole of the zygote, neither cell of the two-cell stage is able to develop into a normal larva. Apparently certain cytoplasmic substances necessary for normal development are heterogeneously distributed in the zygote so that neither half has all of the needed ingredients.

another with regard to their cytoplasmic content. Such differences could influence development.

Recent studies have shown that mitochondria contain DNA (deoxyribonucleic acid) and a system for synthesizing proteins. The zygote contains mitochondria from both the sperm and the egg, and these may be heterogeneously distributed throughout the zygote cytoplasm. It is quite possible that the mitochondria are differentially distributed during mitosis to the two daugher cells. If so, mitochondria may represent an agent by which cells become different from one another.

It is known that the genes themselves are affected by substances in the cytoplasm. In Chapter 14 we shall discuss how cytoplasmic substances turn specific genes "on" or "off." Thus, as soon as the embryo consists of cells that are different from one another in cytoplasmic content, differential gene activity may result, starting the cells down different paths of development.

### Influence of surrounding cells

The position of a cell in relation to its cellular neighbors is critical in determining its fate. One cell, or a group of cells, releases a substance that diffuses out among neighboring cells. This chemical is most concentrated in the vicinity of the cell which produced it and less concentrated farther away. Such a concentration gradient could provide information to the cells regarding their position in relation to the group, and this information could trigger the biochemical events that would determine differentiation.

In embryology a chemical that determines the future differentiation of nearby cell groups is called an *organizer*. Let us look at a specific example, the development of the nervous system. In the late 1920s the biologist Spemann performed a series of experiments which helped to explain why ectoderm cells in the specific region of the neural plate develop into nerve tissue, while other ectoderm cells give rise to the outer layer of skin.

Using early gastrulas, Spemann transplanted a tiny bit of ectoderm from the region which would normally develop into neural tube to the belly region of the embryo. The transplanted pre-neural tube ectoderm no longer formed neural tissue. Instead it joined forces with its new cellular neighbors to form outer skin. Likewise, when belly ectoderm was transplanted to the pre-neural tube region its fate was determined by its new surroundings and it contributed to the formation of neural tube.

When these same experiments are performed on older gastrulas a few hours later in development the results are entirely different. Now, when a piece of pre-neural tube tissue is transplanted into the belly region, it differentiates into neural tissue even though this seems out of place in the midst of the belly skin. And, as might be expected, transplanted belly ectoderm cells proceed to form skin cells even in the midst of the developing neural tube.

Apparently something happened between the early and late gastrula stages which accounted for the difference in results. Spemann solved this mystery by showing that the notochord cells lying just beneath the pre-neural tube ectoderm are responsible for the change. The notochord cells *induce* the ectoderm cells to form neural tube. By producing and secreting a chemical organizer, the notochord cells communicate with the ectoderm cells, stimulating them to differentiate into neural tissue. In the early gastrula the pre-neural tube cells have not yet been subjected to the organizer, so their fate has not yet been determined. Once these cells have been influenced by the organizer during late gastrulation, their fate is decided and cannot easily be altered. When a piece of notochord tissue is transplanted beneath the belly ectoderm of an early gastrula, a second neural tube forms in this region. These and thousands of similar experiments have provided evidence that development depends upon chemical communication among cells and the resulting differential activity of their genes.

### Influence of the external environment

Environmental conditions may also influence the course of development. Such factors as temperature, humidity, light, gravity, pressure, and chemicals in the surrounding environment are known to affect the embryo. It is logical then to suspect that if some cells in the embryo are subjected to slightly different environmental conditions than other cells, they might be affected in different ways. For example the cells on the outside of the gastrula experience different conditions than those lining the inside of the embryo. Such differences may influence the internal workings of the cells so that the genes are affected.

### Interaction of nongenetic with genetic factors

Although nongenetic influences are critical in determining differentiation, we must understand that the genes possess ultimate control. Let us consider a specific experimental example. When ectoderm from the mouth region of an early frog embryo is transplanted to the mouth region of a salamander, the surrounding salamander tissue induces the transplanted cells to develop into mouth structures. However, the mouth that forms is not the characteristic mouth of a salamander. Instead it bears the horny rows of teeth and horny jaws of a frog. Why? As you might guess, the frog cells are not *competent* to form structures specific to salamanders. Frog cells bear genetic instructions only for making *frog* structures.

### Building an embryo

We have seen that variations in cytoplasmic substances, exposure to organizers diffusing from cellular neighbors, and differences in environmental conditions can initiate differences among cells. Once two cells become different from each other by virtue of these influences, their internal chemical composition is affected. Sooner or later the genes in the nucleus are also affected. Certain genes are activated while others are repressed, resulting in the production of different types of proteins. Cellular differentiation, then, is an expression of genetic activity, and genetic activity in turn is influenced by a variety of factors within and without the cell. Although all cells of an organism possess the same set of genes, differential gene activity results in variation in chemistry, behavior, and structure among cells. In this way, an embryo comes to be composed of more than 200 types of cells, each exquisitely designed to perform its specific functions.

## ENVIRONMENTAL INFLUENCES UPON THE EMBRYO

We all know that a baby's growth and development are influenced by the food we give him, the air he breathes, the disease organisms that infect him, the chemicals or drugs he receives, and other factors. But how many of us stop to think that a baby's life begins at the moment of conception, and that his *prenatal* development is also affected by these environmental influences? In fact life before birth is even more sensitive to environmental changes than it is in the fully formed baby.

About 5 percent of all babies born alive, or 175,000 babies per year, arrive with a defect of clinical significance. Such birth defects account for about 15 percent of deaths among newborns. Birth defects may result from genetic or environmental factors or a combination of the two. In this section we shall discuss the environmental conditions that affect the well-being of the embryo. (Genetic defects will be discussed in Chapter 16.)

Substances or conditions that intrude upon the developing embryo from

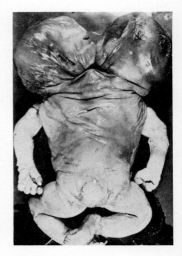

**FIGURE 13-21**

When some genetic or environmental factor prevents or interferes with normal development, an abnormal fetus may be produced. Two-headed monster. (McGraw-Hill Encyclopedia of Science and Technology, 3d ed.)

the outside environment may cause significant damage at one period in development, yet appear to be harmless during a later stage. Timing is important. Each developing structure has a critical period during which it is most susceptible to unfavorable conditions. Generally this critical period occurs early in the development of the structure, when interference with cell movements or divisions may prevent formation of normal shape or size, resulting in permanent malformation. Since most structures form during the first 3 months of embryonic life, the developing baby is most susceptible to environmental conditions during this early period. During a substantial portion of this time the mother may not even realize that she is pregnant and may therefore take no special precautions to minimize potentially dangerous influences.

## Nutrition

The diet of a pregnant woman is one of the most important environmental influences on the health of her developing child. Normal growth and development depend upon an adequate supply of raw materials. Growth rate is mainly determined by the rate of net protein synthesis by the cells. During the first 3 months of development it is especially critical that the mother's diet provide an adequate amount of essential amino acids, as well as other nutrients.

A study of infants born to undernourished women in a major American city showed twice the number of defects as those found in the general population. The brain and other portions of the nervous system are especially vulnerable to the effects of malnutrition, and permanent damage resulting in reduced ability to learn may occur. Recent studies showed that infants who had been deprived of essential nutrients during prenatal life possessed as few as 40 percent of the normal number of brain cells. Experimental studies using rats have shown that rat pups borne by malnourished mothers have 25 percent less norepinephrine in their brain neurons. (Norepinephrine transmits nerve impulses from one neuron to another in the brain.) Dopamine, another neurotransmitter, was also present in reduced quantity.

## Chemical influences

Anything that circulates in the maternal blood—nutrients, drugs, or even gases—may find its way into the blood of the fetus. Some drugs, as well as other types of agents, are known to be capable of interfering with normal development. They are known technically as *teratogens*.

*Vitamins.* Though proper nutrition demands an adequate intake of vitamins, overdoses of certain fat-soluble vitamins taken during pregnancy have been shown to be harmful to the developing embryo. Vitamins are generally prescribed for the pregnant woman by her doctor in order to ensure a sufficient supply of these basic nutrients to meet the demands of the embryo. Occasionally a pregnant woman concludes that if one vitamin capsule is good, four or five should be better, and the embryo is subjected to more vitamins than it can handle. As pointed out in Chapter 4, the fat-soluble vitamins are not easily excreted and may build up to toxic levels. For example, an adequate amount of vitamin D is essential to proper bone development in the fetus, but excessive amounts may result in a form of mental retardation. Oversupplies of vitamins A and K have also been shown to be harmful.

**Drugs.** Drugs are normally prescribed on the basis of weight. The dosage appropriate for a 120-pound woman may be 120 times too much for a 1-pound fetus. Some drugs tend to reduce or interfere with the rate of metabolism, and when taken at a critical time in development may result in permanent malformation.

In the liver of an adult, enzymes are present which act on certain drugs, converting them into other chemical forms. Such enzymes are often absent in the immature liver of the fetus, and so the drug may act differently upon his tissues than it does on adult tissues. The most dangerous time for a pregnant woman to ingest drugs is during the first few weeks of pregnancy.

Some very common drugs are suspected of causing birth defects. Among them are certain antihistamines, including a few which may be purchased without a prescription, certain antibiotics, and possibly even ordinary aspirin. Experiments have shown that when female rats are fed large amounts of aspirin, half their offspring are born with some type of birth defect, such as cleft palate. More recent studies have confirmed that aspirin does inhibit the growth of human fetal cells. Kidney cells are particularly susceptible. Aspirin appears to inhibit prostaglandins, which are especially concentrated in growing tissue. One pediatrician has suggested that if aspirin were a new drug, the government would not approve its being placed on the market.

One problem with proving that a drug is a teratogen is that its effects may be harmful on a subclinical level. That is, in the amounts generally used, no gross defect may result. Yet the fetus may be damaged in some subtle way; for example, an enzyme system may be disturbed. It is difficult to prove a cause-and-effect relationship. Furthermore, it may be almost impossible to relate a defect in a baby to some drug that the mother took 7 or 8 months earlier and has probably forgotten about. The situation is further complicated by the distinct possibility that two or more drugs at "safe" dosages may interact in teratogenic fashion.

*The thalidomide tragedy.* The tragic experience with the drug thalidomide focused worldwide attention on the possible harmful effects of drugs on development. Thalidomide, first developed and used in West Germany in 1957, was considered a mild sedative. Tested on rats and mice without ill effect, it was thought to be completely safe. Many pregnant women relied on this drug to relieve the nausea and vomiting common in early pregnancy. No one at first related its use to a sudden rising incidence of a birth defect called *phocomelia,* a condition in which babies are born with an extreme shortening of one or more limbs, often with no fingers or toes, and other organ defects (Figure 13-22). In the fall of 1961 at a meeting of pediatricians, attending physicians shared with one another their concern over the recent

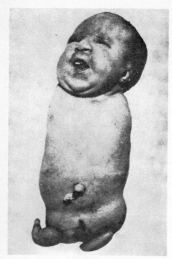

*(a)*

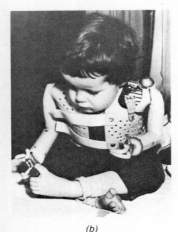

*(b)*

**FIGURE 13-22**
*(a)* Phocomelia due to thalidomide. (C. Scott Russell, Sheffield, England, and *Williams Obstetrics* by Nicholson
J. Eastman and Louis Hellman, Appleton-
Century-Crofts, New York) *(b)* Three-
year-old "thalidomide child"
using power-driven arms.
(OMIKRON)

increase in phocomelia. After much discussion and note comparing, and
even more painstaking inquiries and detective work, it was determined that
thalidomide was responsible for the more than 7,000 grossly deformed
babies born with this condition in 20 different countries. Thalidomide was
never approved for general use in the United States because of the cautious
attitude of a physician, Dr. Frances Kelsey, affiliated with the Food and Drug
Administration. Skeptical of the drug company's recommendation that the
drug be used to stop the nausea of pregnancy, Dr. Kelsey had requested additional information. Before the drug company had satisfied her, thalidomide
had been identified as the cause of phocomelia. A few cases of phocomelia
did occur in the United States as a result of experimental use of the drug by
physicians, and because a few American tourists had brought the drug back
with them from Europe.

Studies performed on thalidomide after the tragic episode have shown
that although the drug does not cause birth defects in rodents, it does cause
phocomelia when given to pregnant monkeys. Under FDA procedure, drugs
had to be tested on two types of animals, usually a rodent and some other
type, such as a dog or chicken. Primates were not generally used because
they are expensive to purchase and maintain. Now that we realize that our
annual death toll of more than 60,000 human beings from birth defects is an
expensive alternative, researchers are using primates more extensively.

You may wonder how thalidomide acts to produce birth defects. There is
evidence that the drug interferes with cellular metabolism in the developing
embryo, inhibiting normal cell division, especially in the developing arms
and legs. Since limb development begins with the emergence of tiny limb
buds during the fourth week of embryonic life, it is not surprising that this
drug is most damaging when taken by the pregnant woman during the
fourth, fifth, or sixth week of pregnancy.

The thalidomide tragedy has served as a terrible lesson from which we
have learned to look with suspicion at all drugs taken during pregnancy.
Many drugs have been removed from the market on the basis of more careful
testing inspired by this experience. Physicians have become more cautious
in prescribing medication for pregnant women. And hopefully, pregnant
women have become aware of the danger of taking any drugs without the
advice of a qualified physician.

*Effects of LSD and other drugs.* Recent studies on the effects of LSD on
development have yielded conflicting results. One study showed that expec-
tant mothers who had taken LSD had three times more birth defects among
their offspring and forty times more central nervous system defects. A later
study, however, failed to confirm these results. At the time of this writing the
matter remains unsettled. The same may be said for the effects of marijuana
and of alcohol. Alcohol is known to be damaging to the nervous system of an
adult user, which would lead one to suspect that there might also be an effect on a developing embryo. In another recent experiment, marijuana smoke
was shown to cause birth defects in rat embryos. Conclusive studies remain
to be done, however.

A word should be said about babies born to hard-drug addicts. Women addicted to narcotics such as heroin, morphine, or even methadone subject their developing offspring to daily doses of these drugs, which pass through the placenta and circulate through the embryo's body. These babies have high prematurity and mortality rates. Those which do survive must be placed on a regimen of prescribed drugs at birth so that they may be withdrawn slowly from dependence upon them. They are, in effect, born as infant drug addicts.

Sometimes women attempt to induce an illegal abortion by use of certain drugs. Such an attempt often backfires, resulting not in successful abortion but in offspring with birth defects.

Hormones administered to a pregnant woman may also interfere with development, while others such as DES (diethylstilbestrol) may cause damage years later (Chapter 5). Some compounds containing progesterone masculinize the fetus.

## Oxygen and other gases

The rapidly metabolizing cells of the developing embryo require a constant, plentiful supply of oxygen. If the mother's supply of oxygen is reduced for any reason, the amount delivered to the fetus is also diminished. The principal way in which the pregnant woman decreases the oxygen supply of the unborn child is by inhalation of carbon monoxide. Cigarette smoke and to a lesser extent automobile exhaust release carbon monoxide into the air which is inhaled. The effects of cigarette smoking on the embryo are discussed in Chapter 9. Suffice it to say here that when a pregnant woman smokes she deprives her unborn child of oxygen. This can result in retarded fetal growth and may cause other subtle forms of developmental damage.

In its more extreme form, maternal carbon monoxide poisoning has been shown to cause gross defects such as hydrocephaly. Laboratory experiments on rats have confirmed that when the embryo is deprived of sufficient oxygen at a critical time in its development, malformations result.

## Pathogens that cross the placenta

Some bacteria and viruses are known to pass through the placenta and infect the developing embryo. Such a prenatal infection may interfere with development and result in malformations, or may cause the fetus to be aborted. Among the pathogens known to infect the developing embryo are those which cause German measles, syphilis, smallpox, chickenpox, mumps, and malaria.

*Rubella (German measles).* The virus that causes rubella is one of the most devastating prenatal pathogens. The German measles epidemic that swept through the United States in 1963–1965 was responsible for an estimated 20,000 fetal deaths and 30,000 infants born with abnormalities.

Once the rubella virus infects the embryo it generally remains, interfering with the normal metabolism and movement of cells. The virus, still highly infectious, can be isolated from the baby for some time after birth. Children born with German measles syndrome often suffer from blinding cataracts, deafness, heart and other organ defects, and mental retardation.

The worst damage is done by the rubella virus during the early weeks of pregnancy when the rapidly dividing cells of the early embryo provide a luxurious environment for the invading virus. For this reason a pregnant woman who contracts German measles during the first month of pregnancy runs about a 50 percent risk of bearing a malformed child. During the second month the risk drops to 22 percent, and during the third month to 7 percent. Because of the high risk of serious damage to the embryo, a woman who contracts German measles during early pregnancy has strong grounds for therapeutic abortion. The recently developed vaccine against German measles should greatly reduce the number of pregnant women susceptible to this damaging virus, as well as the numbers of infected children — from whom adults usually contract the disease.

***Venereal disease.*** More than 40 percent of pregnant women with untreated syphilis transmit the disease to the fetus. An infected fetus may die and be aborted or may be damaged both mentally and physically. About 13,000 cases of congenital (existing at birth) syphilis are still reported annually, and in other thousands of children the disease is thought to be present but undiagnosed. An important part of prenatal medical care is testing the pregnant woman for possible syphilis. Such routine testing has greatly reduced the incidence of congenital syphilis in recent years.

Many years ago it was discovered that if a pregnant woman has gonorrhea, her baby's eyes can be infected as he passes through the vagina during the birth process. Such infection can result in blindness. To prevent this possibility, attending physicians routinely place either silver nitrate drops or penicillin in the eyes of a newborn.

### Other prenatal influences

Many other environmental influences affect the well-being of an unborn child. Recently it has been shown that a fetus can hear sounds of music, sonic booms, and other loud noises. The heartbeat of the fetus may be speeded in response to loud sounds which do not appear to bother the mother.

Ionizing radiation was one of the earliest teratogens to be identified. The incidence of birth defects and leukemia is higher among children of mothers subjected to radiation during pregnancy. Obstetricians no longer perform routine pelvic x-ray examinations on pregnant women because of the possibility of harmful effects.

Mechanical injury to the mother, such as falling, usually does not harm the fetus because of the shock-absorbing amnion which surrounds it. If the injury to the mother is severe enough to harm the fetus, spontaneous abortion often follows.

**1** Development is largely epigenetic, although the blueprints for an organism are preformed in the DNA of the zygote.

**2** Development is a balanced combination of growth, morphogenesis, and cellular differentiation.

**3** Cleavage takes place as the human embryo passes along the oviduct to the uterus.

**4** The blastocyst forms in the uterus and begins to implant itself in the thick uterine wall.

**5** During gastrulation the three basic germ layers are formed, each of which gives rise to specific adult tissues.

**6** The neural tube is formed by a process of growth and precise cellular movement.

**7** The placenta serves as the organ through which materials are exchanged between mother and embryo.

**8** The amnion serves as a shock absorber and provides a fluid medium in which the embryo develops.

**9** Each structure of the embryo develops according to a precise time sequence, so that two embryos of the same age are identical with respect to their stage of development.

**10** Labor may be divided into three stages, with the actual delivery of the baby occurring during the second stage.

**11** Human milk is especially tailored for the nutritional needs of the human infant. Breast feeding affords several other advantages, one of them being the rapid recovery of the uterus.

**12** Development is regulated by the interaction of genes with cytoplasmic factors, chemicals released by surrounding cells, and factors external to the cells.

**13** By controlling environmental factors such as nutrition, cigarette smoking, and drug intake, a pregnant woman can help ensure the well-being of her unborn child.

**14** The pathogens that cause rubella and syphilis are among those which can infect a developing embryo and cause birth defects.

**FOR FURTHER READING**

Demarest, Robert J., and John J. Sciarra: *Conception, Birth and Contraception,* McGraw-Hill, New York, 1969. Contains excellent illustrations of human development and the birth process.

Guttmacher, Dr. Alan F.: *Pregnancy and Birth,* New American Library, Signet Books, New York, 1962. A readable, informative paperback.

Kerr, Norman S.: *Principles of Development,* Concepts of Biology Series, William C. Brown Company, Publishers, Dubuque, Iowa, 1967. A brief description of developmental processes and their regulation, including accounts of important pertinent experiments.

Montagu, Ashley: *Life before Birth,* New American Library, New York, 1964. An excellent paperback, recommended highly for every prospective parent. Discusses the environmental factors that may influence the development of the unborn child.

Patten, Bradley M.: *Human Embryology,* McGraw-Hill, New York, 1968. An in-depth account of human development, well written and well illustrated.

Rugh, Roberts, and Landrum B. Shettles: *From Conception to Birth,* Harper & Row, New York, 1971.

# 14

# THE BLUEPRINTS OF LIFE

*Chapter learning objective.* The student will be able to (1) describe the general course of information flow within the cell from DNA to finished protein, (2) describe in general terms the chemical nature of the various nucleic acids and their main known functions and to explain* mutation in molecular terms, (3) describe the nature of viruses and viral infections, (4) give the general structure and function of ribosomes, and (5) speculatively explain* cancer, hormonal responses, and cellular differentiation in terms of nucleic acids and genetic control mechanisms.

*ILO 1* Describe the experiments of Hämmerling and Brachet upon *Acetabularia*.

    *a.* Show how they support the hypothesis of nuclear control of cellular characteristics.

    *b.* Show how they also indicate the existence of an essentially nonreplicating messenger substance which mediates between the nucleus and cellular characteristics.

    *c.* Analyze these experiments in terms of the scientific method (see Chapter 2).

*ILO 2* Give the general sequence of information transfer from DNA to finished protein.

*ILO 3* Describe or diagram the basic chemical structure of a nucleic acid strand, and distinguish chemically among DNA and the varieties of RNA.

*ILO 4* Given the base sequence of one strand of DNA, predict that of a complementary strand of DNA *or* of RNA.

*ILO 5* Give the known causes and a chemical basis of mutations, describe their general effects, and briefly discuss the social and ethical problems raised by artificial mutagens.

*ILO 6* Explain* the origin of mutations by nucleotide miscopying, relating this to the relationships among DNA, RNA, and protein synthesis.

*ILO 7* Describe the structure and molecular function of ribosomes.

*ILO 8* Describe the structure and life cycles of the three main varieties of virus in terms of nucleic acids and their functioning.

*ILO 9* Describe the hypothesis of Jacob and Monod, giving their experimental bases and employing correct terminology.

*ILO 10* Develop hypotheses similar to those of Jacob and Monod with application to embryonic differentiation and hormone action.

---

* In this context, "explain" has a precise meaning.

One of the puzzles of biology has always been the way in which characteristics are transmitted from generation to generation. Take a hen's egg, for example. What is there to see, once you have broken it open, but nearly formless yolk and white? But 3 weeks' incubation sees these mayonnaise ingredients transformed into a peeping creature that can breathe, see, and learn. You or I couldn't accomplish *that* deliberately, but the hen accomplishes it without effort and in the course of nature. How?

## GIANT CELLS

One of the difficulties of cell biology has always been the minute size of its subjects. Fortunately, a number of kinds of giant cells exist to brighten the lives of investigators.

### Introducing Acetabularia

In the imaginations of the more romantically inclined biologists, the little seaweed *Acetabularia* resembles a mermaid's wineglass. Less imaginatively, it has been described as looking like a little green toadstool measuring, at most, 2 or 3 inches in length. Although it is a typical alga of tropical seas, it also occurs in some subtropical waters if they are both shallow and somewhat rocky.

Nineteenth century biologists discovered that this insignificant underwater plant consists of a single giant cell. Small for a seaweed, *Acetabularia* is gigantic for a cell. It consists of (1) a rootlike holdfast, (2) a long cylindrical stalk, and (3) at sexual maturity, a cuplike cap. The nucleus is found in the holdfast, about as far away from the cap as it can be. In due course, the nucleus divides meiotically and its progeny of pronuclei swim up

**FIGURE 14-1**
*Acetabularia,* collected near the Florida Keys. (Roy Lewis, III)

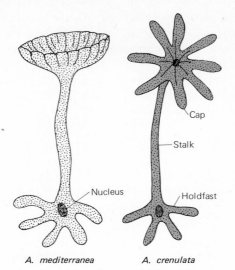

**FIGURE 14-2**
How *A. mediterranea* and *A. crenulata* differ.

Cap

Stalk

Nucleus

Holdfast

*A. mediterranea*          *A. crenulata*

the stem into the cap, where they become the pronuclei of sex cells. These will be released upon maturity to swim away in search of partners. Although there are several species of *Acetabularia,* with caps of different shapes, all species function similarly.

### Hämmerling's and Brachet's experiments
If the cap of *Acetabularia* is removed experimentally just before reproduction, another one will grow after a few weeks. Such behavior, common among lower organisms, is called *regeneration.* This fact attracted the attention of investigators, especially Hämmerling and Brachet, who became interested in the relationship that might exist between the nucleus and the physical characteristics of the plant. Because of its great size, *Acetabularia* could be subjected to surgery impossible with smaller cells. These investigators and their colleagues performed a brilliant series of experiments that in many ways laid the foundation for much of our modern knowledge of the nucleus. In most of these experiments they employed two species of *Acetabularia, A. mediterranea,* which has a smooth cap, and *A. crenulata,* with a cap broken up into a series of fingerlike projections. Since we will become quite familiar with *Acetabularia* in the next few pages, let us establish some nicknames—*A. med* and *A. cren* will do henceforth.

Now the kind of cap that is regenerated depends upon the species of *Acetabularia* used in the experiment. As you might expect, *A. cren* will regenerate a *cren* cap, and *A. med* will regenerate a *med* cap. But it is possible to graft two capless plants of different species together. Through this union, they will regenerate a common cap which has characteristics intermediate between those of the two species involved. Thus, there is something about the lower part of the cell that controls cap shape.

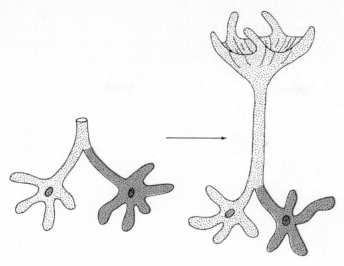

**FIGURE 14-3**
When the holdfasts of the two species of *Acetabularia* are grafted together, a cap of intermediate shape appears.

***Stalk exchange.*** It is possible to attach a section of *Acetabularia* to a holdfast that is not its own by telescoping the cell walls of the two into one another. In this way the stalks and holdfasts of different species may be intermixed.

First, we take *A. med* and *A. cren* and remove their caps. Now, we sever the stalks from the holdfasts. Finally, we exchange the parts.

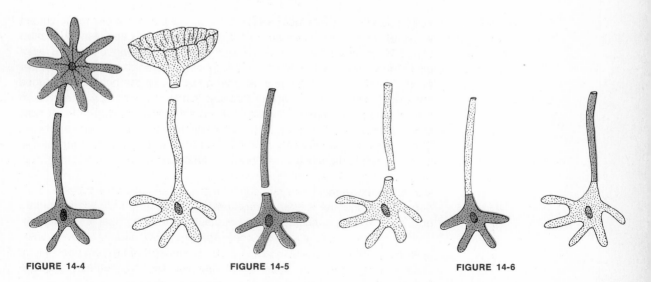

**FIGURE 14-4**          **FIGURE 14-5**          **FIGURE 14-6**

What happens? Not, perhaps, what you would expect! The caps that regenerate are characteristic, not of the species that donated the holdfasts, but of those that donated the *stalks!* (Figure 14-7.)

However—if the same caps are removed still *again, this* time the caps that regenerate *will* be characteristic of the species that donated the holdfasts. That will continue to be the case no matter how many more times the regenerated caps are removed (Figure 14-8).

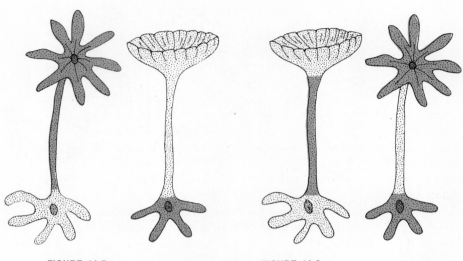

FIGURE 14-7                    FIGURE 14-8

From all this we may deduce that the ultimate control of the cell is vested in the holdfast, for from now on, no matter how often the caps of these grafted plants are removed, *they are always regenerated according to the species of the holdfast.* However, there is a time lag before the holdfast gains the upper hand. The simplest explanation for this delay is that the holdfast produces some cytoplasmic temporary messenger substance whereby it exerts its control, and that initially the grafted stems still contain enough of that substance from their *former* holdfasts to regenerate a cap of the former shape. But this still leaves us with the question of what it is about the holdfast that accounts for its dictatorship. An obvious suspect is the nucleus.

**Nuclear exchange.** If the nucleus is removed and the cap cut off, a new cap will regenerate. *Acetabularia,* however, is usually able to regenerate only once without a nucleus. If the nucleus of an alien species is now inserted, and the cap is cut off once again, a new cap will be regenerated—characteristic of the species of the nucleus! If more than one kind of nucleus is inserted the regenerated cap will be intermediate in shape between the species that donated the nuclei.

There is only one reasonable explanation for these observations: the control of the cell exerted by the holdfast is attributable to the nucleus that is located there.

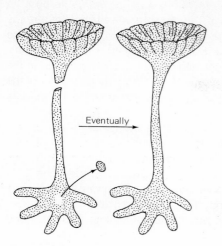

FIGURE 14-9

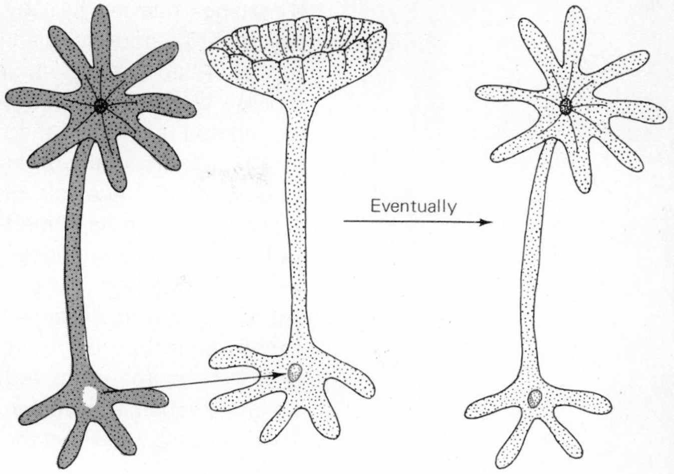

FIGURE 14-10

## The control of the cell
Summing it all up:

1  Ultimate control of the cell is exercised by the nucleus. In the end, the form of the cap is determined by the presence of the nucleus in the holdfast. In the long run, the only thing that can successfully compete for control with a nucleus is another nucleus.

2  Some control of the form of the cap is exercised by the nonnuclear parts of the cell, presumably the cytoplasm or something in it. But since that substance can exercise control for just one regeneration, it must be limited in quantity and unable to reproduce itself without the nucleus. Perhaps it is perishable as well.

3  The source of the messenger substance must be the nucleus.

## MIESCHER DISCOVERS NUCLEIC ACIDS

In the middle of the nineteenth century, Friedrich Miescher, a Swiss chemistry student at the University of Tubingen, began an investigation into the chemical makeup of the cell nucleus. To do that he needed to obtain a reasonably concentrated preparation of cell nuclei that he could employ almost as a chemical reagent. Because it has one of the proportionately largest nuclei of all cells, the white blood cell was an obvious choice for raw material. Fortunately, such corpuscles were available in quantity.

A very rich, though unpleasant, source of white blood cells is the liquid drainage of chronic infections, called pus. It may be regarded as a cellular battlefield strewn with the cadavers and dying specimens of struggling white blood corpuscles and invading bacteria. In the days preceding aseptic surgery virtually all healing was accompanied by infection and hence by pus. Making a virtue of a necessity, physicians of those days

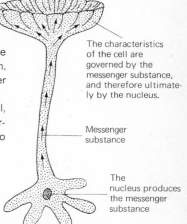

The characteristics of the cell are governed by the messenger substance, and therefore ultimately by the nucleus.

Messenger substance

The nucleus produces the messenger substance

FIGURE 14-11
The conclusions we must draw.

sometimes referred to it by the name (strange to us) of "laudable pus." In any case, the mid-Victorian hospitals of Tübingen afforded Miescher a ready supply. History does not record the reactions of his colleagues as he brought evil-smelling dressings back to the laboratory, extracted the pus, and treated it by incubating it overnight with gastric juice. The pepsin in the gastric juice naturally digested all the protein, leaving only the nuclei of the cells. Chemical analysis disclosed that the nuclei contained a phosphatic substance which he named *nuclein.*

Miescher subsequently continued his work in the Swiss city of Basel, where he switched to a pleasanter material for investigation, salmon sperm. The sperm cell is essentially just an animated nucleus. Furthermore, sperm can be squeezed easily from the male salmon in breeding condition. This improved source permitted him to determine that nuclein is a weak acid, which he renamed *nucleic acid.* Today we call it deoxyribonucleic acid, or DNA. Now we know that there are two basic kinds of nucleic acid, DNA and RNA.

## DNA

DNA may be considered a polymer comprised of many units called *nucleotides* linked in a long chain. In its repeating structure, it is somewhat comparable to proteins or polysaccharides. Each of these nucleotides consists in turn of three chemical parts: (1) a phosphate group, (2) a pentose or five-carbon sugar (deoxyribose, in the case of DNA), and (3) an organic, nitrogen-containing base which may be a *purine* or a *pyrimidine.* The sugars and phosphates combine to form the backbone of the DNA molecule:

. . . sugar-phosphate-sugar-phosphate-

sugar-phosphate-sugar-phosphate . . .

The organic bases are arranged approximately at right angles to the phosphate-sugar backbone. There are four kinds of bases:

adenine &#125; purines    cytosine &#125; pyrimidines
guanine             thymine

These bases are attached to the sugar groups of the backbone, but any base can be linked to any sugar, and they can occur in any order, for example:

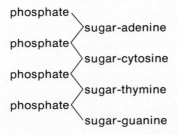

phosphate
      sugar-adenine
phosphate
      sugar-cytosine
phosphate
      sugar-thymine
phosphate
      sugar-guanine

Now although these bases may occur in any order, they do not occur in random proportions to one another in the cell nucleus. In the early 1950s a professor at Columbia University discovered what may be termed *Chargaff's law:* In any tissue of any organism, the proportion of adenine is approximately equal to the proportion of thymine. Furthermore, the proportion of cytosine is approximately equal to the proportion of guanine. Other proportions are not fixed.

Thus, if we were to analyze the DNA of organisms as diverse as frogs and fruit flies, we would always find the same proportional representation of bases: adenine = thymine, guanine = cytosine, adenine ≠ guanine or cytosine, thymine ≠ guanine or cytosine. Furthermore, though frog adenine = frog thymine, frog adenine ≠ fruit fly thymine. Different organisms have different proportions of these bases, *but always the complementary bases are equal in amount within the organism,* and significantly, this is true no matter which tissues of an organism are compared with one another, providing only that they came from the same individual.

Somewhat later, an American and a British investigator, Watson and Crick, working together at Cambridge University, proposed an explanation of Chargaff's law and much other previously mysterious data. According to them, chains of DNA exist in pairs linked together all along their length by precisely pairing bases. Thus, adenine will pair with thymine, cytosine will pair with guanine. These, they said, were the only possible combinations; adenine, for instance, will *not* pair with guanine or cytosine.

They went on to show that generally two strands of DNA are linked in a *double helix* (or coil) by means of their bases:

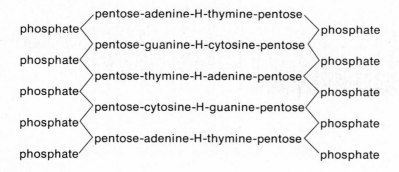

One can predict the composition of one strand if he knows the composition of the other. This is something like the relationship that exists between a photographic negative and its print.

In mitotic interphase the helix separates into its component strands, which disjoin somewhat as an opening zipper does. Under the control of enzyme systems each strand then forms a strand complementary to itself from free molecules. In each case, a new double strand comes into existence, and *two double* strands of identical base composition are formed.

Abundant and compelling evidence indicates that the material in which the cell's hereditary information is stored is DNA, and that DNA is largely

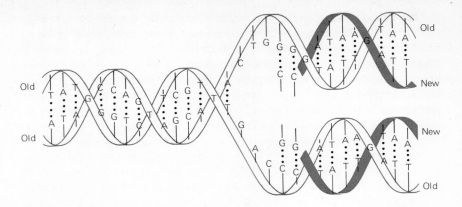

**FIGURE 14-12**
DNA replication.

confined to the nucleus. All nuclei, including that of *Acetabularia* which we have considered, control their cells by virtue of the information content of their DNA. *The information content of DNA is registered in the sequence and arrangement of its bases.* A particular item of such information is known as a *gene.* Each gene is a sequence of perhaps hundreds of nucleotides. We recognize the presence of a gene primarily by its obvious effects—whether it produces blonde hair or brown eyes, for example. The control of these obvious characteristics is exercised by the ability of the DNA to make proteins—especially enzymes—that dictate the ability or lack of it to manufacture the coloring matter of hair or eyes, or to perform any life process whatever.

## RNA

DNA is not the only nucleic acid known. The messenger substance of *Acetabularia,* for example, can be shown to be another nucleic acid, RNA, which resembles DNA in general chemical makeup but differs from DNA in two fundamental ways. The pentose sugar of DNA is deoxyribose, but the pentose sugar of RNA is ribose, which accounts for its name, ribose nucleic acid (or ribonucleic acid). Another important difference is that RNA possesses the base *uracil* in place of thymine, found in DNA. However, uracil pairs with adenine in much the same way that thymine does. RNA also contains a number of other odd bases, important chiefly in prokaryotic organisms such as bacteria.

Just as a single strand of DNA can build up a complement of itself under the proper biochemical conditions, so a negative image counterpart of DNA can be synthesized as RNA. This is called *transcription.* The pairing of bases takes place similarly but is not quite the same because of the substitution of uracil. Thus:

Cytosine in DNA leads to an opposing guanine in RNA
Guanine in DNA leads to an opposing cytosine in RNA
Thymine in DNA leads to an opposing adenine in RNA
Adenine in DNA leads to an opposing uracil in RNA

In the cells of eukaryotes,[1] DNA occurs primarily (but not exclusively) in the nucleus. RNA is found in large quantities *both* in the nucleus *and* in the cytoplasm. RNA occurs in three major forms:

*Ribosomal RNA (rRNA).* Complex little organelles associated with the endoplasmic reticulum, ribosomes are composed of protein and RNA. They seem to be manufactured in the nucleus, stored in the nucleolus, and perhaps released into the cytoplasm with the disappearance of the cell membrane during cell division.

*Messenger RNA (mRNA).* Messenger RNA is usually an unstable substance. It is manufactured by, and in contact with DNA. mRNA reflects the information content of the DNA coded in the sequence of bases occurring on the DNA strand. Messenger RNA will bear the complement of the genetic code specified on the DNA which has acted as its original template.[2]

*Transfer RNA (tRNA).* This nucleic acid is not in contact with DNA during protein synthesis, although tRNA is initially manufactured by the nuclear DNA. A modern conception of its structure is given in Figure 14-13. It is more or less double-stranded, with a group of three free and reactive bases on one end which comprise a *triplet.* These three bases provide a means of coupling with a strand of mRNA. Sixty-four triplet types are theoretically possible, and each variety of tRNA has a specific affinity for a particular one of the 20 amino acids found in natural protein. Thus a particular triplet sequence is associated with a specific amino acid. This particular amino acid chemically combines with tRNA (this is called *activation*), which then positions it precisely along a strand of mRNA. That strand of tRNA provides the scaffolding for the arrangement of amino acids into proteins. (This is called *translation.*) As we shall see, the ribosome plays a vital part in this process.

When the protein is complete, mRNA is released and can be reused, either to be formed into fresh mRNA with a different information content or to produce another strand of protein of the same kind. These relationships are all summarized in Figure 14-13.

Specific enzymes whose task is the destruction of mRNA exist in the cytoplasm. These enzymes ensure that outdated and inappropriate instructions will not interfere with the present needs of the cell. If more copies of a particular kind of mRNA are needed, they can always be made. In the cells of higher animals the average life of mRNA probably can be measured in minutes or hours. Anyone who has had to file reams of outdated memoranda should appreciate the logic of this arrangement. In effect, the cell burns its memoranda as soon as their instructions have been carried out.

---

[1] Organisms whose cells have nuclear membranes. See Chapter 17 and Appendix B.

[2] Very recent evidence hints that an additional RNA intermediate may exist which transfers information from DNA to mRNA in eukaryotes.

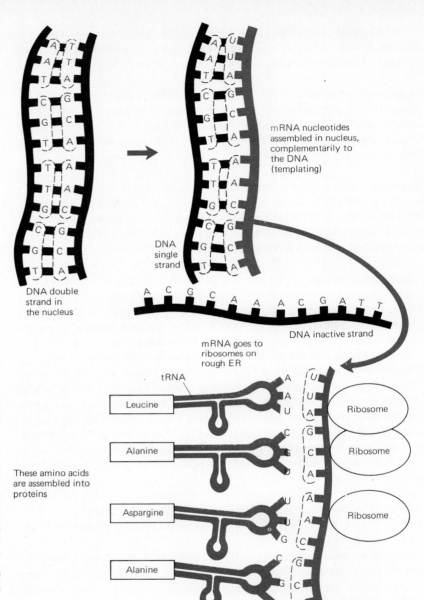

mRNA nucleotides assembled in nucleus, complementarily to the DNA (templating)

DNA single strand

DNA double strand in the nucleus

DNA inactive strand

mRNA goes to ribosomes on rough ER

tRNA

Leucine

Alanine

Aspargine

Alanine

Ribosome

Ribosome

Ribosome

These amino acids are assembled into proteins

**FIGURE 14-13**
Information transfer, DNA to protein. The nucleotide sequence of one strand of DNA is complementary to that of its mate. During protein synthesis the strands separate and the active strand assembles a unit of mRNA whose sequence is complementary to it. The mRNA then becomes associated in the cytoplasm with ribosomes which it binds together into polyribosomes. tRNA fragments are then assembled on the mRNA strand in a sequence complementary to it. Each tRNA bears an amino acid protein fragment. In this way the amino acid sequence of the finished protein is determined. Thus the nucleotide sequence of DNA ultimately determines the amino acid sequence of the finished peptide or protein.

## MUTATIONS

When DNA replicates or passes on its information to mRNA, the complementary bases are not always formed without error. There are a number of ways in which imperfections can take place, and these can be expected to interfere either with the control of the cell or with its heredity, depending on whether DNA or RNA has been erroneously replicated. Such miscopying is called *mutation.*

Mutations usually occur at a very low rate. Most cannot be accounted for by any known mechanism. Nevertheless, radioactive chemicals and other sources of nuclear radiation, many common chemicals, and other influences are capable of vastly increasing the rate of mutation. Under modern conditions most of the radiation we receive from artificial sources is small, and so we are likely to dismiss as insignificant the increased mutation rate from x-rays, television sets, and nuclear weapon tests. On the other hand, we are not dealing with a purely scientific problem when we do anything that affects human welfare. A slight increase in mutation rate worldwide will in fact produce thousands of congenitally damaged people. Are we justified in producing them?

Let us consider a concrete example of how a mutation might take place. (Please understand that the example we are about to employ is only one of several possible mechanisms that produce mutations, most of which are not discussed here.) Turn back to Figure 14-13.

For the sake of simplicity we will discuss changes in the strand of mRNA in this figure, although you understand that inheritable mutations must take place first in the DNA. In all cases, however, mutations *will* interfere with the base sequence and composition of the mRNA. Ultimately mutations change the amino acid sequence and therefore the properties of the protein produced by the mRNA.

Although the genetic information of the cell is registered in the base sequences of the DNA, the most meaningful unit of information on the DNA strand is not so much the base as the group of three bases known as the *triplet*. The triplet is like a word composed of three letters. You could not write a word without letters, but it is the word that is meaningful. It has to be this way for several reasons, but one reason is that tRNA attaches temporarily to mRNA by means of a triplet of bases. This lends importance to the mRNA triplets and, therefore, to the DNA triplets also.

Suppose the base sequence of the lowest triplet of the mRNA strand were to be altered, for example, to CGA. CGA is known to specify the amino acid arginine instead of the alanine that was originally dictated. That would produce a change—perhaps a minor change—in the protein that was being built.

Consider another example. Imagine that the uppermost triplet *loses* a base, such as the first base, uracil. That leaves U and A of the original triplet, but U and A do not constitute a triplet in themselves. Draw a circle around the U and A, and the G of the *next* triplet. That, you see, is what will now act as a *new* triplet, UAG. Now draw a circle around the next new triplet, CAA, and the next, ACG. It is known that UAG is a nonsense triplet, that is, that it has no amino acid equivalent. It would produce a break in the protein strand at this point; in fact it would probably end it. In the event that protein synthesis continued, however, the next new triplet, CAA, specifies glutamine, which formerly did not occur at all in the protein. ACG will produce threonine, another new component. These are major changes. Similar drastic changes may be expected to result from the *additions* of new bases to an mRNA sequence. This is known as a *frameshift mutation*.

One simple change in the base sequence of a DNA strand could affect the entire strand, from one end to the other, and could completely change the character of the protein which the DNA would produce. It has been shown in the case of the hereditary disease sickle-cell anemia, that the cause of the disorder is a defect in the hemoglobin molecule produced by the substitution of just *one* amino acid residue for another which normally occurs in a certain position. Since just one amino acid can make the difference between life and death if it takes place in an enzyme or other strategic protein, you can begin to see how very important mutations are to the life and health of all organisms, particularly since mutations are inherited. Every time a cell prepares to divide, its DNA replicates itself. If the DNA is defective, all copies of it will be similarly defective. All the cells in your body, and those of your offspring, could conceivably share a mutation.

## SMALL WONDERS: THE RIBOSOMES

We have emphasized the importance of ribosomes in protein synthesis, but we haven't explained how they accomplish it. Investigators are only now beginning to form some tentative idea of ribosome structure and function.

### What ribosomes do

You will recall that proteins consist of a linked chain of amino acid units called *residues,* held to one another by a special carbon-nitrogen chemical bond called the *peptide linkage*. The formation of the peptide linkage involves a chemical loss of water, called *dehydration,* and occurs only with enzymatic help.

The cellular synthesis of protein demands a number of enzymatic steps in addition to dehydration. The association of tRNA molecules with their complementary triplets on the mRNA strand requires enzymatic help, as does the establishment of the peptide bond and release of the finished protein. The ribosome probably does that complex sequential enzymatic work, while at the same time it precisely orients the mRNA strand so that it attaches properly to tRNA. Thus, the mRNA is "read" properly so that the right sequence of amino acids is built into the protein. The ribosome also seems to protect the forming protein molecule from the action of certain enzymes which might otherwise tend to disrupt it.

### The construction of ribosomes

Ribosomes are composed of two units, the 50S and the 30S. For our purposes, it is sufficient to know that the 50S particle is bigger than the 30S. If we mix 50S and 30S units under the proper conditions, they automatically unite to form ribosomes. Evidently, the 50S particle contains a trough-like hollow, and the 30S particle fits over it like a lid. Not all ribosomes have these precise measurements. Ribosomes from prokaryotic cells are smaller than those of eukaryotes, except for those found in mitochondria, which resemble prokaryotic ribosomes in size.

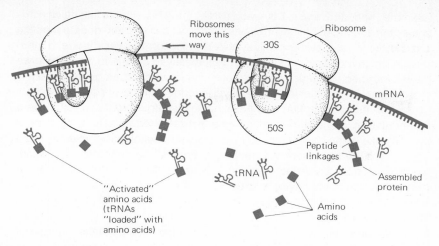

Ribosomes move this way ←

Ribosome
30S

50S

mRNA

Peptide linkages

Assembled protein

tRNA

Amino acids

"Activated" amino acids (tRNAs "loaded" with amino acids)

**FIGURE 14-14**

How ribosomes work—we think. Each ribosome consists of two particles, a 30S and a 50S. These designations refer approximately to their weight and size, so one is larger than the other. Chemically, ribosomes consist of protein (probably several linked enzymes) and RNA, whose function may be to align the mRNA precisely during protein assembly. A strand of mRNA is long enough so that several ribosomes may ride upon it at once, binding themselves together into the temporary structures called *polyribosomes*.

As the ribosomes move down the mRNA strand, that strand threads its way through the tunnel-like cavity between the 30S and the 50S particles. At the same time, tRNA particles bearing amino acids enter the cavity and are assembled into a polypeptide or protein—that is, the amino acids are, and the tRNA molecules are now free to pick up more amino acids in preparation for the next time. The action of the apparatus reminds one vaguely of a zipper.

However, the ribosomes and mRNA do more than just assemble random amino acid sequences. The amino acid sequence of the finished protein has been precisely dictated by the base sequence of the mRNA, which in turn was dictated by the sequence of the bases of nuclear DNA. Perhaps the most apt analogy from human experience would be to consider the ribosome as an automatic assembly line, governed by preprogrammed instructions via computer. And all this wrapped up in structures hardly visible even with a high-power light microscope!

Each of the two particles is composed of from one to three strands of RNA and several separate protein molecules. The RNA seems to serve as a molecular skeleton along which protein molecules are arranged. In a given cell all the 50S particles appear to be chemically identical, but the 30S units may vary somewhat in protein composition. Some scientists think that different kinds of 30S particles may be specialized to work with different kinds of mRNA in the production of various kinds of protein.

As we have tried to indicate in Figure 14-14, ribosomes travel down a strand of mRNA like beads on a string, sweeping up activated amino acids as they go, and producing finished protein chains.

### Ribosomes and antibiotics

Antibiotics of the streptomycin family interfere with the function of the ribosome in such a way as to produce errors in translation, that is, the protein formed is not the same as the specifications coded in the mRNA. Thus a high proportion of the protein manufactured by a streptomycin-treated cell contains the wrong sequence of amino acid residues. In many instances this will render enzymes nonfunctional. Fortunately, bacterial ribosomes are far more susceptible to confusion by antibiotics than are those of eukaryotic organisms like us! Unfortunately, mutant bacteria, whose ribosomes are resistant to streptomycin and related drugs, do exist.

## GENETIC HERETICS

When it first became clear that genetic information was encoded in the base sequences of nucleic acids, the question arose as to which nucleic acid was the fundamental source of that information. It was not at all obvious that the *basic* storage of information was in the DNA. Granted, it was clear that the nucleus controlled most of the life processes of the cell, but it was also true that the nucleus contained more than one substance—histone proteins, for

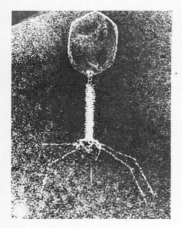

**FIGURE 14-15**
Electron micrograph of a bacteriophage virus. (*McGraw-Hill Encyclopedia of Science and Technology, 3d ed.*)

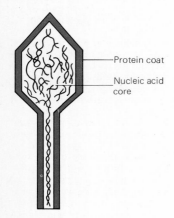

**FIGURE 14-16**
Diagram of a virus.

— Protein coat
— Nucleic acid core

example, and above all, DNA and RNA. There is almost as much RNA present in the nuclei of some cells as DNA. It was eventually decided, on the basis of very good evidence, that DNA did contain the basic genetic information which it passed on to RNA and ultimately to protein in the fashion we have described. Biologists also agreed, though without as good evidence, that information transfer probably never occurred in the opposite direction. To be fair, this assumption was frankly acknowledged from the first to be a matter of faith, and it was given a name that reflected its nature: *the central dogma.* Like many dogmas, it seemed useful and wholesome whether provable or not. We now know of partial exceptions to the central dogma which we suppose must be called genetic heresies. All these exceptions known at present involve the behavior of viruses.

### What is a virus?

Much earlier in this book we defined cells as the smallest units capable of independent life. Of course cells must be given the proper environment in order to live, but they all can produce their own energy from food substances, and except for certain specialized cells, they can reproduce themselves. Everyone has heard of viruses, some of which are responsible for a variety of diseases. Viruses, particles far smaller than cells, are not capable of any metabolic life of their own, nor can they reproduce themselves without the aid of living cells.

### How viruses infect cells

Since virus particles are not really "alive," they cannot swim or actively seek out a host. Chance seems to bring a drifting virus and host cell together. If there are a great many virus particles present, chance favors their finding appropriate hosts. At the time of contact, the virus's "tail" becomes strongly attached to the host's cell membrane. This initial step is known as *adsorption.* Next, the tail seems able to digest a hole in the cell membrane, evidently by some enzymatic means. Then the protein coat[3] of the virus contracts, squeezing the infectious nucleic acid into the cytoplasm of the host cell.

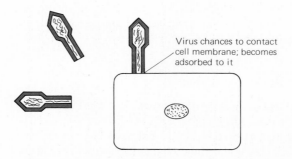

Virus chances to contact cell membrane; becomes adsorbed to it

**FIGURE 14-17**
How viruses attach to a cell.

---

[3] A few extremely tiny viruses have been discovered that lack a protein coat. Many viruses (including the influenza virus) lack tails.

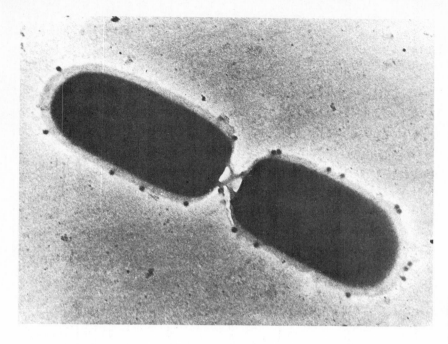

FIGURE 14-18
Bacteriophage viruses adsorbed
on bacterial cell wall. (*McGraw-Hill
Encyclopedia of Science and Technology*,
3d ed.)

Once inside, the nucleic acid does several things: (1) by some unknown means it puts a stop to all activity of the host's nucleic acid; (2) it replicates itself manyfold; (3) it assembles a number of proteins. One of these proteins (in some viruses) is a lysozyme enzyme which dissolves the cell membrane and in due course releases the daughter viruses. Other proteins become the protein coats of new virus particles. These chemical events are precisely timed. It would never do to have the enzymes that burst the cell formed before the virus particles are fully ready to emerge! But when the much-duplicated nucleic acid is fully packaged in new protein coats, a swarm of viruses bursts upon the world from the remains of their dead host, to repeat, if they can, their "life" cycle.

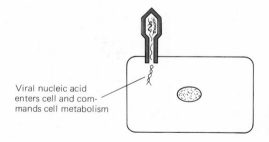

Viral nucleic acid
enters cell and com-
mands cell metabolism

FIGURE 14-19
How viral nucleic acid enters the
host cell.

## Kinds of viruses

Some viruses have a DNA core; others possess an RNA core. Fundamentally their behavior is the same, but there are some important differences. DNA viruses seem to behave in much the same way as the cell's own DNA; that is, they produce an mRNA that is able to manufacture protein. Although the

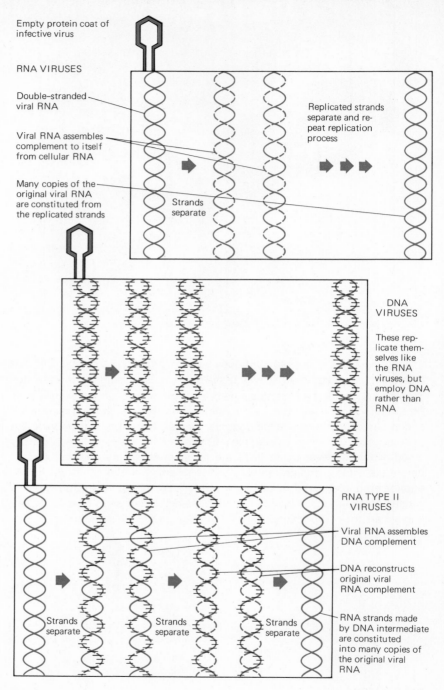

Empty protein coat of infective virus

RNA VIRUSES

Double-stranded viral RNA

Viral RNA assembles complement to itself from cellular RNA

Many copies of the original viral RNA are constituted from the replicated strands

Strands separate

Replicated strands separate and re-peat replication process

DNA VIRUSES

These rep-licate them-selves like the RNA viruses, but employ DNA rather than RNA

RNA TYPE II VIRUSES

Viral RNA assembles DNA complement

DNA reconstructs original viral RNA complement

RNA strands made by DNA intermediate are constituted into many copies of the original viral RNA

Strands separate

Strands separate

Strands separate

**FIGURE 14-20**
RNA and DNA viruses com-pared.

DNA replicates itself in the cell, no viral protein could be formed if mRNA were not produced.

RNA viruses are of two varieties. One type replicates itself by first forming many strands of RNA complementary to itself. The secondary strands then form tertiary strands of RNA identical to the original virus. A recently discovered second type of RNA virus forms an intermediate complement from the components of the cellular DNA. The DNA then forms RNA for the new virus particles. Many cancer viruses appear to be of this type.

Quite recently it has been shown that in animal cells some helical, double-stranded RNA occurs that is *not* dependent upon DNA for its formation, but is self-replicating. The chances are that such RNA actually represents dormant viruses, permanently resident in the cytoplasm of the cell. However, little is known about it as yet.


## THE CONTROL OF GENES

No cell uses all its genes at all times. Even in simple single-celled organisms, the demands of the environment vary. For the common intestinal bacterium *Escherichia coli,* for example, it is a rarity to grow in milk. For that reason, *E. coli* does not usually possess the enzymes necessary to break down and digest milk sugar, lactose.

But it is advantageous for the bacterium to have the capability to make the necessary enzymes, for there is *some* chance (unfortunately) that an intestinal microbe might find its way into milk. But if that is true, how does the bacterium "know" when to turn the appropriate genes on and off, and how does it accomplish this?


### Enzyme induction

The explanation proposed by the French investigators Jacob and Monod is expressed in Figure 14-21. They believed that a series of genes might be functionally associated in an entity which they called an *operon*. An operon, in their view, was a segment of DNA responsible for the production of an enzyme or a series of them. It would actually produce them, however, only when those enzymes were needed. The operon was inhibited from functioning by means of another short segment of DNA which acted somehow like a switch. When that segment, called an *operator gene,* was combined with repressor substance, the operon was "off." When the repressor substance was removed, the operon was "on" and ultimately produced its characteristic enzymes. Genes would also have to exist that produced the repressor substance.

Applying this hypothesis to the *E. coli* bacterium, when lactose was present it removed the repressor substance from the operator gene, probably by chemically combining with the repressor substance. The unencumbered operon then began to work, producing beta-galactosidase and the other enzymes necessary to metabolize lactose. Thus, the very presence of the substrate lactose triggered the formation of enzymes that could break

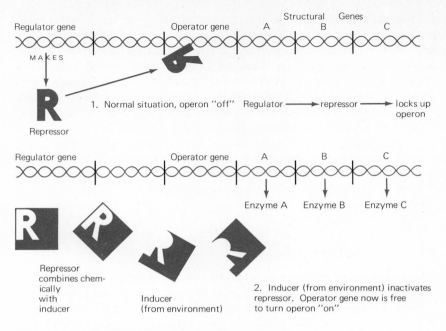

Regulator gene          Operator gene     Structural   Genes
                                       A       B       C

MAKES

R      1. Normal situation, operon "off"    Regulator ⟶ repressor ⟶ locks up operon

Repressor

Regulator gene          Operator gene     A       B       C

Enzyme A   Enzyme B   Enzyme C

Repressor combines chemically with inducer         Inducer (from environment)     2. Inducer (from environment) inactivates repressor. Operator gene now is free to turn operon "on"

3. When enzymes have completed their task and the substrate (inducer) is consumed, the repressor is free to combine again with the operator gene, which then turns the operon "off" again

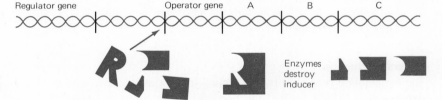

Regulator gene          Operator gene    A       B       C

Enzymes destroy inducer

**FIGURE 14-21**
The operon theory of gene regulation, according to Jacob and Monod.

down that substrate. Such stimulation of enzyme production by the *presence* of a substance is known as ***enzyme induction.***

Recently, the general validity of the Jacob-Monod hypothesis has been demonstrated rather conclusively. It has even been possible to isolate a lactose-associated repressor substance from living *E. coli* cells—no easy task, for there are only about twenty molecules of each such substance in any one cell. Nevertheless, the repressor substance has turned out to be a rather small, slightly acid, and chemically undistinguished protein.

### Gene control in higher organisms

No one has ever been able to demonstrate the existence of a similar mechanism in eukaryotic organisms such as ourselves, but it would seem that some mechanism is necessary to suppress the expression of inappropriate genes in our body cells. A nerve and a muscle cell presumably contain the same genes—yet obviously they are not all expressed, or not expressed in the same way. The genes of some cells are expressed fully only in the presence of appropriate hormones—for example, the growth of the beard produced by male sex hormone. Discovering how this can be explained is a major future task for biology. One current theory is diagramed in Figures 14-22 and 14-23.

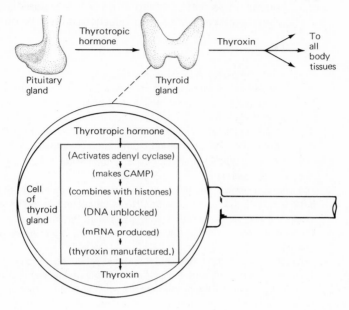

**FIGURE 14-22**
Possible mechanism of thyroid gland control.

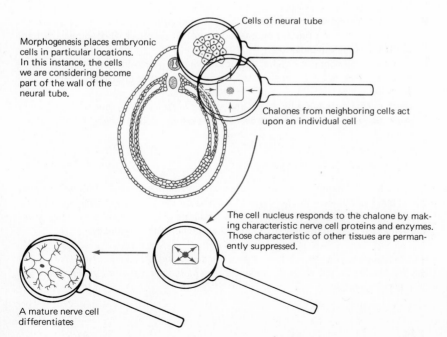

Morphogenesis places embryonic cells in particular locations. In this instance, the cells we are considering become part of the wall of the neural tube.

Chalones from neighboring cells act upon an individual cell

The cell nucleus responds to the chalone by making characteristic nerve cell proteins and enzymes. Those characteristic of other tissues are permanently suppressed.

A mature nerve cell differentiates

**FIGURE 14-23**
Possible mechanism for the control of cellular differentiation. A *chalone* is an intercellular hormone.

Modern research in nucleic acids is proceeding at an explosive rate. While we can expect many of these theories to change, it does seem that now we have the basic concept of the gene well in hand for the first time in the history of biology. But how does the gene—this particular configuration of DNA that produces enzymes, proteins, developmental processes, and ultimately all the characteristics of the organism—how is it passed from generation to generation so that offspring resemble their parents and yet are not identical with them? We explore this question in the next chapter.

**CHAPTER SUMMARY**

1 Experiments involving *Acetabularia* demonstrated that the cell is ultimately controlled by the nucleus through the means of a messenger substance.

2 Miescher discovered nucleic acids in the nineteenth century. More recently, Chargaff and Watson and Crick have clarified their structure. Nucleic acids consist of a pentose-phosphate backbone with a series of bases protruding at right angles to the molecular axis and joined to the sugar residues. DNA, and occasionally RNA, occur in a double-stranded helical form in which the base composition of one strand is complementary to that of its mate.

3 The principal known varieties of nucleic acid are DNA, mRNA, rRNA, and tRNA. Information encoded in the nuclear DNA is transferred to mRNA when the latter is made. mRNA, in conjunction with ribosomes, specifies a sequence of tRNA varieties, which in turn specify a particular amino acid sequence in the finished protein.

4 Mutation is a change in hereditary material produced by changes in enzymes or other proteins specified by the DNA. It follows from this that many mutations represent changes in the DNA or interference with the way in which information is transmitted from the DNA. The addition or deletion of a single base from a section of DNA (or gene) can produce a profound change in the finished protein.

5 Ribosomes consist of two particles, the 50S and 30S. They provide sequential enzymatic activity and precise orientation in protein synthesis. Ribosomes are made of RNA and protein molecules.

6 Viruses usually consist of a nucleic acid core and a protein coat. They reproduce by invading a cell and utilizing its metabolic machinery to synthesize more virus particles.

7 Mechanisms postulated by Jacob and Monod in their operon theory of gene regulation can account for the responses of some simple cells to changes in their environment.

**FOR FURTHER READING**

There is an abundance of books, articles, and learned papers dealing with nucleic acids, but it is not easy to find those that could be mastered by beginning students. The now-defunct *Science Journal* contains a number of useful articles, as does *Scientific American*. Here are a few of the more comprehensible references:

Brachet, Jean, and S. Bonotto: *Biology of Acetabularia,* Academic, New York, 1970.
Haynes, Robert C., and Philip C. Hanawalt: *The Molecular Basis of Life,* Freeman, San Francisco, 1968.

Ptashne, M., and W. Gilbert: "Genetic Repressors," *Scientific American,* June 1970, p. 36.

Rosenburg, Eugene: *Cell and Molecular Biology,* Holt, New York, 1971.

Watson, James D.: *Molecular Biology of the Gene,* 2d ed., Benjamin, New York, 1970.

# 15 INHERITANCE

**Chapter learning objective.** The student must be able to solve the basic types of genetic problems (monohybrid and dihybrid crosses, incomplete dominance, multiple alleles, sex linkage, etc.) and should be able to apply the concepts of Mendel's postulates to these problems. He must be able to describe the mechanisms of inheritance of such human traits as blood types, skin color (polygenic inheritance), color blindness, and chromosomal defects.

*ILO 1* Give Mendel's laws, or postulates, in modern terms, describe their chromosomal or genetic basis, and apply them to genetic problems such as those of ILO 5.

*ILO 2* Be able to explain (in this instance, "explain" has a precise meaning) Mendel's laws in terms of chromosome behavior.

*ILO 3* Recognize a state of genetic linkage, and, given an example, be able to describe the chromosomal basis of sex in human beings and to solve simple genetic problems involving sex linkage.

*ILO 4* Describe the inheritance of such human traits as color blindness, normal skin colors, and blood types. (Some of these require information given in the next chapter for their complete mastery.)

*ILO 5* Solve the following types of genetic problems: monohybrid cross, dihybrid cross, incomplete dominance, multiple alleles (e.g., inheritance of blood types), and sex linkage. (Some of these require information given in the next chapter for their complete mastery.)

*ILO 6* Be able to define the basic terms relating to genetic inheritance, for example, gene, dominance, recessiveness, codominance, chromosome, homozygous, heterozygous, alleles, homologous, genotype, phenotype, and pleiotropism.

Why do you resemble your parents? Even more interestingly, why is it that you do not resemble them exactly? You are not a duplicate of your mother, nor a carbon copy of your father, nor even an exact mixture of the characteristics of them both. This raises the fundamental question of just how the distinctive plans for the immensely complicated living machinery of your body were drawn up and expressed.

## THE NATURE OF HEREDITY

We have already approached the latter question in terms of the DNA of the cell nucleus which contains the blueprints of life. By specifying proteins and developmental processes, DNA specifies you. It follows from this that physical heredity involves passing on that DNA to one's offspring. It is also clear that the DNA of one's spouse is somehow involved as well. By a careful study of simple individual traits, such as eye color or blood type, it is possible to learn how inheritance is accomplished. Today, we call the specifications for these individual characteristics *genes,* and their study we refer to as the science of *genetics.*

Genetics has many practical applications. Animal breeders, for example, need to know something about it to help them improve varieties of cattle, dogs, and even bees. Seedsmen utilize it to produce high-yielding or disease-resistant varieties of vegetables, flowers, and grain. Genetics also has applications to theories of evolution. Our concern here is mainly with the way in which genetics affects us directly, and what, perhaps, we can do about it.

Obviously these considerations tend to focus our attention explicitly upon the genetics of man, but we do not wish to imply that the study of the genetics of other organisms is unimportant. Not only is such knowledge respectable in its own right, but also it casts considerable light on human genetics—which cannot, after all, be readily investigated by breeding experiments. Were it not for the genetic investigations possible with other organisms, genetics would be, like astronomy, a largely observational science, and would doubtless be far less advanced than it is.

### Inheritance—blending or particulate?

Until the end of the nineteenth century most people believed in "blending inheritance," a theory of heredity which held that the heredity of offspring can be expected to be intermediate between those of their parents. To discover whether this theory is valid, it is most convenient to digress from human genetics temporarily and look at the inheritance of color in a genus of garden flower, *Nicotiana,* the four-o'clock plant.

You may recognize the root of the word "nicotine" in *Nicotiana.* That is no coincidence, for it is related to the tobacco plant. Its English name arises from the fact that the flowers of this plant, somewhat resembling those of a petunia, do not open until late afternoon. There are a number of color varieties of *Nicotiana,* and these colors are inheritable. We will consider just three of them: red, white, and pink.

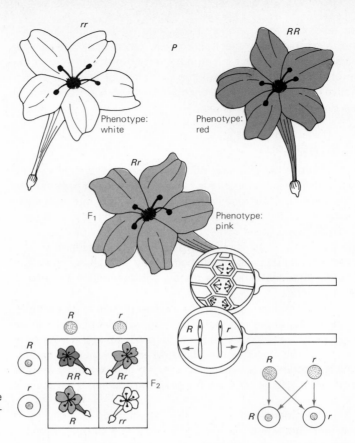

***A simple cross.*** If one were to cross a red nicotiana with a white, taking care to prevent any other fertilization from taking place, the resulting offspring would be all pink—a clear justification, you might think, of the blending theory of inheritance. On the other hand, the blending theory, if it were true, would lead us to predict that if two of these pink flowers were to be crossed, *their* offspring could only be pink. Is this prediction borne out by experiment? No! When pink nicotianas are crossed, one-half their offspring are indeed pink, but one-fourth are red, and the remaining one-fourth are white. This result can be explained only if the pink-flowering plants contain genetic information that is not expressed in them—information that later produces red and white flowers in their offspring. Though masked, those colors were somehow present in the pink-flowering plants; otherwise how could they reappear in their offspring? Indeed, the interaction of these traits of white and red must produce the pink-flowered trait when they are both present at once.

***Dominance and recessiveness.*** Many crosses do not exhibit even the appearance of blending inheritance. In sweet peas, for example, a cross of a red-flowering plant with a white-flowering plant produces offspring whose color is red![1] Even in this case, however, the masked trait for white color covertly persists in the red offspring, for if you cross two of these genetically impure second-generation plants, three-quarters of the offspring will be red, but one-quarter will be white. Now for some terms:

1  In any cross involving contrasting characteristics, the offspring may exhibit one or the other trait. The trait which is expressed is known as the *dominant* trait.
2  If neither contrasting trait is clearly dominant, we refer to them both as incompletely dominant, or *codominant*.[2]
3  The trait which is not expressed, or which is masked by its dominant counterpart, is known as the *recessive* trait.

Even though the masked recessive trait actually could not be seen in the mixed or *hybrid* plants, it reappeared when those hybrids were crossed. This shows that the recessive trait was not lost, but persisted, even though its expression was suppressed by the presence of the dominant characteristic. Since in four-o'clocks neither color is clearly dominant or recessive, we recognize the condition as incomplete dominance.[3]

### How to use genetic symbols

Figure 15-1 shows one way in which the inheritance of flower color in *Nicotiana* can be expressed and predicted by means of genetic symbols. Genetic symbols stand for genes. Any symbol may be used to represent a gene. In Figure 15-1 we represented the gene for "red" by *R*, and the gene for "white" by *r*. We could just as easily have used *R* and *W*, or 1 and 2, for that matter. Most geneticists favor the use of an uppercase letter for the dominant of a pair of characteristics, and a lowercase letter for its recessive counterpart, but there is nothing sacred about this system.

Here the parental generation (*P* in the figure) gives rise to a "first filial," or $F_1$, generation of pink hybrids. The $F_1$ is genetically impure, for each of them has *different* genes for color on the two chromosomes of the pair that bear

---

[1] Even in the sweet pea, dominance is not quite complete. Though your eye would never notice it, a sensitive electronic instrument can detect a slight lightening in the red color of the hybrid flower. As we shall see, hybrids often possess a bit less of the dominant characteristic than if they were genetically pure. This fact is of great medical importance.

[2] The terms *codominant* and *incompletely dominant* are not exactly synonymous. In codominance, both traits are expressed. In incomplete dominance, neither trait is fully expressed.

[3] For simplicity we have selected examples of genes that are clearly dominant, recessive, or codominant. Unfortunately, dominance relationships are not always clear-cut. Many or most genes express themselves in a way that is strongly influenced by whatever other genes may be present, that is, the *genetic background*. Breeding the famous tailless Manx cats, for example, produces some offspring which are tailed and some which are tailless. At first one might suppose that the tailless condition was dominant to the tailed. But if tailed cats are taken from the general population of cats on the Manx island and crossed with one another, tailless offspring are not uncommon. The most likely explanation is that other genes are able to influence the degree of dominance of the tailless gene, or to be more correct, its *penetrance*, so that in the presence of some genes it behaves as a dominant gene, and in the presence of others, as a recessive. In addition to variable penetrance, genes often exhibit variable expression, in that the trait they produce varies in severity in different individuals. Variable expression of genes is often due to the influence of other genes.

them. Such a state of genetic impurity is known as *heterozygosity.* The parents were genetically pure for color, or *homozygous.*

When meiosis takes place before gametes are formed, all the chromosomes (see Chapters 3 and 12) of the cells preparing for meiotic division line up along the equator of the cell. They are arranged in pairs, each pair of chromosomes bearing a set of genes in its DNA that govern the same traits. For example, in a pink nicotiana there would be at least one pair of chromosomes bearing genes for flower color. One chromosome of the pair would possess a gene for red color; its mate would have a gene for white color. The genes for color would occupy corresponding positions on the two chromosomes. The same thing would be true of all the hundreds of other genes occurring on each of these two, and other pairs of chromosomes would be similarly constructed. Genes governing the same trait and located in corresponding places on each member of a pair of chromosomes are known as *alleles.* The two members of each chromosome pair bear allelic genes. They are known as *homologous chromosomes,* for homologous chromosomes are members of a single chromosome pair.

Chromosomes, you will recall, become visible only in cell division, since they are really packages of DNA prepared for shipment to the poles. During interphase they lose their individuality in the nucleus. At that time if the genes contained in the chromosomes are not repressed, they are busy synthesizing protein. During the course of meiosis all homologous chromosomes become separated so that each gamete ends up with only one representative of each homologous pair. Thus each gamete can possess only one of these color-bearing chromosomes and therefore only one gene of a kind governing color. Thus a sperm or an egg of a pink four-o'clock plant is either *R* or *r,* but *not RR, rr,* or *Rr.*

The sperm cell of a flowering plant is in its pollen grain. Should the pollen of one $F_1$ plant have the opportunity to fertilize an ovum of another $F_1$ plant there is no way of telling what kind of zygote will result, for both kinds of sperm have an equal chance of fertilizing both kinds of egg. A *zygote,* or fertilized egg, could end up with the genes *RR, Rr,* or *rr,* depending on the genes possessed by the gametes that united to produce it. Of every four zygotes, one can reasonably be expected to be *RR,* one *rr,* and two *Rr* in genetic makeup, or *genotype.* There are no additional possibilities. The physical appearance, or *phenotype,* of each of these will also follow a $1:2:1$ ratio, for *RR* will be red, *Rr* pink, and *rr* white. If *R* were clearly dominant over *r,* as it is in sweet peas, then both *RR* and *Rr* would be red, and *rr* white. The combined proportions of *RR* and *Rr* would be three, and that of *rr,* one. $F_2$ sweet peas would fall into a $3:1$ phenotypic ratio, if sufficient numbers were present to provide a good sample.

### Using the Punnett square

The Punnett square is probably the best aid to working genetic problems that has ever been invented. An example of its use is found in Figure 15-1. On the top horizontal axis of the square one writes the genes possessed by

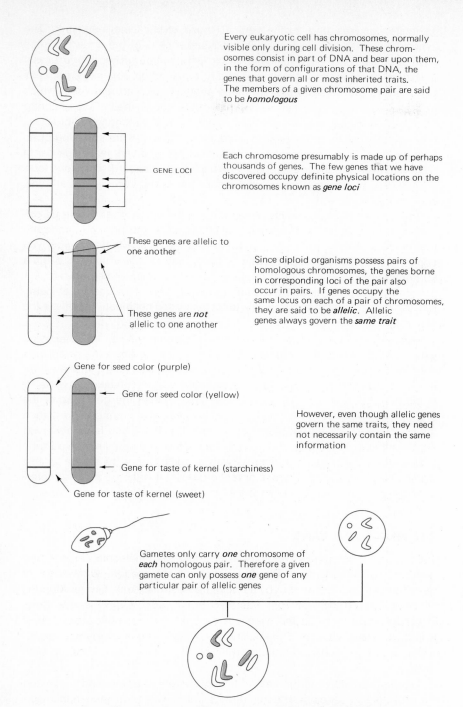

Every eukaryotic cell has chromosomes, normally visible only during cell division. These chromosomes consist in part of DNA and bear upon them, in the form of configurations of that DNA, the genes that govern all or most inherited traits. The members of a given chromosome pair are said to be *homologous*

GENE LOCI

Each chromosome presumably is made up of perhaps thousands of genes. The few genes that we have discovered occupy definite physical locations on the chromosomes known as *gene loci*

These genes are allelic to one another

These genes are *not* allelic to one another

Since diploid organisms possess pairs of homologous chromosomes, the genes borne in corresponding loci of the pair also occur in pairs. If genes occupy the same locus on each of a pair of chromosomes, they are said to be *allelic*. Allelic genes always govern the *same trait*

Gene for seed color (purple)

Gene for seed color (yellow)

However, even though allelic genes govern the same traits, they need not necessarily contain the same information

Gene for taste of kernel (starchiness)

Gene for taste of kernel (sweet)

Gametes only carry *one* chromosome of *each* homologous pair. Therefore a given gamete can only possess *one* gene of any particular pair of allelic genes

When the gametes combine to form a zygote, it, and the resulting embryo, will have homologous pairs of chromosomes, but of each pair, one member will be MATERNAL in origin, and one PATERNAL in origin. Each pair will bear allelic genes.

**FIGURE 15-2**
Homologous chromosomes and allelic genes.

the male gametes, and on the vertical, those of the female gametes. The zygote combinations are written in the internal cells of the square. In this case, a cross of two organisms hybrid for just *one* trait (***monohybrid cross***) is shown. Just one pair of alleles, *R* and *r*, is placed on *both* axes, because both parents are heterozygous for these genes. If you can remember which genes are dominant, you can easily predict the phenotypes of the offspring from the Punnett square—a $1:2:1$ ratio in the case of a monohybrid cross involving pink four-o'clocks. Bear in mind, though, that we are speaking of an *average* ratio. These exact ratios would occur only if a very large number of offspring were produced—a couple of bushels, perhaps. If only four four-o'clock flowers survived from such a cross, they might very well be all pink, or all white, or all red.

The Punnett square predicts more than just the ratios which would be observed in the case of large numbers of offspring; it also predicts probabilities in individual offspring. If you had one of the $F_2$ seeds before you, you could not predict its phenotype with certainty, but you *could* say that it has a 50 percent chance (2 out of 4) of being pink. Its chance of being white is 1 out of 4, or 25 percent, and the same could be said of its chance of being red. But the fact that it is most *likely* to be pink does not mean that this is the way it will turn out.

The Punnett square is useful in deducing genotypes when the phenotype exhibits a dominant characteristic. For instance, a red-flowering sweet pea plant could be either *RR* or *Rr*. Which is it? You could cross it with a white-flowering plant, *rr*. If so much as *one* of the offspring of that union is white (*rr* is the only possible genotype of a white flower), then the genotype of the original plant in question *had* to be *Rr*. If, however, all offspring are red, then the genotype of the parent plants could only be *RR*—if you have observed large numbers of offspring. If there were only a few, you could not be so sure. Crossing a heterozygous individual with a homozygous recessive in this fashion is known as a ***backcross*** or ***test cross,*** for it is the best way of determining heterozygosity. The phenotypes resulting are in a $1:1$ ratio.

|   | *r* | *r* |
|---|-----|-----|
| *R* | *Rr* | *Rr* |
| *r* | *rr* | *rr* |

**FIGURE 15-3**
A backcross.

## THE MENDELIAN LAWS

The roots of much modern biology extend into the nineteenth century, and genetics is no exception. The foundation of our modern knowledge of genetics was laid at that time by an obscure Austrian monk, Gregor Mendel, who lived in the town of what was then Brünn, Austria, and is now Brno, Czechoslovakia. Prior to the twentieth century, "natural philosophy," as it was then called, was not a profession but a hobby; only clergymen, physicians, wealthy people, or others with some leisure time were able to pursue it.

Mendel was active in the natural history society of Brünn, and presented his findings in a series of research reports which were published in the society's journal. His findings were revolutionary, for they overthrew the old blending theory of inheritance in much the fashion that we already have discussed. Despite this, they were almost universally ignored. One can only speculate on the reasons for this. One possibility is that scholars were obsessed with the new theory of evolution propounded by Charles Darwin and could spare no attention for monkish scholarship that bore no obvious relationship to it. An example of the attitude is afforded by a communication from the evolutionist Ernst Haeckel to the embryologist Wilhelm His. "We have better things to do in embryology than to discuss the tensions of germinal layers and similar questions," said Haeckel, "since all explanations must of necessity be of a phylogenetic [evolutionary] nature."

Whatever the reasons for this neglect, Mendel did not gain the recognition rightly due him for more than 30 years, when biologists first rediscovered these principles, and then rediscovered his papers during a prepublication literature search. By then, Mendel was deceased.

Mendel worked with garden peas, which exist in a number of distinct varieties differing in such characteristics as height, flower color, seed coat color, and seed coat texture. By good fortune, all the traits studied by Mendel occurred on different chromosomes. Pea plants also are normally self-fertilized, so simple surgery of the male flower parts makes the plant incapable of being fertilized at all except by artificial means. In this way, illegitimate offspring cannot occur, for the flower can no longer fertilize itself and cannot be fertilized naturally by an unknown plant.

## Mendel's postulates

Mendel's breeding experiments led him to certain conclusions which later scholars restated as Mendel's laws, or the postulates of inheritance.

1 Heredity is transmitted by unit factors (now called genes) which exist in pairs.
2 When gametes are formed, each gamete receives only one gene of each pair (the law of *segregation*).
3 When two alternative forms of the same gene are present in an individual, only one of the alternatives is usually expressed phenotypically (the law of *dominance*).
4 If one considers two or more independent characteristics in a cross, such as flower color and seed coat texture, each characteristic is inherited without relation to other traits. All possible combinations of independent characteristics thus will occur in the gametes (the law of *independent assortment*).

We have already considered examples of all but the last of these laws. In order to explain the principle of independent assortment we will leave the plant kingdom and begin with the first of many examples drawn from human genetics.

## Independent assortment

We now know that Mendel's fourth law applies only to traits which are carried on nonhomologous chromosomes. If different genes are carried on the same chromosome they will tend to be inherited together. However, if they occur on *different* chromosome pairs they will be inherited independently.

To illustrate this principle let's consider two pairs of human traits. The

**FIGURE 15-4**
Curling tongue, a dominant trait.

ability to curl the tongue into a tube is a dominant trait; the absence of this talent is recessive. Attached or absent earlobes are recessive, free earlobes are dominant.

If a man homozygous for curly tongue (*CC*) marries a woman without this trait (*cc*), all their children (*Cc*) may be expected to have this ability, since it is dominant and they are heterozygous for it. Similarly (but quite unrelated), if he has free earlobes (*EE*) and she is homozygous for attached earlobes (*ee*) the children will have free earlobes (*Ee*). One could write the children's genotype for both sets of traits as *CcEe*. Since the two sets of traits are due to genes found on separate chromosome pairs, they will be inherited (or *assorted*) independently of one another.

Let us suppose that—improbable as it is—some of these children were to marry others who have genotypes identical to their own. A variety of children could be expected to issue from their unions. For reasons of convenience let us suppose that the total number of children produced was 16.

As you can see by reference to Figure 15-5, four kinds of gametes are possible with respect to these genes—that is, since each gamete will have one of each pair of alleles, a total of four gametic genotypes is possible, given the genetic makeup of the $F_1$ generation. If, by chance, the chromosome bearing gene *E* is sorted by the meiotic apparatus into the same gamete as the chromosome bearing the gene *c*, the resulting sperm or egg will have the genes *Ec*. Similarly, if the chromosomes bearing *e* and *C* are assorted together, the gamete will have the genes *eC*. Other gametes would be *EC*, and still others would be *ec*. *There are no other possible combinations than these four.* Since allelic genes are always borne on chromosomes that are *separated* in meiosis, a gamete can ordinarily have no more than *one* copy of each allele. Since there are four possible kinds of gametes, working out a cross of that kind will require a box of 16 squares. Count the phenotypes. If you do so properly, the proportions, or ratios, of phenotypes to one another that you will predict among the offspring will fall into a 9:3:3:1 ratio.

## POLYGENIC INHERITANCE

Some traits are governed in their expression by more than one pair of allelic genes, known as *polygenes.* Sometimes such polygenes are located on different chromosomes and therefore assort independently. Not many such traits are known, for they are difficult to investigate. It is likely, however, that many human characteristics are inherited in this fashion and will eventually be proved to be so. One of the better-attested instances of polygenic inheritance is a human characteristic—skin color.

It is well known that if a person of extreme white or blond complexion marries a person of equally extreme black complexion, their offspring will be intermediate in color. If that were all there was to it, it would be an obvious

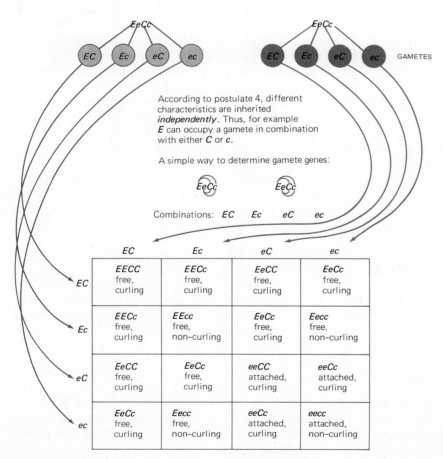

FIGURE 15-5
Independent assortment.

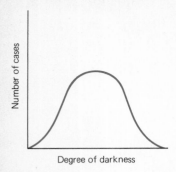

**FIGURE 15-6**
Shape of the frequency distribution for the progeny of the cross *AaBbCc* × *AaBbCc* (somewhat smoothed).

case of codominance. To test that assumption, let us observe the offspring of the union of two such intermediate persons. The assumption of simple codominance predicts a 1:2:1 ratio among their offspring similar to that which we observed in the four-o'clock monohybrid cross. However, what we will observe instead (if their family is sufficiently large, or if we observe offspring in several such families) is a complex distribution not easily expressed by any ratio, ranging from blond to very black, with the majority of intermediate color.

The explanation for this seems to be that human skin color is governed by at least three independent allelic pairs of genes, each pair located on a different chromosome. Thus, the cross black × white is *AABBCC* × *aabbcc*. The $F_1$ offspring can be only *AaBbCc*. What about the $F_2$? We have worked out some of the genotypes of the $F_2$ for you in Figure 15-7. It is up to you to determine their phenotypes, which is not difficult. *aabbcc* would be a blond so light that his blood shows through his skin. *AABBCC* would be very dark indeed. Intermediate degrees of darkness can be predicted by counting the number of capital letters in each genotype. Most of us, including most who are called "blacks," would fall into one of the intermediate genotypes.

Even in the case of a marriage between two mulattos, the chance of bearing a child darker than the darker parent is rather small. In the marriage of a "white" person to a "black" person (definitions here are imprecise) it is extremely unlikely that any legitimate child could be born that would be darker than the darker parent. But, contrary to what one sometimes reads, the chance *is* somewhat greater than zero, although admittedly minute.

When a person with some African ancestry marries a person with none, the question arises of the probable skin color of their children. In such a case it can be predicted with certainty that no dark offspring will be born; at least none will be darker than the darker parent.

Some human genetic characteristics besides skin color bear a social, "racial" meaning, and these can have an inheritance independent of skin color. Many persons light enough to be considered "white" may exhibit facial and hair characteristics commonly associated with "black" persons, and vice versa.

Many human traits which appear to be inheritable but which have a confusing pattern of inheritance eventually may be shown to be produced by polygenes. Still others may be strongly influenced by polygenes but not entirely attributable to them. Height, body build, susceptibility to certain mental and physical diseases, and what we call "intelligence" may be such characteristics.

## LINKAGE

Not all genes assort independently. Many form groups tending to be inherited together, or together with some major characteristic such as sex. As you might suppose, such association is usually due to the genes being borne on the DNA of a common chromosome.

Consider the cross: male AaBbCc × female AaBbCc

| Male gametes | Female gametes |
|---|---|
| ABC | ABC |
| ABc | ABc |
| AbC | AbC |
| Abc | Abc |
| aBC | aBC |
| aBc | aBc |
| abC | abC |
| abc | abc |

|  | ABC | ABc | AbC | Abc | aBC | aBc | abC | abc |
|---|---|---|---|---|---|---|---|---|
| ABC | AABBCC |  |  |  |  |  |  |  |
| ABc |  |  |  |  |  |  |  |  |
| AbC |  |  |  |  |  |  |  |  |
| Abc |  |  | AABbCC |  |  |  |  |  |
| aBC |  |  |  |  |  |  |  |  |
| aBc |  |  |  |  |  |  |  |  |
| abC |  |  |  |  |  |  |  |  |
| abc |  |  |  |  |  |  |  | aabbcc |

| Number of capital letters in each genotype | Number of cases |
|---|---|
| 6 | _____ |
| 5 | _____ |
| 4 | _____ |
| 3 | _____ |
| 2 | _____ |
| 1 | _____ |
| 0 | _____ |

**FIGURE 15-7**

Polygenic inheritance. Most geneticists believe human skin color to be an example of polygenic inheritance – i.e., governed by three separate, incompletely dominant sets of genes on three separate chromosomes. We shall call these gene pairs A, B, and C for those tending toward dark color, and a, b, and c for their light alleles.

| Phenotype | Genotype |
|---|---|
| Extremely black | AABBCC |
| Mulatto | AaBbCc |
| Extremely blond | aabbcc |

Since dominance is incomplete, skin color can be estimated by counting the number of capital letters in the genotype. A mulatto with five such capital letters, e.g., AAbBCC is darker than one with four, e.g., aaBBCC. AABbCc, on the other hand, is the same color as aaBBCC.

## Sex linkage

The sex chromosome pair is not homologous in the same sense that the 23 other (*autosomal*) chromosome pairs are (see Chapter 12). Male and female sex chromosomes are different in shape and obviously carry different genes for most of their length. Only a short segment of both chromosomes seems to be homologous and presumably carries allelic genes.

The way in which the sex chromosomes determine sexual development is not altogether clear, but it appears that the fundamental sex of the human embryo is female, by which we mean that it will develop into a female if it receives no hormonal stimulation. You will recall from Chapter 12 that a female possesses two X chromosomes, while a male possesses an X and a Y chromosome. Normally, the presence of a Y chromosome converts the original ambiguous gonad into a definite testis. The cells of the testis give off a brief burst of male sex hormone in response to the male sex-determining genes in early development. This produces a male body and commits the gonad to develop into a male sex gland, the testis. In the ab-

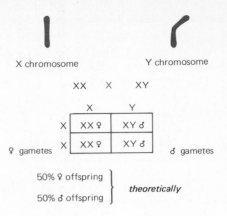

X chromosome    Y chromosome

XX    X    XY

|  | X | Y |
|---|---|---|
| X | XX ♀ | XY ♂ |
| X | XX ♀ | XY ♂ |

♀ gametes                    ♂ gametes

50% ♀ offspring  ⎫
                 ⎬  *theoretically*
50% ♂ offspring  ⎭

**FIGURE 15-8**
The inheritance of sex.

sence of a Y chromosome or if the Y chromosome does not function properly, the male hormone is not produced, and the fetus automatically develops into a female, as it were by default.

Careful analysis of the above will indicate that the X chromosome does not necessarily contain sex-determining genes for femaleness. They could occur on any chromosome. Sex-determining genes must, however, occur on the Y chromosome, for it is solely responsible for the difference between girls and boys. It follows from this that the X chromosome is free to bear many genes that have nothing to do with sex, particularly since the X chromosome occurs in both sexes. By its very nature though, the Y chromosome is far less free to carry any genes not directly related to sex, except perhaps on the short segment that is homologous with the X chromosome.

Genetic research has discovered many such genes on the X chromosome. These govern color vision, blood clotting, development of tooth enamel, blood type, and many other characteristics. In human beings, the Y chromosome is not definitely known to bear any gene other than those related to sex determination.

***Color blindness.*** Let's look at one of the more common X-linked traits—a form of red-green color blindness. Its normal allele, which is dominant, occurs, of course, on the X chromosome.

**FIGURE 15-9**
Sex-linked genes are usually borne on the X chromosome. This condition may be shown by the notation X'. The normal allele is denoted by X. Here are several representative crosses involving sex linkage.

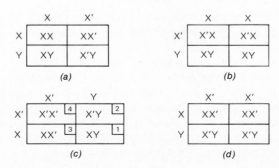

|  | X | X' |
|---|---|---|
| X | XX | XX' |
| Y | XY | X'Y |

*(a)*

|  | X | X |
|---|---|---|
| X' | X'X | X'X |
| Y | XY | XY |

*(b)*

|  | X' | Y |
|---|---|---|
| X' | X'X' [4] | X'Y [2] |
| X | XX' [3] | XY [1] |

*(c)*

|  | X' | X' |
|---|---|---|
| X | XX' | XX' |
| Y | X'Y | X'Y |

*(d)*

It is convenient to indicate an abnormal linked gene by the notation $X'$, which also indicates the chromosome on which it occurs. The normal allele is simply written $X$. Let us consider a girl who is heterozygous for red-green color blindness, $XX'$. If the abnormal gene is not found on *both* the X chromosomes, the girl will have normal sight, since this form of color blindness is recessive. But suppose we consider the case of a *man* who has a gene for color blindness on his single X chromosome. Since there is no homologous X chromosome present in his genotype, there can be no normal allele to oppose the expression of that gene for color blindness, and he will indeed be color-blind (such a condition is termed **hemizygous**). It is said that a noted cardinal suffered from red-green color blindness and sometimes wore his red robe out on the street. To him it seemed conservatively colored; it matched the pine trees outside his study almost exactly!

Consider some examples of the inheritance of this trait. In Figure 15-9*a* a girl heterozygous for color blindness marries a normal fellow. Since he has normal vision, he cannot have any genes for color blindness, for if he had them, they would be expressed. Of their offspring, it is likely that 50 percent will possess the gene but that it will be expressed in only 25 percent of the cases, and they will all be male.

What kind of cross would produce color-blind girls? One cross that would yield them would be that of a heterozygous woman and a color-blind man. In this instance, one-half the girl babies (No. 4) and one-half the boy babies (No. 2) would be color-blind. That is why color-blind girls are uncommon; yet they do occur (Figure 15-9*c*).

***Other X-linked traits.*** "Our poor family seems persecuted by this awful disease," wrote Queen Victoria in her journal, "the worst I know." The disease to which Her Majesty referred was hemophilia. It evidently originated with a mutation in one of her X chromosomes, and has been passed on to almost every family of European royalty. It was even a contributory factor in the Russian Revolution. Prince Alexander suffered from hemophilia. The monk Rasputin, a quack, claimed to be able to preserve his life. Rasputin gained great influence over the Czarina and, through her, over the Czar. Rasputin's disastrous public policies led directly to the revolution.

*Hemophilia* is a disorder of the clotting mechanism of the blood produced by the absence of a necessary globulin. If the globulin is artificially provided by injection, the sufferer may live indefinitely with no more inconvenience than if he had diabetes. In the absence of this treatment hemophilia is almost invariably fatal in early youth, though in exceptional cases the patient may reach reproductive age. Typically, death is caused by a minor wound or bruise, with the person bleeding to death either internally or externally. Should the hemorrhage take place in the brain, death is quick. Formerly the hemophiliac hardly dared to shave—but then, few grew old enough to shave.

An X-linked recessive gene produces the disorder and is inherited in a fashion similar to that of red-green color blindness. As with color blindness, it is much more likely to occur among males than females, but a very few cases of homozygous female hemophiliacs are now known. No respecter of persons, hemophilia also occurs in those who are not of royal blood. It should not be confused with the numerous disorders of the clotting mechanism which superficially resemble it but which are not genetic, or not X-linked.

A few dominant X-linked traits are known to exist. One of these specifies a developmental defect in tooth enamel so that it becomes shot through with tiny cracks and eventually turns brown. Since this trait is dominant, even a heterozygous girl would exhibit the phenotype.

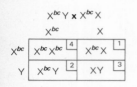

**FIGURE 15-10**
A number of characteristics have been shown to be X-linked. They are inherited along with the chromosome that bears them. Therefore, they tend to be inherited together.

***Mutual linkage of X-linked traits.*** Suppose for a moment that a woman who is heterozygous for both brown tooth enamel and color blindness marries a man who also bears these traits on his single X chromosome. Admittedly, this combination is hardly likely, which is one of the problems in studying human genetics; people, after all, do not marry for the purpose of genetic research, but fruit flies are much less fussy and can be induced to mate with almost any genetic monstrosity to which they are introduced. Thus, though such a cross is unknown among people, similar ones have been done hundreds of times with fruit flies, and the results are predictable (see Figure 15-10).

For our example, it is assumed that in the woman both mutant traits occur on the same X chromosome. This will produce, as you can see from Figure 15-11, a female chromosomal makeup (known technically as the *karyotype*) in which only one gene for color blindness occurs, along with only one gene for brown tooth enamel (No. 1). Since color blindness is recessive, the phenotype does not reflect it. The dominant brown enamel is, however, expressed. No. 4, also female, will have brown teeth and will be color-blind. No. 2, a male, will share these shortcomings. Only No. 3, a male karyotype, will be completely normal.

It should be obvious that the genes for color blindness and brown tooth enamel in our example stay together; they are *not* independently assorted. In fact the inheritance of multiple genes borne on a single chromosome is very much like that of a *single* gene. The condition where genes are borne on the same chromosome is termed one of *linkage. Linked genes tend to be inherited together.* After all, genes are just particular configurations of DNA, and we know that DNA is passed from ancestral cells to their progeny in the packages called chromosomes. It follows that if two or more genes happen to occur in the same chromosomal package, they will tend to be handled together by the mitotic and meiotic apparatuses, and thus will tend to be inherited together.

### Autosomal linkage

Genes also tend to be inherited together if they occur together on a non-sexual chromosome (*autosome*). Let us take another example from human genetics for which there is ample evidence. Some persons secrete proteins associated with their blood type in the saliva. The trait can be detected even before birth by an examination of the cells sloughed off by the embryo in the amniotic fluid. (The process by which this is done is discussed in the next chapter.) The gene which produces this trait is dominant and is called the secretor gene; we shall call it S. The recessive allele is s.

We next introduce a dangerous autosomal hereditary disease involving progressive paralysis and the wastage of muscles, known as myotonic dystrophy, m. The dominant normal condition we shall call M. Since the locations, or *locuses,* of the S, s and M, m genes are found on the same chromosome, they are said to be "linked."

Suppose that a couple, suspecting that their unborn child may be dystrophic, approaches a physician for advice. By careful investigation of them and their relatives he is able to deduce that their genotypes and the probable genotypes of their children are as shown in Figure 15-11.

If these genes were independently assorted, the male, who has the genotype SsMm, would give rise to *four* genetically distinct kinds of gametes. As it is, he produces but two. The possible combinations of his gametes and those of his wife are such that only four genotypes are possible in the offspring, instead of the eight that independent assortment would produce. Notice that in this couple's particular case, if a dystrophic child is born, he *must* be a nonsecretor. The physician has an easy task if the fetus is a secretor—the parents can rest reasonably assured that the child will be normal. But risk increases if the child is a nonsecretor. Even so, it still has a 50 percent chance of being normal.

**FIGURE 15-11**
Genes located on the same chromosome tend to be inherited together; any genes on a nonhomologous chromosome are inherited independently of them.

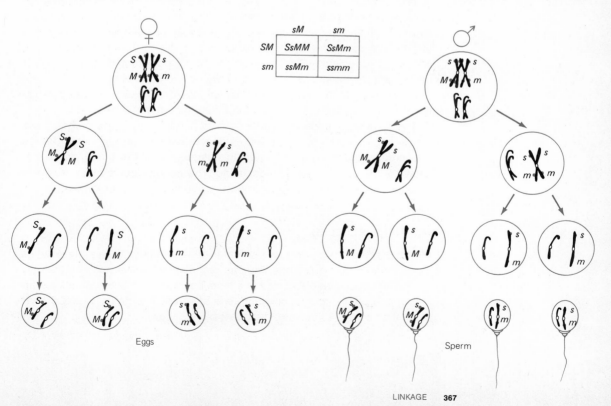

|    | sM   | sm   |
|----|------|------|
| SM | SsMM | SsMm |
| sm | ssMm | ssmm |

Eggs

Sperm

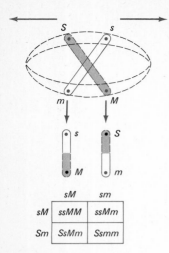

**FIGURE 15-12**
Crossing-over. The greater the distance between linked genes, the greater their crossover frequencies.

|     | sM   | sm   |
|-----|------|------|
| sM  | ssMM | ssMm |
| Sm  | SsMm | Ssmm |

## Crossing-over

Thus far you may have gained the impression that linked genes *always* stay together. As a matter of fact, though they usually do keep company, there are exceptions, and a few fetuses (about 10 percent, from the above example) will not conform to what we would predict on the basis of genetic linkage. For example, some from the very cross we just studied might turn out to be dystrophic secretors, an "impossible" combination of traits. How can they be accounted for? The solution to this puzzle was determined with the microscope by watching the behavior of chromosomes in meiosis. You recall that when gametes are formed, homologous chromosomes are always separated. But imagine those chromosomes *lying across* one another as meiosis begins. If that happens, they may exchange parts, as shown in Figure 15-13. Suppose that had occurred in the body of the female member of the couple whose case we have been considering. In that event, it would turn out *opposite* to our original prediction and *only* a secretor baby would become dystrophic. However, we would emphasize that the experience of geneticists indicates that there would be only about a 10 percent chance of that happening in this case.

## Genetic maps

The farther away from one another the locuses of different linked genes are on the chromosomes, the more frequently they tend to cross over. Thus by counting enough instances of crossing-over, one can estimate the relative distances between linked genes. It is even possible by using this principle to construct a map of chromosomes which gives the order and relative distances of all known genes on them. Genetic maps have been constructed for fruit flies in which hundreds of genes and their loci on chromosomes are known with precision. Similar maps are being constructed for mice. Since people are not experimentally bred, however, we know of only a few linkage groups in man. By computerizing hospital records, however, it should in time be possible to create increasingly detailed human gene maps.

The concept of genetic linkage paid off handsomely in 1972 when Fieve, Fleiss, and Mendlewicz presented evidence that one form of manic-depressive emotional illness is inherited in company with the blood group $Xg^a$, known to be an X-linked characteristic. The conclusion is clear: the gene responsible for this form of manic-depressive illness is also X-linked.

## MULTIPLE ALLELES

*Multiple alleles* exist when a particular gene, occupying one given locus, exists in more than two forms. Thus an organism can have no more than two of a given set of multiple alleles at a time, but more than two such alleles are known to exist. An important example is afforded by a group of blood characteristics known as the ABO series of blood types.

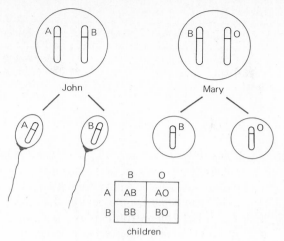

|   | B | O |
|---|---|---|
| A | AB | AO |
| B | BB | BO |

children

**FIGURE 15-13**
Multiple alleles. Here, more than two forms of the same gene exist. In this cross, each parent produces only two kinds of gamete, but between them both, three kinds of gametes are formed. Four genotypes occur among their offspring.

## The ABO series of blood types

The ABO series of blood types is only one of a variety of blood types known in man and in lower animals. It was the first to be discovered as a result of early experiments in blood transfusion, which were all too frequently fatal. As finally determined, the explanation for the fatalities proved to be this:

**1** Blood, as you may know, contains red blood cells suspended in a fluid called plasma. Plasma, it was discovered, contains gamma globulin proteins known as *antibodies,* capable of participation in a variety of immune and allergic mechanisms.

**2** The surface of red blood cells (and of other cells) consists of a membrane with certain antigenic characteristics. For our purposes we will simply regard *antigens* as substances capable of a characteristic chemical reaction with antibodies. An antigen plus an appropriate antibody forms a chemical complex.

**3** There are some similarities between antibodies and enzymes. One example is their *specificity.* A particular antibody will react with only one or a few specific antigens and with *none other.* Antibodies and antigens that are capable of such mutual reaction are said to be *complementary.*

*Landsteiner's law.* In the case of blood cells when complementary antibodies and antigens come into contact, red blood cells clump together (*agglutination*) or break open (*hemolysis*). Both these reactions can take place in laboratory glassware and presumably also in the body of a living person. However, agglutination is the reaction most likely to occur in laboratory glassware, while hemolysis is the characteristic response in vivo. Hemolysis results in the release of hemoglobin in the plasma. This in turn can produce kidney damage that may lead to death. Such reactions do not ordinarily take place in one's own bloodstream, because, as the hematologist Landsteiner discovered, the *antibodies of the plasma are never complemen-*

Antigens, when they occur, are characteristics of the blood cell membrane

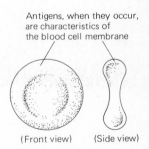

(Front view)  (Side view)

Red blood cells

Antibodies, when they occur, naturally are present in the blood plasma

**FIGURE 15-14**
ABO antigens and antibodies.

**Table 15-1**

| Phenotype | Genotype | Antigens, cells | Antibodies, plasma | Notes |
|-----------|----------|-----------------|--------------------|-------|
| A | AA, AO | A | Anti-B | Dominant |
| B | BB, BO | B | Anti-A | Dominant |
| AB | AB | A and B | None | Codominant |
| O | OO | None | Anti-A and B | Recessive |

tary to the antigens of the cells (**Landsteiner's law**). There is no such assurance when the bloods of two individuals are mixed as in a transfusion.

If blood is transfused into a patient whose plasma contains complementary antibodies hostile to the cells of the donor, hemolysis of these cells is bound to result. The reverse can also happen. If one transfuses blood into a recipient and the donated *plasma* is incompatible with the recipient's cells, trouble will eventually result. Initially, though, such transfusions may be harmless since the half pint or so of potentially dangerous plasma is rapidly diluted in the recipient's blood volume to harmless levels. The trouble is that not all transfusions are limited to a pint or so—in fact, most of them are not. Thus it is likely that sufficient incompatible plasma could be transfused in such a case to produce hemolysis in the patient. For this reason, the old concept that certain blood types make one a "universal donor" or "universal recipient" is now outdated. Blood transfusions are now almost always matched exactly to the recipient's blood type.

Perhaps this will become clearer by example. Imagine a transfusion of type A blood into a type B recipient. Table 15-1 shows that type A blood has type A cells. Type B blood has anti-A antibodies which would immediately attack the type A blood cells. Furthermore, if *enough* type A blood were to be transfused into the recipient, eventually there would be enough anti-B antibodies in his circulatory system to begin to attack his type B blood cells.

**The genetics of the ABO series.** Most immune mechanisms involving gamma globulin appear to be acquired. For example, one develops an immunity to typhoid fever as a result of exposure to the disease, or one develops an allergy as a result of exposure to the causative agent—ragweed pollen, for example. The antibodies of the ABO series seem to form partly in response to environmental influences, but fundamentally they are inherited.

Table 15-1 gives the basic facts of inheritance for both the antibodies and the antigens of the ABO series. Three alleles are known for the ABO series, which by their interaction can produce blood types A, B, AB, and O—depending upon the exact combination of genes in the particular allelic pair possessed by an individual. A normal person can possess no more than two of the ABO alleles. Type O gene is recessive to all other genes, and A and B exhibit codominance with respect to each other. Thus two genotypes can give rise to type A, and two to type B. AB and O each have only one possible genotype.

Occasionally parents suspect that a hospital has awarded them the wrong baby. Sometimes law courts must determine blood relatives when probating an estate. And often the paternity of an illegitimate child must be determined for purposes of assuring its support. In all such instances it is obviously desirable to have an objective means of determining biologic relationship. The genetics of the ABO series, together with some other aspects of human genetics, provides us with an approximation of such a means.

Consider the cross of a type *A* man with a type *AB* woman (see Figure 15-15, top). In such a case two genotypes are possible for the man, *AA* and *AO* (see Table 15-1). The genotype of the woman can only be *AB*. To determine the man's genotype, one must predict the phenotypes observable among the offspring in both the possible cases. If only type *A* and *AB* offspring were produced, the father could be *AA* or *AO*, since these types would result in *either* case. However, a type *B* child could result only if he were *AO*, or — and this eventuality must unhappily always be kept in mind — if the child were illegitimate and the man was not in fact its father. This example and some others can be found in Figure 15-15.

Theoretically, these kinds of genetic studies can be used only to rule *out* possible candidates for paternity, for while the suspect may have an appropriate blood type, so may the governor, the mayor, and the mailman. If the field of candidates is definitely limited, however, genetics is useful. Indeed, many more independently assorting blood types (about 80) are now known than just the ABO series, though most of them are of little importance in blood transfusion. The pattern of heredity produced by all these genes col-

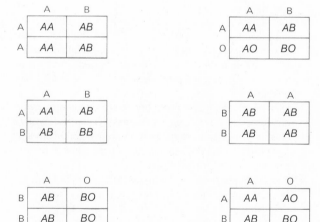

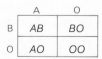

**FIGURE 15-15**
Some representative paternity problems.

lectively is almost as distinctive as a fingerprint. In practice, the guilty or responsible party can usually be positively determined.

Please understand, however, that a rare gene exists that can suppress the phenotypic expression of the B and probably also the A allele. Persons with that gene would be type O phenotypically, even if their genotype is *AA, AO, BB, BO,* or *AB.* What a complication in a paternity case!

## PLEIOTROPISM

The existence of that suppressor gene raises an important point, which we have thus far avoided because of the complications that it introduces. The expression of genes is very rarely as simple as our carefully chosen examples would perhaps lead you to believe. For one thing, genes drastically modify one another's expression, as we have just seen. For another, most genes have multiple effects. This condition is called *pleiotropism.*

For example, among many mammals evincing pigment reduction, such as Siamese cats and white tigers, the same gene that is responsible for the pigment reduction also affects the nervous pathways in the brain responsible for sight. The consequence of these abnormal pathways is the "cross-eyed" condition often observed in these animals. It is likely that all or most genes in all organisms have such multiple effects.

**CHAPTER SUMMARY**

1 "Blending inheritance," the idea that offspring are an intermediate mixture of their parents, is untrue. Traits may be suppressed in the heterozygous condition, but even so they persist and can be demonstrated in the $F_2$ by crossing the $F_1$ siblings with each other.

2 Inheritance is accomplished by means of hereditary units called genes which govern individual characteristics. Genes are particular configurations of DNA which give rise to distinctive enzymes, thus controlling the biochemistry and ultimately all characteristics of the organism. The genetic composition of an organism is its genotype. It is expressed physically in the overt characteristics known as the phenotype.

3 In solving genetic problems genes are denoted by symbols. The possible combinations of genes may be conveniently visualized by means of the Punnett square. This method shows not only the proportions of different phenotypes to be expected among the offspring, but also the chance that a particular offspring may have a given genotype or phenotype.

4 Gregor Mendel, a gifted naturalist, discovered four genetic principles, or laws: (*a*) the concept that inheritance is particulate, (*b*) the law of segregation, (*c*) the law of dominance, and (*d*) the law of independent assortment.

5 Genes assort independently only if they are located on nonhomologous chromosomes. Sometimes independently assorting genes govern the same characteristics (polygenic inheritance).

6 Many genes having nothing to do with sex are carried on the X chromosome. The Y chromosome, on the other hand, appears to be almost exclusively sexual. If more than one trait is carried on the X chromosome, these traits are not independently assorted but tend to be inherited together.

7  As in the case of those on the X chromosome, autosomally linked genes tend to be transmitted together. However, crossing-over often leads to some unexpected results. Since crossover frequency varies directly with distance apart, it is possible to use crossover statistics to prepare linkage maps showing the order and relative spacing of genes on chromosomes.

8  Allelic genes sometimes exist in multiple form, as in the ABO series of blood types. Cells bearing antigens complementary to plasma antibodies do not normally occur within an individual. The mixture of complementary antibodies and antigens in transfusion may produce hemolysis and other grave consequences. Genetically, O is recessive to both A and B, but A and B are codominant with respect to each other.

9  Blood group inheritance may be used to rule out possible candidates in disputed paternity and similar suits.

**FOR FURTHER READING**

Reisman, Leonard E., and Adam P. Matheny: *Genetics and Counselling in Medical Practice,* Mosby, St. Louis, 1969.

Sinnott, E. W., L. C. Dunn, and T. Dobzhansky: *Principles of Genetics,* 5th ed., McGraw-Hill, New York, 1958.

Stern, Curt: *Human Genetics,* 2d ed., Freeman, San Francisco, 1960.

Strickberger, M. W.: *Genetics,* Macmillan, New York, 1968.

# 16

# HUMAN GENETICS

. . . genetics is the science of the future;
more significantly, it bids fair to become
*the* science of the 21st century.
—*John B. Breslin, S. J.*

**Chapter learning objective.** The student must be able to give the principal genetic causes of birth defects and enzyme deficiency diseases (inborn errors of metabolism), together with specific examples, with emphasis on mutations and the disruption of biochemical processes. He must also be able to summarize and to criticize several arguments involving the ethical and social implications of genetics and of eugenics.

*ILO 1* Give several examples of inborn errors in metabolism, and describe their genetic basis (e.g., phenylketonuria, sickle-cell anemia).

*ILO 2* Describe the inheritance of the Rh factors, and outline the mechanism of Rh disease.

*ILO 3* Describe the chromosomal basis of such abnormalities as mongolism (Down's syndrome), Turner's syndrome, and *cri-du-chat*.

*ILO 4* Define consanguinity and give its principal genetic implications.

*ILO 5* Discuss controlled genetics, genetic engineering, and eugenics as ethical, social, and scientific problems, giving their primary implications and proposed applications.

The ability to work genetics problems is not especially important in itself. However, genetic problems *can* help your understanding of genetic principles, which are extremely important in our modern world. The social, political, and ethical implications of genetics are becoming increasingly important; this will be even more true in 5 to 10 years. You will be able to make intelligent personal and political judgments about genetics only if your knowledge of the subject is sufficiently thorough to aid you in discerning sound scientific ideas from pseudoscientific rubbish.

## INBORN ERRORS OF METABOLISM

The lack of a single crucial enzyme in a cell can result in the failure of a chemical reaction to take place, since normally such reactions occur only in the presence of appropriate enzymes. The result depends upon the enzyme that is missing or inactive. If the enzyme is essential in the metabolism of the pigment melanin, for instance, then its lack will result in the congenital absence of color that we call albinism—a relatively harmless defect. But not all enzymatic defects result in such minor handicaps. Almost any step of the thousands that occur in all the countless biochemical pathways of the cell is a potential cause of disaster if it should be blocked. In some instances, the most minor change in the amino acid composition of an enzyme is enough to block the course of the reaction which that enzyme would otherwise catalyze. In other instances, the changed enzyme still functions, but often less efficiently. In still others, detours and alternate pathways exist whereby the cell can bypass its roadblocks, though usually at some cost in efficiency or economy. Every instance of a genetic defect which can be investigated biochemically has been traced to a cause of this nature. It seems virtually certain that all genetic defects, including those expressed through development, will some day be shown to depend upon one or more specific enzymatic deficiencies, or on similar defects of structural proteins.

### Alkaptonuria

Writing near the start of this century, an English physician, Sir Archibold Garrod, described a number of congenital diseases as "inborn errors of metabolism," an expression which has persisted ever since in the vocabulary of genetics and medicine. Even before the concept of the gene had been fully developed, and far before our modern understanding of enzymes and nucleic acids, Garrod essentially hit upon the modern view of genetic defects. By careful observation and deduction he recognized several genetic defects as being due to metabolic disorders. These genetic defects are now known to result from the congenital absence or inactivity of certain enzymes. One relatively harmless recessive disorder which he discussed is the disease *alkaptonuria*.

Persons with alkaptonuria excrete normally colored urine that turns very dark upon exposure to air. Late in life dark pigmentation develops in the cartilages and in some cartilaginous external body parts, such as the external ears and tip of the nose. Ultimately, crippling arthritis may occur.

These symptoms seem to result from the presence in the blood of the substance *alkapton,* which accumulates because the enzyme that breaks it down in normal people is not functional in those with alkaptonuria. Alkapton itself is a normal product of the metabolism of the amino acid *phenylalanine* and therefore is present in small amounts in us all.

Since Garrod's time many more metabolic disorders besides alkaptonuria have been shown to depend upon the lack of a single enzyme.

### Phenylketonuria

The amino acid phenylalanine is also involved in the disease phenylketonuria, usually abbreviated PKU. PKU is a recessive characteristic involving a defect in the metabolism of phenylalanine. It often (*not* always) leads to idiocy in untreated persons homozygous for the trait. In PKU, the enzyme responsible for the conversion of phenylalanine to tyrosine is functionally absent. Unable to follow the normal pathways, phenylalanine accumulates in the blood and tissues and eventually is converted to a compound called *phenylpyruvic acid,* which is then excreted in the urine as a waste product. Its presence in the urine indicates PKU. Before the excess phenylpyruvic acid circulating in the blood is excreted, it appears to damage the growing nervous system. An amino acid imbalance caused by an excess of phenylalanine also appears to contribute to the course of the disease.

PKU is usually treated by greatly reducing the phenylalanine content of the diet, for without phenylalanine, the toxic phenylpyruvic acid is not formed and the amino acid imbalance is less severe. The diet of a person with phenylketonuria must be carefully managed by an experienced nutritionist, for phenylalanine is an essential amino acid; it therefore must be provided in amounts great enough for incorporation into body proteins, but not so great as to produce mental retardation. The adult with phenylketonuria is little inconvenienced by phenylalanine or its by-products, so that the onerous and expensive diet can usually be abandoned during adolescence. Most infants are now given blood tests routinely for phenylketonuria. The chances of detecting this disease before it does substantial harm are now excellent.

### Sickle-cell anemia

Found predominantly in Africans or persons with some African ancestry, sickle-cell anemia is a hereditary disorder of hemoglobin molecules. It involves an improper substitution of just one amino acid residue out of the several hundred in a normal hemoglobin molecule. In sickle-cell anemia many of the red blood cells are sickle-shaped, fragile, low in hemoglobin, and inefficient in oxygen transport. Untreated, the condition is often fatal by late adolescence.

Technically, the sickling trait is recessive, since heterozygotes exhibit only a mild, harmless anemia. They can be detected, however, by the ability of many of their red blood cells to assume a sickle shape when chemically deprived of oxygen in the laboratory. From a biochemical point of view, the sickling gene and its normal allele really exhibit codominance. The red blood cells of a heterozygote, therefore, contain *both* normal and abnormal hemoglobin. The exact proportion of abnormal hemoglobin in the red cells of heterozygotes is governed by a quite different set of modifying genes, which determine the degree of phenotypic expression of the sickling gene.

These benevolent modifying genes are especially prevalent in African people living in malarial districts. Although the malaria parasite is able to subsist all too well in a normal blood cell containing only normal hemoglobin, it finds sickle-cell hemoglobin uncongenial. Even the mixture of normal and sickle-cell hemoglobins found in heterozygotes seems unpalatable to the malaria parasite. Thus those heterozygous for sickle-cell anemia are protected to some extent against malarial disease and yet are otherwise healthy, for enough normal hemoglobin is present to allow their blood usually to function quite well in oxygen transport.

Not all Africans have the sickling gene. Its area of greatest prevalence appears to be the very malarial Congo Basin, but pockets of the gene occur around the Mediterranean, in India, and even in Southeast Asia—all malarial districts. In a highly malarial region the partial protection that the gene confers appears to be so advantageous that natural selection preserves the gene, even though in the homozygous condition it is almost invariably fatal—at least it was before the advent of modern medicine. In places such as the United States, where malaria is rare, one may expect the eventual extinction of the gene, for reasons to be discussed in the next chapter.

A somewhat similar disease of the red blood cells, *thalassemia,* occurs in a number of Mediterranean countries, particularly Greece and Italy. Because thalassemia also occurs frequently in the more heavily malarial districts it is suspected of conferring protection against malaria in somewhat the same way as sickle-cell anemia. In the homozygous state, thalassemia produces extremely small red blood cells and a form of anemia which has usually been fatal. Like sickle-cell anemia, thalassemia is attributable to an abnormal form of hemoglobin.

A number of other abnormal forms of hemoglobin are known to occur in human beings, usually with only mildly harmful effects, or with no obvious harm at all. Abnormalities of the hemoglobin provide an excellent example of the chemical nature of mutation.

## THE Rh FACTORS

With the Rh factors we begin to take up a number of congenital disorders whose biochemical origin is not clear-cut, and which do not seem to be attributable to any simple error of metabolism. Indeed, the Rh factors are not really defects at all, nor are any of them abnormal in any useful sense. Still, they are responsible for a congenital disease.

The Rh series of blood types has been known only since the 1940s. There are several somewhat codominant varieties, but all may be thought of as producing either a *positive* or a *negative* phenotype. For our purposes we will lump the several alleles into two groups—*R*, which produces positive phenotypes, and *r*, which produces negative. *R* is dominant. Thus a person with Rh-positive blood has the genotype *RR* or *Rr*. The negative phenotype can only be *rr*.

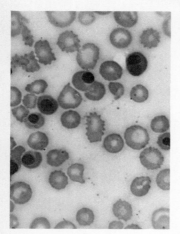

**FIGURE 16-1**
Erythroblastosis fetalis—"Rh dis-
ease" of infants. (Carolina Biological
Supply Company)

A person with Rh-positive blood has antigens associated with the blood cell membranes, and, as you might expect, *no* anti-Rh antibodies in the plasma. An Rh-negative person possesses no Rh antigens, but, contrary to what you might think, he has no anti-Rh antibodies in his plasma, either. That is, he has none naturally. But if he were to receive a transfusion of Rh-positive blood, he would quickly develop anti-Rh antibodies, which would then hemolyze the cells of the *next* transfusion of Rh-positive blood. The first such transfusion would, however, be relatively harmless in itself.

Another way in which an Rh-negative person can become sensitized to Rh-positive blood is through pregnancy with an Rh-positive fetus. Though there is, of course, no direct connection between the circulatory systems of mother and fetus, nevertheless capillary defects do occur from time to time which permit the bloods and antigens of mother and fetus to mix slightly. In time, the Rh-positive blood cells of the fetus may stimulate the production of anti-Rh antibodies in the mother's plasma. Much greater mixing may occur in childbirth. These antibodies penetrate the placental capillary walls easily and begin to destroy the red blood cells of the fetus, particularly immediately following birth. The hemoglobin thus released into the infant's circulatory system damages the kidneys; usually more important, it breaks down there, forming the bile pigment *bilirubin,* which produces a severe jaundice. Bilirubin in excessive amounts can severely damage many areas of the brain, producing a condition known as *kernicterus,* which in turn can lead to paralysis, idiocy, or death. All these symptoms are accompanied by a more or less severe anemia. The entire condition is known as *erythroblastosis fetalis.*

Only a cross between an Rh-positive man and an Rh-negative woman can produce this result, since only such a father can transmit a positive genotype to his offspring. If he is heterozygous, half the babies may be negative themselves and, therefore, in no danger. Indeed, Rh disease does not always develop, even when circumstances appear most to favor it. Rh-negative women have been known to produce as many as a dozen Rh-positive infants, all perfectly healthy! This, however, is exceptional. Usually the maternal antibodies require at least two or three pregnancies to develop fully, but eventually they come into existence and cause trouble.

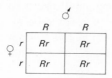

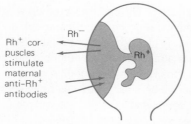

Rh⁺ corpuscles stimulate maternal anti-Rh⁺ antibodies

Rh⁻

Rh⁺

You should work out this cross: Rh⁻ ♀ X heterozygous Rh⁺ ♂. Does it matter if Rh⁺ ♀ X Rh⁻ ♂?

**FIGURE 16-2**
How Rh disease begins.

You may wonder why we have not mentioned the case of an Rh-negative fetus borne by an Rh-positive mother. Why doesn't the fetus, in such a case, develop antibodies that attack the maternal red blood corpuscles? The explanation is simple: the fetus simply is not able to develop most antibodies until some time after birth. In scientific terms, the fetus is said to be *immunologically incompetent.*

The effects of erythroblastosis fetalis can often be minimized by *exchange transfusion,* in which the fetal blood is removed and replaced with fresh Rh-negative blood, which is naturally immune to the effects of anti-Rh antibodies. It is even possible to accomplish this *in utero,* before the baby is delivered. Recently it also has become possible to prevent the development of anti-Rh antibodies in the mother. This is accomplished by injecting the new mother with a concentrate of anti-Rh antibodies contained in a preparation of gamma globulin obtained from persons with strong anti-Rh antibodies. The presence of these artificially injected antibodies prevents the mother from developing her own by a negative feedback mechanism. Since the baby has been born just before the time of injection, these antibodies cannot harm it. Of course the treatment is ineffective if the mother already has developed the antibodies, so it should be given after the first baby of a susceptible mother, and at all subsequent births.

## CHROMOSOMAL DEFECTS

A number of birth defects are known to be caused by chromosomal abnormalities. Although in a sense abnormal karyotype (chromosomal makeup) is the known cause of many such defects, in another sense their cause is still unknown, for it is not clear how the abnormal chromosome makeup produces an abnormality in development.

### Some sex chromosome abnormalities

A remarkable array of abnormal sex chromosome karyotypes is known. Most of them were discovered in fruit flies before they were observed in human beings, but now many human occurrences have been described. To use the original terminology, abnormal fruit flies whose karyotype was XXX or XXXX were known as *superfemales,* for they possessed feminine characteristics in exaggerated form. Similarly, XYY flies were called *supermales.* This comic strip terminology has fallen into disfavor, however, for the "super" flies were in fact slow to develop, sickly, and often more or less grossly defective. In human abnormalities of these types, it is more proper to call each by a specific name—Fröhlich's syndrome, Turner's syndrome, and the like, or simply by its karyotype—XYY, for example.

*Turner's syndrome.* A normal woman possesses two X chromosomes, as previously discussed. Nevertheless two do not seem to be required for life. After all, men have only one X chromosome. Therefore it is not really surprising that some women have only one X chromosome, a condition whose karyotype is indicated by XO. Such a person, though phenotypically female, has undeveloped or somewhat degenerate sex organs, is short, and has rudimentary breasts. Women with an XO karyotype are also often mentally retarded. This condition is called *Turner's syndrome.*

Many cells in the normal female body contain projections or condensations of the nucleus known as "drumsticks" or "Barr bodies," after their discoverer. It seems that these represent condensed and inactivated X chromosomes. Thus, even in normal females, only one X chromosome per cell is genetically active. If that is so, why should there be any phenotypic difference at all between the XX and XO karyotypes?

This puzzle may be related to the fact that apparently it is not always the same X chromosome that is inactivated. It is believed that in some cells of the normal female body one of the X chromosomes is in an inactive state; in others, the *other* X chromosome may be inactive. Thus there are always enough cells around with an X chromosome bearing the dominant normal allele to ensure that it will be expressed.

There is a great deal of evidence for this view, which was first proposed by the British geneticist Mary Lyon. A hereditary deficiency of the enzyme glucose 6-phosphate dehydrogenase (G6PD) in red blood cells is suspected of protecting the carrier against malaria somewhat as sickle-cell anemia does. Unfortunately, a person with G6PD deficiency is also likely to have his red blood cells abruptly hemolyze if he eats broad beans *(Vicia fava)* or inhales their pollen, takes certain common drugs, or accidentally eats the chemical naphthalene, widely used as a moth repellent. This characteristic is X-linked, and, as you might think, severe cases are usually found in males. However, the trait exhibits codominance, and even female heterozygotes have some susceptibility to poisoning from these ordinarily fairly harmless agents. The really interesting fact, though, is that half the red blood cells of heterozygous females have G6PD deficiency and half are normal—indicating that half the cells in the bone marrow that give rise to red cells are using one X chromosome, and half are using the other![1]

In Turner's syndrome, a large proportion of the fetal cells may initially "turn off" the X chromosome, in accordance with Lyon's hypothesis, and yet possess no functional mate to that chromosome. The readjustments necessary to compensate for this situation may lead to abnormalities in development. It may also be that early in the development of the human embryo the products of *both* X chromosomes are indeed necessary to produce normal female sexual differentiation.

***Disorders of the Y chromosome.*** Persons with the XYY karyotype may be considered normal males, they are fully fertile, but they are larger than the average, somewhat inclined to emotional disorders, and of slightly lower intelligence on the average than the rest of the populace. Since many crimi-

---

[1] There is evidence that the inactivated X chromosome may bear *some* active genes. The genes for an important blood globulin, immunoglobulin (IgM) are borne on the X chromosome. Females with two X chromosomes have twice as much IgM as males, who have only one. Those with an XXX karyotype have a triple concentration of IgM. Therefore, for this gene, both X chromosomes are active.

nals seem to possess this karyotype, there has been some disposition to attribute their undesirable behavior directly to the extra chromosome. Yet many normal persons also possess the XYY karyotype. Thus it cannot be said with any certainty that the XYY criminal could not have made up his mind to be a professor, a politician, or a businessman instead!

Almost every imaginable abnormality of the sex chromosome karyotype has been described—XXX, XXXX, XXY, XXXY, XXYY, etc. Such persons are frequently abnormal sexually and are often feebleminded. Quite often genetic or chromosomal abnormalities of all kinds are associated with feeble mindedness, however diverse other effects may be. It is suggested on the basis of very good evidence that in mice only about 3 percent of the genetic information of the cell is expressed in the tissues of the heart or kidney, but fully 12 percent is expressed in the brain. Presumably this reflects the tremendous complexity of that organ, which in man is at its most complex. Because of its complexity, it is likely that the brain is uniquely vulnerable to almost any disturbance of normal developmental processes. We know from other studies conducted on mice that genetic interference with a single crucial developmental process early in embryonic development ultimately can lead to multiple, serious defects in the finished organism. Small imbalances in biochemistry or development caused by multiple chromosomes probably damage the brain in some such way also. We appear to pay a considerable price in developmental risks for the specialization of the human brain.

*Origin of sex chromosome defects.* There are a number of ways in which such defects as XO, XXY, XXX, and XYY might arise, but most of the possibilities involve defects of the meiotic mechanism.

For example, if an abnormal meiosis were to produce an egg in which the X chromosomes had failed to disjoin, it would possess two of them. If it was fertilized by an X-bearing sperm, the result would be XXX; if by a Y-bearing sperm, the result would be XXY. Or if an egg received no sex chromosomes at all, it would produce an XO or a YO zygote. YO embryos appear to die in the two-cell stage of development.

Similar causes probably underlie the production of the autosomal abnormalities which we are about to consider.

### Autosomal abnormalities

*Autosomes* are chromosomes other than sex chromosomes. These are involved in some abnormal phenotypes, but fewer of these are known than is the case with sex chromosomes. Paradoxically, this appears to result from the greater severity of autosomal defects. Since both sexes have the same number of autosomes, there is never any need to shut any of them off. No autosomal equivalent of a Barr body is known. In fact, both members of all pairs of autosomes seem to be required for life, and the absence of a member of even one homologous pair seems to be enough to ensure early abortion of an embryo.

One known chromosomal defect involves the *loss* of chromosomal material. It is the *cri-du-chat* syndrome, rather recently discovered by French investigators. In this disease, the baby is microcephalic—that is, he has an abnormally small head and brain. Mentally, he is an idiot, and his cry sounds something like that of a cat, which is what the name indicates in French. In these infants a particular chromosome, No. 5, is found to be shorter than normal. This chromosome seems actually to lack some of its genetic material. The missing information appears to be necessary for normal mental development, and its lack produces the symptoms of the disease.

Multiple autosomes are rare in living babies, and only a few varieties of *trisomy* (three homologous chromosomes) are known that are compatible with life. These trisomies, in which a trio rather than a pair of autosomes exists, involve a surplus instead of a shortage of genetic material. They produce several physical and mental defects, perhaps by subtly disrupting or imbalancing metabolic pathways during early development. The commonest and best known of the trisomies is *trisomy 21,* otherwise known as mongolism and correctly known as *Down's syndrome.*

**FIGURE 16-3**
Down's syndrome typically produces severe mental retardation, but programs of physical therapy and other training can help sufferers to lead a far more meaningful life than is generally realized. (Association for the Help of Retarded Children and Manhasset Club Life Magazine)

"Mongolism" is a poor name for Down's syndrome because the disease has nothing to do with the Mongolian race and does not always result in idiocy. Originally it was called mongolism because those afflicted with it have somewhat slanted eyes. Additional symptoms include a peculiar crease on the palm of the hand; enlarged, protruding tongue; protruding lower lip; saddle-shaped nose; thick, short extremities; short stature; varying (usually severe) degrees of mental retardation; and, frequently, obesity. It appears to be produced by a triploid No. 21 chromosome; hence the name "trisomy 21."

The person with Down's syndrome does not always have three *separate* No. 21 chromosomes. Sometimes he has only two, but a third No. 21 chromosome is attached to another chromosome. This situation usually arises from a chromosomal defect in the mother, harmless to her, known as *translocation.* Translocation can be detected by microscopic examination of the mother's tissues (usually the white blood cells). If it is present, she must decide whether she wishes to risk having children. It is not a good risk. For that matter, Down's syndrome may be detected prenatally by a technique known as amniocentesis. A small portion of the amniotic fluid surrounding the fetus is withdrawn through the mother's abdominal wall by means of a large hypodermic syringe. Some of the embryonic skin or bladder lining cells slough off and float free in the amniotic fluid. These cells are withdrawn along with it. After being centrifuged, these cells are then grown in tissue culture. Their chromosomes are examined microscopically, and the presence or absence of trisomy 21 is determined. When it is present, many women consider therapeutic abortion. Today, science can do much to prevent the conception and birth of these unfortunate people. Modern means of birth control and chromosome analysis make this practical.

In mongoloids with three distinct No. 21 chromosomes the parents are often chromosomally normal. The risk of having such babies increases with age, becoming substantial for mothers over 40 years of age. The reasons for this are not known, but it is one of the many hazards encountered by older women who bear children. It should be recalled that the second meiotic division of ova does not take place until fertilization—in these cases more than 40 years after the first!

Before the development of antibiotics, most patient's with Down's syndrome died of respiratory infections, such as pneumonia, by age 7 or 9. This was then considered almost a diagnostic characteristic of the disease. Now these patients often can be kept alive into late middle age.

Though many mongoloids are trainable to some extent, and some few, whose condition has perhaps been misdiagnosed,[2] have nearly normal intelligence, the vast majority never achieve more than one-fourth to one-half normal intelligence and constitute a decided social problem. Their care may cost $1.75 *billion* annually by 1975.

A few other, quite rare, trisomic diseases resembling mongolism are

---

[2] Some patients with Down's syndrome appear to be *chimeras,* that is, some of their body cells have triple No. 21 chromosomes and others have a normal karyotype. A number of the more intelligent persons with Down's syndrome have proved upon investigation to be genetically heterogeneous.

known. Trisomy of most chromosomes results, however, in the death of the fetus in the uterus and its subsequent expulsion in spontaneous abortion. In fact, lethal genetic and chromosomal defects appear to cause the premature termination of 25 percent of all pregnancies, frequently so early in gestation that the mother mistakes the abortion for an unusually heavy flow of menstrual blood.

## CONSANGUINITY, OR MARRIAGE BETWEEN RELATIVES

Marriage between relatives is referred to as consanguinity. In all known societies throughout history, cultural and religious precepts[3] have forbidden marriage between very close relatives (brother-sister, parent-child), though a few exceptions have been made for royalty. First-cousin marriages are more common, despite laws in most states forbidding them. In the United States marriage between first cousins accounts for a little less than 1 percent of all marriages.

Taboos against first-cousin and closer marriages are genetically sound, although they probably originated for social reasons. Certainly, not *all* such marriages are disastrous. Yet the risk of bearing a child with a major congenital abnormality of genetic origin is 70 percent greater if the parents are first cousins than if they are unrelated. The death rate for children of consanguineous marriages is between three and four times higher than among a comparison population. Almost half the offspring of brother-sister marriages are more or less grossly handicapped by genetic disease.

No mystery enshrouds the increased risk of malformed offspring from consanguineous marriages. The problem can be traced to a greater chance of sharing harmful genes. All of us no doubt shared the same ancestors, hence the same genes, in the distant past. But with time the pool of human genes became diverse as a result of mutations, so that today each of us is genetically unique. Only identical twins have identical genes. Each of us carries an estimated eight harmful recessive genes. If you marry a nonrelative, the chance is remote that you will select someone who shares any of the same harmful genes. There is little chance that any of your offspring will be unfortunate homozygotes, receiving a double dose of recessive, harmful genes. However, if you select your first cousin as your mate, you will have a common set of grandparents, and so in all probability one of every eight of your genes will be identical. Your offspring will be homozygous for one-sixteenth of all their gene traits. Second cousins share one thirty-second of their genes.

Consider some specific examples. Only 1 of every 70 persons in the general population carries a recessive gene for albinism (an inherited lack of pigmentation). Thus, if you have no family history of albinism, the chance that you are a carrier for this disorder is 1 in 70. If you marry someone with

---

[3] See, for example, the commands of God to the Israelites in Deuteronomy 27:22, Leviticus 18, etc.

no family history of albinism the chance that your spouse will be a carrier for albinism is also only 1 in 70. Therefore, the chances that both of you would be carriers is $1/70 \times 1/70 = 1/4,900$. And then the chance that your child would inherit both recessive genes would further decrease the risk to $1/4 \times 1/4,900 = 1/19,600$. Not much to worry about! However, if you were to marry a first cousin the odds would change considerably. The risk that you are a carrier is still 1/70, but the risk that your cousin is a carrier also is no longer 1/70 but rather, 1/8. Now the chance that your offspring will be albino is $1/70 \times 1/8 \times 1/4 = 1/2,240$. The risk of bearing an albino child is now eight times greater.

Here is another example. The chance that you are a carrier for cystic fibrosis is 1 in 25. If you marry a nonrelative, the chance that your spouse will be a carrier is also 1 in 25. However, if you select your first cousin as a mate the chances again increase to 1 in 8. The likelihood that you will have a child with cystic fibrosis is three times greater than if you married a non-relative.

Studies of the incidence of consanguinity among the parents of children suffering from rare genetic defects confirm the risk. Up to 50 percent of children suffering from infantile amaurotic idiocy have parents who are first cousins. Up to 59 percent of all albinos are products of first-cousin marriages. Consanguineous marriages produce up to 21 percent of those with congenital total color blindness. Many similar statistics are available for other genetic disorders.

It is only fair to point out that consanguineous marriages also increase the probability of bringing out *desirable* heterozygous genes in the offspring. This possibility, however, does not justify the vastly increased risk of producing a child who is mentally retarded or has some serious physical malformation. With all the nonrelatives in the world to choose from, one should not have to select his mate from among his kin. Kiss your cousin, if you must, but it isn't wise to marry her.

## WHAT CAN BE DONE ABOUT HUMAN GENETIC ILLNESS?

We have mentioned a few representative human genetic and chromosomal disorders. There are dozens of them we have not mentioned. In some of these conditions the patient can be helped to some degree by treating the phenotype and supplying its lack (euthenics). In such cases, the vital product for which genetic competence is absent—for instance, insulin in congenital diabetics—can sometimes be administered artificially; or a substance that cannot be tolerated may be withheld, for instance, phenylalanine in PKU. Some half-dozen genetic disorders can be treated with massive doses of B vitamins. Like PKU, the disease *galactosemia* can be treated by withholding a food substance that cannot be metabolized—in this case, lactose, a disaccharide sugar occurring in milk products. The vast majority of genetic disorders, however, are like cystic fibrosis—there is little, in our present state of knowledge, that can be done to alleviate suffering or prevent death. What hope is there for eventual remedy?

## Treating the genotype

Genetic defects are often due to the functional absence of a particular enzyme from the body's cells. Usually, the patient must be homozygous for that gene, because if the enzyme-producing gene is present on even one of a pair of homologous chromosomes, enough enzyme is produced by that one to sustain normal life. For example, in the *Tay-Sachs syndrome,* a disease particularly prevalent in Jews of East European origin, the enzyme hexosaminidase A is missing, leading to degeneration of the central nervous system, idiocy, and death in infancy or very early childhood. Heterozygous carriers can be detected by testing their blood for the amount of the enzyme that is present. In heterozygotes only one chromosome makes the enzyme, so carriers have only 50 percent of the normal amount. That is enough for them, though. Heterozygotes for Tay-Sachs disease are vigorous and normal. (The blood test, incidentally, provides such people with the basis for birth control decisions.)

If a disease like this results from the *lack* of an enzyme, it is probably because one or a few DNA bases have been added or deleted on that section of the chromosome. Can the defective section of DNA be replaced, or can a functional copy be provided in addition to it? No, not yet, but there is promise that one day this will be possible.

One potential means is by virus infection. An important characteristic of what are called "temperate" viruses is that they can be induced to carry genetic information into their host cell in their DNA, which the host can use to overcome an enzymatic defect. This has been demonstrated in bacteria. It is likely that it has application to human beings as well. Some day, a benign virus infection might be used to carry the missing genetic information into the cells of a baby with PKU or Tay-Sachs disease.

Somewhat similarly, zygotes might be genetically repairable by fusing them in the laboratory with human or animal cells bearing the genes they lack. The repaired zygote could then be reinserted in the uterus of its mother, or perhaps eventually raised in laboratory glassware.

A note of caution is in order. First, genetic therapy by the *addition* of DNA capable of producing active enzymes would work only in recessive diseases where those enzymes were *absent*. It seems very unlikely that it would be effective against dominant disorders, or disorders such as sickle-cell anemia, where the problem is not an absent enzyme, but an abnormal protein that is present in large quantities. Such proteins must be *replaced,* which involves the removal or suppression of the defective gene—a vastly more difficult task. It is also easily possible that, concomitant with the introduction of a desirable gene, undesirable ones might also be inserted in the DNA, *producing* disorders. Some forms of cancer seem to be due to viral interference with the normal cellular genotype. Third, chromosomal defects have no simple genetic basis, and it is unlikely that any currently proposed technique would be effective with them or with any genetic characteristic involving *development* rather than adult phenotype. Still, we do anticipate substantial progress in the correction of at least some genetic defects in the near future.

A further problem, of course, is that we have no guarantee that genetic surgery or other intervention will be used solely for good and not for evil. Genetic therapy could be used to correct hereditary disorders; but there is a fuzzy line between this and *genetic engineering,*[4] the alteration of human beings and their heredity for reasons unrelated to disease. You could imagine, for instance, a Hitler who would breed a race of super-soldiers or a caste of human ants to work in factories. It might even be possible to introduce genes by means of virus infections into an enemy population that would produce genetic disease and imbecility in most of their next generation. Some suggest, for this reason, that all research in the correction of genetic defects be halted. Do you agree?

## Eugenics

All this bears upon the science of eugenics. *Eugenics* involves the selective breeding of organisms, especially human beings, so as to favor the preservation and spread of desirable genes in the population. *Dysgenics* is the opposite—favoring deleterious or disease-producing genes.

Human life, as it is now lived, would seem to be quite dysgenic. Radiation and the myriads of chemicals we are exposed to—both of natural and artificial origin (such as some food additives and DDT)—ensure an increasingly high mutation rate in the human population in addition to whatever mutation rate might occur without known cause.

Modern medical science, on the other hand, has ensured the preservation of the lives of many persons with inherited disorders such as phenylketonuria, diabetes, hemophilia, and galactosemia. Many of these people marry and reproduce, passing on genes that in former years would have been eliminated by the early death of their possessors. It is possible to prove mathematically that under these circumstances unfavorable genes will eventually spread widely (though slowly) throughout the human population. It is proposed by some that the carriers of these genes should be required by law to be sterilized. This could be done by tubal ligation, or vasectomy, however much it might constitute a violation of their civil rights. This is known as *negative eugenics,* for by such practices you are removing undesirable genes. (Negative eugenics can, of course be practiced voluntarily, with great social benefits and much to the advantage of parents, who without benefit of genetic counseling might produce grossly defective offspring.)

*Positive eugenics* is sometimes also advocated. Its proponents envision that persons with genetic traits held to be desirable might be bred like prize cattle. It need not necessarily involve their physical contact—it might not even disturb their existing marriages! They need only contribute their eggs and sperm; these gametes could be allowed to unite in vitro, that is, in laboratory glassware. The early embryo could then be implanted in the uterus of a professional host mother to mature. According to proponents of this view, we could eventually breed super-geniuses with the strength of Hercules who

---

[4] Some authorities define genetic engineering as any deliberate genetic alteration.

might live for several centuries free of all disease. They say further that if one nation started such a program, all others would have to follow suit or be submerged by an enemy master race that this time really *would* be a master race.

Please do not think these ideas are the far-fetched products of fevered brains; they have been seriously proposed by supposedly responsible political and scientific leaders and are, definitely, finding an audience. Perhaps you yourself may find them reasonable. Nevertheless, they are open to a number of serious objections:

1   The same mathematics that indicates the spread of harmful genes in a population also indicates that the spread of those genes is slow. Conversely, any attempt to stop the process will also be slow to show results, particularly if the gene in question is recessive, as most of them are. We have no assurance that any governments initiating such programs would continue to exist long enough to obtain any worthwhile results by them (see Chapter 17).

2   The genetic basis of many disorders, particularly most mental disorders, is questionable.

3   It is not always possible to predict what characteristics may be useful to future generations. If our descendants should desire to colonize the moon or be compelled to abandon industrial society because of exhausted natural resources, many of the traits now deemed undesirable might be essential to their welfare. G6PD deficiency, for instance, is handy in the absence of effective antimalarial programs. The deliberate genetic homogenization of the human race would tend to restrict our future options.

4   It is not even possible to say with certainty what characteristics are useful to us in the present. Many psychologists claim that mild schizoid traits are an essential part of the creative personality. A program to eradicate schizophrenia (which may be, in part, of genetic origin) could well result in the creative impoverishment of the human race.

5   Highly bred mental and physical supermen may exhibit undesirable emotional or character disorders along with those we hold to be desirable, or they might prove unsatisfactory in some other unpredicted way. The fact that great intelligence is rare implies that it may be in some way pathologic! Though this is, admittedly, an argument from ignorance, our ignorance of genetics (particularly of mental traits) is profound. Would it not be premature to start such programs at this time?

6   On the other hand, were such a program to succeed, the ruling caste, thus created, would probably reduce the remainder of mankind to slavery or, at best, compete with them so successfully that worthwhile careers would no longer be open to "normal" people.

7   The creation of specialized or even "subhuman people" by eugenics is theoretically just as feasible (or unfeasible) as the creation of supermen. Most of us would hold this to be undesirable because our culture biases us against it. From an objective viewpoint, if such were possible, our ideals of the genetic superman may be just as perverted as those that would lead to the perfection of human termites. Sub- or superhuman, would such "people" be human at all? If not human, would their creation be a betrayal of humanity?

Other criticisms are possible, and some may occur to you that we have not listed. By the end of the century most of our readers will probably still be alive, and indeed will help to comprise the age group most actively involved in running our society. You will be voting, and, we hope, engaged even more deeply in the political process. Before then, many of the possibilities we have discussed in this chapter may well be realities, and many other genetic advances that we have not foreseen probably will have come to pass as well. Considering the vast potential both for good and for evil that these developments possess, we can only hope that you will continue to learn, to think, and to commit yourself morally in these areas throughout your lifetime.

1 A number of recessive genetic diseases known as "inborn errors of metabolism" are attributable to missing or inactive enzymes needed to catalyze specific biochemical reactions. Among them are alkaptonuria, phenylketonuria, and albinism.
2 Sickle-cell anemia and thalassemia are recessive blood cell disorders traceable to chemically abnormal hemoglobins. Sickle-cell hemoglobin differs from normal hemoglobin in one amino acid residue. The trait, though essentially lethal, is an advantage in the heterozygous condition, since the presence of the abnormal hemoglobin in the erythrocytes of heterozygotes seems to inhibit the malaria parasite.
3 Erythroblastosis fetalis is a severe hemolytic anemia produced in the newborn Rh-positive child by antibodies produced in the body of his Rh-negative mother to his blood cells. It can often be treated by exchange transfusion, but is best prevented by the artificial injection of anti-Rh-positive antibodies in the form of gamma globulin immediately following the birth of all children to such mothers.
4 Multiple or missing sex chromosomes give rise to a variety of disorders in the persons so afflicted. Proportionately far fewer disorders involving autosomes are known, but one of the most important congenital diseases known, Down's syndrome, is attributable to a trisomic karyotype of the No. 21 chromosome.
5 A number of approaches to genetic disease are theoretically possible. It is sometimes feasible to supply the missing genetic product, as in the case of congenital diabetes or hemophilia. The birth of a genetically damaged fetus can often be prevented, either by abortion (following amniocentesis or other means of assessing genetic risks) or by eugenic measures involving the prohibition of marriage of consanguineous individuals or those known to carry specific harmful genes. The future may bring genetic treatments whereby hereditary defects can be specifically remedied by adjustment of the DNA. The ethical and social implications of some of these matters are grounds for real concern.

The understanding of genetic principles can be tested by solving problems employing those principles. Furthermore, practice in problem solving may produce insights into the basic concepts of genetics not easily obtained in any other way. Nevertheless, the ability to solve genetic problems is not an end in itself. It is useful only to the degree that it may contribute to an understanding of genetic concepts and the ability to apply these concepts to the real world.

These problems are to be regarded as illustrations only. A greater variety are to be found in the study guide which accompanies this book.

1 The fatal congenital disease, xeroderma pigmentosum, expresses itself by abnormal pigmentation of the skin, which if exposed to sunlight becomes lethally

**FIGURE 16-4**
White forelock, a dominant gene. The inheritance of this trait can be traced through four generations of the woman's family.

cancerous at an early age. Recent investigations have shown that in these patients the DNA is not able to repair itself after exposure to ultraviolet light. In normal persons the DNA is capable of self-repair to a large extent. In the heterozygous condition, the xeroderma gene is harmless but produces a typical type of skin freckling. If two such heterozygous, freckled persons marry, what phenotypes can be expected among their offspring?

**2** Two varieties of xeroderma pigmentosum are known. In one type, the condition is accompanied by mental deficiency, and in the other, it is not. Recent studies employing tissue culture indicate that when cells taken from patients exhibiting the two varieties are fused, the resulting compound cell does not exhibit the biochemical abnormalities associated with the disease; that is, as far as xeroderma pigmentosum is concerned, it is normal. Propose an explanation for this experimental finding.

**3** A streak of white in an otherwise dark head of hair is known as white forelock. It is dominant to the normal allele. If a woman with white forelock marries a normal man and their first child is normal, what is her genotype? (See Figure 16-4).

**4** Two second cousins marry. Blood tests disclose that though phenotypically normal, they both have only partial activity of the hexosaminidase A enzyme in their blood serum. Such a biochemical characteristic marks them as heterozygous carriers of Tay-Sachs disease, an invariably fatal recessive trait associated with the total absence of hexosaminidase A and with fatty degeneration of the nervous system. Homozygotes die long before they reach reproductive age.

(*a*) It is now possible to diagnose the Tay-Sachs syndrome by amniocentesis, a somewhat dangerous procedure. In this couple's case, is it worth the risk?

(*b*) Draw a family tree showing four generations of their immediate ancestors. Assuming that the gene entered their family in their mutual great grandfather, show all the crosses and genotypes that their ancestors possessed.

**5** Two persons heterozygous for albinism marry. The lady is able to curl her tongue (as discussed in Chapter 15), an ability she shares with her father but not with her mother. If tongue curling is dominant, what is the chance that their first child will be an albino capable of curling his tongue? Albinism is a recessive trait.

**6** A color-blind man marries a woman who is afflicted with brown tooth enamel (review Chapter 15). One of their daughters, a girl with brown teeth, marries a normal man. What is the chance that their first son will be color blind and will have brown teeth? That he will be completely normal?

**7** An unpronounceable enzyme that we shall call HGP is missing in some cases of the Lesch-Nyhan syndrome, and has abnormal characteristics in others. When homozygous, the gene responsible produces a bizarre form of mental deterioration in which the patient persistently bites himself. It is fatal in childhood, is X-linked, and is recessive.

Mary comes from a family of six in which she is the only girl. Two of her brothers have died from the effects of the Lesch-Nyhan disease, but the rest are phenotypically normal, as are her parents. Jim, her fiance, is normal. Before they marry they wish to know the chance that their children will exhibit this disorder. What would you tell them?

**8** Work out the blood type cross $ABRr \times BOrr$. What are the phenotypes of the progeny? Now work out $ABRr \times BORR$. What are the phenotypes of the progeny? If a man of genotype $BORR$ is married to a woman of genotype $ABRr$, can he be the father of her B-positive baby? If it were A-negative?

**9** Baldness in man is thought to be due often to a single pair of allelic genes. The genotype $B_1B_1$ is bald in either sex. $B_2B_2$ produces a full head of hair. $B_1B_2$ is bald if the bearer is male, but normally hairy if female. (These phenotypes are usually expressed in later life.) What phenotypes can be expected among the offspring of a bald lady and a hairy man?

10 The urine of some persons has a distinctive odor following the eating of asparagus. This appears to be a recessive trait. If a person with aromatic urine marries a person heterozygous for this trait, what phenotype *proportions* will their offspring theoretically exhibit?

11 There are several varieties of X-linked color blindness. **Protanopes** cannot see red, for example, and **deuteranopes** sense red but cannot distinguish it from green. Assuming these genes to be X-linked, nonallelic, and recessive, what offspring would be observed from a cross between a normal man and a woman whose mother was a protanope, and whose father was a deuteranope?

12 Manic-depressive mental illness of one type (bipolar) is X-linked and recessive. If a manic man marries a woman whose father was color blind, what is the chance that their male child will be a color-blind manic? Their female child?

13 If a color-blind man who is also manic marries a normal woman, and their daughter marries a normal man, what is the chance that their male grandchild will be a color-blind manic? (Ignore crossing-over.)

14 Emphysema can be caused by the homozygous presence of a recessive gene producing a deficiency of the enzyme alpha 1-antitrypsin. Even in the heterozygous condition this gene predisposes one to emphysema *if he smokes*. Explain this fact in terms of what you know about the biochemical basis of gene action.

15 In a very small percentage of cases, women heterozygous for red-green color blindness exhibit this defect phenotypically. Explain this finding in terms of Lyon's hypothesis. (*Hint:* when, during development, does X inactivation take place?)

16 "Two out of five. There will be no more children, because the Walshes have learned that the deafness of Mary Ann and Nancy was caused by hereditary genes." (From a medical article in a popular magazine.)

(*a*) Is this disease dominant or recessive?

(*b*) What are the probable genotypes of the parents?

17 The nature of *dominant* genes which produce biochemical defects is somewhat mysterious. Can you propose a hypothesis to account for their existence?

18 It has recently been discovered that an X-linked disorder, Fabry's disease, is associated with a deficiency of the enzyme alpha-galactosidase. But even in patients exhibiting the disease, about 20 percent of the enzyme's normal activity is present. Investigation disclosed that two alpha-galactosidase isoenzymes exist in normal cells, one of which is missing in Fabry's disease. The remaining weaker enzyme accounts for the residual activity. What degree of enzyme activity would you predict in a woman heterozygous for Fabry's disease? Would the Lyon hypothesis affect your answer?

19 In all mammals except man, the enzyme lactase ceases to be produced by adulthood, so that adult animals are unable to digest milk sugar. Of course, most mammals have no need to drink milk after maturity. But in Europeans and certain other cattle-raising people the lactase enzyme persists through life. What kind of mutation could account for this odd racial characteristic?

20 A codominant mutation occurs in hamsters, which, when homozygous, produces a naked condition of the skin (Figure 16-5). Heterozygous hamsters have coats of hair that are thinner than normal. Predict the phenotypes and proportions of the offspring of a naked and a sparsely furred hamster.

21 Another codominant mutation occurs in human beings, congenital joint laxity (Fig-

ure 16-6). The heterozygote is "double-jointed," and the homozygote is usually cripped by poor joint articulation. What phenotypes will occur among the offspring of two "double-jointed" persons?

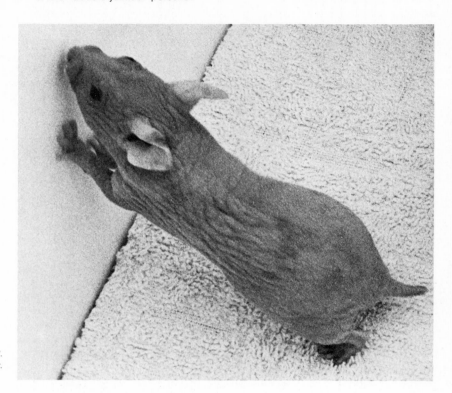

**FIGURE 16-5**
The naked hamster. (Drs. M. F. W. Festing and M. K. Wright, and *Nature,* Mar. 10, 1972)

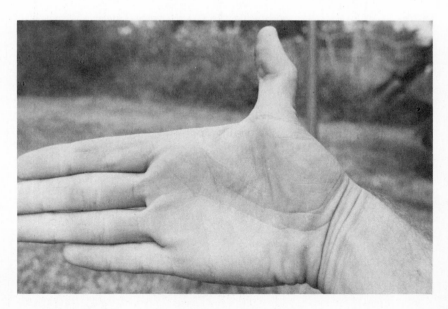

**FIGURE 16-6**
Congenital joint laxity.

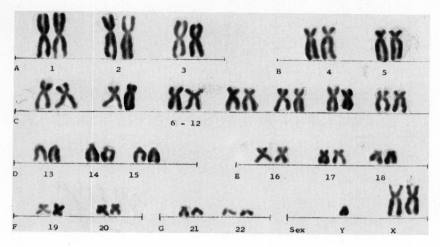

FIGURE 16-7
Karyogram of an individual with Klinefelter's syndrome. This is an XXY, which is phenotypically male, but sexually somewhat undeveloped. (Dr. Gilbert Echelman)

We have listed a number of articles from *Scientific American,* but you should also see more recent issues of this magazine. Another excellent source is the magazine *Hospital Practice,* particularly the 1969 and 1970 volumes. If you are near a medical school or a hospital with a good professional library, you should consult that magazine. *The Journal of the American Medical Association* frequently carries papers on human genetics. Finally, a number of specialized genetic journals exist, but their contents are generally too technical for the beginner to master.

## FOR FURTHER READING

Anderson, V. Elving: "Chromosomes and Human Behavior," *American Scientific Affiliation,* June 1969, p. 48.

Cole, William: "The Right to Be Well-Born," *Today's Health,* January 1971, p. 42.

Friedmann, Theodore: "Prenatal Diagnosis of Genetic Disease," *Scientific American,* November 1971, p. 58.

Lerner, I. Michael: *Heredity, Evolution and Society,* Freeman, San Francisco, 1968.

McKusick, Victor A.: "The Mapping of Human Chromosomes," *Scientific American,* April 1971, p. 104.

————: "The Royal Hemophilia," *Scientific American,* August 1965, p. 88.

Moody, Paul A.: *The Genetics of Man,* Norton, New York, 1967.

Winchester, A. M.: *Human Genetics,* Merrill, Columbus, Ohio, 1971.

Macalpine, Ida, and Richard Hunter: "Porphyria and King George III," *Scientific American,* July 1969, p. 38.

"Man into Superman: The Promise and Peril of the New Genetics," *Time,* Apr. 19, 1971, p. 33.

# 17

# THE ORIGIN OF LIVING SYSTEMS

*Chapter learning objective.* The student must be able to complete an examination on the process of organic evolution including its history as a concept, modern evolutionary synthesis, terminology, and evidence bearing upon its validity.

*ILO 1* Discuss the history of evolutionary thought, listing the most important contributors and summarizing their ideas.

*ILO 2* (*a*) Give the Hardy-Weinberg law, discuss its significance and consequences in terms of population genetics and evolution, and (at the option of the instructor) solve simple problems in population genetics.

(*b*) Define evolution in genetic terms, discuss the relationship between the Hardy-Weinberg law and evolution, and describe the dependence of evolution upon genetics.

*ILO 3* Summarize the "modern" concept of evolution, discussing its mechanisms (mutation, genetic recombination, genetic drift, natural selection, reproductive isolation) and the adaptive results. Cite specific examples and illustrative experiments.

*ILO 4* Summarize the general features of:

(*a*) The currently popular theory of the spontaneous origin of macromolecules and other organic compounds necessary for life's beginning.

(*b*) The symbiotic theory of the origin of eukaryotic cells.

*ILO 5* List and critically discuss the various types of evidence bearing upon evolution (e.g., evidence from comparative anatomy, geology, etc.), giving the specific application of each.

Among all the cultures of man known to anthropologists there is some explanation for the origin of the world, of mankind, of other organisms, or of what we would call the universe. Man's curiosity about his origins has demanded explanations, and these explanations continue to be modified today.

## TWO CONCEPTS OF ORIGINS

Such explanations tend to fall into one or the other of two broad categories: special creation or evolution. Various admixtures and modifications of these two concepts exist, but it seems impossible to imagine an explanation of origins that lies completely outside the two ideas.

Both views are as old as Western civilization, or older. It is well known, for example, that the book of Genesis in the Bible describes the creation of the animate and inanimate world in 6 days, with emphasis upon the personal intervention of God in the creation of a single ancestral pair of human beings. Less well known to us today are the evolutionary speculations of ancient Greek philosophers which coexisted with their ideas of supernatural creationism. At first their ideas had little impact. They reentered the mainstream of philosophy only during the Renaissance. Creationism, for its part, has persisted to this day but has, since Darwin's time, become a minority view.

## THE NATURE OF EVOLUTION

In its broadest sense the word *evolution* simply means change. However, biologically and historically *organic evolution* refers to the gradual devel-

*"Maybe you're descended from a lemur, but I'm not descended from any lemur!"*

**FIGURE 17-1**
(William Miller, Copyright 1972, *The New Yorker Magazine*)

opment of complex organisms from simple beginnings. Most early evolutionary writers were concerned with this aspect of its meaning. In this sense, evolution is the idea that all forms of life developed gradually from very different and often much simpler ancestors, and that all lines of their descent can, in the end, be traced back convergently to a common ancestral organism. Most people today understand the term "evolution" in this way.

There is another sense, however, in which the term evolution may be used. This aspect is less concerned or even unconcerned with speculation about the actual course of change in populations of organisms going back to their ultimate origin. Rather it involves a consideration of the *mechanism* whereby relatively minor genetic changes take place in populations. This form of evolution is sometimes referred to as *microevolution* and is recognized even by most creationists. Adam and Eve could, at the most, have represented the equivalent of only two races, and one must consider the enormous range of human variation present today!

The earlier portions of this chapter will be concerned mainly (but not exclusively) with evolution in this latter sense—that is, with microevolution. The other question—the extent to which it has in fact taken place—will be discussed at the end.

## DEVELOPMENT OF EVOLUTIONARY THINKING

Although the idea of evolution is ancient, older views do not resemble modern concepts of evolution at all closely, and probably had little influence upon the originators of the modern theory of evolution.

### Lamarck, Darwin, and Wallace

The name most commonly associated with theories of evolution is that of Charles Darwin (1809–1882), who does indeed deserve recognition as the father of almost all modern evolutionary thinking. Darwin, however, did not inhabit an intellectual vacuum. His predecessors and contemporaries made important contributions to his thinking.

Darwin's most important predecessor was probably the Frenchman Lamarck. All too often Lamarck is remembered merely as someone who proposed an unsatisfactory theory of how evolution proceeded, but he was also the first to tackle seriously the question of *how* evolution took place. Those writing before him for the most part toyed with some evolutionary ideas, merely proposing that evolution did take place. Lamarck proposed a theory of use and disuse of organs combined with the inheritance of acquired characteristics. For example, ancestral giraffes, persistently stretching their necks in search of food would, over the course of many generations, produce our present long-necked giraffes. Disuse of the eyes over a long enough period of time might similarly produce blind cave fish. Thus Lamarck thought that individual adaptive response to the demands of the environment on the part of an individual organism was somehow transmitted to succeeding generations.

The first to develop a plausible alternative to this explanation was

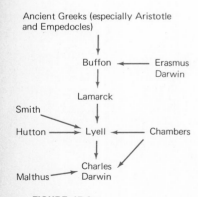

**FIGURE 17-2**
The historical development of evolutionary thought.

Charles Darwin. Darwin, partly as a result of a voyage of exploration and animal collecting on the HMS *Beagle,* eventually proposed the theory that bears his name today. Together with a fellow naturalist, A. R. Wallace, who independently had come to precisely the same conclusions, he published a summary of his thought.

## The mechanism of evolution

As Darwin came to view it, all organisms tend to produce more offspring than their environment can support. This in turn leads to a competitive struggle among living things with similar ecologic requirements. Add to this the need to deal with natural enemies such as predators and parasites, and a basically hostile physical environment, and you will have the basic conception of the Darwinian struggle for survival. Since all organisms differ from one another to some extent, those in the past that were best fitted to succeed and survive by virtue of those variations did succeed and survive. Others did not, or at any rate did not reproduce themselves as successfully. The characteristics that ensured the survival of parents were inherited by their offspring. Those offspring, in turn, varied among themselves, and those best fitted for the demands of their environment were preserved by *natural selection,* i.e., the ability of the environment to favor or to discriminate against organisms. If a plant or an animal has the ability to survive under a given set of conditions, it will survive; if not, it will perish. If a particular organism possesses a competitive advantage over its fellows, it will prevail over them. Recently some rats exposed to the rodenticide *warfarin* developed a resistance to the chemical. This presumably occurred because only those rats that possessed superior resistance survived to produce the present generation of resistant animals. The end result would be systematically to bring species into closer harmony with the demands of their habitat with every passing generation.

Darwin saw nature itself selecting those most fit to pass on their heritage to the next generation (not, of course, in a conscious or intelligent fashion). Although Spencer, not Darwin, coined the phrase "survival of the fittest," nevertheless it expresses the Darwinian idea quite well. The organism that is strongest, toughest, subtlest, most enduring, or that can fly fastest, fly farther, secure more mates, or best fulfill whatever other demands its environment puts upon it—that organism would survive, reproduce, and ultimately prevail.

Thus the four main points involved in the mechanism of evolution that Darwin proposed are overproduction of offspring, variation among those offspring, inheritability of that variation, and natural selection of the variants.

Though Darwin and Wallace did not originate the concept of evolution, they did propose a plausible mechanism whereby it might take place. Theirs was the first credible theory of evolution.

## GENETICS AND MICROEVOLUTION

The modern concept of microevolution is not so much a contradiction of the mechanisms Darwin postulated, as a logical extension and development of

Imagine a population of organisms which is panmictic for two allelic genes: A, whose frequency is p, and a, whole frequency is q. Their combined frequency is obviously 1, so we may write:

(1) $p + q = 1$, so $p = 1 - q$; $q = 1 - p$

Now a gamete bearing gene A may combine with another one such to form an AA zygote. Similarly, $A \times a \rightarrow Aa$; $a \times A \rightarrow aA$ and $a \times a \rightarrow aa$. The frequency of each combination is the product of the frequencies of its component genes, as follows:

|  | $p(A)$ | $q(a)$ |
|---|---|---|
| $p(A)$ | $p^2(AA)$ | $pq(aA)$ |
| $q(a)$ | $pq(Aa)$ | $q^2(aa)$ |

Adding these up, the totality of all genotypes is:

(2) $p^2 + 2pq + q^2 = 1$

where $p^2$ = frequency of AA
$2pq$ = frequency of Aa
$q^2$ = frequency of aa

Now in the *next* generation, the present Aa organisms will be able to produce *either* A or a gametes, that is, half the gametes will be A and half a from this source. The AA organisms will, of course, only yield A gametes; the aa only a gametes. Thus the total frequency of all A and a gametes may be expressed as follows:
(Let p′ stand for the new frequency of A; q′ for the new frequency of a)

(3) $p' = p^2 + \frac{1}{2}(2pq)$
$= p^2 + pq$
$= p^2 + p(1 - p)^*$
$= p^2 + p - p^2 = p$

(4) $q' = q^2 + \frac{1}{2}(2pq)$
$= q^2 + pq$
$= q^2 + q(1 - q)^*$
$= q^2 + q - q^2 = q$

*Therefore,* if left undisturbed, gene frequencies do not change from generation to generation in a panmictic population, regardless of dominance or recessivity.

* See equation (1)

his ideas. The modern evolutionary principles most essential to our understanding are those of the *gene pool* and the *Hardy-Weinberg law.*

### Gene pool and the Hardy-Weinberg law

A population of organisms in which genes combine essentially at random is said to be **panmictic.** Even human populations are panmictic for certain genes; that is, mating is at random with respect to them. People are not greatly concerned about blood types, for example, and therefore marry without taking them into account. We therefore are panmictic with respect to blood types. However, mating among human beings is not panmictic with respect to some genes—those governing skin color, for example.

A panmictic population may be considered as a pool of allelic genes, any one of which has a chance of combining with any other, at least when borne by an individual of the opposite sex. What that chance is will depend upon the frequencies of the genes involved, and upon no other factor.

It can be shown that in a freely interbreeding population that mates randomly with respect to two allelic genes, the genotypes of the population may be described by the expression $p^2 + 2pq + q^2$, where $p$ = the frequency of the dominant allele, $q$ = the frequency of the recessive allele, $p^2$ = the frequency of the homozygous dominant genotype, $2pq$ = the frequency of the heterozygote, and $q^2$ = the frequency of the recessive homozygous genotype (see Figure 17-3).

One can demonstrate also that, regardless of dominance or recessiveness, the relative frequencies of allelic genes do not change from generation to generation if they are left undisturbed.

Discovered in the early twentieth century by Hardy and by Weinberg independently, the law summed up by the formula $p + 2pq + q^2$ is known by their name. It is true only in the absence of mutation, natural selection, genetic drift, or selective migration. Without those perturbing forces genetic frequencies in a freely interbreeding population do not change significantly from generation to generation. Thus the Hardy-Weinberg law establishes a base line from which evolutionary departures must take place.

### Changes in gene frequency

The Hardy-Weinberg principle demonstrates beyond possibility of serious challenge that allelic genes will retain their original frequency in a population no matter how many generations elapse and regardless of degree of dominance. This being the case, at first it might appear that the Hardy-Weinberg principle renders microevolution impossible, for if genetic frequencies within populations cannot change, how can the organisms that comprise those populations evolve? The resolution of this paradox depends upon an examination of the conditions under which the Hardy-Weinberg law can be expected to hold true.

*Mutation.*  Quite clearly, genetic frequencies would change anyway if one kind of gene changed into another form of the same allele. This in fact does happen in all known populations, although usually very rarely.

*Natural selection.* Another influence that tends to perturb the genetic equilibrium of populations is natural selection. You will recall the case of sickle-cell anemia from Chapter 16. In this disorder an abnormal form of hemoglobin occurs in the red blood cells of those homozygous for the trait. In the absence of treatment the condition is almost invariably fatal. Yet a heterozygote suffers from no more than a mild, harmless anemia. In East Africa, where the trait probably originated, the heterozygote enjoys some protection against the malaria parasite. So prevalent is malaria in the Congo Basin and similar areas, that someone heterozygous for sickling apparently has a considerable advantage over a person of normal genotype. But not everyone with sickle-cell anemia lives in the Congo Basin. The trait also persists in many persons of African origin in the United States, now nearly malaria-free. Obviously the sickle-cell heterozygote has no advantage in Chicago. Natural selection may be expected to remove this gene from the American population, given enough time in which to do so, for here it constitutes a disadvantage. (Modern methods of medical treatment may, however, preserve the gene despite the workings of natural selection.)

*Selective immigration or emigration.* Theoretically, selective immigration or emigration can change genetic frequencies, since it represents the arrival or departure of organisms possessing one genotype at a greater rate than others in the population, and thus changes their proportional representation in the gene pool.

*Genetic drift.* Genetic frequencies can remain constant only if the population is fairly large. Otherwise, random events will tend to cause changes. If a population consists of only a few individuals, predators could destroy the only representatives of a particular genotype and miss the others purely at random. Such an event would be most unlikely in a large population. The production of changes in gene frequency by random events is known as *genetic drift.*

*Founders and bottlenecks.* Genetic drift is substantial in small populations with limited gene pools, and may be expected to be greatest in the smallest populations. If a new habitat is accidentally colonized by a small number of organisms, the only genes that will be represented among their descendants will be those few which they chanced to possess. Thus, isolated populations may have very different gene frequencies than those characteristic of the species elsewhere. The disproportionate effect exerted upon a population by a limited number of ancestors is termed the *founder effect.* Like genetic drift, it can produce great changes in gene frequency even in the absence of natural selection (Figure 17-3).

In some species, very few individuals survive some critical stage in their life cycle. For example, among houseflies in Northern areas, only a few representatives survive the winter months, and they give rise to most of the summer population. In principle, this is similar to the founder effect. Only a few individuals, which are perhaps not truly representative of the genetics of

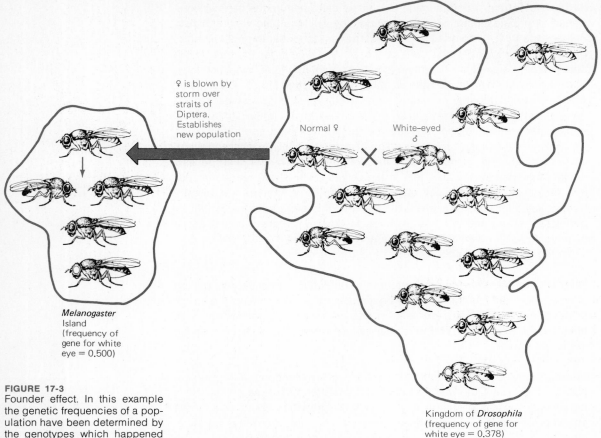

♀ is blown by storm over straits of Diptera. Establishes new population

Normal ♀          White-eyed ♂

X

*Melanogaster*
Island
(frequency of
gene for white
eye = 0.500)

Kingdom of *Drosophila*
(frequency of gene for
white eye = 0.378)

**FIGURE 17-3**
Founder effect. In this example the genetic frequencies of a population have been determined by the genotypes which happened to be possessed by its founders, but which were not characteristic, in frequency, of the ancestral population as a whole.

the population from which they came, will give rise to the entire future population. They will by chance exert a disproportionate influence over its prospective genetic frequencies. Since one can think of this phenomenon as a periodic squeezing out of some of the genes in a gene pool in random fashion, it is termed the **bottleneck effect**. Like genetic drift and the founder effect, it can change gene frequencies even in the absence of natural selection.

Thus the Hardy-Weinberg situation will prevail only under certain conditions, which may be listed as:

**1** Absence of mutation
**2** Absence of selection
**3** Presence of panmixis
**4** Absence of selective immigration or emigration
**5** Large population

Changes in gene frequencies can be produced only by departures from

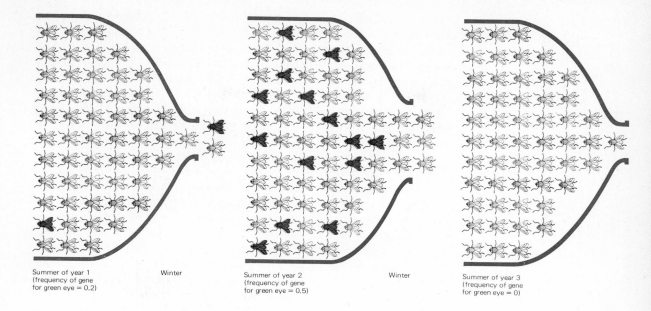

Summer of year 1
(frequency of gene
for green eye = 0.2)

Winter

Summer of year 2
(frequency of gene
for green eye = 0.5)

Winter

Summer of year 3
(frequency of gene
for green eye = 0)

**FIGURE 17-4**
Bottleneck effect. Since only a small population of flies survives the winter, its genotypes determine the genetic frequencies of the entire succeeding summer population.

the basic Hardy-Weinberg conditions, but under natural conditions such departures nearly always do take place.

Mutations are always occurring, as is natural selection. Panmixis seldom prevails, and many collections of organisms in nature are divided into small, more or less isolated gene pools separated by geographic or topographic barriers. Thus in most instances, microevolutionary changes in genetic frequencies are inevitable in natural populations of organisms.

## Examples of natural selection

In recent years a number of examples of natural selection that have come to light in the laboratory or in the field have afforded an incomparable opportunity to study this microevolutionary mechanism in action. Among these we may list the development of DDT resistance in insect populations exposed to this pesticide, industrial melanism in moths, and the development of resistance to drugs among disease bacteria. Space permits the discussion of only one of these examples.

*Protective coloration and industrial melanism.* A subtle example of natural selection is the development of protective coloration. Protective coloration exists when an organism blends in with its surroundings in such a way as to make it hard to see. This has the effect of protecting it from its predators, or in the case of those which *are* predators, of keeping their victims from noticing them until it is too late.

Many examples of protective coloration and mimicry readily come to mind. Stick insects resemble sticks so closely that you would never guess that they are animals—until they start to walk. The chicks of ground-nesting birds are usually colored to blend in with the surrounding weeds and earth so that they simply cannot be discerned from a distance. Some leaf-hoppers resemble leaves not only in color but in the pattern of veins in their wings. There are praying mantids that resemble flowers—pity the poor bee that comes to visit! Crab spiders are often colored like the flowers they inhabit—in fact, they seek out flowers whose colors are similar to their own. The suggestion is obvious that evidently such protective coloration has been preserved and accentuated by means of natural selection. How might these adaptations originate?

In the past there were many moths whose colors blended in with the light-colored lichens on the trunks of trees. Unfortunately, air pollution kills lichens so readily that they are often considered an index of it. Prior to about 1850 in England, one scarcely could find a dark-colored specimen of the moth, *Biston betularia,* and they were extremely rare in museum collections made before that date. Nowadays it is hard to find any of the formerly predominant light-colored specimens, at least in industrialized portions of that country. *Industrial melanism* is the name used for the development of such dark coloration in populations exposed to air pollution. It is not just dirt, but an increase in the dark pigment melanin in their bodies (see Figure 1-10).

When air pollution kills light-colored lichens on tree trunks, light-colored moths stand out against the dirty dark tree bark. Under these circumstances it is no longer an advantage, as it was formerly, to resemble the light-colored lichens that no longer exist. Instead, birds eat those light-colored moths, and only those that happen to be darker survive well under the changed conditions. So here we have a well-documented case of protective coloration occurring just since the industrial revolution began to produce substantial air pollution in the nineteenth century. Is natural selection responsible? Light-colored moths released experimentally in polluted districts are known to be eaten more readily by birds than dark ones, and the reverse is true in non-polluted districts. Many moths in the vicinity of large cities now even have melanistic caterpillars. There is a bright side. Since air pollution control was instituted in the 1950s in certain British districts, the proportion of light-colored moths is again increasing.

*Mimicry.* When an organism closely imitates the appearance of a model to which it is unrelated, it is called mimicry. One example is *Batesonian mimicry,* wherein a harmless or palatable species resembles one that is dangerous, obnoxious, poisonous, or revolting. There are many examples. A harmless fly may resemble a bee so closely that even a biologist would hesitate to pick it up in his hand. The well-known monarch, or milkweed, butterfly is poisonous to birds and mammals because of the toxic substances it absorbed while feeding on poisonous milkweed plants as a caterpillar. The monarch has many insincere imitators among butterflies, which closely resemble it in color, but which are entirely edible.

Natural selection has not turned a fly into a bee or even one species of butterfly into another in these instances, but it apparently has maintained a resemblance that gives its possessor almost as much protection as the

(a)

(b)

model, for as soon as predators learn to associate the distinctive markings of the model with its undesirable characteristics, they will tend to avoid all similarly marked animals.

### Genetic recombination

Almost all members of a species have some means of transmitting genes to one another, or from one generation to another, in such a way that new combinations of genes can result. This is true even of viruses and bacteria. Without genetic recombination offspring would resemble their parents precisely, and the only means of genetic change from one generation to another would be mutation. Even favorable genes would come to predominate in a population of such organisms much more slowly than if they could reproduce sexually.

Often new combinations of genes are more favorable than old. It has long been recognized that when two genetically uniform inbred strains of organisms are crossed, the offspring of the union display *hybrid vigor,* that is, they are usually more vigorous than the parents.

## OTHER ASPECTS OF MICROEVOLUTION

Clearly, it is not enough to speak of mutations and the change of gene frequencies as accounting for all the major differences among organisms. By themselves, these processes tell us little about the mode of formation of new species, and still less about the production of higher taxonomic categories.

### Speciation

A *species* is a reproductively isolated population of organisms incapable of genetic exchange with other species. It follows from this definition that the key to the development of a species (or *speciation*) is the development of its reproductive isolation so that it cannot interbreed with other species. It is thought that there are two main ways in which such a situation may arise.

**FIGURE 17-5**
(*a*) Upper left: The distasteful *Heliconius erato* butterfly of South America. Upper right: Its comimic, *Heliconius melpomeme.* In this form of mimicry, two unpalatable species are similarly colored, which helps predators to learn to leave them both alone. Bottom left: Woodruff Benson obscured the coloration of some specimens of *H. erato* by painting them with dye; others (bottom right) he left alone. After releasing the altered and unaltered butterflies, he found that the altered ones suffered heavier predation, showing that the color pattern tended to protect this species of insect from the attacks of its enemies. (Woodruff Benson and Science) (*b*) The harmless bee fly closely resembles a bee.

One is through the chance effects of long physical isolation and independent development. This may be called *allopatric speciation.*

The other principal means of speciation is *sympatric speciation,* in which two populations occupy the *same* territory. Sometimes such sympatric species can be detected only by virtue of the fact that they will not interbreed. How might this come about?

Imagine two diverse habitats in the same location. A versatile organism is able to live in them both. However, certain mutants are capable of living in one of them—call it habitat A—better than the normal members of the population. It is obviously to their advantage if no genes from the parent population find their way into the gene pool of this new group, which shall now be called population A. Meanwhile, a population B is developing, specialized for life in habitat B. It is to their mutual advantage that they not interbreed, and natural selection can be expected to favor those genotypes that are incapable of it.

## Isolating mechanisms

In the process of speciation, organisms employ various mechanisms to achieve reproductive isolation from one another. These are known as *isolating mechanisms.*

Isolating mechanisms that interfere with mating are known as *prezygotic isolating mechanisms,* because they act before the zygote can be formed and in fact prevent its formation. For instance, fruit flies exhibit a definite courting behavior that is species-specific. Unless both partners behave in just the proper instinctive fashion, they will not mate successfully. Part of the behavior has recently been shown to be a love song—a series of buzzes of just the right pitch and rhythm performed by the male. If he doesn't have the right beat, that ends the encounter right there. These differences in love song keep some *Drosophila* species apart. Even geneticists often cannot tell all the species apart in any other way, and apparently the flies can't either. Similar though more elaborate mechanisms isolate various species of crickets, birds, and many other animals which have instinctive, highly stereotyped patterns of courtship behavior which must be performed precisely in order for mating to take place.

Probably the most obvious prezygotic isolating mechanisms are physical barriers to mating—more or less gross inappropriateness of structure, so that reproductive parts simply will not fit one another. But even among species that differ physically, it is rare that an attempt at mating actually takes place.

Presently two kinds of postzygotic isolating mechanisms are known. Perhaps the most obvious of these is embryonic lethality, in which the zygote or embryo does not develop properly and is aborted. The other more subtle mechanism is that of hybrid sterility. In this case two species can mate and the offspring may be healthy or even unusually vigorous; but if the offspring is sterile, then in the long run genetic mixture between the species does not truly take place. One such case will immediately occur to you—that of the mule, the offspring of a horse and a donkey.

(a)

(b)

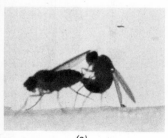

(c)

**FIGURE 17-6**
Courtship and mating in the fruit fly, *Drosophila.* (*a*) The male follows the female and vibrates his right wing. (*b*) If all goes well, the male licks the genitalia of the female while continuing to vibrate. (*c*) Copulation follows. (H. C. Bennett-Clark)

### Stabilizing selection

Toward the end of the nineteenth century an ornithologist named Hermon Bumpus collected a group of dead sparrows, killed in an exceptionally severe snowstorm. He compared these dead birds in nine different ways with those that survived the storm, using such characteristics as wingspread and body weight as the bases of his comparison.

Bumpus discovered that the dead birds tended to be abnormal in many of their characteristics, that is, they represented the extreme ends of the normal range of variation in a sparrow population. Bumpus concluded that there is a more or less standard body build suitable for a bird with the life style of a sparrow. Though extreme deviants from this standard may get along all right when conditions are not rigorous, extreme stresses such as this blizzard periodically arise which tend to weed out the unsuitable phenotypes, together with, presumably, the genotypes that produce them.

While Bumpus's data have been subjected to several reanalyses (one cannot duplicate the observations until a comparable blizzard bird-kill recurs), his basic thesis seems to stand. The phenomenon is known not as Bumpus's law (thank goodness) but as *stabilizing selection*. It is, in a sense, antievolutionary, for it tends to maintain a standard phenotype in a population of organisms.

As long as the environment of a species does not undergo long-term changes, natural selection will tend to stabilize the genetic composition of populations. Should the environment change, however, or should the organism find itself able to expand its range into a new kind of environment or life style (*ecologic niche*), then and only then will natural selection produce microevolutionary changes.

### THE ORIGIN OF LIFE

We must now leave the discussion of evolutionary mechanisms to consider the broader question of the actual history of life upon our planet. Here we enter a realm of inference more than of direct observation; a realm in which there has always been more room for debate, discussion, and even controversy.

According to current evolutionary theory, life on earth originated by spontaneous generation as described below. According to this view, spontaneous generation is no longer possible because life itself has changed the primitive conditions of the atmosphere and the earth.

### Origin of organic molecules

Theorists believe that the primitive atmosphere consisted of a mixture of nitrogen, carbon dioxide, methane, ammonia, and perhaps even hydrogen gas. These constituents could have reacted with one another in the presence of volcanic heat or lightning (energy sources) to form rather complex organic compounds—amino acids, carbohydrates, and even nucleic acids. Investigators have constructed experimental models of what they think primitive earth conditions were, and when the inorganic substances mentioned above

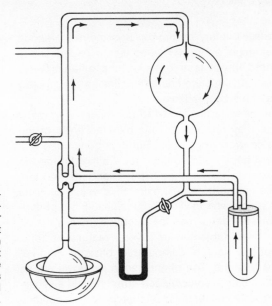

**FIGURE 17-7**
Apparatus for investigating hypothetic early chemical evolution. Mixture of gases and water is boiled in the flask at lower left and subjected to energy from electric sparks or ultraviolet light in the large round vessel in the upper right. Organic molecules or other products are collected in the vessel at the lower right.

were permitted to react under these conditions, a surprising array of organic compounds formed, including some macromolecules—that is, giant molecules such as protein fragments.

### Origin of cells

If such organic molecules could form spontaneously under present-day conditions, they would quickly be consumed by bacteria or slowly oxidized by oxygen. But neither bacteria nor oxygen is thought to have existed on the earth at that time. Macromolecules could have arranged themselves to form tiny bubble-like structures called *coacervates,* which have actually been observed to form in laboratory experiments. These coacervates possess catalytic activity and share some properties of living cells. Nucleic acids might have duplicated themselves, and eventually by chance some could have developed the ability to synthesize needed enzymes. Natural selection would, of course, have favored those aggregates which developed such abilities. Coacervates may eventually have become somewhat similar in structure to modern viruses.

But the cell, even of the simplest known forms of life, is an extremely complex little structure. Many find it difficult to see how organic molecules could have become associated in such a fashion to produce even the simplest of them.

These first living things would have to have been heterotrophs, nourishing themselves upon the organic molecules about them. Once the supply of organic molecules (especially ATP) began to diminish, however, survival would depend upon the ability to capture energy from organic molecules (glycolysis) and finally upon the ability to capture energy from sunlight (photosynthesis). Once photosynthetic producers evolved, oxygen would eventually be produced, rendering possible the development of the modern aerobically respiring consumer class of organisms.

## Origin of eukaryotes

Evolutionists think that the earliest cells were prokaryotes. You will recall from Chapter 3 (see also Appendix B) that prokaryotic cells lack nuclear membranes as well as other membranous organelles such as mitochondria, ER, chloroplasts, and Golgi bodies. Two ways have been proposed for the evolution of eukaryotes. The older view hypothesized that membranous organelles arose by multiple inpocketing and infolding of the cell membrane in an ancestral prokaryote.

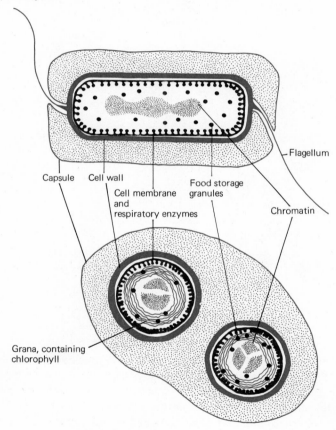

Capsule    Cell wall

Cell membrane
and
respiratory enzymes

Food storage
granules

Flagellum

Chromatin

Grana, containing
chlorophyll

**FIGURE 17-8**
A typical bacterium (above) and
a blue-green alga.

More recently set forth, the *endosymbiotic theory* suggests that mitochondria, chloroplasts, and perhaps even centrioles and flagella may have originated from a cooperative union among prokaryotes. Thus mitochondria are seen as former bacteria, chloroplasts as former blue-green algae, and centrioles and flagella as former something elses. The theory further stipulates that each of these partners brought to the union something that the

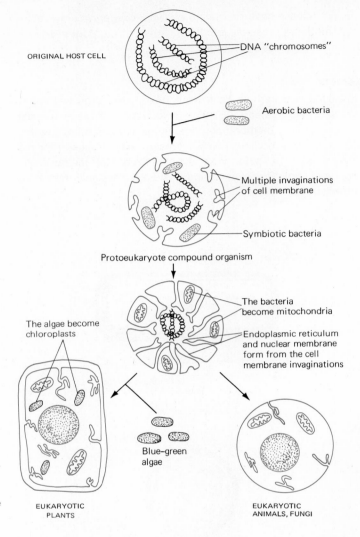

ORIGINAL HOST CELL

DNA "chromosomes"

Aerobic bacteria

Multiple invaginations
of cell membrane

Symbiotic bacteria

Protoeukaryote compound organism

The bacteria
become mitochondria

Endoplasmic reticulum
and nuclear membrane
form from the cell
membrane invaginations

The algae become
chloroplasts

Blue–green
algae

EUKARYOTIC
PLANTS

EUKARYOTIC
ANIMALS, FUNGI

**FIGURE 17-9**
The endosymbiotic theory of the
origin of eukaryotes.

others lacked. For example, mitochondria provided the ability to employ ox-
idative metabolism, which was lacking in the original host cell.

The principal evidence in favor of the endosymbiotic theory is that mitochondria and chloroplasts do possess *some* (but *not* all) of their own
genetic apparatus distinct from that of the cell's nucleus. Thus they have
both their own DNA and their own ribosomes. In the case of chloroplasts
there is as much DNA as is possessed by the average virus particle. Moreover, the DNA and ribosomes of these organelles are rather similar to those
found in prokaryotes.

The generally accepted course of evolution of eukaryotes is illustrated in
Figure 17-10. For a summary of the diverse organisms found on our planet
today consult Appendix B.

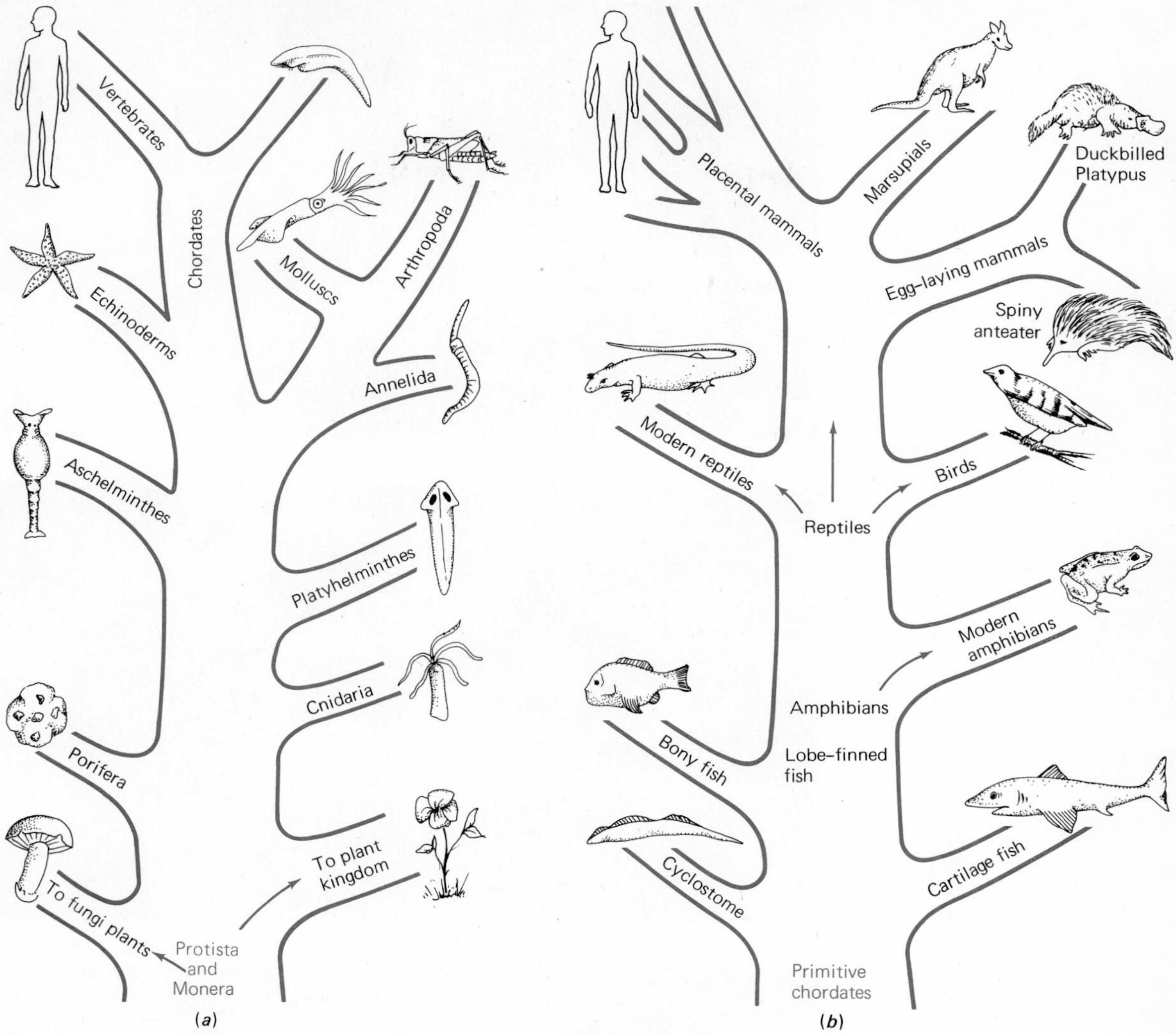

Vertebrates
Chordates
Echinoderms
Molluscs
Arthropoda
Annelida
Aschelminthes
Platyhelminthes
Cnidaria
Porifera
To plant kingdom
To fungi plants
Protista and Monera
(a)
Placental mammals
Marsupials
Duckbilled Platypus
Egg-laying mammals
Spiny anteater
Modern reptiles
Reptiles
Birds
Modern amphibians
Amphibians
Lobe-finned fish
Bony fish
Cyclostome
Cartilage fish
Primitive chordates
(b)


**FIGURE 17-10**
(*a*) The general course of animal evolution (a majority view). (*b*) A vertebrate family tree.

## EVOLUTIONARY THEORY AND THE CONCEPT OF CREATIONISM

So far, we have discussed the mechanisms by which small genetic changes are made in populations. We are now ready to turn to the concept that much more extensive genetic changes have occurred in organisms since whatever origins life may have had upon the earth. The ***theory of evolution*** states that all organisms have gradually developed from a common, simple ancestral type. The theory is based on data from such areas as morphology, biochemistry, the fossil record, and observed microevolutionary processes. Some of the evidence that supports the theory of evolution follows:

(a)

(b)

**FIGURE 17-11**

A bombardier beetle in action. (*a*) An ant approaches the captive beetle, and (*b*) attacks it. (*c*) The beetle mixes a peroxide and a catalyst in a special chamber, and sprays the ant in the face with the boiling, noxious fluid. Since neither the peroxide nor the catalyst is useful by itself, the creationist asks from what beginnings the beetles' present mechanism could have evolved. (An evolutionist might suggest that their original function may have been quite different.) (T. Eisner and D. Aneshansley and *Science*)

(c)

**Microevolution.** Many cases of naturally occurring microevolution have been observed. For instance, bacterial populations have become resistant to antibiotics, flies and mosquitos have become resistant to pesticides, and industrial melanism has occurred in moths. The domestic rabbit and the house mouse are subspecies that have evolved during the time of modern man. A few plant species have evolved during recent years. If microevolution has been witnessed in a relatively short span of years, it is quite plausible that over millions of years many microevolutionary changes have accumulated to produce evolution. Furthermore, where can one draw the line between micro- and macroevolution? If a species can evolve, why not a whole genus, order, class, or phylum, given enough time?

**Morphologic resemblances.** Comparative anatomy has shown interesting patterns of structural similarity between organisms that can best be explained by relative degrees of genetic kinship. For example, all vertebrates have the same pattern of circulation, nerves, muscles, bones, and other organs, and structural complexity gradually increases as one moves from agnathans to mammals. Even more striking is the fact that these organs develop in the embryo in much the same manner in all vertebrates. Vestigial, or useless, structures also provide evidence for evolution. All vertebrate embryos develop pharyngeal pouches, which give rise to gills in fish and amphibians but either disappear or are modified to form different structures in birds, reptiles, and mammals. Our wisdom teeth are vestigial structures thought to be holdovers from our vegetarian past. Each of us has a complete set of muscles for wiggling our ears, a useful ability in many animals but of little interest in man. The presence of the same anatomic, embryologic, and vestigial structures suggest that all vertebrates have a great many traits in common and must therefore have a common ancestry.

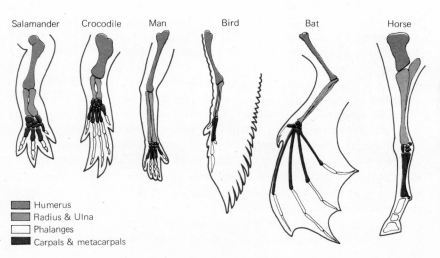

Salamander   Crocodile   Man   Bird   Bat   Horse

■ Humerus
■ Radius & Ulna
☐ Phalanges
■ Carpals & metacarpals

**FIGURE 17-12**
Comparison of anterior limb bones of six vertebrates. Those who believe in evolution interpret similarities of limb bones in vertebrates to suggest common ancestry. Creationists simply affirm that the Creator chose to use a common pattern. Both interpret the evidence according to previous beliefs. (John N. Moore and Harold Slusher, *Biology, A Search for Order in Complexity,* Zondervan, Grand Rapids, Mich., 1970)

*Biochemistry.* All organisms have the same fundamental biochemical mechanisms. All employ DNA, the citric acid cycle, cytochromes, etc., and it seems inconceivable that the biochemistry of living things would be so similar if all life did not develop from a common ancestor group.

Furthermore, the blood sera of morphologically similar animals are immunologically similar—an observation that indicates similarity of protein structure. The sequence of the 300 amino acids in hemoglobin is identical in man and in the chimpanzee. In the less closely related gorilla two of the amino acids in the sequence are different. Monkey hemoglobin differs in the sequence of twelve amino acids. Since DNA codes protein synthesis, protein similarity is a strong indicator of genetic similarity. Further biochemical evidence is accumulating that supports the taxonomic relationships that have been previously established.

*Genetics.* Mutations are rarely directly responsible for evolutionary change. Rather, as unfavorable alleles are continuously weeded out by selective pressures of the environment, the production of new mutations simply keeps up a genetic variability within the population. The new mutations are usually, in turn, weeded out and replaced with other variants. Evolutionary changes are based on the ability of a population to produce recombinations (i.e., to have sufficient genetic variability) that allow enough individuals to survive under the stress of a changed environment. Thus evolutionary change does not rely on a high rate of mutation. However, the question arises whether enough mutations are likely to occur to produce the necessary genetic variability. Researchers who studied rates of mutation occurring within selected species in recent years have estimated that at least 500 million useful mutations that are reproduced in the offspring occur during the evolutionary life of a species (i.e., 50,000 generations in a population of 100 million individuals). Of these, perhaps as few as 500 would be necessary to provide the genetic basis for evolution.

Chromosomal mechanisms can also supply genetic variability. For instance, more than one-fourth of the species of flowering plants are polyploids of their closest relatives. That polyploidy is in fact a mechanism for plant evolution has been demonstrated experimentally. One researcher synthesized a species of hemp nettle by interbreeding two species, each with eight pairs of chromosomes. The offspring contained sixteen pairs of chromosomes and turned out to be identical with a naturally occurring species that also contained sixteen pairs. The synthesized species would not interbreed with either of the species from which it had originated. Rather, it reproduced true to its kind, producing plants with sixteen chromosome pairs, and was capable of crossing with the natural 16-chromosome species. This is an obvious, directly observable instance of evolution at the present time. There is also evidence that some animals, such as the catastomid family of fish, arose by means of polyploidy.

Furthermore, many organisms that seem to be related on the basis of other similarities also show close similarities in the shape, size, or markings of their chromosomes.

*Fossil record.* When scientists explore fossil beds in sedimentary rocks, they often find that the fossils were deposited in a striking sequence. In the deepest and oldest strata they find the most primitive fossils. Buried in successive layers from the bottom to the top are a progression of fossils from the simplest to the most complex.

Careful studies have indicated that approximately one foot of sediment is deposited every five thousand years. By measuring the depth at which fossils are buried in a deposit, scientists are able to arrive at an estimate of their age.

A more recent and more accurate method for determining the age of fossils is radioactive dating. Radioactive elements decay into stable products at specific, constant rates. This decay rate is not altered by such factors as temperature or pressure. One of the most useful systems is the uranium-lead method. When igneous rocks were first formed, some contained separate crystals of uranium 238. The uranium immediately began to decay at a constant rate to form lead 206. The half-life of uranium 238 is 4.5 billion years, which means that half the atoms in a particular sample will be converted to lead during that time period. By measuring the ratio of uranium 238 to lead present in a rock sample, scientists can compute the age of the rock.

The potassium-argon method is based upon the decay of radioactive potassium 40 into argon and calcium. Potassium 40 has a half-life of 1.3 billion years. For archeologic artifacts or fossils less than 30,000 years old, a radiocarbon method is utilized. The half-life of carbon 14 is 5,568 years.

The earliest known fossils are ones resembling bacteria; they are believed to be about 3 billion years old. A more or less continuous record of fossils has been identified from about 600 million years ago to recent times. Whole sequences of fossil organisms can be traced throughout geologic time, visibly undergoing evolutionary change.

*Distribution.* Organisms are not found in all the habitats that could support them. Related organisms tend to be found together in geographically contiguous areas. Furthermore, since such related organisms are often of a very distinctive type, it seems virtually certain that they shared a common ancestry.

The island continent of Australia, for instance, has its own distinctive group of organisms that apparently have evolved there in relative isolation from the rest of the world. Marsupial mammals (that is, those like the opossum that have a pouch for carrying immature young) seem to have reached Australia early in evolutionary time when Asia and Australia were linked. Thereafter, placental mammals evolved throughout the rest of the earth and slowly replaced marsupials except in island continents such as Australia and South America which were isolated by water from the new types of mammals.

In Australia, marsupials remained the dominant mammal type, and their evolution proceeded in many different directions, fitting them for many different life styles very much like those enjoyed by the placental mammals

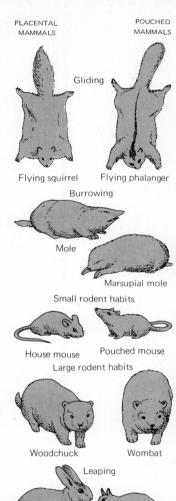

PLACENTAL MAMMALS  POUCHED MAMMALS

Gliding

Flying squirrel  Flying phalanger

Burrowing

Mole

Marsupial mole

Small rodent habits

House mouse  Pouched mouse

Large rodent habits

Woodchuck  Wombat

Leaping

Hare  Hare–wallaby

**FIGURE 17-13**
Convergent evolution among mammals. Two groups of mammals not closely related (placentals and marsupials) have similar ways of life and occupy similar ecologic niches. It will be noted that for every occupant of a given niche in one group there is a counterpart in the other group. This correspondence is not restricted to similarity of habit but also includes morphologic features. (C. P. Hickman, *Integrated Principles of Zoology* 3d ed., Mosby, St. Louis, 1966)

elsewhere. Thus in Australia we find marsupials which correspond to our placental wolves, bears, rats, moles, flying squirrels, and others.

To sum it up, the vast majority of biologists consider the evidence to be overwhelmingly in favor of evolution, that is, that the diversity of organisms is best and most simply explained in terms of evolution. Most scientists, while readily conceding that some of the hypotheses about particular events may have to be modified as new evidence is found, accept the concept of evolution as one of the most fundamental theories in biology.

***Creationism.*** A few scientists even today remain unconvinced, however, holding the view that evolutionary theory does not satisfactorily explain all the facts, and that the divine *creation* of organisms is at least as probable. This view, called *creationism,* is generally ignored in science textbooks on the grounds that it is not a scientific explanation. Thus far, at least, most of the concepts surrounding creationism have not been of the kind accessible to the techniques of scientific inquiry. Consequently, creationism is generally held to be an unfalsifiable hypothesis. In the words of the American Association for the Advancement of Science, " . . . the statements about creation that are part of many religions have no place in the domain of science and should not be regarded as reasonable alternatives to scientific explanations for the origin and evolution of life."

Nevertheless there have been vigorous demands in the past few years by some creationist groups[1] that the creationist view be given "equal time," so to speak, at least in elementary biology texts. You might find the creationist view quite interesting for several reasons. The principal reason was stated clearly by John Thomas Scopes, of the famous Scopes "Monkey Trial."

> Education, you know, means broadening, advancing, and if you limit a teacher to only one side of anything the whole country will eventually have only one thought, be one individual. I believe in teaching every aspect of every problem or theory.

But there are other reasons, too. For one thing, consideration of creationist arguments should help considerably to delineate the nature of science. Is creationism scientific? If not, why not? Is evolution scientific? If so, why so?

Finally, we cannot imagine that the cause of truth is served by keeping unpopular or minority ideas under wraps. Today's students are much less inclined than those of former generations to unquestionably accept the pronouncements of "authority." Specious arguments can only be exposed by examining them. Nothing is so unscientific as the inquisition mentality that served, as it thought, the truth, by seeking to suppress or conceal dissent rather than by grappling with it. Therefore, we will briefly state, for those who

---

[1] Creationism is not just a single point of view, though space does not permit us to record all the variations within the movement. Most creationists accept microevolution, for example, but believe that missing data, plus some actually contradictory data, make proof of long-term evolution impossible at this time. As they see it, the Creator made all the major varieties of life at one time, but some organisms have since become extinct and others have undergone slight change. Others take this viewpoint further and assert that evolution is, in effect, the means employed by God in creation. This is *theistic evolution.*

are interested, several major theses of the creationist position and a few of the questions raised by this dispute. In general, the majority of creationists support their view with most or all of the following arguments:

1  Evolutionary theory rests on the concept of uniformitarianism, which assumes that events proceeded in the past at the same rate as they do today. But uniformitarianism can be neither proved nor disproved. Because man was not around to record events that supposedly occurred before his time, we cannot prove that mutations occurred at the same rate in the past as they do in the present. We cannot be sure that sedimentary rock has always been deposited at the same rate that is observed today, and thus we cannot calculate with certainty the age of a fossil in sedimentary rock.

2  Morphologic and biochemical similarities may reflect an economy of design on the part of the Creator; whereas, major differences would indicate different types of organisms made for different purposes. Moreover, organisms that are morphologically similar are *not* always biochemically similar. Evolutionist arguments about similarities are sometimes convincing but cannot constitute proof; they are only more or less plausible.

3  Many supposedly vestigial structures have minor or even major functions. The human appendix, listed as a vestigial remnant of a cecum, more fully developed in our vegetarian ancestors, appears to be a sort of intestinal tonsil and may be important in the early establishment of immune mechanisms. At one time the thyroid gland was considered vestigial and was removed by surgeons for reasons we would now consider frivolous.

4  Nonharmful mutations occur seldomly and at random. The number of very specific mutations needed to make one radical change in an organism is large. Thus the probability is minute that mutations could lead to the development of new phyla.

5  Dating fossils is not easy. While it is doubtless true that, in general, younger sedimentary rocks are deposited on top of older ones, the original relationship is easily disturbed by geologic events, and this disturbance is not always easy to detect. Even dating by the technique of radioactive decay is not above criticism. How can we be sure that a given rock contained none of the products of radioactive decay to begin with or that none of them (such as argon, a gas) has subsequently leaked away? In any case, results of the various dating methods tend to conflict. If they are valid, they should agree, at least roughly. Even *general* evolutionary sequences can be disturbed by unreliable dating. In human "evolution," for instance, it now appears that the most recent *Australopithecus* is geologically younger than the oldest known *Homo*. How, then, can the existence of *Australopithecus* be considered evidence for the evolution of *Homo*, as it is usually supposed to be?

Such fossil evidence as exists might be explained by the view that organisms were created in a timed sequence. Another view is that organisms were destroyed in a universal flood and that the segregation of distinct communities of fossil organisms in separate strata may be explained ecologically. A marine community, for example, might be the first to be buried by a catastrophic sedimentation, followed by a swamp community, followed by a terrestrial lowland community washed over both, etc.

6  In the final analysis, evolution is inherently incredible. All geologic time is actually too short to develop man from an organic soup by microevolutionary processes without the purposeful intervention of a Creator.

Now that you have read, and presumably thought, about evolution and the creationist arguments against the theory, you should be prepared to consider a few questions about the logical nature of evolutionary theory as compared to the creationist view.

1  Reviewing Chapter 2, what is the difference between an unfalsifiable and an untestable hypothesis? Is creationism unfalsifiable, or is it merely untestable? What of evolution?
2  Which viewpoint actually makes use of the evidence that we have?
3  What is the validity of some of the factual issues that are raised? For example, does evolutionary theory in fact rest upon the principle of uniformitarianism? What if sedimentary processes did take place more slowly or rapidly in the past? What if the rate of deposition was very rapid indeed? Which is more important to evolution: mutation rate, genetic recombination or selective pressure? Is a high mutation rate really indispensable to evolutionary change?

And, if you are interested in pursuing these questions further, turn to the references listed in the bibliography.

## CHAPTER SUMMARY

1  The term evolution can be used and understood in more than one sense. One of these meanings refers to the change of gene frequencies within populations, and is more precisely termed microevolution. In a broader sense, evolution is used to refer to the concept that all life is commonly descended from one or a few ancestors and is therefore genetically related.
2  Recognizable evolutionary theories in the modern sense were first propounded in the eighteenth and nineteenth centuries. The principal theory which has survived to the present day was that proposed by Darwin and Wallace.
3  Darwin and Wallace did not originate the concept of evolution, but proposed a plausible mechanism whereby it might take place. Their view comprised four main points: overproduction of offspring, variation among those offspring, inheritability of that variation, and natural selection of the variants.
4  The modern concept of evolution continues to be based upon the idea of natural selection, but emphasizes changes in gene frequencies within populations. Such changes will not take place, according to the Hardy-Weinberg law, if mating is at random (panmixis), if there is no selection against a genotype, if there is no selective immigration or emigration, and if the population is relatively large. Since these conditions are rarely met, the usual situation is for changes in gene frequency to take place in most populations of organisms over a period of time.
5  An interfertile population of organisms that cannot exchange genes with other such populations is termed a species. The origin of species, that is, speciation, involves the development of this genetic isolation. In addition, speciation usually involves the genetic divergence of populations, resulting in the division of a single ancestral population into populations specialized for different modes of life (ecologic niches) or for life in different geographic areas, or both.
6  Stabilizing selection tends to stabilize genetic frequencies within a population and to preserve a standard type of organism if the environmental conditions remain the same. If, however, environmental conditions change systematically, genetic frequencies will tend to change also.

**7** According to current theories, life may have originated on the earth from naturally occurring macromolecules by a kind of spontaneous generation. Eukaryotic cells may have originated from a symbiotic union of several prokaryotes.

**8** The principal alternative to the concept of evolution as a theory of origin is special creation. Creationists question the degree to which evolution actually has taken place and the fundamental assumptions underlying evolution, such as uniformitarianism. Most evolutionists contend that all life is related to some degree and shares common ancestors. Evidence of several types is presented.

**FOR FURTHER READING**

A mountainous accumulation of literature has grown up on the subject of evolution. We have tried to provide only some of the more readable and popular evolutionary works here. Additional references are easily obtained in the card catalog of any good library. We have taken more pains to obtain a fair-sized listing of creationist literature, since this is not readily available, and what is available is often irresponsible. Creationist titles are starred (*). Finally, there are a few of the more important historic studies of the development of evolution as a scientific theory, and its impact upon the modern world.

Aulie, Richard P.: "The Doctrine of Special Creation," *The American Biology Teacher,* April and May, 1972.

Clark, R. E. D.: *Darwin, Before and After,* Paternoster Press, London, 1958.*

Darwin, Charles: *The Voyage of the Beagle,* Doubleday, Garden City, N.Y., 1962 (originally published 1845 and 1860).

Davidheiser, Bolton: *Evolution and the Christian Faith,* Presbyterian and Reformed, Nutley, N.J., 1969.*

DeCamp, L. Sprague: *The Great Monkey Trial,* Doubleday, Garden City, N.Y., 1968.

Dobzhansky, T.: *The Genetics of the Evolutionary Process,* Columbia, New York, 1970.

Eaton, Theodore: *Evolution,* Norton, New York, 1970.

Eisley, Loren: *Darwin's Century,* Doubleday, Garden City, N.Y., 1958.

Greene, John C.: *Darwin and the Modern World View,* Louisiana State University Press, Baton Rouge, 1961.

Irvine, William: *Apes, Angels and Victorians,* McGraw-Hill, New York, 1955.

Kenyon, Dean, and Gary Steinman: *Biochemical Predestination,* McGraw-Hill, New York, 1969.

Kerkut, G. A.: *Implications of Evolution.* Pergamon, New York, 1960.

Klotz, J. W.: *Genes, Genesis and Evolution,* 2d ed., Concordia, St. Louis, 1970.*

Macbeth, Norman: *Darwin Retried,* Gambit, New York, 1971.

Moore, John, and Harold Slusher (eds.): *Biology—A Search for Order in Complexity,* Zondervan, Grand Rapids, 1970.*

Salisburg, Frank B.: "Doubts about the Modern Synthetic Theory of Evolution," *The American Biology Teacher,* September 1971.

Wallace, Bruce: *Chromosomes, Giant Molecules and Evolution,* Norton, New York, 1966.

Zimmerman, Paul, et al. (ed.): *Darwin, Evolution and Creation,* Concordia, St. Louis, 1959.*

# 18 BEHAVIOR

**Chapter learning objective.** The student must be able to summarize the modern ethologic view of the behavior of organisms, with emphasis upon the concepts of innate behavior, learning, and the formation of societies and their interaction.

*ILO 1*  Define and distinguish between tropisms and taxes.

*ILO 2*  Using appropriate examples, summarize the role of releasers in the expression of simple and complex programmed behavior.

*ILO 3*  Describe biologic clocks and migratory behavior, and give some of the mechanisms known or suspected to be responsible.

*ILO 4*  Summarize the contributions of heredity, environment, and maturation to behavior.

*ILO 5*  Give an operational definition of learning.

*ILO 6*  List six major types of learning, giving an example of each.

*ILO 7*  Given examples, distinguish between social and other aggregative behavior.

*ILO 8*  Given a description of an animal society, identify the cooperative result of the mutual action of the participating organisms, the suppression of aggression, and the modes of communication employed by the interacting animals.

*ILO 9*  Present the concept of a dominance hierarchy, giving at least one example, and speculate upon its possible general adaptive significance and social function.

*ILO 10*  Define territorial behavior, distinguish between home range and territory, be able to recognize territorial behavior when given an example, and give three theories relating to the adaptive significance of territoriality.

*ILO 11*  Contrast insect and typical vertebrate societies with respect to plasticity of behavior, general complexity, and specialization of roles.

*ILO 12*  List and define three distinctive properties of human society.

It has only recently been generally recognized among life scientists that the behavior of organisms is a legitimate part of biology. Behavior is as important a part of an animal's adaptations as any physical characteristic it may possess. Anyone can appreciate the importance of a female rat's mammary glands to her successful reproduction. But her ability to retrieve straying young and return them to the nest is just as important. In this chapter we shall examine both the main types of behavioral adaptations found among organisms and their possible significance for mankind.

## TROPISMS AND TAXES

As almost everyone knows, plants grow toward the light. A houseplant placed near a window must be regularly turned or it will grow in a lopsided fashion. Plants exhibit many such consistent and apparently automatic responses to stimuli in their environment. For example, the roots of plants grow toward a source of water, toward the force of gravity, or *away* from light. The clinging tendrils of peas, cucumbers, and many other vinelike plants respond to the touch of solid objects by curling around them.

Such automatic growth responses resulting from fairly simple stimuli are called *tropisms.* They have been investigated by a number of eminent naturalists, including Charles Darwin and, more recently, the botanist Frits Went. Many of these studies have employed the young seedlings of oats as experimental organisms. Generally, the future stem and leaf sheath of the seedling (known as the *coleoptile*) has been the portion subjected to experiment. These studies have led to an explanation of the mechanism of many plant tropisms.

The upper tip of the coleoptile is necessary to its further growth; if this tip

**FIGURE 18-1**

**FIGURE 18-2**
Positive thigmotropism is exhibited by the tendrils of this "wild cucumber" vine, which curl about any solid object of the proper size.

is removed, no more growth takes place. Perhaps the tip produces some substance that is essential to the growth of the entire coleoptile. It is easy to test this hypothesis. If a severed coleoptile tip is restored to the stalk from which it was taken, the stalk will resume growth. However, no such effect can be demonstrated if some impermeable substance such as glass, plastic, or mica is interposed between the tip and the stalk. Presumably these materials block the stalkward flow of whatever essential substance the tip produces.

Further support for the growth-substance hypothesis may be gained if a *permeable* barrier is interposed between the stalk and tip. Our hypothesis

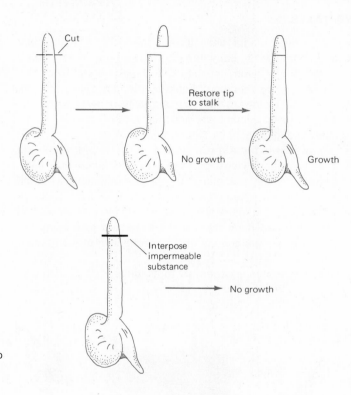

**FIGURE 18-3**
The coleoptile tip is essential to its growth.

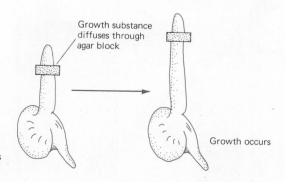

**FIGURE 18-4**
The growth factor diffuses through a permeable barrier.

would lead to the prediction that under these circumstances, growth should take place. The jelly-like substance *agar* can be utilized in this way, and when it is so used, growth does indeed take place.

Now if a growth-producing substance passes through the agar block, it follows that the block becomes impregnated with that substance if it is left in contact with the tip for a while. The block *itself* should thereafter be able to produce growth; that is, it should act much as the tip does. A control block of plain agar should have no such properties. Or, to take still another suggestive test, if an agar block impregnated with such a substance is placed off-center on *one side only* of a decapitated coleoptile, growth should take place on that side only, producing a bend in the stalk.

All these predictions relating to a growth-substance hypothesis have been experimentally confirmed. Such experiments led to the formulation of the **auxin theory** of plant growth. Briefly stated, this theory holds that plant

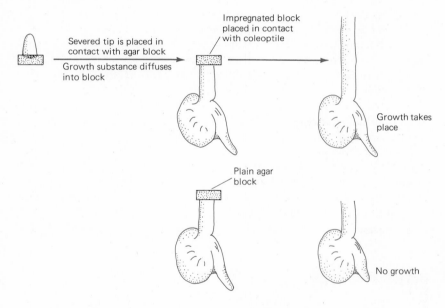

Severed tip is placed in contact with agar block

Growth substance diffuses into block

Impregnated block placed in contact with coleoptile

Growth takes place

Plain agar block

No growth

**FIGURE 18-5**
An agar block impregnated with growth substance can act as a temporary substitute for the tip itself.

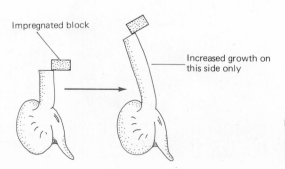

Impregnated block

Increased growth on this side only

**FIGURE 18-6**
When the growth substance is applied to one side only, growth takes place only on that side.

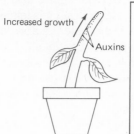

Phototropism: growth response to light

Geotropism: growth response to gravity

Thigmotropism: growth response to solid objects

Hydrotropism: growth response to water

**FIGURE 18-7**
The auxin theory of tropism.

growth is regulated by hormones known as *auxins,* which respond to environmental stimuli. How can the auxin theory explain our early example of a plant growing toward the light (positive phototropism)? Auxins respond to light by becoming most concentrated on the side of the stem *away* from the light, and therefore causing the plant to grow *toward* it.

Similar experiments have shown that other tropisms also are controlled by hormone action. Tropisms clearly take place without volition on the part of the plant. To be more precise, there is no need to postulate volition in order to explain the response. Thus when we say that plants "love" the sun, we are speaking figuratively.

Acellular slime molds are not plants. They are many-nucleated organisms resembling giant amebas, and their responses to stimuli are not solely growth responses. They can be raised on moist blotting paper and fed flakes of raw oatmeal which they attack, envelop, and digest. It has been shown that slime molds move in response to chemical substances given off by food—by flowing toward these substances. Even bacteria are able to do this.

**FIGURE 18-8**
Tropistic behavior in a non-cellular slime mold, *Physarum.* This unusual pet can be fed with oatmeal flakes. In (a) pseudopods of the slime mold approach a flake, and in (b) they engulf it.

(a)          (b)

Many animals far more complex than plants, slime molds, or bacteria respond to stimuli in much the same way as these simple organisms. Flatworms, for example, will congregate on a piece of raw meat left in a stream overnight. They need no other stimulus than the chemical cues given off by the meat. If a flatworm is placed in the apparatus shown in Figure 18-10, and meat extract is placed in one of the bottles, the flatworm will swim into the arm of the trough that receives water from the bottle containing meat extract.

**FIGURE 18-9**
Chemotaxis in bacteria. In this highly magnified view, bacteria cluster about the opening of a capillary tube containing a chemical that is attractive to them. (Dr. Julius Adler and *Science*)

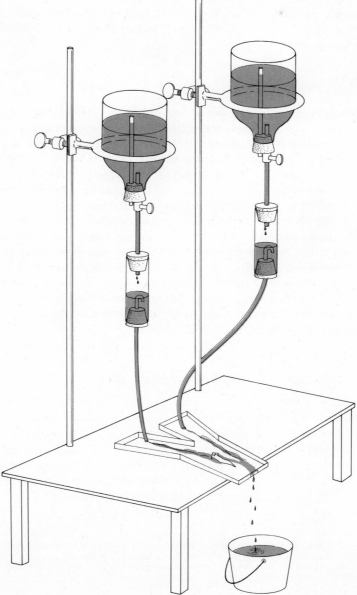

**FIGURE 18-10**
Simple maze for studying chemotaxis in a flatworm.

*(a)*

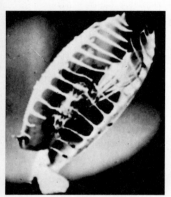

*(b)*

By adjusting the lighting and the rate of flow from both bottles so that all other stimuli are equal on both sides of the animal, it is possible to show that it is responding to the meat extract alone.

What is the difference between these simple animal responses to stimuli, called *taxes* (or in the singular, *taxis*), and those of the plant? A distinction is difficult, for except that they are neural and muscular rather than hormonal and tropic in their mechanism, there is not a great deal of difference. But a tropic response to environment is all that a plant can produce. The potential behavioral repertoire which the possession of a nervous system opens up to more complex animals is far greater and more sophisticated than anything available to plants—whose primary response to challenge, after all, is in the way in which they grow.[1]

## WHAT IS BEHAVIOR?

Most behavior is the external response of an organism to its environment. It ranges from the simplest tropisms, taxes, and other orientational responses to complex manifestations in the elaborate societies of insects and man. The scientific study of behavior is restricted to that which can be observed or deduced directly from observations. A pet dog rolling on its back in front of its master is displaying love, as most people would see it. The behaviorist might classify the same behavior as "submission," but even that is an interpretation that many would avoid. An objective analysis of behavior must avoid, as much as possible, attributing human emotions or motivations to the organisms studied. Viewed in this light, behavior must be considered, as we have already noted, to be an *adaptive characteristic,* that is, something which fits an organism or a population to the demands of its environment, just as teeth, claws, or wings do. It is from this point of view that we shall deal with behavior in this chapter. The scientific study of behavior is termed *ethology* or, popularly, *animal behavior.*

## ACQUIRED AND INHERITED BEHAVIOR

Like any other characteristic of living things, behavior has both genetic and acquired components. The number and form of the muscles in the forearm, for example, is genetically determined. However, the degree of development of these muscles is acquired, depending principally upon nutrition and exercise, and it may be expected to differ in blacksmiths and accountants. Similarly, the ability to *use* these muscles is an innate behavioral capacity, but the specific skills of a watchmaker, typist, or truck driver are acquired.

---

[1] Some rapid plant responses, such as those of guard cells, the Venus flytrap, and the sensitive plant, are due largely to turgor (pressure) changes within cells, but these responses are exceptional. Many such plant responses may, however, be mediated electrically, in much the same way as an animal nervous impulse.

## Behavioral genetics

Behavior is essentially a property of the coordinating mechanisms of the body, that is, of the nervous and endocrine systems. The capacity for behavior is therefore subject to whatever genetic characteristics govern the development of these systems. But there is a broad range of plasticity in behavior, from simple tropisms and taxes that can be little modified by learning, to complex intellectual tasks in which learning is paramount. The more narrowly stereotyped a system of behavior is, the more obvious is its genetic control. For example, in bee colonies dead larvae usually are removed from their waxen cells and discarded, a trait known among ethologists as the "hygienic" trait. This behavior actually has two components: (1) the removal of the wax cap of the cell, and (2) the removal of the dead larva. Some breeds of domestic honeybees do not exhibit any hygienic behavior. They leave the dead larvae to rot. Appropriately, this contrasting trait is called "unhygienic." The genetics of hygienic behavior in honeybees was investigated in 1964 by W. C. Rothenbuhler.

Rothenbuhler crossed hygienic and unhygienic bees. The $F_1$ generation was unhygienic, so the hygienic trait was evidently recessive. Backcrossing these hybrid unhygienic bees with hygienic ones, he obtained four behavioral phenotypes in approximately equal proportions:

1. Worker bees that would neither uncap the cells of dead larvae nor remove the corpses even if the experimenter opened the cells for them.
2. Workers which both uncapped the cells and removed larvae.
3. Workers that opened caps of cells but left larvae untouched.
4. Workers which did not uncap cells but which would remove larvae if they were uncapped by the experimenter.

These results can be explained easily by the hypothesis that the ability to uncap is controlled by a single pair of allelic genes, and that the ability to remove is controlled by an independently assorted pair. (Review Chapter 15 on genetics, if you need to.) Presumably the appropriate neural pathways can develop in the nervous systems of the insects only if functional copies of the genes are present and unsuppressed by abnormal alleles.

Although there are as yet few other demonstrations as elegant as Rothenbuhler's, the genetic control of many other stereotyped behaviors seems undeniable in a broad range of animal types comprising several phyla. Consider some additional examples. Among ants the corpses of workers are routinely removed from the nest by living ants. The way (and apparently the only way) a dead ant is recognized is by the presence of the chemical products of decay in or on its body. If these chemicals are experimentally daubed on a living ant, it will be unceremoniously carried squirming to the refuse heap no matter how many times it struggles back to the nest—at least until the odor wears off. It seems obvious that with diligence the genetics of this behavior could be demonstrated. Laboratory rats, kept in cages with no opportunity to burrow for more than 100 generations will still dig typical rat

(c)

(d)

**FIGURE 18-11**
Some plants are irritable in somewhat the same way that animals are. The Venus flytrap is able quickly to engulf insect prey, partly through quick changes in the turgor of certain cells. (General Biological Supply House)

Let $U$ = no uncapping
$u$ = uncapping
$R$ = no removal of dead larvae
$r$ = removal
$U, R$ dominant
$u, r$ recessive

$UURR \times uurr$

$Uu\ Rr$ F$_1$

(phenotype: no uncapping, no removal)

$UuRr \times uurr$ (backcross)

|  | UR | Ur | uR | ur |
|---|---|---|---|---|
| ur | UuRr | Uurr | uuRr | uurr |
|  | no uncapping no removal | no uncapping removal | uncapping no removal | uncapping removal |
|  | 1 | 1 | 1 | 1 |

**FIGURE 18-12**
Inheritance of hygienic behavior in honeybees.

burrows when allowed to do so out of doors.[2] Many temperate-zone frogs will react to a lowering of water temperature by attempting to bury themselves in the mud at the bottom of ponds and streams, thus protecting themselves against freezing. There are some Puerto Rican frogs which have presumably had no "use" for this behavior since the Wisconsin glaciation, more than 10,000 years ago. Nevertheless when their temperature is artificially lowered, these frogs make typical burrowing motions.

The genotype defines a range of behavior of which the organism is capable; but the way in which it is phenotypically expressed depends upon the interaction of the organism with its environment—particularly that interaction which we call learning. Put another way, the behavior itself is not inherited as much as the capacity to develop it is.

## Releasers

Complex stereotyped behavior such as the corpse removal practiced by ants is usually triggered by some specific stimulus, known as a *releaser.* It would be hard to argue that ants are exhibiting intelligent or purposeful behavior as they bear off their vigorously struggling colleague to the disposal heap. Releasers are also found among vertebrates. Behavior triggered by these releasers in any organism can be shown to be adaptive, in that it usually works to their advantage. However, such behavior can also be shown to be nonpurposive. At least the organisms involved do not necessarily have purpose in mind, as we might, when they perform their actions.

---

[2] However the laboratory environment does seem to select out different genes than the natural environment, so that strains of rats adapted to laboratory conditions are losing their reliability as authentic models of animals in a state of nature.

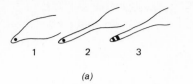

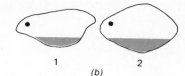

(c)

**FIGURE 18-13**
Releasers. (*a*) Models of bills of herring gulls, effective in eliciting pecking by young birds. Long, thin model bills (as in 2) are more effective as symbols than short, thick ones, and the models are most effective if the red pigmented areas near the tips have the form of several stripes. (*b*) A male stickleback will court an inanimate model of a fish having a swollen underside (1) suggestive of a female with eggs. Also, a male will attack a model having a red underside (2) suggestive of another male in breeding condition. (Paul B. Weisz, *Science of Biology*, 4th ed., McGraw-Hill, New York, 1971) (*C*) Many butterflies have conspicuous pigmented areas on their wings, suggestive of the eyes of an owl. Such eyelike patterns probably serve to protect the insects from predation by owl-fearing birds. (David E. Carter)

## Releaser chains

In the herring gull, the sight of a bill spot releases pecking in the chick. The peck of the chick releases a further response in the adult, that of regurgitation of food, which the chick then devours. Regurgitation does not ordinarily occur without the proper stimulus from the chick. The consummation of the act results from a chain of two consecutive released behaviors. Many much more elaborate chains are known, in which the entire sequence is complex, yet each step is an example of simple released behavior. Very sophisticated stereotyped types of behavior such as the bower building of bowerbirds or the comb construction of bees, are probably of this sort. Despite superficial similarities to certain human behavior, such sequences are rigidly predetermined, do not involve intelligent thought, and really resemble the sort of things we do in only one respect: they are adaptive, tending to promote the survival of the individual or the group.

Released behavior differs from conditioning in that it is not learned.[3] It differs from taxis in that it is not usually simple or orientational. In fact, were it not for released behavior, lower animals would probably be unable to display any complex behavior at all. The only way that an insect with a brain the size of a bee's, for instance, could possess the elaborate society that it does possess, is for the instructions for that society to be genetically inherited and preprogrammed. There is room in the head of a bee for only the most limited learning—that which is required by the immediate demands of its environment, and none other.

---

[3] It is becoming evident that some behavior previously thought innate is actually learned in part.

**FIGURE 18-14**
This brilliantly colored oleander moth fed while a caterpillar upon oleander leaves. Poisonous substances from the leaves entered the body of the insect, rendering it poisonous for life to vertebrate predators. Such a defense relies heavily on the ability of the predator to learn to associate the bright color of the moth with its obnoxious qualities.

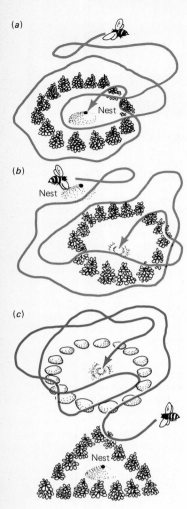

(a)

Nest

(b)

Nest

(c)

Nest

**FIGURE 18-15**
Tinbergen's *Philanthus* experiment. When the ring of pine cones is moved to position (b) from position (a), the wasp behaves as if her nest were located in their center. That it is the arrangement of the cones rather than the cones themselves to which the wasp reacts is shown by the substitution of a ring of stones for the cones in (c). The learning ability of *Philanthus* is very limited. (N. Tinbergen, *The Study of Instinct*, Clarendon Press, Oxford, 1951)

An illustration from the research of Niko Tinbergen will help to clarify this. The bee-killer wasp *Philanthus* captures bees, stings them, and places the paralyzed insects in burrows excavated in the sand. *Philanthus* then lays an egg on her victims, which are devoured alive by the larva that hatches from that egg. From time to time *Philanthus* returns to her hidden nest to reprovision it until the larva becomes a hibernating pupa in the fall.

When *Philanthus* covers a nest with sand, she takes precise bearings on the location of the burrow before flying off again to hunt. There is no way in which the location of the burrow could be preprogrammed in the wasp. How to dig it, how to cover it, how to kill the bees—these behaviors appear to be innate. But since a burrow must be dug wherever a suitable location occurs, the location of the burrow *must* be learned after it is dug.

Tinbergen misled the wasp in this fashion: he surrounded the burrow with a circle of pine cones, on which the wasp took her bearings. Before she returned with another moribund bee, Tinbergen moved the circle of pine cones. The wasp could not find her burrow—the cones no longer surrounded it. Only when the experimenter benevolently restored the cones to their original location could *Philanthus* bring home the bacon successfully!

Notice how efficiently the wasp learned what she needed to learn. A biologist would have difficulty in pinpointing a tiny location so efficiently in so short a time. And yet that was *all* she could learn, in that context, at least. Her BFQ (Burrow-finding Quotient) must have been at least 100, but her IQ?

As we shall see, the very existence of preprogrammed behavior is hard to demonstrate in human beings. We owe the complexity of our behavior to a *generalized* ability to learn. *Philanthus's* intelligence is as narrowly specialized as her stinger.

### Displacement behavior

One way or another, all behavior is motivated, whether by an external stimulus or by some internal state of the organism. Ethologists generally employ the more precise term *drives* to denote the physiologic states of internal tension that lead to action. Thus a hungry rat is impelled by one drive, a pugnacious rat by quite another drive, and an amorous or frightened rat by still others. It is inevitable that on occasion two or more of these drives may interact or actually conflict.

As everyone knows, it is entirely possible to be angry and frightened at the same time. Fortunately for the endocrine system, this poses no great problem, for the physiologic responses necessary to prepare the body for either fight or flight are hardly distinguishable. The two *behaviors* are certainly different, however. When conflict looms, which action shall be taken? It is not at all certain that people actually reason out an answer to that, but it is virtually certain that animals do not. The conflicting drives can produce very hesitant behavior, wavering between attack and defense. Even when attack is dominant, it is usually tempered by some admixture of caution; contrariwise, submissive behavior is rarely abject. Many animals possess a series of signals whereby they are able to subtly communicate the balance of such conflicting drives to their opponents. Presumably, such behavior af-

FIGURE 18-16
Apparent displacement behavior in a swan. Not quite daring to attack a rival, the bird picks up leaves and twigs from the ground.

fects the emotional balance of the opponent, who in turn communicates that balance to *his* opponent. The result is a complex and shifting situation, constantly teetering on the verge of explosion until one of the opponents flees or launches an attack. The whole process can be observed by watching the preliminaries of a cat or dog fight. Other things being equal, the encounter may be carried by the animal with the strongest aggressive drive without any need for actual combat.

When opposing drives are precisely balanced, an animal is likely to behave in a completely irrelevant manner. When both fear and aggression are balanced, a gorilla will nibble on a leaf, a bird may peck at the ground, a man may scratch his head. When opposing drives release irrelevant behavior, it is known as *displacement behavior.* Displacement behavior forms a significant part of many courtship rituals, which, as everyone ought to know, are often tense situations even among people, and which may involve the opposition of many potential releasers and drives among animals.

Please do not confuse displacement behavior with *redirected behavior.* In displacement, behavior is transformed. In redirection, behavior remains the same but finds another, often inappropriate, object. A caged ostrich may furiously bite the cage walls, since he is unable to reach the hated keeper. But although frustrated as to object, the behavior remains aggressive.

### Biologic clocks
Some behavior is unmotivated by obvious external cues, and follows a regular daily or longer cycle. For example, sparrows are more active in the day than in the night. Such behavioral cycles are not the only ones known. Many physiologic processes seem to follow an intrinsic rhythm. Human body temperature, for example, follows a typical daily curve. Activities generally carried out by day are *diurnal;* those carried out at night are *nocturnal.* No one knows how all daily (circadian) activities are governed. Many obvious

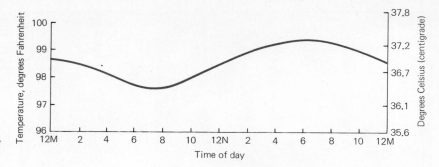

**FIGURE 18-17**
Daily variation in human body
temperature.

hypotheses have been excluded, for example, the alternation of light and
darkness, although these do govern daily activity in some animals, for instance sparrows. It is possible that the interaction of a number of biochemical processes produces the timed accumulation of certain substances to
critical levels, and that these substances, whatever they may be, are responsible for the action of the "biologic clock," which governs behavioral and
physiologic rhythms.

Biologic clocks can be reset. Anyone who has traveled long distances by
jet plane is likely to get his days and nights mixed up when he arrives at his
destination. Eventually, however, his body adapts to the new schedule. Many
of the trials of new parents are concerned with getting the baby's biologic
clock on the same schedule as theirs!

### Migration

Long-distance travel that has no obvious immediate motivation such as
hunger may be observed among many animals. Birds often fly south even
*before* the weather turns cold, or food becomes in short supply. Salmon swim
into fresh water toward the end of their life cycles. Monarch butterflies fly
southward, and *the next generation* flies north in the spring. Such characteristic travel by a species is known as *migration.*

The adaptive significance of migratory behavior is not always clear. It is
obviously to the advantage of birds to fly south in the winter, but why eels
must migrate to the Sargasso Sea to spawn is a mystery. In some instances
animals migrate as a result of maturation. In others, explicit environmental
cues trigger the process. In migratory birds, for example, changes in day
length trigger the characteristic restless behavior called *Zugunruhe—*
migratory restlessness. The bird evinces increased readiness to fly, and flies
for longer periods of time.

Of course the disposition to travel is not enough to account for migration,
since true migration always involves a specific end for the journey. The
direction of travel is also important, and this raises the general problem of
animal navigation. In many instances, we must frankly admit that we do not
know the cues employed by many animals to negotiate their migratory
journeys. Among birds, however, a combination of celestial (sun and star)

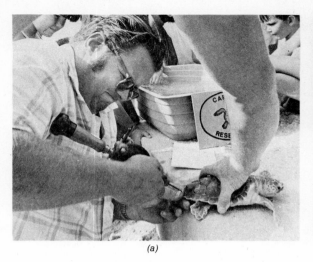

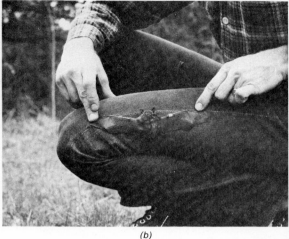

(a)

(b)

plus geographic navigation appears to be used. Some birds may also be sensitive to the earth's magnetic field. Green turtles may smell the islands on which they lay their eggs. Salmon, similarly, probably can "smell" the stream of their birth and direct their journey at least partly in this way. Many aspects even of bird migration still are not understood (for example, navigation in fog), and even less is known about that of other organisms. Here we still have much to learn.

(c)

FIGURE 18-18
Migration is usually studied by tagging animals, in the hope that the tags, when found, will be returned, together with a report of where they were found. (a) Tagging a marine turtle. (*Clearwater Sun, Fla.*) (b) A banded bat. (c) A typical bat band. (Dr. Bob Martin)

## The development of behavior

In some animals behavior matures in a more or less rigid sequence. Among honeybees, for example, division of labor among the workers is determined by age. Ordinarily, a nurse bee is younger than a guard bee or forager. Among African hornbill birds, eggs are laid about 24 hours apart in a tree cavity. When the chicks hatch, they are walled up in the cavity by the parents, who then feed them through a crack in the wall. The wall efficiently repels predators, and when it shows signs of wear is diligently repaired by the chicks. When the chicks mature, some behavioral switch is pulled that causes them to break down the wall in order to escape the nest. The trouble is that one sibling is a day older than the other. This results in the ludicrous spectacle of one bird building the wall as fast as its brother tears it down. Usually, they get out only when both are mature.

In animals such as man the innate preprogrammed behavior commonly called *instinct* is difficult to demonstrate. Infants have a very limited instinctive repertoire. The sucking reflex would be one example, but it is quite possible that we may possess much more complex innate behavior that matures later in life. The problem is to distinguish it from learning.

How many proud (or anxious) parents have consulted some such list as the one shown in Figure 18-19 to determine whether their child is exceptionally brilliant, as they hope, or retarded, as they may fear. Child psychologists believe that in general, given a reasonably normal environment, babies are

**FIGURE 18-19**
The development of posture and locomotion in infants. (Mary M. Shirley, *The First Two Years: A Study of Twenty-five Babies,* vol. II, *Intellectual Development,* The University of Minnesota Press, Minneapolis, 1933 and 1960)

able to develop these kinds of motor behavior without explicit teaching although trial-and-error learning may enter into the process. Similarly, it has been shown that parent birds do *not* teach their young to fly. The flightlike movements that young birds exhibit just before leaving the nest are apparently more in the nature of exercise. The ability to fly is inborn in them as, in a sense, is the ability to walk in us. But what do we really mean by such a statement?

***Physiologic readiness.*** Before any pattern of behavior can be exhibited, an organism must be physiologically ready to produce it. Breeding behavior does not ordinarily occur among birds or most mammals unless steroid sex hormones are present in their blood at certain concentrations. A baby cannot walk unless its reflexive and muscular development permits it. A bird cannot fly unless its nervous coordinating mechanisms are ready. And yet these states of physiologic readiness are produced by a continuous interaction with the environment. The level of sex hormones in a bird's blood may be caused by seasonal variations in day length. The coordinating mechanisms of the bird's nervous system are developed (in part) by reason of use. The baby's muscles develop in response to exercise. Without the trial and error involved in learning how to walk, walking would be retarded or perhaps prevented. Perhaps the best example of such interaction between readiness and environment is afforded by the white-crowned sparrow,[4] which exhibits considerable regional variation in its song. This bird, even if kept in isolation, eventually will sing a poorly developed but recognizable white-crowned-sparrow song. However, if it is allowed to grow up under the care of its parents for the first 3 months of its life, when it matures it will sing in the local "dialect" characteristic of its parents or foster-parents. If such learning does not take place in those 3 months, it never will.

There is evidence that an infant less than 16 weeks old can perceive motion but does not understand the significance of form, color, or the size of objects it sees until it is older. There is almost no way to determine whether this psychologic development is "innate" or "learned," since the infant is, after all, constantly exposed to visual stimuli and opportunities for visual learning.

**The function of play.** Play is a very important aspect of the development of behavior. It serves as a means of practicing adult patterns of behavior, of perfecting means of escape, prey killing, and even sexual conduct. The main thing that distinguishes true play from the real thing (at least among lower animals) is that the behavior is not actually consummated. Thus a kitten pounces upon a dead leaf but of course does not kill it even though it administers a typical carnivore neck bite. When interacting with a littermate, the same kitten may practice the disemboweling stroke with its hind claws, but the littermate is not intentionally injured in the process. Puppies will playfully mount one another as a part of their social play, but genital penetration does not usually occur.

Play also seems important in establishing social bonds in some species, and in establishing dominance hierarchies in others, especially among primates. Among such animals dominance relationships may be sorted out among members of a play group before maturity, and without the necessity of actual fighting.

The picture that these and other examples leave with us is something like that revealed by modern embryology for the physical development of the organism—genes specify a range of response, but the actual realization of these potentialities is evoked environmentally as the organism matures.

---

[4] Different species of birds vary greatly in the extent to which their songs are learned or innate.

(a)

(c)

(b)

**FIGURE 18-20**
(a) Play can function in determination of dominance relationships; (b) the formation of social bonds between mother and child; (c) and in the development of motor skills. (b, Lion Country Safari, Inc., West Palm Beach, Fla.)

## Learning

Ethologists and psychologists recognize several varieties of learning, among which we may list conditioning (classical and operant), trial and error, insight, imprinting, and habituation. Learning is defined usually in terms of measurable results. We assume learning has taken place when, as a result of environmental stimuli, *behavior* has changed in a persistent or permanent manner. The ethologist is not primarily concerned with what an animal may think or feel, since there is no unbiased direct way to measure thoughts or feelings. In fact, thoughts or feelings as we know them may very well not exist in many simpler animals.

*Conditioning.* In the 1920s the Russian physiologist Ivan Pavlov pioneered the study of *conditioned reflexes.* Pavlov initially fed meat or meat powder to dogs and at the same time rang a bell. The stimulus of the food produced salivation and secretion of gastric juice. Such a response is to be expected and is what might be called innate behavior—a reflex, probably a very simple one (Chapter 10). Pavlov's epochal discovery, however, was that if he rang the bell and then gave meat in close sequence for a long enough period of time, eventually he could produce the same reflexes (salivation and gastric secretion) with the bell alone. Eventually the learned stimulus of the bell took the place of the unlearned stimulus of the meat. Ethologists term the physiologic stimulus (such as the food in this case) the *unconditioned stimulus.* The *conditioned stimulus* is the normally neutral occurrence paired with it—in this case, the bell. The *response* is simply the reaction that is produced by both or either of the two stimuli (here, the secretion of gastric juice). If the unconditioned stimulus is withdrawn after a time the animal ceases to respond to the conditioned stimulus. This process is termed *extinction.*

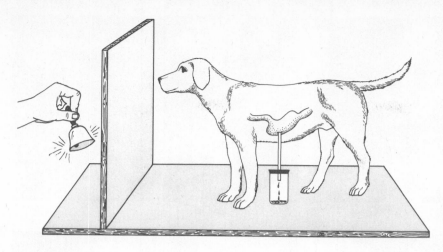

**FIGURE 18-21**
Classical conditioning a la Pavlov.

It is worth noting that conditioned stimuli need not involve awareness of learning. The very same experiment Pavlov performed on dogs can also be performed on human volunteers with much the same results. Yet the secretion of saliva and gastric juice are not ordinarily under volitional control.

Pavlov and subsequent researchers have greatly extended the application of these principles to a variety of physiologic reflexes, and to the movement of parts of the body that *are* normally under conscious control. To take one example of a modern study, the internal sphincter of the bladder opens and produces the desire to urinate when the pressure of urine in the bladder reaches a certain critical point or stage. The same result can be produced by inserting a tube into the urethra of a volunteer, and introducing water into the bladder until a desire to urinate is felt. Once the proper pressure is determined, a bell can be rung automatically at that point. Shortly, a desire to urinate is felt whenever a bell rings, regardless of the fullness of the bladder. One hopes that in such a case the process of extinction will proceed rapidly following the experiment.

We, too, salivate along with Pavlov's dogs when confronted with the stimulus of the dinner bell, or the sight, smell, or even mental image of food—at least when we are hungry. As C. S. Lewis once wrote, "He that but looketh on a plate of ham and eggs to lust after it, hath already committed breakfast with it in his heart (or in his glands)."

Please direct your attention to two features of classical conditioning: (1) the conditioned stimulus *precedes* the unconditioned stimulus, and (2) the organism is a passive participant. Its behavior is elicited; it does not arise spontaneously.

Another form of conditioning is also known which has neither of these characteristics. It is known as *operant conditioning.* In operant conditioning, the animal responds not to some external cue such as a bell, but to its own internal needs and drives, and its own spontaneous behavior produces the conditioning stimulus, better known as a *reinforcement.*

If a rat, monkey, bird, child, or elephant is placed in a cage and provided with a lever, button, pedal, or something of the sort to push, and if pushing

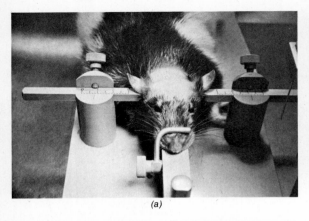

(a)

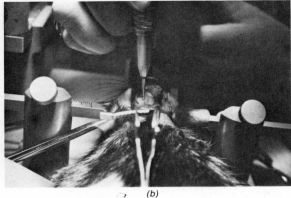

(b)

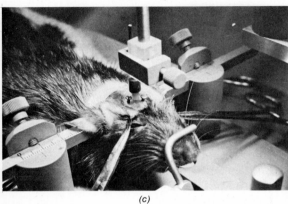

(c)

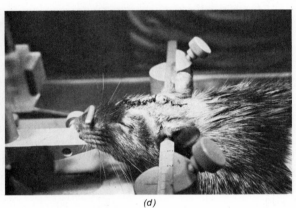

(d)

**FIGURE 18-22**
Operant conditioning. Here a rat (a) is anesthetized and prepared for surgery. (b) The skull is exposed and broached with a dental drill. (c) With the aid of a stereotaxic device, an electrode is precisely implanted in the pleasure center of the hypothalamus, and the incision (d) is sewn up. (e) In about a week the rat can be trained to depress a lever so as to deliver a stimulating electric current directly to the pleasure center of its brain.

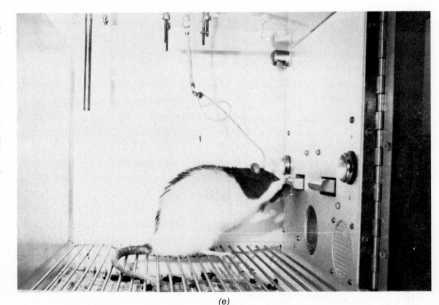

(e)

that lever produces a reward such as food, in time the subject will learn to push it in order to produce the reward. Initially, the rewarded action will occur as a coincidental aspect of the animal's random exploratory behavior, but as the association between the rewarded action and its reinforcement becomes explicit, the behavior of the animal will become steadily more predictable and it will tend to perform the rewarded action consistently. (Reverse experiments are also possible, in which the animal must do something in order to avoid some unpleasant stimulus such as an electric shock.)

Since the organism is an active participant in the operant learning process, far more complex and sophisticated experimental designs are possible, and the resultant learning is of a far higher order than is that of classical conditioning.

Animal species vary widely in the degree of complexity of the tasks that can be learned in this fashion. Monkeys, for example, can master far more complex manipulative tasks than rats. Human beings surpass all other organisms in such learning and are able to respond to far more subtle rewards and punishments. The programmed textbook and various forms of computer-assisted instruction are adaptations of instrumental conditioning to formal human education.

**FIGURE 18-23**
Operant techniques are widely used in animal training. (Miami Seaquarium)

Conditioning has many applications to human psychology and learning, and even to medicine and pure biology. For example, some researchers claim to be able to reduce gastric secretions in ulcer sufferers, relieve excessively high blood pressure, and otherwise alleviate many disorders resulting from abnormal physiologic reflexes which can be retrained by conditioning techniques. Single bundles of muscle fibers can even be brought under conscious control by these methods.

Biologists can determine the ability of animals to sense stimuli, and can estimate the degree of their sensory sensitivity by attempting to pair a conditioned stimulus in the sensory mode which they wish to test (such as sight) with an unconditioned stimulus. For example, a fish may be trained to nudge a bar every time a light flashes. The fish is fed as a reward for doing so. By varying the color of the light flash, one can objectively determine which colors a fish can see. By a modification of the experiment, the ability of the fish to tell colors apart can also be estimated.

*Trial and error.* Another learning mode closely related to operant conditioning involves trial and error on the part of an animal. Many organisms can be induced to find their way through mazes by providing some reward at the end of the maze. Mammals often need no reward save the satisfaction of their innate exploratory drive. If an animal is placed in such a maze, and a reward such as food is placed at its exit, the animal will eventually find its way out of the maze by trying different routes at random. After it has been placed in the maze several times, its number of wrong choices will diminish, and in consequence, it will travel the maze more quickly. In time, it will learn the maze perfectly and will make few or no further mistakes. If placed in a different maze, however, it must begin the learning process all over again.

Many comparative studies have been done using such techniques. For example ants learn mazes about half as rapidly as rats, certain lizards learn them about as quickly as rats, and people learn them more slowly than you might expect—only slightly faster than rats do! While such results are interesting, they do not faithfully reflect the general mental abilities of the organisms involved. Some mental capabilities may reflect the specialized mode of life that an animal leads. It seems likely, for example, that the ability to learn complex mazelike systems of tunnels would be most important to a burrowing, penetrating animal with secretive habits like the rat.

*Insight.* Some animals seem to possess the ability to put the elements of a problem together in a unique fashion, supplying from some inner resource whatever portion may be missing to generate a solution. To take a simple example, a dog can be placed in a blind alley which it must *circumvent* in order to reach a reward. The difficulty of the problem seems to be that the organisms must move *away* from the reward in order to get *to* it. At first, if the dog has no experience in coping with such a situation, it will only fling itself at the barrier nearest the food. Eventually, by trial and error, the frustrated dog will find its way around the barrier and reach the reward. A baboon placed in the same kind of situation is likely to see the solution immediately. There is clearly a large difference in the behavior of these two animals. Insightful behavior has frequently been observed in chimpanzees. They have piled boxes on top of one another to obtain food, fitted sticks together to reach a banana, led an experimenter to an appropriate location, and even climbed on him in order to obtain a reward. They did not always remember their insights, however, and sometimes could not do the same task twice.

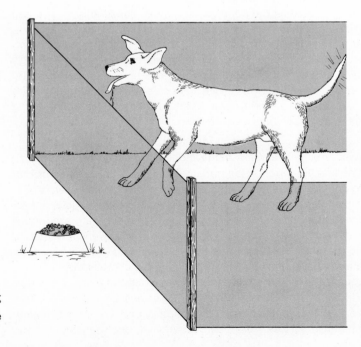

**FIGURE 18-24**
Limitations of insight in a dog. The dog cannot "understand" that he must walk away from the food in order to get to it.

Watching such behavior in another ape did not necessarily transfer that behavior to the watcher.

Even insightful behavior depends upon past experience with trial-and-error learning. Investigators who have dealt with insightful behavior have noted that it is both more likely to occur and more likely to be remembered if it grows naturally out of the previous experience of the organism. For instance, if chimpanzees are first permitted to play and experiment with sticks and sticklike objects, they are much more likely to be able to solve the banana-and-stick insight problem, since by trial and error they have become familiar with some of the properties of sticks.

Insightful behavior is most highly developed in human beings. The provision of sufficient past learning experience appears to be important in the development of this most important behavior in human beings as well as in chimpanzees.

*Imprinting.* The process by which certain stimuli elicit behavior patterns during a critical period of an animal's development is known as *imprinting*. The most commonly cited example of this occurs when the young of many birds learn to recognize their mother during a brief period soon after hatching. Thereafter they will accept no other adult bird. Punishment, for some reason, seems to facilitate the process. If an experimenter makes the proper noises (characteristic of mother birds of that species) he may be able to get the chicks to accept a moving model or even himself as their mother. The cues which produce this irreversible behavior apparently depend upon sound and sight. Among graylag geese, for instance, the experimenter must not only cluck like the mother goose but also crouch down and waddle about on the grass so as to fall within the general apparent size limits of what the chicks will accept as a mother. If this is done successfully, the chicks will thereafter reject even their own mother in favor of the experimenter. These cues release the specialized learning behavior of imprinting. Imprinting at first relates only to the complex of mother-young relationships, but in later life it apparently transfers to the sexual domain as well. One who attempts to imprint young birds is likely to become the object of their sexual advances when they become mature.

Recently it has been demonstrated that with diligence it is sometimes possible to reverse unnatural imprinting and to get young birds to follow their natural mother or another female of their own species. But natural imprinting is not reversible. A partial explanation seems to lie in the fact that the mother bird communicates by voice with her chicks even before they hatch from the egg.

Something like imprinting establishes the tie between mother and offspring among many mammals, too, but here it operates basically upon the mother. The mother of some species of hoofed mammals will initially accept her offspring for only a few hours after birth. If they are kept apart past that time, the young are thereafter rejected by the mother. Normally this behavior results in the mother distinguishing her own offspring from those of others. Apparently the cue that is employed is olfactory, for the maternal bond is not established if the nasal membranes are anesthetized for an hour or so following birth, and strange babies will be accepted if anointed with her amniotic fluid or wrapped in the skin of her offspring.

Behavior of this general type is probably widespread among birds and mammals. Even among monkeys and apes, the mother-offspring bond must be properly established if the offspring is to achieve normal sexual and social behavior in adult life. Even human mothers may experience something akin to imprinting. It has long been a practice in the United States to separate the new mother and her infant immediately after delivery. Recently researchers compared the maternal behavior of women who had not been initially separated from their offspring with that of those who had. Their preliminary findings indicated that those not initially separated from their infants exhibited more consistently maternal behavior toward them.

*Habituation.* Although not usually thought of as a form of learning, habituation does fit the formal definition. We are not normally conscious of the touch of our clothing after we have put it on. Even a horrendous stench becomes more bearable in time. We unconsciously screen background noises out of our consciousness when conversing, and so on. Learning to ignore constant or extraneous stimuli is termed *habituation.*

*Human and animal learning.* What relation does animal learning bear to that of human beings? Every form of animal learning has also been demonstrated in human beings, so there are obviously extensive points of contact. Nevertheless, what is going on in your head at this moment far transcends anything of which any animal is capable. Indeed it seems that our animal-derived models of learning are not sufficient in themselves to explain the capabilities of the human mind. Nevertheless there is a common pattern that fits all learning, and it is this: all higher, that is, more complex organisms tend to repeat those patterns of behavior which are instrumental in gaining rewards or escaping punishments. What you are doing now may bring you happiness through a good grade, praise from your friends, or intellectual pleasure. Also it avoids several obvious kinds of pain, such as the loss of self-esteem that might result from a lack of diligence! But while we may be able to explain your motivation partially in terms such as these, we have not penetrated to the root of learning itself.

## SOCIAL BEHAVIOR

When animals belonging to a common species interact in an adaptive fashion, we refer to that behavior as social behavior. If they are mutually attracted together, the group that they comprise is termed a *society.* Examples of a society, from this point of view, would be a hive of bees, a tribe of Indians, a flock of birds, a pack of wolves, and a school of fish. Not every group of animals, however, forms a society.

On a hot day flamingos may congregate in the shade. The flamingos in such an instance are only incidentally grouped. Really, they are attracted to

(a)

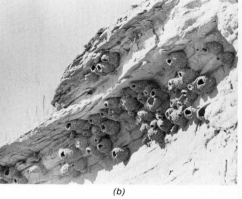

(b)

(c)

**FIGURE 18-25**

Social animals. (a) Baby bats on a cave roof. (Bob Martin) (b) Cliff swallow nests. (Frank R. Martin) (c) Herd of zebra. (Busch Gardens) (d) Family of rhinoceroses. Animal in foreground is rolling to rid itself of parasites. Photographed in Africa. (Lion Country Safari, Inc., West Palm Beach, Fla.)

(d)

**FIGURE 18-26**
An aggregation of flamingos in the shade.

the shade. To be sure, the flamingo is a social animal, but at this particular time they form a nonsocial group known as an *aggregation.* Aggregations also occur among nonsocial animals. Ladybird beetles hibernate in sheltered places in vast swarms weighing several pounds. On a cold fall day, sheltered outside walls frequently collect a large covering of houseflies in such aggregations. The organisms are all attracted by the same set of favorable environmental conditions, but they really do not interact socially with one another.

The minimum social contact for many species of animals is the sex act. As someone has said, familiarity breeds contempt, but breeding is not achieved without familiarity! For many species which do not otherwise exhibit much social behavior, fertilization and perhaps the rearing of young are almost the only forms of sociality. Let us consider the sex act as a basic example of social behavior, for the elements to which it can be reduced are also the least common denominators of most social behavior.

Like other social relationships, the sex act is adaptive in that it promotes the welfare of the species. It requires *cooperation,* the *suppression of aggressive behavior,* and a *system of communication.* Among many spiders, for example, mating is preceded by a ritual courtship on the part of the male, the effect of which is to produce temporary paralysis in the female. While

(a)

(b)

(c)

**FIGURE 18-27**
(a) In many higher animals the sexual relationship forms the basis for the cohesion of the society. (Lion Country Safari, Inc., West Palm Beach, Fla.) (b) In others, the sex act is almost the only social relationship which they possess. In these lizards, even parental care is absent, the eggs (c) being deposited in a suitable place and then abandoned to hatch as catch can. Pencil is included for size comparison.

she is thus enthralled, the male inseminates her. Should she recover before he makes good his escape, he becomes the main course at his own wedding feast.

This behavior conforms to the criteria of social interaction. In order to ensure cooperation, the aggressive behavior of the female must be suppressed. To manage this, the male produces paralysis by means of certain releaser cues, provided in his courtship dance, which therefore constitutes a means of communication from him to the female he is courting. To be sure,

**FIGURE 18-28**
Among many animals elaborate social relationships may surround the parental care of offspring. (Joe Van Wormer)

the spider's morbid love life is rather extreme, though *not* unique. Higher animals are not usually so bloodthirsty in their sexual behavior, although the courtship rituals of many birds seem designed to suppress aggressive behavior on the part of the male, whose territory the female must naturally invade before mating can take place.

### Why animals form societies

In a pioneering work, *The Social Life of Animals,* W. C. Allee showed that many species are far more resistant to noxious environments in groups than alone. Schools of fish have been shown to be less vulnerable to predators than single fish because large numbers tend to confuse predators. Flocks of birds may be able to find food better than single individuals. Insects are able to construct elaborate nests and raise young by mass production methods when they cooperate. A pack of wolves or a pride of lions hunts far more successfully by cooperative means than individual animals could hunt by themselves. It seems clear that social behavior is often a biologic advantage.

### Animal communication

The ability to communicate is an essential prerequisite for social behavior. Without it, how could organisms cooperate? Animal communication differs significantly from most human communication in that the former is not symbolic. As you read these words, information is being conveyed to your mind by words; yet the words themselves are not the information. They stand for it. The relationship between the word "cat" and the animal itself is a learned one. A Japanese would not recognize it, nor would an Arab.

This is not to say that *signs* in some sense are not employed by animals. In a way all releasers are signs. Releasers are not learned, but true symbols are. It has recently been shown that the pupil of the eye in human beings dilates in certain emotional situations such as sexual interest or excitement. Without realizing it people respond to such subtly transmitted cues. In experiments, a photograph of a girl's face with the pupils retouched to appear greatly dilated was far more attractive to male subjects than one in which they had been shown as hostile pinpoints. Dilation of the pupils seems to serve as a communication sign of the releaser type in human beings, and one which may bear little relationship to symbolic language. Although some body communication is culturally determined learned behavior, a large part of it is truly universal and appears to be physiologically determined.

(d)

FIGURE 18-29
Parental care is especially complex in the chimpanzee. (a) The male parent (rear) normally makes little contribution to the care of his own offspring. Indeed, among chimpanzees it is a wise father that knows his own baby. (b and c) Mother chimpanzee awakening sleeping infant in order to nurse. (d) Young infants (less than 18 months) are transported by clinging to the leg of the mother. (Lion Country Safari, Inc., West Palm Beach, Fla.)

(a)                                                  (b)

**FIGURE 18-30**
(a) Male giraffes "neck" in a stylized form of combat. (Lion Country Safari, Inc., West Palm Beach, Fla.) (b) Male sticklebacks adopt a ritualized fighting posture. (After Niko Tinbergen)

Among animals, signals are generally transmitted involuntarily in an accompaniment of the physiologic state of the organism. If certain emotional or mental states are present, they will be transmitted even if other members of the species are nowhere about. A bird cannot "help" giving an alarm call when it sights a predator, or the appropriate cry if the bird is likely to alight or take off. The pronghorn antelope flashes its white rump patch when alarmed, whether another antelope is watching or not. Cats and dogs communicate the exact balance of their aggressive and defensive tendencies to their opponents without, it seems certain, any desire to do so. Certainly, opponents in a human poker game have no desire to communicate their feelings.

Of course, flocking, courtship, or territorial behavior involves the cooperation of other organisms, but the very presence of a potential spouse or rival produces an endocrine and emotional state that will produce certain behaviors if the releasing stimulus is provided. The signs whereby animals communicate are also themselves determined physiologically. Among human beings, the closest equivalent to animal communicating mechanisms would be a physiologic response which someone else can "read," such as a blush.

Do animals ever employ symbols? The matter is controversial. Most ethologists feel that even dogs do not respond to spoken language as we do, but deduce the behavior we command from an astute reading of human

(a)

(b)

**FIGURE 18-31**
Animal communication. (a) This male anole lizard issues a territorial challenge by expansion of his brilliant red dewlap. (b) The male sharptail grouse issues visual invitation to romance for the benefit of nearby hens. (Frank R. Martin)

facial expression, voice intonation, and bodily attitude. On the other hand, chimpanzees have been taught to speak a very few words meaningfully, and apparently can use sign language appropriately. Whether apes employ symbolic language in nature is unlikely, although the potential seems to exist.

Animals employ a number of modes of communication. The singing of birds is an obvious example of auditory communication, here serving to announce the presence of a pugnacious territorial male. Lions roar, wolves howl, crickets chirp, and frogs croak. And some animals communicate by scent rather than sound. Antelopes rub the secretions of facial glands on conspicuous objects in their vicinity. Dogs urinate on posts, fire hydrants, hi-fi speakers, and the like. Many carnivores leave piles of dung about their home range, distinctively scented by their anal glands. Many insect societies are coordinated by olfactory cues and other chemical messages called *pheromones* (see further on in this chapter, under Insect Societies). The gentleman baboon assesses the lady baboon's sexual readiness by the brilliance of the coloration on her buttocks. The lifted tail of the skunk conveys an unambiguous message to anyone who has cause to associate it with the release of his unmistakable perfume. None of these patterns of behavior is symbolic language in the human sense. Even the so-called "language of the bees," the dances performed by returning workers to convey the location of food supplies and nesting places, seems to be genetically determined behavior.

It is worth emphasizing that we do not necessarily inhabit the same sensory world as other animals. For us the dominant sense is sight; for dogs it is smell; for bats it is hearing; for catfish it is taste. All these latter have had to be inferred. The existence of the bat's sonar system has only recently been demonstrated, as has the fact that whales "sing," and that dolphins possess fairly elaborate sound communication. For years it was unsuspected that rats emit an ultrasonic cry at the conclusion of copulation, whose meaning, apparently, is to break off contact. Most of the communication of rats with one another employs frequencies of sound which we cannot hear. The most elaborate systems of communication may pass by us completely if they are not visual in nature.

The mode of communication employed by a given organism is affected by its ecology. Birds living in dense foliage communicate far more efficiently by song than by sight; but prairie birds such as turkeys convey many messages by visual display. Others like the peacock make the best of both.

**FIGURE 18-32**
Other animals may inhabit a very different sensory world from ours. The common black-eyed susan looks very different to us (a) than it does to a bee (b) because our eyes are not equipped to discern the striking ultraviolet-absorbing pigment of this flower.

(a)

(b)

### Dominance hierarchies

In the spring, a paper wasp nest may be founded cooperatively by a number of females which have survived their winter hibernation. During the early course of construction a series of squabbles among the ladies takes place, in which the combatants bite one another's bodies or legs, and (rarely) sting. In the end, one of the young queens gets the upper hand over all the rest, and thereafter she hardly ever is challenged. The paramount queen spends more and more time on the nest, and less and less time out foraging for herself. She takes the food she needs from the others as they return, and if they don't like it, they can leave; some do.

The paramount queen now begins to take an interest in raising a family—her family. Since she is almost always at hand, she is able to prevent other wasps from laying eggs in the brood cells by rushing at them, jaws agape. At the same time, she cannot be stopped from laying all the eggs she wants, since she has already demonstrated that she cannot be successfully challenged. Furthermore, if any of the other females *do* manage to lay, she eats their eggs. *Her* eggs are undisturbed! Those subordinates that stay experience a definite regression of the ovaries, and eventually become sterile workers. The young that have been raised in the nest grow up as sterile workers from the outset, although if the paramount queen dies, they retain the potential of becoming reproductively competent.

A careful analysis of this arrogant behavior will disclose that the paramount queen can bite any other wasp without serious fear of retaliation. There is another wasp, however, which while usually not a queen, can nevertheless bite any wasp she chooses (other than the queen) without fear of retaliation. Thus, though the queen can bite any wasp in the nest, all other wasps are not equal in their relationships with one another. One can arrange the wasps into a definite *dominance hierarchy,* an arrangement of status which regulates aggressive behavior within the society.

Queen ⟶ Wasp A ⟶ Wasp B ⟶ · · · ⟶ Wasp E ⟶ Wasp F

Once it is established, far less time is wasted in fighting than would otherwise be the case. In fact there is almost no actual fighting, for subordinate wasps, upon challenge, generally exhibit submissive poses which inhibit the aggressive behavior of the queen toward them. Consequently few or no colony members are lost through wounds sustained in fighting.

Dominance hierarchies are advantageous to all animal societies, and may occur almost universally. They have been observed among birds such as chickens (where it is sometimes termed the *peck order*), in many mammals, in some reptiles, and in various social and subsocial insects. Dominance hierarchies are not always arranged in a simple straight-line sequence, however. Dominance relationships are frequently established among animals on the basis of their *first* contact. If the *alpha* (dominant) animal A happens to be ill or emotionally upset on the day when it first meets animal D, it may lose place to D, even if D is a generally subordinate individual. This can result in a complicated situation, yet it is the rule rather than the exception in most animal societies. In such a case, the alpha animal is considered to be the individual with the greatest *overall* dominance, even if it is subordinate to D (see below).

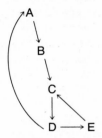

Dominance hierarchies exemplify the inhibition of aggression, which, as we have seen, is a necessary feature of social behavior. Hierarchies require a system of communication, since a subordinate animal must signal its unwillingness to contest the dominant. Like the wasps in our example, most social animals seem to have certain stereotyped poses that *tend* to inhibit aggression from dominant individuals. These are not always easy to interpret, however. For many years it was believed that dominant and submissive poses in wolves were the reverse of what they in fact turned out to be. This misconception had the effect of thoroughly muddling early observers who attempted to sort out dominance relationships in wolf packs.

In some animals dominance is a simple function of aggressiveness, which is itself often influenced directly by sex hormones. Thus if a hen receives testosterone injections, her place in the dominance hierarchy shifts upward. If she is spayed, the reverse takes place. Of all chickens, the cock is the most dominant. In rats, estrogen injections reduce dominance. Recent tests on rhesus monkeys have shown than when males are dominant, their testosterone levels are much higher than when they have been defeated. Thus, not only can estrogen reduce dominance and testosterone increase dominance, but dominance may increase testosterone. It isn't always easy to unscramble the situation!

An extreme form of the dominance hierarchy occurs among certain labrid coral reef fishes. Like many fishes, these are capable of sex reversal. What is odd is that the most dominant individual is always male, and the remaining fish within his territory are always female. If the male dies or is removed, the most dominant female will become the new male. Should anything now happen to "him," the next ranking female will become the new sultan of the harem. Why these male chauvinist fish behave as they do is not altogether clear, but their behavior does serve to propagate preferentially the genes of the most dominant individual, since "he" can mate with *all* the others.

In other animals, however, males and females have separate dominance systems, and in many monogamous animals, especially birds, the female takes on the dominance status of her mate by virtue of their relationship.

Dominance hierarchies sometimes occur consistently in the relationships of one species to another. Great blue herons, for example, are automatically dominant over most other species of herons. Similarly, blue jays are generally able to rout smaller song birds. Such inherited statuses minimize conflict between species and presumably operate to their mutual advantage.

Even human beings seem to develop dominance hierarchies when placed together in unstructured social situations. However, in normal society, many individuals possess an economic or social status linked with a job or administrative position which they would never be capable of attaining in the "natural" course of events. The introduction of numerous variables, such as the sudden inheritance of great wealth, confuses the dominance relationships among people.

### Pairing and territoriality

Most animals have a geographic area that they seldom or never leave. Such an area is called a *home range.* Since the animal has opportunity to become familiar with everything in that range, it has an advantage over both its predators and prey in negotiating cover. Furthermore, the resident animal becomes familiar with all resources of the area, such as the location of water, salt licks, fruit trees, and the like. In some, but not all, animals a portion of the home range is defended against other individuals of the same species, and sometimes against individuals of other species. Such a defended area is called a *territory.* The tendency consistently to defend such a territory is known as *territoriality.*

Territoriality is most easily studied in birds. In most birds males choose territories at the beginning of the breeding season. This is a behavioral consequence resulting from high levels of sex hormones in the blood. Fighting takes place between the males of adjacent territories until territorial boundaries become fixed. Generally, the dominance of a cock varies directly with his nearness to the center of his territory. Thus, close to "home" he is a lion. When invading some other bird's territory, he is likely to be a lamb. The interplay of dominance values among territorial cocks eventually produces a neutral line at which neither is dominant. That line is the territorial boundary. Bird songs announce the existence of territory and often serve as a substitute for violence. Furthermore, they announce to eligible females that a propertied male resides in the territory. Typically, male birds take up a conspicuous station, sing, and sometimes display visually to their neighbors and rivals. It has been shown that the red shoulder patches of a red-winged blackbird are important in establishing his territory. Rivals are less likely to take him seriously if these are experimentally obscured.

The arrival of a female establishes a new relationship called *pair bond.* The pair bond is a stable relationship between animals of the opposite sex which ensures cooperative behavior in mating and the rearing of the young. When a female arrives, she is initially treated as a rival male, but through the use by both male and female of instinctive appeasement postures and gestures, the initial hostility is dissipated and mating takes place. Such sexual appeasement behavior may be very elaborate and gives rise to mating dances in some birds.

A male flicker possesses a black mustache-like marking under the beak. This is lacking in the female. If a happily married female flicker is captured, and such a mustache painted on her, her mate will vigorously attack her as if she were a rival male. He will accept her again if it is removed. The releaser mechanisms involved in the establishment and maintenance of the pair bond are often remarkably detailed.

Since mating cannot take place until all the necessary types of behavior have been released in both partners, and since at least some of the releasers are usually unique to the species, courtship rituals can be very effective sexual isolating mechanisms.

Many kinds of territoriality occur among animals; not all of them can be shown to be sexual in origin. In some cases, particularly in carnivorous birds and mammals, territory may be coextensive with home range. In such cases it may serve to conserve the food supply. In some animals, such as bears, territory is not so much a geographic area as certain trails, food caches, etc. Among such animals territories may seem to us to interpenetrate, but they do not actually do so, if the animals avoid one another's preserves. In sea birds territorial behavior is often restricted to nesting sites on a rock or island, and serves to establish nesting sites and pair bonds. Some social animals, such as hyenas, possess tribal, or collective, territories which are vigorously defended against other tribes.

We will guess that territoriality among animals is adaptive for a variety of reasons. Probably the most important reasons are that territoriality tends to reduce conflict, to control population growth, and to ensure the most efficient utilization of environmental resources by encouraging migration and spacing organisms more or less evenly throughout a habitat.

### Other social bonds

Animals are held together by a variety of social bonds besides the pair bond. Among most mammals and birds at least a temporary bond is established between one or both parents and their offspring. Elaboration of these bonds appear to be partly responsible for the cohesion of some insect and vertebrate societies such as wolf packs, which are essentially families. Some societies employ bonds not easily derived from the parent and reproductive relationships, as in the primate societies of the baboon, chimpanzee, or man. In these, social bonds are reinforced by cooperative interaction, such as mutual grooming, bodily contact, and play.

### Elaborate societies

So far we have for the most part dealt with rather simple social interactions among organisms. Such interactions may be quite different from what we would observe in elaborate societies, such as insect or human societies, although there would also be points of contact. While it is difficult to define precisely, we shall consider an elaborate society one in which there is considerable division of labor not directly connected with the care of the young.

*Insect societies.* Although many insects cooperate socially, such as tent caterpillars which spin a communal nest, the truly social insects are generally held to be bees, ants, wasps, and termites. Subject to the strictures which we already have noted, much insect social behavior appears to be genetically determined. Opportunities for learning are rigidly circumscribed, and indeed the insect brain is probably not capable of sophisticated learning. Insect societies are held together by an elaborate system of releasers, and tend to be quite rigid in consequence. Many social insects seem to secrete social hormones called pheromones, which accomplish such tasks as suppressing the ovaries of worker honeybees without recourse to the dominance hierarchies which we saw among wasps, or alerting an ant hill to the presence of an enemy. (An "alarm substance" is given off from a special abdominal gland by excited or terrified workers.) Virtually all the intercommunication within insect societies is comparable to the hormonal communication within a single organism, and the social coordination achieved approaches the behavior and singleness of purpose of an individual organism rather than the rather diffuse goals of human society.

*Vertebrate societies.* Because of far greater range and plasticity of potential behavior, vertebrate societies tend to spring from a different behavioral base than those of insects. But, as we have seen, the social relationships of vertebrates can be almost as highly stereotyped as anything that insects have. Still, vertebrate societies are far less rigid, far more adaptable to changing needs, and far more "intelligent" in their general approach to life Paradoxically, perhaps, they are also far less complex. Vertebrate societies usually contain nothing comparable to the physically and behaviorally specialized castes of termites or ants, but they do not need them, since the individual can learn his task and his place. What is more, except for man, vertebrates are not as specialized in their tasks as insects. A beehive may be considered a kind of superorganism, but is is hard to interpret a wolf pack or a herd of red deer in that fashion.

## HUMAN IMPLICATIONS OF ANIMAL BEHAVIOR

It is remarkable that more parallels can be drawn between human and insect societies than between human and other vertebrate societies. Of all the vertebrates, apparently only man exhibits behavior of sufficient plasticity to produce the more or less rigid specialization and division of labor that elaborate human society requires. If insects achieve this end by their genetic programming of behavior, man achieves it by nongenetic control of behavior, which is nevertheless transmitted from generation to generation. This continuity of behavior is called *culture.* It is learned behavior, but it *is* transmitted from other organisms. The medium of its propagation is not the information contained in DNA but the information contained in language.

Since the transmission of information is never free from error, and since information is also subject to deliberate innovation, culture is even more subject to change than is genetic information. To continue the analogy, in insect societies the bodies of the component organisms are often specialized to reflect the division of labor in it; in human societies specialized roles are learned, and in general are performed not by adapting the body to the task, but by the invention of tools and other tangible examples of so-called material culture. Thus the principal distinguishing mark of human societies is their reliance upon symbolic language, culture, and the use of tools. Do these behaviors also occur among the various lower animals?

We have already seen that communication by learned symbols does not seem to take place among animals in a state of nature; they have no language in the human sense of the term—not even in rudimentary or simplified form. Bees may be able to transmit direction and distance of food by means of largely innate signaling systems, but they have no provision for the development of a system of algebraic notation. Chimpanzees may be taught some human speech sounds and sign language, but they do not seem to employ this or anything like it in a state of nature.

Since culture depends upon language, it would seem on the face of it that lower animals cannot possess this human trait either. To be sure, some analogies to culture do exist in some animal societies. Among jackdaws (European blackbirds), adults seem to teach younger birds in some way—for example, that a man with a gun is dangerous. Differences in food preferences exist among various populations of urban and rural rhesus monkeys in India, and also among various groups of Japanese macaque monkeys who occupy much the same environments. Such examples seem to argue for some kind of nongenetic transmission of learned behavior in these animals. In Scandinavia and the British Isles, entire populations of tits (a kind of bird) have learned to break through the caps of milk bottles and drink the cream from the top—and they learned this *after* World War I, even though such milk was widely available before the war. Evidently this behavior did not spread by imitation in the human sense. Apparently some tit genius originated the habit, and since tits tend to feed where they see other birds feeding, each was rewarded individually, and the habit spread. But these protocultural behaviors are not the same as human culture.

Tool use is the only touchstone of human society that remains for us to consider. Does tool use occur among animals? Technically, it does. A number of instances are known or suspected. Among the Galapagos Islands' finches, one species is adapted to a woodpecker-like ecologic niche. Its bill,

FIGURE 18-33
The loggerhead shrike. This passerine (perching) bird has relatively weak feet, but is nevertheless a predator, as evidenced by the hooked beak. Unable to hold its prey as it tears at it with its beak, the feathered bandit impales its victims on long thorns, which serve as vises to hold the food in place.

however, is unspecialized and does not resemble that of a woodpecker, so it must use the spine of a cactus grasped in its bill to dig grubs out of holes and crevices in bark. The California sea otter uses rocks to break open the shells of abalones and other marine organisms on which it feeds. Chimpanzees extract ants and termites from their nests by inserting moistened straws which the insects grasp in their jaws, making it easy to pull them out. Certain predatory African birds break the shells of eggs with rocks in order to eat the contents. Some digging wasps tamp the earth plug of their burrow down using a pebble held in the jaws. Most of these instances of tool use have not been investigated rigorously, but it is likely that the majority are genetically predetermined. Those that are learned are apparently usually learned by imitation.

There can be no question that ethology has provided us with many insights into human behavior and social relationships. We already have mentioned programmed learning in this connection; to this we may add behavior modification, greater understanding of the structure and function of the brain and nervous system, the cause and alleviation of psychosomatic disorders, the treatment of phobia, various methodologies for dealing with schizophrenia, the mode of action of many drugs, and others. But the greatest thrust of ethologic endeavor has been an attack upon the problems of human society.

Experiments and observations made upon lower animals are never perfectly applicable to human beings. Nevertheless, conclusions drawn from experiments on other animals can generally be applied to man when the

FIGURE 18-34
A beaver dam. The beaver is capable of elaborate construction, but his only tools are his teeth, paws, and tail. (U.S. Forest Service)

questions investigated are physiologic in nature. Experiments on mammals are most applicable to man because of the physiologic similarity between other mammals and man. In the case of behavior, there is far less similarity. Is the social life of a rat remotely comparable to that of a human being? Even the societies of chimpanzees and dolphins, though far more complex than those of rats, do not resemble human society at all closely. Extrapolations from the behaviors of rats and baboons to that of human beings are subject to far more pitfalls than we are used to dealing with in other areas of biology. Caution is called for!

The most eminent ethologists currently disagree regarding such signal questions as whether man is a territorial animal by nature or by cultural training; whether his aggressiveness is "innate," and if so, to what extent it may be modifiable; and how much crowding he may tolerate without psychologic or physiologic harm. Despite sensational publications purporting to solve these problems, assured answers seem to be premature at this time. Pressing though these questions may be, a recklessly conceived set of "solutions" may well be worse than no solutions.

**CHAPTER SUMMARY**

1 Behavior consists of the active external relationship of an organism to its environment. Put more simply, behavior is anything that an organism *does*.

2 Simple orientational behavior of plants is termed tropism, that of animals is called taxis. These differ principally in that tropism is basically a growth response, but taxis involves nervous and muscular action.

3 Behavior itself is not so much inherited as is the capacity for behavior. The actual development of behavior is a product of a complex interaction between heredity and environment. The characteristic learning capacity of some organisms is much more narrowly circumscribed than that of others.

4 A releaser is a specific stimulus that triggers given behavior more or less automatically. Released behavior is basically unlearned, and though it is often a response to a simple stimulus, it may be quite complex. Releasers and the responses they evoke are sometimes arranged in complex chains.

5 Most organisms have an active rather than passive relation to their environment. Internal drives presumably reflect some physiologic need, on the level either of gross function or of neural mechanism. These drives lead to the increased probability of specific kinds of actions such as eating.

6 When drives conflict, an organism tends to exhibit irrelevant displacement behavior. In ritualized form, displacement behavior does, however, play an important part in courtship displays and other stereotyped social behavior. Displacement behavior should be distinguished from behavior that is redirected as a result of frustration.

7 Most organisms possess internal timing mechanisms of unknown nature which are known as biologic clocks.

8 Many organisms migrate. The adaptive significance, triggering stimulus, and navigation mechanisms employed are known only in a few cases.

9 Certain behavior may develop in a trial-and-error fashion, or as a result of the maturation of neuromuscular systems, without the need for specific learning.

10 Learning is operationally defined as a persistent change in behavior resulting from environmental stimuli. Several categories of learning are customarily distin-

guished: classical conditioning, operant conditioning, trial and error, insight, imprinting, and habituation.

11 Social behavior is intraspecific interaction. All social behavior requires a means of communication, suppression of aggression, and mutual attraction.

12 Animals form societies because it is adaptive for them to do so, from the standpoint either of the organism or of the total population of which they are members. In a society organisms interact with one another. In aggregations such interaction is incidental, and organisms are drawn together coincidentally by a common environmental stimulus.

13 Animal communication involves the transmission of releasing signals but does not utilize (as far as is known) symbolic language in the human sense.

14 Dominance hierarchies formalize aggressive behavior and suppress its expression.

15 Organisms often inhabit a home range, from which they seldom or never depart. This range, or some portion of it, may be defended from members of the same (or occasionally different) species. Defended areas are called territories, and the defensive behavior is territoriality. Often territorial defense is carried out by display behavior rather than by actual fighting.

16 Territoriality exists in a number of forms and may have a number of adaptive significances in different organisms. Like the dominance hierarchy, in general it formalizes and suppresses aggression. Territoriality also tends to ensure the fullest utilization of the resources of the environment by spreading organisms more or less evenly throughout it.

17 Complex vertebrate societies employ pair bonds, parent-offspring bonds, and other bonds, such as those established by mutual grooming, to ensure their cohesion.

18 Insect societies tend to be rigid, with the role of the individual narrowly defined by innate behavioral tendencies. Division of labor is often ensured by modification of the bodies of the society members. Often insect societies are quite complex.

19 Vertebrate societies are far less rigid than insect ones. Though innate behavior is important in them, in general the role of the individual is learned. With one exception, vertebrate societies are far simpler than those of the more highly social insects.

20 Human society is by far the most complex of all vertebrate societies. It is almost uniquely characterized by the possession of language, culture, and tool use.

21 Although ethology has shed light upon many problems of human individual and social behavior, since the social dimensions of lower animals differ radically from those of human beings. Conclusions drawn on the basis of animal experimentation and observation must be applied with caution to human social problems.

**FOR FURTHER READING** In addition to the specific references given below, we particularly recommend *Natural History* magazine as a source of naturalistic studies of the behavior of wild animals. Many worthwhile articles occur also in *National Geographic Magazine*. Psychology texts should be consulted for a much fuller treatment of learning than we have been able to provide here.

One area of ethology that deserves more intensive treatment than we have been able to provide is the study of animal societies. We have included a number of works in this bibliography that are essentially specialized studies of particular animal societies. The student and instructor will find these a fruitful source of material for term papers, discussions, and special assignments.

Barnett, S. A.: *The Rat, a Study in Behavior,* Aldine, Chicago, 1963.

Blakemore, Colin: "The Limits of Learning," *New Scientist,* Apr. 20, 1972, p. 145.

Chauvin, Remy: *Animal Societies,* Hill and Wang, Inc., New York, 1968.

Fox, Michael W.: "Neurosis in Dogs," *Saturday Review,* Oct. 28, 1972, p. 58.

Hess, Eckhard: "Imprinting in a Natural Laboratory," *Scientific American,* August 1972, p. 24.

Jonas, Gerald: "Visceral Learning," *The New Yorker,* Aug. 19, 1972, p. 34, and Aug. 26, 1972, p. 28.

Manning, Aubrey: *An Introduction to Animal Behavior,* Addison-Wesley, Reading, Pa., 1967.

*The Marvels of Animal Behavior,* National Geographic Society, Washington, 1972.

Myerriecks, Andrew: *Man and Birds,* Bobbs-Merrill, Indianapolis, 1972.

Premack, Ann James, and David Premack: "Teaching Language to an Ape," *Scientific American,* October 1972, p. 92.

Schaller, George B.: *The Year of the Gorilla,* University of Chicago Press, Chicago, 1964.

Tinbergen, Niko: *Curious Naturalists,* Doubleday, Garden City, 1968.

Van Lawick-Goodall, Jane: *My Friends the Wild Chimpanzees,* National Geographic Society, Washington, 1967.

Wilson, Edward O.: "Animal Communication," *Scientific American,* September 1972, 1953.

# 19

## OUR SPACESHIP EARTH

***Chapter learning objective.*** The student must be able to complete an examination concerning ecologic levels of organization, the various modes of organismic nutrition, the various types of asocial mutual interaction among organisms, the physical and biotic factors that influence organisms, the energy relationships among members of food pyramids, chains, and webs, and the ecologic and chemical effects of biocides and other toxicants. Using the correct ecologic terms, he must also be able to summarize the concept of succession in relationship to productivity and the problems of wildlife and habitat conservation.

*ILO 1* Define biologic adaptation, giving an example from both the plant and animal kingdoms of structural, functional, and (in animals) behavioral adaptations to the environment.

*ILO 2* Put the "law of the minimum" in your own words, giving at least one original example of its operation and relating it to the concept of carrying capacity and to the distribution of communities.

*ILO 3* Define "ecosystem," give its necessary components, and describe the principal chemical and trophic relationships among the components.

*ILO 4* Describe the cycling of substances through the biosphere, and also the pathways and distribution of chlorinated hydrocarbon biocides, giving their principal ecologic effects.

*ILO 5* Summarize the environmental problems posed by the presence of heavy metals and radioactive substances in the biosphere.

*ILO 6* Define a community. Give the major characteristics and summarize the course of a sere.

*ILO 7* Distinguish among the main varieties of ecologic succession and climax communities.

*ILO 8* Give the principal relationships of the characteristics of successional and climax communities to conservation goals and problems.

## OUR LIFE SUPPORT SYSTEM

The spaceship, of course, is the planet Earth. It is striking that this immense sphere of rock, water, and air can be called fragile but it *is* both fragile and small compared to the scale of the universe. It is the only home we have; the only place in the galaxies that we *know* to be able to support life. Viewed from space, as we have begun to view it, the earth is small indeed, and it is vulnerable. In an earlier age when the world seemed vast in relation to man's numbers and activities, no one imagined that we endangered its existence. But all that has changed.

### Earth: a closed system

Every spacecraft has a life-support system. For a trip to the moon, for instance, oxygen must be carried along in tanks for the astronauts to breathe, and some means must be provided for removing the carbon dioxide produced by human metabolism. On a longer journey, such as to Mars, some means of reconverting the carbon dioxide to oxygen would probably be provided. Science fiction writers have pictured vessels that might be built to reach the planets of other stars. Perhaps they would be miles long and contain the population of a small city. Because of the immense distance, generations would live and die before the goal was reached. Oxygen sufficient for such a journey could not possibly be carried in tanks but could be produced by living green plants. The plants also would be eaten by the passengers and their domestic animals. Human and organic wastes, from paper bags to corpses, would be broken down by bacterial or chemical means and their constituent elements returned to the soil or hydroponic farms, in which the

**FIGURE 19-1**
Our spaceship earth. (NASA)

all-important plants grew. During the journey, no puff of air, no ounce of water, could be wasted. The passengers would have to be thrifty past all previous meanings of the term. All metals, ceramics, plastics, and industrial chemicals would have to be endlessly remanufactured and repurified for use over and over again. Self-sustaining and self-contained, powered perhaps by thermonuclear fusion energy for light and heat, such a ship could travel for millennia, and theoretically could drift from one edge of the Milky Way to the other. What a triumph of human ingenuity, technical skill, and social engineering it would be! Imagine how delicately balanced its mechanisms, how careful and conscientious its crew would have to be.

Yet such a spaceship already exists, and on an even vaster scale. Whether it will continue much longer is debatable. The crew is ignorant and careless. It damages and consumes the life-support system and even wantonly destroys it. When reproached, crew members justify their conduct on the grounds that they do not understand what they are doing! You belong to the crew. This chapter is intended to help you understand your spaceship.

### Some basic concepts of ecology

In the nineteenth century few people thought in terms of spaceships, but even then some saw that the earth could be compared to the economy of a well-run estate or household, called *oikos* by the Greeks. Ernst Haeckel combined this word with another Greek root, *logos,* meaning, among other things, a study or body of knowledge. The word that resulted was **ecology,** the study of nature's household. In more modern terms, it is the systems approach to nature. The ecologist studies the systems of the earth as a whole, and he also studies the *relationships* among its component organisms, and between them and their physical environments.

*Adaptation.* Organisms are *adapted* (fitted) for the interaction they have with their environments. Countless specific adaptations are known to biologists. Light-colored animals, for example, often are adapted to blend

**FIGURE 19-2**
(a) A bizarre adaptation. The African baobab, or elephant tree, loses its leaves in the dry season to reduce water loss, but stores large quantities of water in the tissues of its swollen trunk. (Pat Gill) (b) The fennec, an African desert fox. The large ears probably act as radiators from which excess body heat is lost.

(a)

(b)

visually with light-colored surroundings, dark-colored animals with dark sur-roundings. As a result they are not easily detected by predators, or if they are predators, they are not easily seen by their prey. Sometimes, in temper-ate and arctic climates, they are dark in the summer and white in the winter.

To take other examples, warm-blooded animals living in cold climates usually have thicker fur or feathers than those living in warm climates. Cold-climate animals generally have shorter extremities also, exposing less of the body to possible frostbite and reducing the total surface area of the body available for the loss of body heat by radiation. Cold-climate animals also tend to be larger, for a large animal has relatively less surface area than a small one, and therefore tends to lose less body heat. Desert animals tend to have special means of conserving moisture—the kangaroo rat is so efficient at this that it does not ordinarily need to drink. It subsists instead on the water formed in the course of metabolism:

$$C_6H_{12}O_6 + 6O_2 \longrightarrow 6CO_2 + 6H_2O$$

The kidneys of this rat excrete so little water that the urine is semisolid.

Many extreme adaptations also occur among plants. The cactus has no leaves so that water loss can occur only from the stem, which has far less surface area. It possesses great masses of spongy water-storage tissue. In these and other ways it is adapted to water scarcity. The water hyacinth, which floats in water, has air-filled tissues to keep it afloat and to hold its green parts *out* of the water. Alpine plants are tiny and have small leaves to resist the cold dry winds of mountaintops. Swamp plants often have large leaves but have so much trouble getting nitrogen from the acidic soil that they sometimes have specialized organs to capture insects and consume them for the sake of their amino acids.

Not all adaptations to the environment are structural or physiologic. Some are behavioral. For example, many reptiles regulate their body temperatures quite precisely by orienting their body at just the proper angle with respect to the rays of the sun. Very complex behavioral adaptations are exhibited by social insects and those higher vertebrates which form societies among themselves.

In these and countless other ways plants and animals are precisely fitted to the characteristics of light, temperature, water, soil, and other conditions typical of their surroundings. Even creatures living in moderate conditions are well adapted to them. A cactus would have trouble competing with the native vegetation in a rain forest or in the Adirondack Mountains of New York State, for in order to conserve water it must sacrifice the efficient photo-synthesis it could enjoy if it had leaves. Lacking them, cacti grow slowly and are at a competitive disadvantage where more "normal" plants grow well. In the desert, of course, the shoe is on the other root.

Organisms also adapt to one another. The yucca moth is exquisitely adapted to the yucca plant which it pollinates. The larvae of the moth live on yucca seeds but do not consume them all. Enough are always left to ensure the reproduction of the plant. The adult moth accomplishes the pollination as it lays the eggs which later hatch into larvae that eat some of the seeds it has helped produce.

***The law of the minimum.*** Communities of organisms are not the same everywhere and are not uniformly distributed over the face of the earth. Table 19-1 shows some of the major recognizable communities, called *biome types,* that occupy our land masses. The table is not complete. A thorough treatment would occupy this entire book. But however complex the distribution of these communities may be, it is clearly not haphazard; it is, in fact, governed by the interaction between the requirements of organisms and the provisions of the environment, especially temperature and moisture.

(a)  (b)  (c)  (d)  (e)

**FIGURE 19-3**
(*a*) A typical inhabitant of the taiga, an elk. (Frank R. Martin) (*b*) Road building and attempted habitat restoration in the taiga. (Alyeska Pipeline Service Co.) (*c*) Virgin Amazon rain forest. (John Sharp) (*d*) Burning tropical forest in Liberia. (Food and Agriculture Organization) (*e*) Devastated tropical rain forest, destined to be converted to agricultural use. (Dr. A. Gomez-Pompa and *Science*)

**Table 19-1 Some major biome types***

| Biome | Distribution | Typical organisms | Characteristic threats | Notes |
|-------|--------------|-------------------|------------------------|-------|
| Tundra | Northernmost circumpolar community of any importance; essentially no equivalent in Southern Hemisphere. Similar communities on high mountains of all latitudes | "Reindeer moss," sedges, heather, etc. Mosquitoes and some other insects in summer. Owl, ptarmigan, grizzly and polar bears, arctic fox, musk ox, caribou, lemming, stoat, weasel, snowshoe hare, and similar forms. Few or no reptiles or amphibians | Mechanical abrasion of slowly regenerating tundra; road building, oil pipelines | Very cold; usually low rainfall. Kept wet by low rate of evaporation plus poor drainage caused by permanently frozen soil known as the *permafrost* layer |
| Taiga | Northern Europe, Asia, and North America, but in areas of more moderate temperature than the tundra | Coniferous evergreens, some deciduous trees, (chiefly in subclimax† situations). Small seed-eating birds plus their predators, such as hawks. Fur-bearing carnivores such as mink, fisher, martin. Elk, bear, puma, tiger (Siberia only), wolverine, wolves, etc. | Lumbering, unregulated hunting and trapping, and, in some areas, agricultural development | Thick blanket of dead needles overlying acidic soil poor in mineral nutrients. Short growing season |
| Deciduous woodlands | Variable. Some are probably subclimax† stages of evergreen forests. Central and southern Europe, eastern North America, western China, Japan, New Zealand, etc. Tend to develop in temperate climates of moderate rainfall | Varied, familiar vertebrate and invertebrate fauna. Deciduous, hardwood trees and shrubs of many species, e.g., beech, oak, maple, cherry. Most fauna found near or on the forest floor | Plowing, which mixes top and subsoil, converting it to agricultural land. High human population densities with spilling over of suburbs onto prime forest or agricultural land | Historically, industrial civilizations have tended to develop in this biome; in many areas, especially Europe, very little left of the original community at this time |
| Grasslands | Continental interiors, especially temperate situations with rather low rainfall, e.g., the North American Midwest and the Ukraine | Dominated by grasses. Large herbivores such as bison, antelope, cattle. Jackrabbits, rodents such as prairie dogs, wolves. Rich, diverse array of ground-nesting birds | Original community destroyed by any and all agricultural development, for which this habitat is eminently suitable. Now almost none of original community left. In places, massive erosion produced by overgrazing and overexploitation | Soil characteristically quite rich in mineral content, which may form a calcified layer some feet below the surface. Useful for grazing and, where rainfall permits, for grain crops |

**Table 19-1. Some major biome types\*** (*Continued*)

| Biome | Distribution | Typical organisms | Characteristic threats | Notes |
|---|---|---|---|---|
| Deserts | Continental interiors; sparse rainfall. A few (e.g., the Gobi), temperate in climate; the majority, subtropical or tropical. Tend to lie in "rain shadows" of mountains | Drought-resistant vegetation such as sagebrush, cacti, euphorbias, and even some kinds of algae. In moister deserts, fauna may be numerous but tends to be nocturnal. Numerous reptiles and mammals; some birds | Threatened in some places by irrigation and residential-industrial development, but so far, damage is local | One of the more variable biomes, the variation depending mainly on rainfall rates. Some deserts are essentially devoid of life. Irrigation frequently accentuates already high mineral content of soil to saline levels |
| Tropical rain forest | Tropical areas of high rainfall, e.g., Congo and Amazon Basins. Grassy, open woodland called *savannah* produced by more moderate rainfall or a pronounced dry season. Some temperate rain forests, usually coniferous | Extremely rich flora and fauna representing almost all phyla. Habitat dominated by multiple stories of broad-leaved evergreen trees. Most fauna and considerable epiphytic flora concentrated in the canopy or treetop zones | Lumbering, ill-conceived agricultural development, depletion of fauna (e.g., monkeys, leopards, etc.) by trophy hunting and collecting for the pet trade | Most soil nutrients tied up in the living bodies of trees and other vegetation. When the natural vegetational cover is removed, a process of soil impoverishment and oxidation often turns the soil to a stony substance (*laterite*). Original climax community slow to regenerate |

\* We may distinguish a number of major zones of terrestrial life, each of which may contain many smaller communities but is dominated by one or two characteristic community types overall. These great life zones are called *biome types*. This tabulation of the major biome types is much simplified, and many communities of great local importance have been left out altogether.

Biome distribution is complicated in large part by mountains, because the tops of mountains have colder climates than the bases. Some tropical mountains have arctic summit climates; in climbing them one passes through first a rain forest, then a habitat resembling a temperate forest (e.g., in India this region would be dominated by rhododendron-like plants), then a zone of coniferous trees, and finally a tundra-like area below the snows or glaciers of the top. The steepness of the slopes and dryness of the air cause vegetation to possess adaptations retarding moisture loss. Montane animals (e.g., Rocky Mountain bighorn sheep or the chamois) may have special adaptations permitting them to clamber about steep slopes. Its topography makes the mountainous habitat ill suited to human settlement for the most part, but it is widely endangered by mining, mineral exploitation and refining, logging, fires, and sometimes by overgrazing.

† "Subclimax" is discussed on page 481.

Organisms are not merely adapted to aspects of their environment in the sense that they must *withstand* them. Many aspects of their surroundings are, after all, *required* by living creatures. Certain minimum amounts of nutrients, light, water, warmth, carbon dioxide, and oxygen are required by most organisms, though the details and proportions of these needs differ among them. It follows from this that living things must receive at least the minimum amount of all their needs in order to survive. To illustrate, an orange tree from Florida could be transplanted to South Carolina in the spring and might thrive throughout the summer. The winter, however, would kill it. Yet the *average* annual temperature of South Carolina is probably within the range of tolerance for a citrus tree. Receiving excess provision of its other needs, such as mineral nutrients or water, could not make up for the fact that the orange tree is unable to tolerate the below-freezing temperatures that prevail

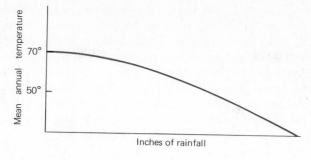

FIGURE 19-4
The interrelationship of temperature and rainfall conditions tolerated by a hypothetical species of tree.

in South Carolina, even if they occur for just a few days or weeks out of the entire year.

The influence of the physical and chemical surroundings of an organism can be modified by the presence of other organisms and their adaptations. In nature, they always are. Consider the shade produced by forest trees, to which many shrubs are specifically adapted. It is an oversimplification to speak in terms of a single limiting factor or a few which govern the distribution of organisms or of the communities which they comprise.

Imagine a species of tree, which if all its other requirements are satisfied, will grow only in a climate whose average annual temperature exceeds 50 degrees Fahrenheit. Since thus far only *one* of the tree's requirements is under discussion, it could be visually represented by a bar graph or even a picture of a thermometer. But it is quite likely that the tree could not withstand as low a temperature if its water needs were not ideally met. There should, in other words, be an interaction between inches of rainfall and low temperature tolerance. This interaction may be represented by a two-dimensional graph (Figure 19-4).

A plant usually requires more water in a highly saline soil than in one with a lower salt content. Salinity, then, is a third factor which may interact with temperature and water supply in governing the occurrence of this species of tree (Figure 19-5). Any point above the graphic surface of Figure 19-5 represents a combination of salinity, rainfall, and temperature which can be tolerated by the tree.

But, as we have already intimated, the *totality* of the environmental influences that impinge upon the organism is much greater than these three. Add to them the need for trace elements in the soil, the intensity of sunlight, the

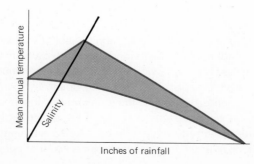

FIGURE 19-5
The interrelationship of temperature, salinity, and rainfall conditions tolerated by a hypothetical
species of tree.

kinds and frequencies of competing organisms, parasites, and predators, the seasonal distribution of rainfall, and much else, and you still will have failed to define completely the life requirements of the species. We do not believe that we exaggerate when we assert that a complete list of all potential limiting factors is probably greater than human imagination can conceive, and the complexity of their possible interaction is beyond human comprehension. A graphic representation of all of them could be constructed only in what a mathematician might call an $n$-dimensional hyperspace. In the end, a complete understanding of ecologic factors affecting the distribution of organisms will probably be approximated only through the use of sophisticated computer techniques, and then only for certain communities of major importance. Such a complete summary of the requirements of an organism virtually defines its life style, or *ecologic niche.*

Limiting factors affect not only the distribution of organisms but also their numbers in particular areas—their population density, in other words. In general, one may imagine organisms increasing to such a density that one of their requirements comes to be in such short supply that no newcomers can be accommodated. Ordinarily, they may be expected to "run out" of one supply before they run out of any others, so it is customary to speak as if there were just one limiting factor upon the population of any organism. Which one it is will be determined by which one happens to be in shortest supply. Returning to the hypothetic example of our species of tree, under some conditions of salinity the limiting factor might be rainfall; under others it might be temperature. The nature and absolute value of the limiting factor would not be the same in all parts of the tree's range of distribution, so in some areas its population density would be greater than in others. The number of a particular kind of organism that can be supported in a given habitat is known as the *carrying capacity.*

### Energetics and ecosystems

Bumper stickers are sometimes seen that bear the question "Have you thanked a green plant today?" You may have wondered why in the world you should! The idea behind the slogan is that green plants (or more specifically, their photosynthesis) form the basis of essentially all life on earth. To review this process briefly, carbon dioxide and water are combined by chloroplasts in green plants. The products of the reaction are carbohydrates, which may be converted by the plant into all other food materials, and oxygen. Even plants must respire. The food they produce reacts with the oxygen they also produce. The relationship between plants and animals, then, is not just one of food and eating, but of energy. Energy produced by the sun is captured by plants and released by animals and also by plants themselves. The living world is driven by thermonuclear solar energy, just as an interstellar spaceship might be driven by thermonuclear power. The photosynthetic *producers,* the *consumers,* the *decomposers,* and their physical environment make up a self-sustaining system requiring only the input of light energy to drive it. Such a self-sustaining biologic system is called an *ecosystem.* It is the life support system of our spacecraft: life itself.

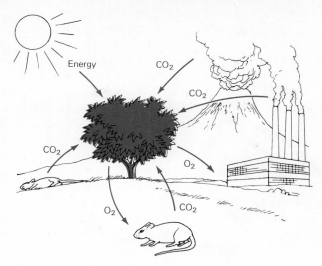

## Ecochemical cycles

As you learned in Chapter 3, organisms require inorganic as well as organic nutrients. The phosphorus of ATP, the magnesium of chlorophyll, the iron of cytochrome or hemoglobin, and the nitrogen of protein are a few examples of mineral elements found in vital compounds required by organisms in their bodies. These cannot be produced from the carbon, hydrogen, and oxygen of photosynthesis. It should be obvious that in a closed ecosystem these inorganic substances must follow cycles from one organism to another, or sooner or later they would become so scarce that life would be impossible.

*The nitrogen cycle.*   Many mineral cycles are poorly understood. Perhaps the most familiar cycle is the nitrogen cycle. The nitrogen cycle is important because several vital classes of biologic compounds, chief among which are protein and nucleic acids, require nitrogen.

As you know, proteins are made up of linked amino acid units known as *residues.* Each amino acid is really a fatty acid bearing an *amino group,* $NH_2$. Neither amino acids nor proteins can be made without nitrogen. For most purposes amino acids may be considered to originate in plants. Plants make them from nitrogen in the soil that has already been brought into chemical combination, that is, *fixed nitrogen.* Most plants require nitrogen in the form of nitrate, $NO_3^-$, although it has recently been shown that many plants absorb enough of the minute quantities of ammonia that are present in the atmosphere to comprise a significant portion of their nitrogen budget. Not only can organisms make proteins from amino acids, but they also can make nucleic acids from them. Without a source of nitrate all this is impossible. It is easy to see that the existence of all life on earth demands that plants have access to nitrate, for without it there could be no proteins and none of the blueprints of life, DNA and RNA.

Nitrate is produced by certain chemosynthetic bacteria known as *ni-*

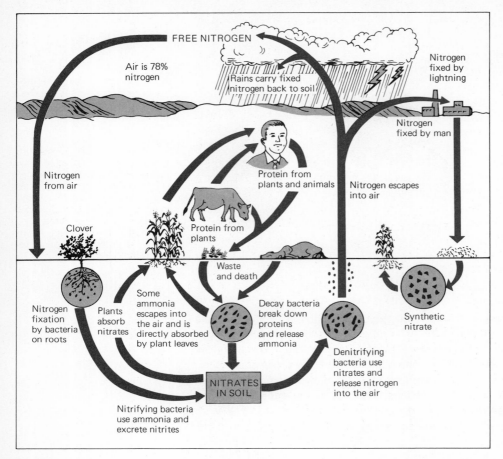

**FIGURE 19-7**

The nitrogen cycle. (After John Y. Moore and Harold L. Slusher, *Biology: A Search for Order in Complexity,* Zondervan, Grand Rapids, Mich., 1970)

Within the figure:

FREE NITROGEN

Air is 78% nitrogen

Rains carry fixed nitrogen back to soil

Nitrogen fixed by lightning

Nitrogen fixed by man

Nitrogen from air

Protein from plants and animals

Nitrogen escapes into air

Clover

Protein from plants

Nitrogen fixation by bacteria on roots

Plants absorb nitrates

Some ammonia escapes into the air and is directly absorbed by plant leaves

Waste and death

Decay bacteria break down proteins and release ammonia

Synthetic nitrate

Denitrifying bacteria use nitrates and release nitrogen into the air

NITRATES IN SOIL

Nitrifying bacteria use ammonia and excrete nitrites

*trifying bacteria,* which oxidize nitrite, $NO_2^-$, to nitrate, $NO_3^-$. The oxidation furnishes them with the energy by which they live. Nitrite, in its turn, is produced from ammonia by other bacteria which oxidize ammonia as their source of energy. Ammonia in *its* turn is produced by still other bacteria which reduce amino acids to ammonia, water, and carbon dioxide—for amino acids are the products of the decay of plant and animal proteins such as result from corpses and feces. Furthermore, a large amount of an animal's protein intake is biologically oxidized in its cells. This involves the removal of the amino group, which then usually is converted either to urea or to uric acid and excreted in the urine. Some bacteria convert urea or uric acid to ammonia.

*Denitrifying bacteria* produce free nitrogen gas from fixed nitrogen as the ultimate result of their work, reversing the action of the nitrifying and nitrogen-fixing bacteria. Denitrifying bacteria are anaerobic. Apparently they live chiefly in the deepest reaches of the soil near the water table. They employ nitrate or nitrite as electron acceptors in place of oxygen. The action of the denitrifying bacteria would produce a net deficit of chemically combined nitrogen in the biosphere were it not for the action of the nitrogen-fixing bacteria.

Most soil nitrate comes not from amino acids but from the direct combination of nitrogen and oxygen, or from the formation of ammonia from nitrogen. Though some of this fixation takes place through the agency of lightning or high-temperature combustion, most of it is produced by *nitrogen-fixing microorganisms* which employ for the purpose a specialized and little-understood enzyme known as *nitrogenase.* The most important of these microorganisms live in special swellings, or *nodules,* on the roots of leguminous plants such as beans or peas. It is estimated that these mutualistic bacteria can fix nitrogen at 100 times the rate at which other, less important nitrogen organisms are able to. Legumes often are planted to introduce ammonia, and ultimately nitrate, into nitrogen-poor soils, but they cannot, of course, make up other deficiencies such as those of phosphorus or potassium. In fact, nitrogenase contains, and therefore requires, some odd trace elements. Even legumes will not grow in areas where these elements are deficient, such as certain parts of Australia.

To sum it up, then, the four main groups of microorganisms involved in the nitrogen cycle perform, respectively, the following functions: (1) produce ammonia from proteins, urea, or uric acid, (2) oxidize ammonia—ultimately to nitrates, (3) defix nitrogen by liberating it from nitrate, and (4) fix nitrogen directly and turn it into nitrate. Ecologists believe that virtually all nitrogen now in the earth's atmosphere has been fixed and liberated many times by these organisms. In light of this, it is remarkable that at any one time, there are probably only a few pounds of the enzyme nitrogenase on the entire planet. Those few pounds have sustained the biosphere for millennia.

Many of the agricultural problems of the world are due to nitrogen-deficient soils. Such deficiencies can be remedied (in a manner of speaking) either by planting legumes and plowing them into the soil, or, far less desirably, by applying nitrogen in the form of ammonia or nitrate that has been manufactured industrially.

Blue-green algae (which, like bacteria, have no nuclear membranes) share with the nitrogen-fixing bacteria the ability to fix nitrogen, using the energy of sunlight to do so. These algae may add considerably to the nitrogen budget of rice paddies, fish ponds, and other aquatic habitats. In the presence of large amounts of phosphorus, such as is provided by sewage effluent, blue-green algae sometimes gain a competitive advantage over the green algae, which cannot fix their own nitrogen. Thus, in the presence of large amounts of phosphorus all algae tend to thrive, but the green algae are limited by the amount of nitrate present whereas the blue-green algae are not. Thus even the removal of all organic matter from sewage is not enough to keep aquatic algae from growing into great concentrations called *blooms,* and in such blooms the blue-greens predominate. These blooms contribute to the undesirable changes of eutrophication (see Chapters 20 and 21) and damage aquatic ecosystems in other ways. For example, they are not very suitable for fish food. The more phosphate that sewage effluent contains, the worse the problem will be.

*Cycling of toxicants.* A number of artificial toxic compounds have been distributed throughout the ecosystem and tend to cycle through it. Among these substances are the "hard" chlorinated hydrocarbon insecticides (better termed *biocides,* since their toxic effects are not limited to insects), the polychlorinated biphenyl (PCB) plasticizers, and some heavy metallic ions such as mercury and lead.

*Biocides.* The widely used chlorinated hydrocarbon biocides such as DDT and dieldrin pose a threat to terrestrial and even aquatic ecosystems, even though most of them are almost completely insoluble in water. Many of these compounds are not readily broken down by the metabolism of normal organisms and are not readily excreted either. They tend to accumulate in body fat. By far most of the food metabolized by organisms is excreted as carbon dioxide, water, and nitrogenous waste products. But the hard pesticides are not excreted, or at least not excreted as readily, even though they are taken in with food. A large portion of them remains in the body of the animal which has consumed them. Thus in a sprayed lawn, the cricket that eats the grass has more biocide than the grass did; the toad that eats crickets has more than the crickets it eats, and so on. An ultimate predator, such as a hawk, concentrates in its body an appreciable portion of the biocides absorbed by many acres of vegetation. This process is called *biologic magnification.*

In many terrestrial communities birds function as ultimate predators. They are perhaps the most sensitive of all vertebrates to the effects of chlorinated hydrocarbon pesticides (especially DDT), which appear to damage their

**FIGURE 19-8**
A toxicant pyramid. Most of the biomass consumed on each trophic level is excreted, but DDT and similar fat-soluble biocides are excreted much less rapidly. Thus, they tend to concentrate progressively in the later members of all food chains, becoming most concentrated and damaging in the ultimate predators.

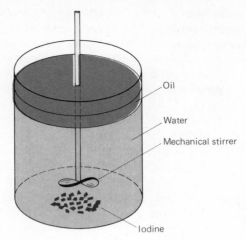

Oil

Water

Mechanical stirrer

Iodine

**FIGURE 19-9**
How a substance almost insoluble in water can become concentrated in another medium.

reproductive physiology in several ways. For example, apparently in part through its interference with $Na^+/K^+$ ATPase, DDT causes eggshell thinning in birds. Some predatory birds saturated with the compound have been observed to lay eggs *without* shells! Even if such eggs could be hatched and the chicks survive, parent birds usually refuse to incubate them and often eat them instead.

The effects of chlorinated hydrocarbon toxicants are not limited to terrestrial communities, even though most of them do not dissolve well in water. To appreciate the transportation of these chlorinated hydrocarbons in the aquatic realm, imagine a simple experiment with iodine, water, and mineral oil. The iodine is placed on the bottom of a vessel, and water is added. On top of the water we place a layer of oil. The iodine is quite insoluble in water but very soluble in oil. It must pass through the water in order to get to the oil. Very soon the water is saturated with iodine, giving it a light yellow color which never deepens. However, as hours pass, the iodine in the water enters the mineral oil, which absorbs it avidly. There is now less iodine in the water, and more dissolves to replace it. Still more is absorbed by the oil. More dissolves in the water. This process will continue until almost all the iodine is dissolved in the oil, even though it had to be dissolved first in the nearly incompatible water.

Like the iodine in our experiment, chlorinated hydrocarbons tend to be soluble in fat, although not in water, and many of them are not chemically destroyed in nature to any appreciable extent. Thus, although they are present in only parts per billion or even trillion in surface water, they accumulate rapidly in the fats and oils of living things in that water. Very significantly, they occur in the oil droplets which help diatoms and other microscopic aquatic plants (phytoplankton) stay afloat in the upper regions of bodies of water—the topmost few feet into which light for photosynthesis can penetrate.

DDT, other chlorinated hydrocarbon pesticides, and the PCB plasticizers have been shown to depress the photosynthesis or growth of some marine phytoplankton when present in amounts similar to or even much below those which now occur in many natural waterways. A significant portion of the photosynthesis of the world is carried out by marine phytoplankton.

It is remarkable how far DDT can be carried from its point of application by wind and by water. Easily detectable quantities of DDT occur even in antarctic organisms. It is possible that there is now no living creature on the face of the earth whose tissues are completely free of at least small quantities of DDT. The long-term effects of such small concentrations are largely unknown at present, but there is some reason to think they may be carcinogenic (cancer-producing) or genetically damaging.

Besides the water route, chlorinated hydrocarbons are transported in several other ways. Scientists have shown that DDT can be blown about in the air or transmitted through the water in particulate form—free or adsorbed on other particles, such as clay. One of the most important means of dispersal is simply that of being blown away in the form of dust. Most DDT applied from the air is lost in this fashion and ends up at points remote from its application. DDT is also somewhat volatile and can be lost and dispersed by evaporative processes. In the end, such chlorinated hydrocarbons which are not absorbed by living organisms probably come to rest through geologic processes of sedimentation. Chlorinated hydrocarbon particles are adsorbed on clay silt particles which accumulate on the bottoms of lakes, rivers, and bays, along with the toxicants they contain. Often these biocides find their way into the bodies of bottom plants or bottom-dwelling animals and reenter the ecosystem in that fashion. We are fortunate, indeed, that the chlorinated hydrocarbon biocides leave the biosphere at all.

Other biocides may be locally important, e.g., in specific instances of heavy fish mortality called "fish kills," but most of them are chemically degraded under natural conditions and therefore do not accumulate as the chlorinated hydrocarbons do. *Polychlorinated biphenyls* are a class of substances added to plastics to improve their molding characteristics. They are also widely used in paints. They behave in much the same way as DDT in global ecosystems and similarly are universally distributed. The ecologic effects of the PCBs are at present unknown but are under strong suspicion. There is reason to believe that mammals such as seals are more easily poisoned by them than are the fish and invertebrates on which they may feed. Furthermore, since predators may be more likely to eat prey sickened by PCBs than healthy prey, PCBs are even more likely to be transmitted upward through food chains.

Not all the harmful effects of biocides (or, for that matter, of other pollutants) are obvious. Some evidence indicates that at far less than lethal concentrations they may disrupt the breeding and danger-avoidance behavior of fishes. Such effects ultimately will reduce the fish population or render it extinct. Though this is little studied, biocides also may affect the behavior of other organisms.

Ironically, perhaps, one of the more serious effects of most biocides arises from the ability of some few organisms to develop resistance to them. For technical reasons, this is more likely to occur in rapidly breeding creatures (e.g., guppies, mosquitoes, or flies) than in those that breed slowly (e.g., eagles, ospreys, or man). Thus it has been observed that a few kinds of rapidly breeding fish seem to develop great resistance to at least some chlorinated hydrocarbons. These fish, when taken from their polluted habitat (such as Southern rice fields and cotton irrigation ditches, which are notoriously loaded with DDT) and placed in an aquarium with normal fish of the same species, may actually excrete enough DDT to kill their nonresistant companions. The more slowly breeding game fish do not develop such resistance and would not be edible if they did.

Pesticide resistance also occurs among terrestrial animals. DDT-resistant flies, mosquitoes,[1] and even mice now occur globally and require such high doses of pesticides to kill them that almost everything else in their habitat is exterminated, including the natural enemies of the pests. These enemies are more vulnerable to DDT, both because their populations are small and because of biologic concentration of the toxicant. Outbreaks of damaging plant mites have been traced to the elimination of predatory insects by the DDT that was intended to control another pest. DDT and almost all other pesticides actually can *create* very serious pest problems in this way. The development of such resistance is not limited to DDT, but occurs with almost all pesticides. Even rats resistant to the rodenticide warfarin have been discovered.

Though many nations have outlawed or are in the process of outlawing many "hard" (nondegradable) pesticides such as DDT, tremendous damage already has been done, and the elimination of these biocides is proceeding

---

[1] Several species of mosquitoes in California are now resistant to practically all appropriate pesticides.

**FIGURE 19-10**
Biologic control of the imported cabbage worm. *A. panteles* wasps are released in cabbage plots. These parasitoids lay their eggs in the worms, which, upon hatching into larva, devour the pests alive. (USDA)

**Table 19-2. Some pest control alternatives**

| Method | Procedures employed | Advantages | Disadvantages |
|--------|---------------------|------------|---------------|
| Resistant crop plants | Genes conferring resistance to pest organisms or diseases are bred into cultigens, which are domestic plants and animals. | When effective, cost is low and nothing need be done to the pest organism. Use of chemicals is avoided. | Pests tend to develop strains capable of attacking resistant crops. Developing new resistant strains to counter this may be slow—10 to 15 years is the average required time. |
| Natural enemies | Parasites and predators of the pest insect are discovered, artificially bred, and released into the pest population. They may be left to maintain themselves, or continued artificial saturation of the pest may be practiced. | Once established, natural enemies may require no further attention, and to a degree, their population automatically adjusts itself to the prevalence of the pest insects. Use of chemicals is avoided. | The relationship of pest to enemy population is cyclic in such a way that there is often unacceptable damage to crops before the pest is controlled by its enemies. The search for natural enemies may be very long, expensive, or unsuccessful, and sometimes a natural enemy becomes a pest in its own right. Usually, enemies only destroy a small portion of the pest population and are themselves vulnerable to the application of pesticides. |
| Bacteria and viruses | Bacteria and viruses are artificially bred and applied to pests somewhat as insecticides are. A variant is to isolate the bacterial toxins and spray them on the pests. | If well chosen, these disease organisms will infect *only* the target pest; they can be very specific. To some degree, they may persist in the pest population, and sometimes may render other control unnecessary. Use of chemicals is avoided. | Culture methods and dispersal techniques are not as yet well understood. Theoretically, pests may develop resistance to these agents as they do to conventional insecticides. |
| Pheromones and attractants | Sexual attractants or odor of host plant is manufactured artificially and usually used as bait for traps or to attract specific insect to poison. In some cases may be used to disrupt reproductive behavior or to interfere with the ability of males to locate females. | Since these substances attract the target pest, they are very specific and can seek out insects even when their population is as yet sparse. Minimize the use of biocides. | Although future research will probably overcome practical difficulties, the effects of these compounds on insect behavior are subtle, complex, and sometimes negligible. |
| Synthetic hormones | Insect hormones, especially "juvenile hormone," are manufactured artificially and applied as insecticides are, with precise timing. They disrupt life cycles of insects and in some cases may kill them outright. | Insects cannot possibly develop resistance to their own hormones, and it is hoped these hormones will not affect other organisms. This method can be made very specific for the target pest by careful scheduling of application, to coincide with vulnerable parts of the life cycle. | As yet these compounds are rather untested and may not be as harmless to other organisms as is presently believed. Despite this, the technique is regarded as promising. |

**Table 19-2.** (*Continued*)

| Method | Procedures employed | Advantages | Disadvantages |
|---|---|---|---|
| Sterile male release | Large numbers of male insects are reared artificially and sterilized by radiation or chemicals. Released into the pest population they compete with fertile males, causing the females with which they mate to lay nonviable eggs. | Complete eradication of a pest species can sometimes be achieved by this method. Use of chemicals avoided. | Most effective at *low* population densities, and in situations where the pest population is geographically isolated. Not practical with all insects, especially with those that accept multiple mates. Sterilization procedures can reduce the mating ability of male insects. Insect populations can develop resistance to this method. |
| Genetic manipulation | Strains of a pest organism are artificially bred for genetic characteristics which confer *susceptibility* to other control methods (i.e., pesticides) or which reduce the noxious qualities of the pest. These are then released in large numbers into the pest population. | Sole known technique whereby insecticide resistance may be reversed. Has potential of permitting the use of minimum quantities of biocides or of special biocides to which the genetically manipulated pest only is susceptible. | Large numbers of pest organisms would probably have to be released, and the release would probably have to be frequently repeated. Theoretically it would seem difficult to maintain the desired genes in the pest population, except perhaps in a very few species. Furthermore, this technique is completely untested. |
| Cultural control | Insect life cycles are interrupted by crop rotation or sanitary measures; crops are interspersed to slow the spread of pests, or are grown under conditions or at seasons unfavorable to them. In sum, one chooses farming methods uncongenial to pests. | Involves the minimum disturbance to the environment and may be cheap, since in some instances it would require only minor modifications of existing techniques of crop culture. | Some techniques consume large amounts of labor or may be otherwise uneconomical or impractical under local conditions. Requires sophisticated knowledge not widespread, especially in underdeveloped countries. |
| Integrated pest control | All appropriate control measures are combined on an individual prescription basis. | Maximum economic benefit is obtained with minimum ecologic damage. | Since the control program must be tailored to pest, crop variety, and local conditions on an individual prescription basis, large amounts of the most skilled manpower are required, both for planning and for application. A danger is the possible failure to take broader ecologic considerations into account and to sacrifice the environment as a whole for the sake of short-term economic gains. |

so slowly that even more damage is likely to occur. Furthermore, as long as such toxicants are used in countries with an unenlightened environmental policy, they may be distributed worldwide by natural processes, even to the countries in which their use is illegal.

It is true that the problems connected with persistent pesticides are reduced to some extent when degradable compounds such as the carbamates or organophosphates (which break down in nature) are substituted.

Unfortunately, however, there is no known pesticide that is free from undesirable ecologic consequences. The carbamate Sevin, for example, is sudden death for bees, and bees are vital for pollination of food and wild plants.[2] Since all pesticides kill more than the pests against which they are directed, in the end our pesticide problems can be *solved* only if we can *avoid using them*. We can avoid using them, to be realistic, only if we are content to produce less food. We can produce less food only if we have fewer mouths to feed. In the end, many environmental problems seem to lead us by straight or devious paths back to the population question (Chapter 24).

There are some alternatives to the use of pesticides short of complete abandonment of our crops to the insect world, though none of them will permit as high a level of agricultural production as our overpopulated world demands. Nevertheless, their wide use could reduce our pesticide problems. Table 19-2 presents a brief summary of some of these alternatives.

*Heavy metals.* Recently we have come to realize that chemically heavy metals such as mercury, cadmium, and lead may be dangerous environmental contaminants. Mercury, for example, has been released into aquatic environments by a variety of industrial processes for a number of years. Virtually all such mercury appears to enter into certain organic compounds through the agencies of bacteria and other organisms, a process called *methylation.* Methylated mercury is concentrated along food chains in somewhat the same way as the chlorinated hydrocarbon pesticides are. In toxic doses it produces unpleasant, crippling, or fatal neurologic disorders. Most cases of human mercury poisoning in America have resulted from the misuse of mercury-containing fungicidal seed dressings, but considerable human poisoning resulting from biologic magnification has occurred in the vicinity of the town of Minamata, Japan. A number of persons were sickened there as a result of eating fish and other seafood contaminated by mercury discharged from a plastic-manufacturing plant. As yet the full dimensions of

---

[2] We must distinguish between the acute and chronic toxicity of biocides. We can scarcely improve upon the definitions of Carl Gustafson, who distinguishes between

> . . . acute toxicity, which is immediately evident because of a high death rate or other maiming effect; and chronic toxicity, which is the result of a slow accumulation of the poison in the body and is a sublethal effect. Whenever acute toxicity is high, a pollutant will be readily recognized and appropriate remedial actions will be introduced. If on the other hand, acute toxicity is low, the physiological effects will go unnoticed until the chronic effects make themselves evident. When this happens, the problem may have advanced beyond immediate or easy correction. (*Environmental Science and Technology,* October, 1970, p. 814.)

Cigarette smoking affords a familiar example of chronic toxicity. No one expects to die as a result of smoking a single cigarette, but over a long period of time the habit is definitely harmful and greatly enhances one's chances of an unpleasant end. DDT, for instance, has about the *direct* toxicity of aspirin, as has been amply demonstrated by people wishing to prove it "harmless." One California pest control operator has been taking it daily in pill form. The long-term effects of persistent pesticides are ecologically damaging, however, and have nearly rendered some species extinct in the wild. Even the pest control operator may find DDT very damaging in the long run.

The organophosphates and carbamates, on the other hand, are far more acutely toxic. A single drop of Parathion on the skin is sufficient to kill a man. It is as toxic as many anticholinesterase nerve gases, and was in fact originally intended to be used as a nerve gas. However, though anticholinesterases are very dangerous to agricultural workers and wildlife that are directly exposed to them during their storage, mixing, or application, they are quickly broken down by microorganisms and other agencies, and so are not as subject to biologic magnification as are the chlorinated hydrocarbons.

the mercury problem are not exactly known. Nevertheless, in several countries, including the United States, the industrial discharge of mercury has been sharply curtailed. Apparently mercury has always been present in ultimate predator fish in near-dangerous amounts. Museum specimens of fish captured in the nineteenth century exhibit mercury content comparable to that found in similar fish today. Even a slight increase in mercury contamination of waters could lead quickly to fish that may be toxic to eat.

Another heavy metal, lead, has been an environmental contaminant through most of recorded history. Today lead apparently enters our bodies chiefly from the air, into which it is exhausted by automobiles. Highway tollbooth guards, among others, may have particularly high blood concentrations because of their occupational exposure. Little is absorbed by our food, so biologic concentration of lead is no great danger to us, although it may be for other organisms. The obvious way to reduce lead intake is to remove it from automobile gasoline, and government pressure is being exerted against refiners to do just that. Lead spoils the catalyst in the afterburners used on automobiles to control air pollution. For this reason automobile manufacturers, now under legislative mandate to reduce automotive pollution, are trying to influence gasoline manufacturers to remove lead from gasolines.

Accidental lead poisoning is definitely a serious environmental problem, particularly (though not exclusively) in slum and ghetto areas. Many children display the behavior known as *pica*—the craving for and eating of nonfood substances such as dirt, soap powder, or paper. If their perverted diet contains chips of lead paint from deteriorating surfaces, death may result, or their nervous systems may be seriously damaged. Simply covering lead paint in an old dwelling with a modern nontoxic product (and not all modern paints are nontoxic) does not stop the chipping and does not solve the problem. The lead-containing paint must be removed.

*Radioactive substances.* Pollution of this type arises from nuclear power plant waste cooling water, wastes from the mining, refinement, and milling of radioactive fuel elements, certain industrial processes, reprocessing of spent nuclear fuel, fallout from radioactive air pollution arising from the explosion of nuclear and thermonuclear weapons, and stray gamma radiation from such sources as dental x-ray machines and even color television sets. In addition, there is some natural background radiation from cosmic rays and minerals. Microscopic water plants may concentrate some *radionuclides* (radioactive elements) 700- to 1,200-fold (along with their stable isotopes). The arctic lichen called reindeer moss is even more efficient. Further biologic magnification may take place in the ecosystem, as with persistent pesticides. (In fact, biologic magnification was first demonstrated with radioactive pollution arising from the early Pacific Ocean nuclear tests.) Radioactive substances may be lethal, sterilizing, cancer-producing, or mutation-producing, depending on degree of exposure.

Speaking cautiously, it appears that pollution of this type will not be significant in the *near* future unless we have nuclear conflict, serious power plant accidents, or the widespread testing of nuclear weapons in the atmosphere. Pollution of other kinds is much more likely to cause the disintegration of much of our environment long before pollution from radioactive substances is significant. For example, the potential thermal pollution from nuclear power plants is ordinarily a much more serious threat than any radionuclides they may release into the environment.

On the other hand, radioactive wastes from nuclear fuel reprocessing constitute a serious *potential* hazard for future generations. Some of these wastes will still be dangerous hundreds of years from now. Let us hope that our descendants will be civilized, wealthy, and conscientious enough to maintain the tanks and caverns full of radioactive wastes that we are bequeathing to them. Perhaps the advent of thermonuclear (fusion) power, if it is developed, will spare them some of that worry—providing that *it* does not produce radioactive waste, as it may.

## THE PASSENGERS AND THEIR QUARTERS

Thus far we have mainly discussed some details of the function and malfunction of the life-support system of our spaceship earth, and have given rather little attention to the life that the passengers must lead on board, except for a brief summary of the principal biomes. Now it is time to examine the internal economy of the communities and interrelationships of the primary producers, consumers, and decomposers that make them up.

### The community concept

As we have seen, the basis for all life on earth is the energy of sunlight trapped by chlorophyll and bound up in the chemicals we eat. It follows from this that the distribution and abundance of all life are governed by green plants. Almost every part of an animal serves its need to eat in some way. Animals are usually precisely adapted to their food organisms, whether these are plants or other animals. Thus, specialized animals usually accompany their particular host plants. Primary consumers, together with their plants and the predators that in turn feed upon them, comprise a *community*—a definite, fairly stable association of organisms bound together by a set of ecologic relationships with one another and with their common physical environment.

### Succession

Communities do not ordinarily spring into existence fullblown, in their final form. Generally, the community that first comes to occupy a given habitat is quite different from the final community found there. Communities typically undergo a series of maturational changes, known as *succession,* before the final steady state is reached. Succession may be more formally defined as the serial replacement of a community by another until a final state is reached.

***An example of succession.*** In his fascinating book, *The Molds and Man*

(University of Minnesota Press, 1965), C. M. Christensen writes:

> Many practical grain men once believed, and a lot of them still believe, that grain has "an urge to heat and germinate in the spring." That is, about the time that corn and wheat are being planted in the fields the grain in its dark storage bin feels the eternal call of spring, begins to breathe heavily, like an ardent bull, and sweats and heats. Granted that stored grain often heats and spoils in the spring, this romantic explanation of the process is complete nonsense; the same heating occurs in bins of turkey feed, baled cotton and wool, and manure piles, materials that by no possible stretch of the imagination could feel the mysterious call of spring, or respond to it if they did feel it.
>
> The answer to the heating problems when insects are absent, and often indeed when they are present but when the temperature goes above that endurable by the insects, was found to be in the storage fungi. Some of these fungi, mainly in the *Aspergillus glaucus* group, will invade grain slowly when the grain has a moisture content of only 14–15 per cent. They do not grow rapidly enough to produce any detectable heating at that moisture content, but they produce moisture as one of their products of growth, and so increase the moisture content of the grain. Once the moisture content is increased to 15–16 per cent, *Aspergillus candidus* and *A. ochraceus* take over. They grow more rapidly, cause some rise in temperature, and release more water. Once the moisture content of the grain reaches about 18 per cent, *A. candidus* grows very rapidly, and is joined by *A. flavus,* which also grows rapidly. These two species like it hot and humid, and *A. candidus* can raise the temperature up to about 130 degrees Fahrenheit and hold it there for weeks, by which time the grain is a hot and stinking mess.
>
> At that point some of the ever present thermophilic or heat-loving bacteria take over, and given conditions for rapid growth, they can raise the temperature to about 170 degrees Fahrenheit. These bacteria cannot survive temperatures much higher than this, but they produce materials that can undergo purely chemical oxidation, and this may carry the temperature up to the point of spontaneous combustion.

In a rather simple fashion, this passage illustrates the phenomenon of *ecologic succession.*[3] Notice that each community *produced* the conditions that permitted the establishment of the next; and that the successor community was better adapted to the changed conditions than the predecessor community which produced them and which it replaced. The stages through which a particular community develops are known collectively as a *sere.*

**Varieties of succession.** Very complex successions of communities of organisms also occur in nature. Though in principle all successions are similar, we may distinguish two main varieties: primary and secondary.

---

[3] We must warn you that, though it is useful as an illustration, this example is typical only of *microsuccessions* in nature—the kind that might take place in a rotting log or a bit of dung. The community mentioned was composed exclusively of decomposing organisms, and since the community contained no primary producers, the climax could not possibly be self-sustaining. Furthermore, only a few species made up the community—a situation not likely in nature. Nonetheless, it is a good example of the way in which communities actually prepare the habitat for their successors.

*Primary succession.* The seres of a primary succession take place in newly formed habitats previously unoccupied by life. For example, in 1883 the oceanic volcano Krakatoa, located in what is now Indonesia, erupted (or perhaps we should say "exploded") with the force of several nuclear bombs. Dust ejected by its convulsions temporarily modified world climate and reddened sunsets everywhere.

The volcano left a small lava island behind. Its surface was scoured clean of life. Over the years, however, both animals and plants have colonized the bare lava. For the most part they have been small, light creatures such as spiders that could be blown there by the wind, or flyers such as birds, or plants whose seeds are dispersed by wind or seawater. However there was an additional immigration requirement: the initial plants had to be capable of growing on bare ash or lava. Only after the original colonizers had by their life and death produced some topsoil could new arrivals requiring it prosper. Animals, too, were subject to the same type of limitation. Though the climate was wet and tropical, jungle dwellers could not settle there in the absence of a jungle. Presumably there will be a jungle there some day, but it can develop only by a series of steps in which one community by its life processes prepares the habitat for the following community.

Primary succession on bare rock is generally begun in temperate climates by lichens. These compound organisms have a fungus component which can extract minerals even from bare rock by means of acidic secretions. However, without its algal component the fungus, ordinarily a decomposing organism, would starve for lack of organic food. Lichens need no topsoil; they produce it by their action upon rock.

Moss will germinate and grow in the soil that the lichens produce. But moss grows much faster than lichens and eventually crowds its benefactors out. After perhaps hundreds of years moss produces an accumulation of rotted organic matter called *humus* from its decomposing parts, which, together with the particles of rock loosened by the action of the plants, forms enough soil to allow ferns or grass to invade the habitat. Eventually, after a succession of many communities, a *climax community* will occur. A climax community is one that has no successor. As long as environmental conditions remain the same, the climax community will, in theory, remain unchanged.[4]

---

[4] As a matter of fact, no one can say for sure that a particular community will never change. Even climax communities are subject, in theory, to evolutionary change and may experience successional changes on a time scale too vast for human perception. A number of varieties of climaxes are recognized by ecologists. For example, an *edaphic* climax is produced by special conditions of soil and moisture, and differs from that characteristic of the area as a whole. *Preclimax* and *postclimax* communities are not successional stages, as you might think, but are local variations produced by local differences in exposure, such as whether a hill slope is northerly or southerly. Some ecologists believe that in time all seres in a given area converge to a single type of climax. Others dispute this. No one has lived long enough to be sure. There is even a school of thought that denies the reality of the very concept of a climax community. Those who hold this view state that even in established climax communities, individual organisms are always dying, and that they are not usually immediately replaced by identical substitutes. Instead, a local succession takes place in the spot vacated by their death. Furthermore, a few instances are known of unstable successions in which the terminal community dies off periodically, each time being replaced by an earlier successional stage.

FIGURE 19-11
A cypress "hammock," an example of an edaphic climax. The normal climax vegetation for much of Florida is hardwood forest, perhaps oak or magnolia. However, in localized swampy areas, such as old sinkholes, a localized community dominated by cypress tends to develop and persist. These hammocks are believed to be true climax communities, although they are not typical of the climax community that tends to develop in this area.

*Secondary succession.* Sometimes a habitat may be partially destroyed, for example by completely destructive logging, fire, or bulldozing. In that case, it is believed that if the habitat is left undisturbed the same climax will be attained eventually as in primary succession, but the subclimax communities leading to it may differ from those observed in primary succession, since the sere develops on soil different from that of the earlier stages of primary succession.

In parts of Florida, a severe forest fire will produce conditions favorable for the growth of certain tall leguminous weeds. These in turn permit the germination and growth of palmetto grass, cabbage palm, and yellow pine, more or less in that order. If the habitat is undisturbed (which it seldom is), a climax community of hardwoods, such as laurel oak, or magnolia, may be expected to develop. Today most Eastern American forest communities are decidedly subclimax, because of burning or logging. All agricultural land may, of course, be considered subclimax, since if left to itself it would swiftly harbor a very different community. Ask any gardener who has ever pulled a weed!

A point not to be overlooked is that the entire community changes in the course of succession, not just the dominant plants. A few species of organisms may stay with the community over its entire history, but this is exceptional. As communities approach their climax stages, their component animals and plants not only change but also increase in diversity, so that the climax community typically contains not only different but more species than did its predecessors.

*Implications for conservation.* It is some comfort to reflect that, given time, man's devastation of the natural landscape may be repaired by successional processes. However, certain man-caused changes seem irrepara-

**FIGURE 19-12**
Secondary ecologic succession. The evergreen forest was burned some years ago. Deciduous trees now occupy this area, their yellow leaves contrasting with the surrounding gymnosperms in the fall. (Dr. G. H. Ketchledge)

ble—for example, the extreme soil destruction of moist tropical habitats, called *laterization,* and the permanent poisoning of the soil with heavy metals from smelting. Even heavy erosion may cause irreparable damage. The majestic biblical cedars of Lebanon occur only in scattered sheltered areas today. In more than 2,000 years that climax community, mostly because of erosion resulting from overgrazing, has not reestablished itself. Parts of the Badlands of South Dakota afford an example closer to home. Here, unwise agricultural development led to erosion that rendered the land almost completely sterile.

Even in less severe situations most climax communities cannot be expected to redevelop in less than several hundreds of years. Will we live to see it? A completely logged Southern hardwood forest, cut at the time of the American Revolution, would only now be approaching its original condition, and then only if it had been left completely undisturbed all that time. On the time scale of the individual human life, most such destruction is irreversible.

Some climax species, such as Douglas fir,[5] do not require an elaborate secondary succession in order to recolonize a logged-off area immediately. Others, by far the majority, do require the repreparation of their habitat by subclimax communities. Such forests will not regenerate immediately after logging unless substantial numbers of mature trees are allowed to remain standing, for their seedlings require shade and other conditions that only their original community can provide.

*Artificially maintained subclimaxes.* Particular kinds of wildlife are characteristic of particular seral stages. Earlier in the century Pennsylvania was restocked with deer to replace those originally there which had been hunted

[5] Douglas fir may not be truly a climax species.

almost to extinction. The deer prospered and now are present in such numbers in the mountains of Pennsylvania and New York as to be a hazard to motorists. Most of the forests in those states have been subclimax mixtures that are ideal deer habitat; deer do not do well in the climax forests that are eventually characteristic of that area. As forests mature, there is a tendency in some parts of Pennsylvania for deer herds to decrease. Some sportsmen demand logging to restore their deer, but farmers would be glad to see fewer deer, for they can be annoying agricultural pests. It seems clear that there is a need for agreement on the kind of wildlife desired in a given area in order to formulate a sound conservation policy. The way in which we manage our wildlands obviously will depend in part upon whether we wish to have climax or subclimax species living on them.

Subclimax communities may be maintained artificially for reasons having nothing to do with wildlife. Much Southern longleaf or yellow pine forest is a subclimax community that would, if left alone, eventually be replaced by oak or other hardwood climax. However the pine forests are more valuable for pulp, timber, naval stores, and even grazing than the hardwood community

**FIGURE 19-13**
Succession following open pit phosphate mining in Florida. Occasionally in such cases natural communities may reestablish themselves in time.

would be. Periodic controlled burning of these pine woods destroys hard-wood seedlings but not those of pine, whose terminal buds are more resistant to fire.

*Disturbance climaxes.* Sometimes the presence of alien species makes the attainment of the original climax impossible. For example, in California and some other parts of the American Southwest the dominant community is the *chaparral,* a scrubby, bushy miniature forest that develops under some dry conditions. Several nonnative plants have invaded the area and produced a stable climax which is not at all the same as the original chapparal or grassland that it replaced. Such a climax is called a *disturbance climax.* It is not always easy to recognize disturbance climaxes. Rabbits probably were brought to Great Britain in ancient times. Since rabbits are serious agricultural pests, farmers were happy to see them dying recently from an introduced disease, myxomatosis. With the rabbits gone, quite a number of communities thought to be at the climax stage have begun to develop further. By eating seedlings the rabbits evidently had maintained these communities in a subclimax condition for generations.

**CHAPTER SUMMARY**

1 Ecology is the study of the interactions of organisms with one another and with their environment.

2 Each organism is precisely adapted to life in the community of which it is a member.

3 Organisms must at all times receive at least the minimum provision of their needs. This fact largely accounts for the limitation of both the distribution of species and the densities of populations.

4 An ecosystem is a fairly self-sufficient group of organisms. Producers, decomposers, and consumers interacting with one another and with the nonliving environment constitute the ecosystem.

5 Carbon, oxygen, and various minerals behave cyclically in ecosystems and are continually exchanged among their component organisms.

6 Toxic and radioactive substances also are cycled throughout the biosphere, producing actual or potential damage in all parts of the globe.

7 A community is an integrated group of interdependent organisms living together.

8 Ecologic succession is the serial replacement of one community by another until a final unchanging state is reached. This final state is called the climax community.

9 The exact successional stages and climax attained by communities differ according to their environment. All such stages are known as seres.

10 Ecologists recognize two main types of succession, primary and secondary.

11 The total biomass and species diversity of a climax community are greater than those of any of its predecessors. Desirable wildlife or timber harvest or other use is sometimes associated with subclimax communities, which may therefore be deliberately maintained in that state.

12 Exotic species may form a permanent disturbance climax which prevents the reestablishment of the original climax community.

Buchsbaum, Ralph, and M. Buchsbaum: *Basic Ecology,* Boxwood Press, Pittsburgh, 1957.

Editors of *Scientific American: The Biosphere,* Freeman, San Francisco, 1970.

Garcia, Richard: "The Control of Malaria," *Environment,* June 1972, p. 2.

Holcomb, Robert W.: "Insect Control: Alternatives to the Use of Conventional Pesticides," *Science,* Apr. 24, 1970, p. 456.

Huffaker, C. B.: "Life against Life—Nature's Pest Control Scheme," *Environmental Research,* vol. 3, 1970, p. 162.

Knight, Clifford B.: *Basic Concepts of Ecology,* Holt, New York, 1963.

Nicholson, H. Page: "Pesticide Pollution Control," *Science,* Nov. 17, 1967, p. 871.

Novick, Sheldon: *The Careless Atom,* Houghton Mifflin, Boston, 1969.

Odum, Eugene P., and Howard T. Odum: *Fundamentals of Ecology,* 3d ed., Saunders, Philadelphia, 1971.

Snow, Joel: "Radioactive Waste from Reactors, the Problem That Won't Go Away," *Scientist and Citizen,* May 1967, p. 89.

Whittaker, Robert H.: *Communities and Ecosystems,* Macmillan, New York, 1970.

**FOR FURTHER READING**

# 20

## AQUATIC ECOLOGY

*Chapter learning objective.* The student must be able to interpret the structure and function of aquatic communities in ecologic terms, to summarize the characteristics and component organisms of each major community, and to summarize the principal dangers to the aquatic habitat, proposing means of their conservation.

*ILO 1* Describe at least three ways in which the aquatic habitat differs significantly from terrestrial habitats.

*ILO 2* Describe a planktonic community, summarizing the food-web relationships, their zones of greatest productivity, and the importance of plankton to man and other organisms.

*ILO 3* List several types of organism that belong to the nekton, and explain why they are particularly susceptible to human predation.

*ILO 4* Describe the benthos, and explain its importance in the aquatic ecologic system.

*ILO 5* Summarize the characteristics of an estuarine habitat, contrasting it to other aquatic areas such as the marine habitat, giving its ecologic role and importance.

*ILO 6* Describe ecologic succession in an aquatic environment.

*ILO 7* Describe the principal threats to aquatic communities, estimate the ecologic and economic importance and implications of these threats, and propose measures for the conservation, farming, and rational use of biotic aquatic resources.

To the casual observer, the aquatic habitat looks much the same everywhere, because its watery covering obscures it from our view. Nevertheless, beneath and within this obscuring interface is a wide variety of communities and habitats, comparable in diversity to the terrestrial ecosystems.

## PROPERTIES OF WATER

The life found in water is obviously very different from that on land. The reasons for the differences can be appreciated when we consider how greatly the properties of water differ from those of air.

For example, carbon dioxide is so highly soluble in water that it tends to become highly concentrated in solution. It is sometimes difficult for aquatic creatures to rid themselves of this gas. Oxygen, on the other hand, dissolves less readily in water and is never present in abundance. Thus there are no really warm-blooded animals[1] that breathe water, for not enough oxygen is available to support necessary organic combustion.

On the other hand, aquatic animals can excrete the nitrogenous products of protein metabolism more readily than terrestrial organisms. The parent substance of urea and uric acid is ammonia. Though toxic, ammonia is so soluble and so readily washed away by water that aquatic animals can excrete it directly. This is advantageous to them, because it requires the expenditure of energy to make urea or uric acid, whereas making ammonia actually yields one ATP per molecule.

Water is a remarkable solvent. Almost all polar or ionic compounds, including many of the more common salts, readily dissolve in it. Chlorides, some carbonates, phosphates, sulfates, nitrates, calcium, sodium, potassium, aluminum, and ammonium are found in most waters that have not been specially purified. Rainwater leaches substances from rock and soil and carries them far from their origin. In the end, almost all surface water finds its way to the sea. Evaporating, it eventually becomes rainwater again and repeats the leaching process if it falls upon land. As Solomon wrote, "All the rivers run into the sea; yet the sea is not full; unto the place from whence the rivers come, thither they return again" (Ecclesiastes 1:7). The salts and most other dissolved substances that are carried to the sea do not evaporate but have accumulated there throughout the history of the earth. Even within freshwaters and within the seas, the salinity varies greatly, contributing to the diversity of the aquatic habitat.

Water also can transport substances that are not soluble in it. Some rivers are capable of rolling quite large boulders along their beds, and all watercourses transport considerable quantities of silt and suspended organic matter. Bottom-dwelling aquatic organisms burrow in deposited silt, fight perpetually against burial by it, and even feed upon it.

Water has the highest specific heat of any common substance. That is, it

---

[1] A few species of very fast-swimming fish do maintain body temperatures somewhat higher than that of the surrounding water.

takes a great deal of heat to make water change its temperature; thus it is slow to warm or cool. Consequently, the temperature of aquatic habitats tends to be more constant than that of terrestrial areas. In fact, a nearby large body of water tends to moderate the climate of adjacent lands.

Water is among the few substances that are lighter in their solid than in their liquid form. Ice, therefore, rises to the top of a body of water and floats upon it, rather than sinking to the bottom as might be expected. Were this not the case, some seas and many lakes would be filled with ice to their bottoms, with only a thin layer of liquid water lying on the surface, melted annually by the returning summer. As it is, the tendency of ice to float on the surface permits considerable life to exist beneath the pack ice of polar waters or the annually frozen lakes and rivers of temperate climates.

Water is denser than air. The specific gravities of many aquatic animals are almost the same as that of the water around them. Thus they tend to be buoyed up by it, which permits either a reduction in heavy skeletal structures or growth to great size. Aquatic animals of all phyla are able to attain much greater size in the aquatic realm. The blue whale, for instance, is the largest animal known to have lived at any time.

Water develops a high *surface tension,* which behaves like a thin skin on the surface of any body of water—from a cupful to an oceanful. The surface tension results from the mutual attraction of its component molecules to one another at the air-water interface. Perhaps you have noticed that some objects heavier than water can "float" upon this surface tension. So long as they do not break through it, a needle or even a razor blade can be supported by the surface film if it is placed upon it with care. Some animals, such as the water strider insect, are able to support themselves on the surface film, for they are specifically adapted to this strange life style by

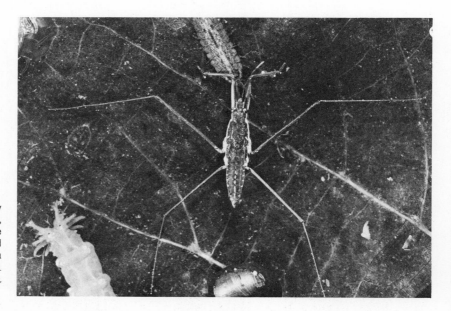

**FIGURE 20-1**
Some freshwater community members. The water strider, *Gerris,* supports itself on surface film. Also shown is the head of a horsefly larva (left) and a water snail in lower center. (C. P. Hickman, *Integrated Principles of Zoology,* 3d ed., Mosby, St. Louis, 1966)

special hairs on their feet which repel water. Many species of algae also tend to collect on the underside of the surface film.

Last, but very significant, water does not transmit light nearly as well as air does. In part, this is a result of the suspended matter universally occurring in natural waters, but it is also a characteristic, to some extent, of the pure substance. Consequently, active photosynthetic organisms can occur only within about 30 meters of the surface, at the deepest. Furthermore not all wavelengths of light are absorbed to equal extents. The red end of the spectrum is absorbed rapidly, so that even in shallow depths light tends to be blue or green in color. These colors are least effective in photosynthesis, for which accessory photosynthetic pigments are only a partial compensation. In all, the aquatic habitat seems singularly unsuited to photosynthesis. Nevertheless, a very large portion of the world's photosynthesis occurs there—in the topmost few feet or sometimes even inches.

The aquatic habitat is a threatened one. Despite vigorous, sometimes even frantic attempts by governmental bodies and conservation organizations to reverse the trend, pollution and other insults to our water resources are rapidly overwhelming their living communities. It is our hope that this discussion of the aquatic habitat will help to suggest not only a cause for action but also the form of action needed to preserve a reasonable facsimile of King Neptune's realm into the twenty-first century.

## AQUATIC COMMUNITIES

It is traditional to divide the aquatic habitat into the major categories of freshwater and marine. Naturally, these habitats have a great deal in common, though in general, the freshwater habitat is subject to greater environmental fluctuations and is in general more stressful to the physiology of its inhabitants. The organisms that occupy these habitats may be divided into the categories of *plankton, nekton,* and *benthos.* Their nature and distribution are influenced by the restrictions of aquatic photosynthesis, upon which they are, for the most part, dependent.

### The plankton

The plankton are organisms which drift free in the water and which depend for motion mostly upon the gross movements of the water around them. It is not that planktonic organisms are incapable of locomotion (although some are), but that their ability to swim is comparatively feeble. Many planktonic organisms are nevertheless capable of extensive vertical migration, either daily or seasonally.

Most plankters (as the individual organisms are called) are small, and the vast majority are microscopic. Nevertheless, some quite large plankters occur. A number of large mollusks (such as squid), coelenterates, and other jellyfish-like creatures, together with certain medium-sized shrimplike crustaceans, are a part of the plankton.

The planktonic organisms are divided into *zooplankton* (animal forms) and *phytoplankton* (tiny plants on which the animal forms are dependent).

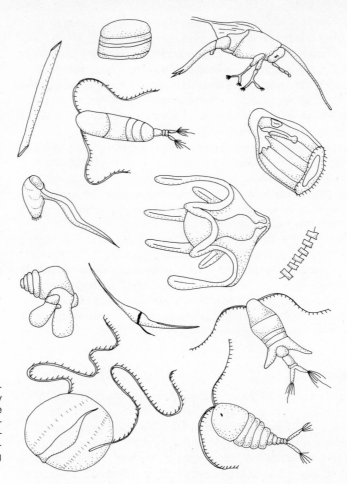

**FIGURE 20-2**
Some representative zooplankton. For the most part these tiny animals feed upon the algae that compromise the phytoplankton, but some are carnivorous. All are adapted to a perpetual struggle against sinking to the bottom.

These photosynthetic primary producers are, for the most part, unicellular or colonial green algae (especially diatoms). Blue-green "algae" also occur, especially in polluted situations. Much as in terrestrial communities, the microscopic primary producers are eaten by primary consumers. In general, these primary consumers are tiny crustaceans called copepods, which may, in turn, be preyed upon by carnivorous copepods or by others of the large plankters.

The familiar pyramidal relationships of biomass and numbers occur in the planktonic community. The phytoplankton is rather sparsely distributed, by terrestrial standards—as compared, for example, with the plants of a grassland. The various consuming organisms are even rarer. The uppermost members occur sometimes at frequencies of one to two per several thousand cubic yards of near-surface water. Such rarity of prey has led to fantastic adaptations on the part of many of the larger planktonic predators. They simply cannot afford to miss any of the rare opportunities they have in which to feed.

The tenuous and shallow web of the planktonic community supports almost all the larger organisms of the open sea and produces much of the oxygen in the earth's atmosphere. Any major interference with it, as by pollution, might ultimately have serious consequences for all living things. The planktonic communities of the freshwaters are less differentiated and less important than those of the sea, but still are of considerable importance in the larger lakes and rivers. Because of the relative shallowness of many bodies of freshwater, their benthic communities, that is, those situated on the bottom, may be more important than planktonic ones to their total ecology.

Like terrestrial plants, the minute phytoplankton are dependent upon a sufficient supply of certain inorganic substances for growth. Many of these, such as sodium and potassium, are almost always abundant and, therefore, do not limit the growth of plankton. Phosphate, nitrate, manganese, and iron occur in only very small amounts in the open sea. Their ultimate source appears to be the land, and they are quickly tied up in the bodies of living things. For this reason, and also because of upwellings, planktonic communities are, in general, most productive near the continents and especially in estuaries such as Tampa Bay or Delaware Bay. Sometimes this productivity is of the wrong kind. A number of planktonic algae, such as *Gymnodinium* and *Gonyaulax* shown in Figure 20-3, give off toxic substances capable of killing fish. When vast concentrations (technically called "blooms") of these

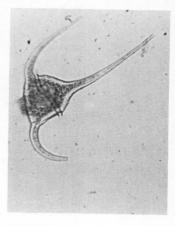

**FIGURE 20-3**
*Gymnodinium,* one of the organisms responsible for outbreaks of "red tide." (X200)

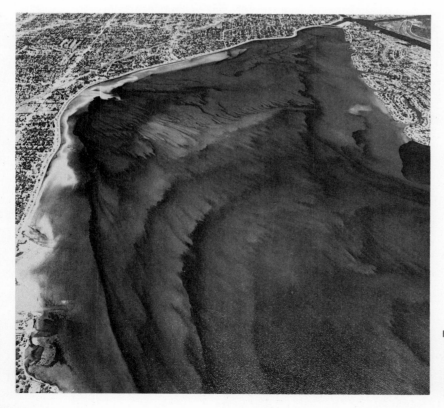

**FIGURE 20-4**
Red tide in Tampa Bay. The oddly patterned cloudiness in the water is produced by countless billions of organisms such as *Gymnodinium.* (SELBYPIC)

microorganisms occur, they actually may turn the water red, for the algae themselves are red-pigmented. Such blooms often are referred to as "red tide." Though they may occur in the open sea, generally they are found in coastal waters, especially estuaries. Figure 20-4 shows a recent (1971) bloom of red tide in Tampa Bay, a typical estuary. (Estuaries are discussed further on in this chapter.)

Red tide is favored by several factors which must coincide in order for truly massive blooms to occur. These seem to be high water temperature, high salinity (usually resulting from low rainfall, especially in estuaries), and the presence of phosphate, nitrate, and vitamin $B_{12}$. The nutrients are often provided by domestic or other sewage, and a hot, dry summer does the rest. Perhaps once in every 10 years in recent history, Tampa Bay has been virtually depopulated of fish by an outbreak of red tide.

The wastage of marine life thus produced must be seen to be believed (see Figure 20-5). If not promptly removed, dead fish may pile up on the shore in stinking heaps 2 or 3 feet deep. It can be a major strain on waste disposal facilities to restore beaches to an attractive condition. The loss of food resources represented by these fish, plus the cost of cleaning them up, plus the loss of tourist and recreation-related revenues may amount to millions of dollars.

The waters of the deeper parts of the ocean tend to stratify, with plant nutrients becoming most concentrated in the lower depths. Part of the reason for this is that dying plankters usually sink to the bottom. Although the disintegration of their bodies releases the nutrients which they have absorbed during their lives, this release occurs in the dark reaches of the abyss, where

**FIGURE 20-5**
Fishes dead from red tide outbreak in Tampa Bay, Fla. Red tide organisms compete with fish for oxygen, mechanically clog their gills, and sometimes secrete a toxin. These fish were cast up on a nearby causeway on July 24, 1971. (SELBYPIC)

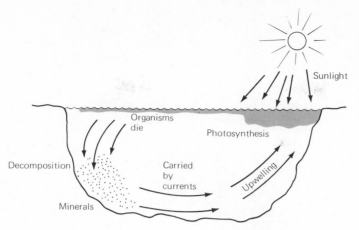

**FIGURE 20-6**
In the open sea organisms die and sink to the bottom. Decomposition releases nutrients, but nutrients are unavailable to the producers up near the surface. Increased wave action carries the nutrients upward (usually in the fall, producing autumn blooms of plankton), or they are carried to the poles or some other area of upwelling where they come to the surface and stimulate productive plankton communities. Phytoplankton cannot grow beneath the depth (*compensation point*) at which the demands of their metabolism exceed their ability to carry out photosynthesis.

those nutrients are not at all accessible to photosynthesizing organisms.

The nutrient-rich lower depths often are separated from the surface waters by a fairly regular decline of temperature. This decrease of temperature with depth is by no means *always* uniform, especially in the larger freshwater lakes, where a fairly abrupt transition, called the **thermocline,** occurs. In most temperate waters the thermocline is established only in the summer. A remixing of the waters occurs as the thermocline is destroyed in the fall by a combination of falling temperatures and increased wave action due to winds. This mixing process produces increased quantities of plant nutrients in the upper, photosynthetically active waters. Autumn growths, or "blooms," of plankton result. Seasonal thermoclines are not universal. Permanent thermoclines exist in the tropics but seldom occur in the polar regions. Relatively minor thermal disturbances (as from electric power plants cooling water) can produce very significant changes in the thermocline structure of even large freshwater lakes and rivers.

The deep, nutrient-rich waters of the oceans are in constant slow movement, in response to wind and tidal, climatic, and other forces. There is a tendency for these deep, cold waters to be carried to the continental shelves and the poles, where they upwell to the surface and replenish the nutrients of the photosynthetic zone of coastal waters and the polar seas. For this reason, the cold polar waters are rather productive, despite the low temperatures and the rather short supply of solar illumination for photosynthesis.

Among the most productive waters, however, are those few in which tropical levels of sunlight are combined with local areas of upwelling caused by some geologic or oceanographic peculiarity. One of the best examples is the Humboldt Current, a great upwelling off the Peruvian and Ecuadorian coast, now the site of one of the world's great fisheries.

Many creatures besides man, among which are thousands of cormorants and other seabirds, exploit these fisheries. The seabirds nesting upon certain nearby islands have deposited millions of tons of excrement upon them. The digestion and metabolism of fish protein by these birds results in

excrement rich in uric acid. That uric acid is transformed, in time, to nitrates, which were formerly a mainstay of the Ecuadorian economy before the development of artificial chemical fertilizers. The cormorants, which formerly generated the guano (manure), are now in a state of decline. A number of possible causes have been suggested; the use of DDT and the competition of Peruvian fishing fleets for the anchovies on which these seabirds feed may be contributing factors.

### The nekton

There is no sharp line of distinction between plankton and nekton. Nekton are organisms (all of them animals) whose powers of locomotion permit them to be fairly independent of the movements of the waters around them. Most large, powerful, and habitual swimmers belong to the nekton. They include representatives of the cartilaginous and bony fishes, certain crustaceans, large squid, and some aquatic representatives of mainly terrestrial air-breathing groups. Whales and porpoises are nektonic mammals; formerly, the ancient ichthyosaurs and mososaurs (large marine reptiles) occupied the ecologic niche now held by whales. Certain large swimming and diving insects are considered by some investigators to belong to the freshwater nekton. However, not every large aquatic animal is a strong swimmer. The immense ocean sunfish, for instance, a weak swimmer, is regarded as part of the plankton.

The food, or trophic basis, of nekton is principally plankton, although in shallower waters, such as those found over parts of the continental shelves, the benthic organisms on the bottom (and on the bottoms of lakes) make a contribution to the nutrition of nekton which may be very important locally. In general, the biomass of nekton is very small in comparison to that of the plankton which supports it. Plankton itself is very sparse by terrestrial standards. Since most nektonic organisms are trophically several steps removed from the phytoplankton, they are really rather uncommon animals, especially in the truly open sea. It is only because the seas are vast that they can support sizable fisheries, and most of these are located over the continental shelves or near reefs that push into the zone of possible photosynthesis near the surface. It follows from these considerations that populations of fish and crustaceans, being relatively small, are very susceptible to pressure from human predation. Many once-prosperous marine fisheries are now in decline (e.g., the menhaden and anchovy fisheries) for no other known reason than overfishing. Marine resources are finite; the sea is *not* inexhaustible.[2]

Most planktivorous nekton, though small in body size, may run in enormous schools. Such small fish as the thread herring and menhaden are good examples; they form the staple food of many large fish and aquatic birds. At the other end of the planktivores is the largest of all mammals, the baleen whale, which strains planktonic crustaceans out of the water in tremendous quantities. Certain very large cartilaginous fish, such as whale sharks, and bony fish, such as the ocean sunfish, are planktivorous also, but most of the larger fishes are carnivorous, as are the toothed whales, seals, and porpoises.

## The benthos

The most highly differentiated aquatic communities are those made up of benthic, or bottom-dwelling, organisms. The diversity of these creatures and their relations is almost beyond belief and certainly cannot be fully described here.

As we have observed, the benthos is the most important, although not necessarily the best developed, of freshwater habitats. Since it forms most of the trophic basis of nekton, it is of the utmost importance to freshwater fisheries. The benthos naturally contains plants as well as animals, at least in very shallow water. In the shallowest water, many of these plants even flower (see Figure 20-8). Some, such as the water hyacinth, float on the water surface. Others, such as water lilies, float only their leaves or upper stems. The stems of water lilies actually extend all the way to the bottom, where they are rooted. These stems usually contain air under pressure, forced down through special air vessels, or hollows, to the roots from the leaves. A number of odd insect larvae comprise a specialized community living around the stems and roots of these plants. Such animals are as dependent upon the air supplied to them by the plants as a diver is upon his air hose, and have specialized organs for tapping the air supplies of the plants on which they are, in a double sense, parasitic.

---

[2] This is to say nothing of the deplorable state of the whaling industry which, through lack of restraint, has pushed several species of great whales to the edge of extinction and may cause the demise of all of them in the next decade or two. Dramatic declines in freshwater and coastal fisheries (such as those of the Great Lakes) are due, in addition, to acute or chronic pollution and to the introduction of exotic predators into these relatively restricted ecosystems.

 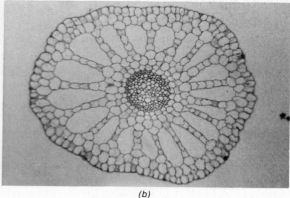

(a) (b)

**FIGURE 20-8**
(a) Water lilies, air-dwelling benthic plants. (b) Microscopic view of a cross-section through the stem of a water lily. Note large air spaces.

Eel grass and fixed algae predominate in the somewhat deeper areas and often have strange, complex, and specialized adaptations—especially in the sea. Some of these algae, the big kelps, are comparable to trees in size. Many marine benthic animals are filter feeders; that is, they filter suspended particles from the water and feed upon them. For the most part, these particles are of planktonic origin. Thus, a substantial number of benthic primary consumers, such as oysters and clams, are basically dependent upon the rich phytoplankton of the continental shelf. Despite that, the most productive benthic areas are also those in which fixed aquatic plants are the most numerous. Coastal or estuarine marshes and shallows may exceed even terrestrial tropical rain forests in the total biomass which they are able to support. The spawning of many fish and commercially important crustaceans takes place here, and most edible mollusks occur nowhere else. Burrowing worms, living on detritus (organic particles suspended in water) in the mud as terrestrial earthworms do, support entire populations of aquatic birds and fishes. Oceanic reefs similarly are of great economic importance and form the ecologic basis of important fisheries (e.g., the North Sea).

Benthic communities vary greatly in response to their physical environments. Depth, bottom composition, tides, and wave action are among the most important factors. Solid support is a rarity on the sea bottom. Most of the sea bottom, by far, is composed of sand or mud in which certain specialized organisms can burrow, but upon which the most important *sessile* organisms (that is, those rooted to one spot) can find no purchase. Sessile and some mobile organisms reproduce by means of free-swimming larvae, which form an important part of the zooplankton. The vast majority of such planktonic larvae find no ultimate resting place and thus die. So pervasive are these larvae, that the echinoderms have special pincers (pedicellaria) covering the exposed parts of their bodies with which they kill and remove any larvae attempting to stake out a homestead. The growth of sessile "fouling" organisms on ship bottoms requires special repellent paint, and periodic painstaking dry-docking and scraping.

(a)

(b)

**FIGURE 20-9**
Coral reef habitats off the Florida Keys. (Roy Lewis, III)

A rocky substrate obviously provides a wealth of cover, solid support, and an abundance of niches for organisms to occupy. A living coral reef is even richer and more diverse, partly because it is largely composed of the elaborately branching skeletons of colonial cnidarian animals which are alive themselves and which provide a wealth of nooks, crannies, and habitats for an abundance of organisms.

## The intertidal community

A special division of the benthos is the intertidal community—those organisms that live between the low-tide mark and the high-tide mark. Creatures living in this area are specialized to tolerate periodic immersion and exposure. They display a distinct zonation in their distribution. Those most tolerant of exposure, mainly semiterrestrial crustaceans, occur at the uppermost limits of the high tide. Barnacles, mussels, certain snails, and particular types of algae are characteristic of the intermediate zones. Echinoderms, hermit crabs, blue crabs, certain fish, and other more thoroughly aquatic organisms live at levels that are exposed only a few times a year by the most extreme "spring" tides.

On rocky shores, intertidal organisms possess clinging adaptations, i.e., mechanisms that prevent the waves from washing them away. On the most exposed rocky shores, only the tightest stickers can survive. Devices employed range from the anchoring threads of mussels known as byssus, through the clinging feet of snails and limpets, to the permanent cement that attaches barnacles to their substrate. An extreme adaptation of considerable interest is that of certain sea urchins and mollusks which can burrow into solid rock or even concrete and steel bridge pylons.

Where the shore is sandy, adaptations must employ active burrowing rather than clinging. The sand launce is a slim bony fish that maintains itself by constant burrowing just below the surf line. A filter-feeding crustacean with somewhat similar habits is the sand "crab," *Emerita.* Southern shores have *coquina* clams, and sandy beaches everywhere swarm with other clams, tiny worms, amphipod crustaceans, and an abundant microscopic fauna and flora known as the *psammon,* inhabiting the film of interstitial

**FIGURE 20-10**
Barnacles, typical intertidal organisms. (Charles E. Carter, Jr.)

**FIGURE 20-11**
A muddy intertidal zone inhabited by various mollusks and worms. (Roy Lewis, III)

water between the sand grains. A unique inhabitant of the supratidal beach is the little flatworm *Convoluta* found in Roscoff, France. This worm is symbiotic with (or perhaps parasitic upon) a green alga which lives in its body tissues. The worm does not seem to feed upon anything but the food manufactured by its algae. The life cycle of the worm is terminated in senility by digestion of the algae and starvation. Perhaps this odd habit may one day form the basis for a replacement of the old cliché about the goose and the golden eggs.

### The estuarine community

*Estuaries* are bodies of brackish water formed behind lines of offshore islands or at the mouths of rivers where they flow into the sea. The loss of speed of river water at that point usually causes it to deposit its silt and other suspended matter, leading to the formation of extensive marshes. Many bays technically are estuaries, as are many rivers with broad mouths, such as the Hudson River. Two varieties of estuary are usually distinguished: *positive* and *negative*. In a positive estuary, more freshwater flows in than evaporates; in a negative, the reverse is true. The water of a positive estuary tends to be fresher than that of a negative. A negative estuary is more susceptible to pollution than a positive.

**FIGURE 20-12**
Certain nutrients such as phosphate, nitrate, manganese, and iron are not plentiful in the open sea. Their scarcity is a limiting factor for plankton. These nutrients are most plentiful near continents. Therefore, the most productive planktonic communities are found near continents, especially in estuaries.

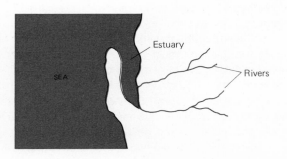

SEA

Estuary

Rivers

Nutrients flow into an estuary from the streams that lead into it and also, at the change of the tide, from the marine habitat. Because of the shallow, marshy nature of most of the area, a thick growth of emergent vegetation such as marsh grass (or "salt grass"), *Spartina,* or a forest of the shrublike mangrove trees tends to develop. This vegetation efficiently traps the nutrients that swirl about its roots. Few habitats on the face of the earth are as productive of biomass as the estuarine. Since the waters are shallow, predators find it difficult to pursue smaller organisms among the tangled grasses and roots. For this reason and because rich plankton and other food occurs there, estuaries are a remarkable nursery for young shellfish and finfish. It is estimated that of all the seafood landed in the United States, fully 60 percent spawns and passes its youthful stages in estuaries.

Tampa Bay, a subtropical estuary, produces at least 10 tons of plant material per acre per year—six times the world average yield of wheat. No wonder ecologists regard estuaries as indispensable parts of the marine habitat. To take just one example, mullet would doubtless become extinct if estuaries and other intertidal marshlands were to be destroyed. Young mullet are so vulnerable to predation that they cannot survive anywhere else. Those who have enjoyed smoked mullet do not need to be told what a loss their extinction would be! But we are not the only predators interested in mullet. Mullet form a vital component of the food chains that lead to such valuable species as tarpon, porpoise, snook, and pelican. Mullet is just one of the many species that cannot survive without this habitat.

*Spartina* occupies most Northern tidal marshes, producing endless acres of salt hay. Further south, mangroves take their place in the estuarine habitat. The black mangrove grows in shallow water, elevated above the surface by prop roots. Oysters grow on these roots, and numerous invertebrates and fishes shelter among them. Raccoons and birds also find rich hunting among the mangroves.

Mangroves play an important role in stabilizing coastlines and in forming land. Their roots tend to trap silt, and their leaves and woody parts may be reduced to peat as they die. During a 30 year-period, mangroves formed about 1,500 acres of new land in Biscayne Bay, Florida, alone. It has been shown that the soil of many areas far inland in the Everglades consists of mangrove peat. Successional processes eventually make mangrove-reclaimed land indistinguishable from the surrounding habitat. The discarded leaves of both *Spartina* and mangroves are reduced by bacteria and other organisms to a rich detritus that rivals plankton as an estuarine food source.

Remarkably, the mangrove is viviparous; it bears its seedlings alive! The seeds germinate in the fruit while still attached to the tree. The young mangrove seedling is released only when it is capable of independent life. After perhaps months of floating, if it is fortunate, the mangrove seedling sinks to the bottom in an area shallow enough to permit it to grow. One wonders how many times a single seedling is the beginning of an island as the new tree grows and spreads out in the shallows.

Mangrove communities in Florida have been seriously threatened by real

(a)

(b)

**FIGURE 20-13**

Mangrove life cycle. The mangrove is a flowering plant (*a*) which produces seeds and fruit (*b*). The seeds germinate within the fruit while still attached to the parent plant (*c*), and the leafless seedlings drop off only when they have attained approximately the size of a pencil (*d*). They may germinate in the mud near the base of the parent tree, or may be carried to a location remote from it. Leaves develop on the upper end of the rooted seedling (*e*). The seedling grows into a small tree (*f* and *g*). In the case of the black mangrove, it sends down prop roots; in the case of the red, *pneumatophores,* breathing roots, develop to bring oxygen down to the roots in the almost anaerobic mud. [The fringe of pneumatophores at the base of a red mangrove may be seen in (*h*).] Oysters, periwinkle snails, and algae may utilize prop roots, stems (*i*), or pneumatophores (*j* and *k*) as places of attachment. Shed leaves and decaying stems provide the basis of extensive food chains in estuaries where mangroves occur (*l*). Mangroves provide a habitat for many higher organisms, e.g., raccoons and shore birds, such as this great blue heron (*m*). (Parts *a, b, f, i, j, k,* and *l,* courtesy of Roy Lewis, III)

(e)

(f)

(j)

(k)

(c)

(d)

(g)

(h)

(i)

(l)

(m)

estate development. Real estate promoters and developers have tended to bulldoze or dredge-fill over coastal mangroves, building a sea wall or bulkhead between the filled land and the remaining subtidal areas. Naturally, this destroys the mangrove habitat. It should be obvious that those wishing to live on Florida seaside properties should be content to have a backyard full of mangroves. However, for the developer, there is less profit in such a course.

## SUCCESSION IN AQUATIC COMMUNITIES

**FIGURE 20-14**
A cross section through a lake. As the lake eutrophicates, it gradually fills. All stages of succession may be observed as one approaches the shore, the more advanced being those closest to the shore. A fully terrestrial community occupies what was once the shallows of the lake. One day, presumably, the entire lake will be converted into a terrestrial community.

Lakes, ponds, and rivers usually do not possess an intertidal zone. Their benthic habitats are much more subject to successional processes than those in marine situations. Such succession is largely dependent upon the deposition of silt and is, therefore, more of a geologically based than a biologically based phenomenon, in contrast to the usual terrestrial successions. Nevertheless, it is just as relentless, if not more so. The woodland pond has been described by one ecologist as a temporary wet spot on the floor of the forest. In the absence of intervention, any freshwater lake or pond will become progressively shallower until it becomes a part of the surrounding

terrain. (This can also happen in some estuaries and bays of the marine habitat.) The pond's aquatic vegetation will change at the same time. Bladderwort and tape grass may be among the earliest stages in this succession. Eventually, water lilies or similar plants may appear. With the establishment of a marsh, a cattail and bulrush community may develop. The climax of this aquatic succession is not aquatic at all, but will be simply woodland, prairie, or quaking bog.

The associated fauna undergoes dramatic changes in keeping with the plant successions. The more "desirable" species of fish are usually associated with the early and intermediate (*oligotrophic*) stages of succession. Successional changes in fish fauna are chiefly associated with rising water temperatures, decreasing oxygen content, and bottom changes affecting egg laying and breeding. These *eutrophic* changes, as they are known, are even more directly associated with nutritional enrichment of the water, and they promote the growth of aquatic plants.

The earliest stages of freshwater succession may be associated in an odd way with nearby terrestrial ecosystems. Trout, especially brook trout, are almost entirely insectivorous. A major portion of their insect food comes from *outside* the stream. Productive trout streams may be almost entirely without aquatic vegetation, but often are closely overgrown with bushes and trees. Insects feeding upon the bushes fall into the stream and feed the trout. Such a trout stream may be ruined by the conversion of the surrounding woodland to pasture. The absence of shade raises water temperatures; grazing adds silt to the water from soil erosion; animal manure encourages undesirable aquatic plant growth; the rain of insects from the bushes ceases. And soon, the trout leave.

## CONSERVATION OF AQUATIC HABITATS

Aquatic habitats are subject to at least as many threats as terrestrial ones. Perhaps the principal threat arises from municipal or industrial pollution, in which flowing water is used to carry off wastes, apparently free of charge. So important is this threat that we have devoted much of Chapters 21 and 22 to it. We will not consider pollution further in this chapter.

### Dams

Dams are constructed for a variety of reasons—the most important being flood control, water storage for industry, drinking, irrigation, and power generation. On the positive side, artificial lakes produced by dams may be important sources of recreation for those who enjoy water sports. Sometimes they can be stocked with fish. They would also appear to be indispensable sources of water for irrigation and for urban areas.

But the price that is paid for these advantages is immense. First of all, the flooding of large areas of terrestrial habitat by the artificial lake formed behind the dam destroys it. It is unlikely that a comparably productive aquatic community will develop in its place, since dammed lakes tend to be

(a)

(b)

**FIGURE 20-15**
Some near-climax plants in an aquatic ecologic succession. (a) Southern bald cypress. Note the "knees," thought to function in root respiration. (b) Sundew, a carnivorous plant that obtains needed nitrogen from the bodies of insects which it entraps in its sticky leaves and digests.
(Carolina Biological Supply Company)

(a)

(b)

**FIGURE 20-16**
Mono reservoir in the Santa Barbara (California) region. This dam was originally constructed to stop silt from flowing into the Gibraltar reservoir. In (a), taken in 1938, silt has completely filled in the area behind the dam, and in (b) successional processes have established a woodland. (a) was taken in 1938, (b) in 1949. Despite the existence of this dam, Gibraltar reservoir was partly filled with silt anyway. (Department of the Interior)

**FIGURE 20-17**
A fish "ladder" permitting spawning fish to bypass the Bonneville Dam on the way to the headwaters of the Columbia River. (Bureau of Sport Fisheries and Wildlife)

too deep for the effective penetration of sunlight, and, more important, they fluctuate greatly in water level. Priceless scenic and other values may also be destroyed by damming. The scenic Glen Canyon has already been lost in this fashion to form the artificial Lake Powell. The flooding of the Grand Canyon by a Colorado River dam is frequently proposed by Southwestern power interests and others.

Dams tend to trap nutrients, preventing them from flowing naturally to the sea through estuaries or deltas. The Nile Aswan Dam has virtually destroyed the Egyptian sardine industry in this fashion, for the plankton that supported the sardines, and, ultimately, many other species of marine life, is no longer supplied with its essential nutrients. Ironically, silt and other materials that would do much good in an estuary simply fill up the lake behind the dam that traps them, rendering it useless, sometimes in much less than a generation.

Dams increase the salinity of river water. Since the artificial lake has a much greater surface area than the stream had before damming, water evaporates from it in greater volume. After having passed through a series of dams, river water can approach the limits of salinity for agricultural use.

Finally, dams interfere with the migration of fish. Despite the provision of fish "ladders" for the use of trout and salmon, it is often too much to expect the fish to swim up them, or the newly hatched fingerlings to pass down uninjured to the sea. Many fish are also frequently destroyed in the turbines of power generators, or damaged by internal gas bubbles produced by sudden pressure changes in spillways.

Such matters have only very recently been taken into consideration in the planning of dam construction, and even today they are often given only lip service. But there is a growing disposition to take a skeptical view of new large-dam construction. If flood control is a consideration, then the building of industry and homes on flood plains should be prohibited. Some flood control can be achieved by small earthen dams or by beaver dams at headwaters, with far less ecologic harm than large dams produce. Alternate sources of power may be considered, and needs for power and water may

FIGURE 20-18
A small headwater dam, Vermont. (Jane Baker)

be lowered by a reduction in the rate of population growth and industrial development, particularly in areas that are inherently unsuitable for heavy settlement, such as much of the American Southwest.

## Channelization

Channelization is a flood control measure sometimes carried out by the Army Corps of Engineers, the Soil Conservation Service, and other agencies. In it, a stream is straightened, most of the bends being removed from its course. Often it is deepened, also, and may be lined with concrete or rock riprap. The increased velocity of the water resulting from the straightened channel would change the character of the communities in the stream even if the steam bottom remained otherwise undisturbed. As it is, however, most

FIGURE 20-19
Crow Creek, Tenn., before (a) and after (b) channelization. (Tennessee Fish and Game Commission)

(a)

(b)

aquatic life is bulldozed, dredged, entombed in concrete, and otherwise deleted from the bed of such a stream. Eventual recovery of a sort may be expected, but since the number of different potential habitats is reduced by channelization, it is unlikely that most channelized streams will ever be restored to any close semblance of their former selves.

It is not that government agencies or others take a malicious delight in the destruction of streams, but that they seek to benefit agricultural interests, construction firms, and other groups that profit from flood control. This goal is commendable as far as it goes. Nevertheless we can look only with horror at projects that, according to Nathaniel Reed, Assistant Secretary of the Interior, permanently reduce local populations of fish, plant life, and ducks by 80 to 99 percent. Reed has stated that if all such projects envisioned in 1972 were to be completed, 25,000 to 60,000 acres of stream habitat would be adversely affected, and from 120,000 to 300,000 acres of terrestrial wildlife habitat would also be destroyed or very severely damaged.

It is time to ask seriously whether the ecologic costs of such projects do not considerably outweigh their benefits.

## Drainage

In the minds of many, the words "swamp" and "useless" seem to be closely linked, but the association is unfortunate. Swamps, marshes, shallow lakes, and other wetlands are really the inland equivalent of estuaries and share some of the estuarine characteristics. They are absolutely essential to the breeding and preservation of many unique species of plants and animals, including the waterfowl shot by sportsmen. What is more, they are frequently an important point of entry for water into an aquifer. (An *aquifer* is an underground water-bearing geologic formation.)

Today swamps increasingly are being filled for residential development or drained for mosquito control. Often they are not drained but merely polluted by use as a dumping ground for garbage and other wastes. When they are gone, it will be scant improvement to contemplate subdivisions, refuse heaps, eroded farmlands, or burning peat wherewith progress has replaced the useless swamps.

## Dredging and filling

Perhaps the most blatantly threatened aquatic habitat is the estuarine. Estuaries are a natural location for large port cities such as Wilmington, Tampa, Baltimore, and New York. Consequently, heavy population density characteristically develops near them, with predictable results. The New Jersey "meadows" were once unspoiled fastnesses of *Spartina*, tidal marshes teeming with waterfowl, other wildlife, invertebrates, and fishes. Today the oxygen content of much of the water in these marshes approaches zero, and the only wildlife easily found is an occasional red-winged blackbird. Extensive industrial development, combined with the waste disposal needs of New York City, has virtually expunged this vast area from the world of life.

Some states, however, are proving themselves more enlightened. By

(a)

(b)

legislative act, the eastern shore area of the Delaware Bay estuary has been preserved from virtually all industrial development. We are hopeful that other maritime states will begin to follow this enlightened example. Industry can almost always find alternatives, although it is granted that those alternatives may reduce profits.

A particularly pernicious threat to Southern estuaries is the practice of dredge-fill real estate development. In dredge operation, sand is sucked by a vacuum apparatus from the bottom of a fairly shallow bay and dumped in another area so as to build up an artificial island or, more commonly, an artificial peninsula. A severance tax is paid to the state for this purpose, but it is seldom anywhere near as much as sand would cost by the cubic yard from a commercial source. Naturally, organisms are buried under the sand dredged up, and other organisms are smothered for miles around by the silt liberated into the water as a by-product of the dredging. It is less obvious that the channels from which the sand has been dredged (the "borrow" areas) are rendered unsuitable as marine habitats. For one thing, to accommodate the boats of the new residents, they are usually made 6 feet or more in depth. Considering the silty water, that puts them out of the range of most effective photosynthesis and so greatly reduces the productivity of the area. Then, too, algae that do drift into these channels die and contribute to their pollution. Artificial channels between "fingers" of dredged land are very susceptible to pollution of all sorts, since usually they are not adequately flushed out by tidal action. However, since all this devastation is concealed by a beautiful blanket of blue water, people pay premium prices for the privilege of living in such "Venices in Florida," where, from their patios, they can gaze out over the sterile waters upon the much-promoted Florida sunset.

In 1971 in the state of Florida, about 100 dredge-fill applications per month were approved by regulatory agencies. As a consequence, fully 20 percent of the Tampa Bay bottom has been filled, with an additional 20 percent dredged to provide that fill. Thus, 40 percent of old Tampa Bay has been lost, probably forever, as a productive aquatic habitat. One recent study, for example, discovered only 0.8 bottom-dwelling animal per square

**FIGURE 20-20**
(a) The *Essayons*, the largest dredge in the world. Completed in 1950, at a cost of $11.5 million the *Essayons* maintains channels in a 5-mile section of the lower Hudson River at a cost of $5 million annually. The ship is 525 feet long, accommodates 147 officers and men, and can carry more than 8,000 cubic yards of silt, to be later dumped into the ocean through 24 trap doors in her bottom. Since the Port of New York would swiftly become unusable without the *Essayons* and her sister ships, such dredging must be classed among those few cases in which the benefits outweigh the obvious ecologic costs. (U.S. Army Corps of Engineers) (b) Dredging for oyster shells, Tampa Bay, Fla. The shells are sold for use in road building and other construction. Damage to the bottom is extensive wherever such dredging is carried out. Here, dredging is of questionable public benefit. (SELBYPIC)

foot of dredged channel, as compared with 238 such animals per square foot in an adjacent undredged area.

Yet Florida and New Jersey are not the worst examples of such practices. The leader now appears to be California, which has destroyed over 67 percent of its estuaries. But other states, at their present rate of progress, will catch up soon, unless they swiftly profit from California's experience.

Estuaries are also threatened by port-maintenance activities and by mining. Commercial sand and gravel interests mine these materials from the bottom of estuaries. Already many Texas estuaries are threatened by the dredge mining of oystershell for roadbuilding and other uses. Offshore oil exploration and pumping pose other hazards resulting from silt and possible oil pollution.

There is no known instance of the successful restoration of a destroyed estuary to anything like its former productivity.

### Aquatic farming

**FIGURE 20-22**
Harvesting milkfish in Bataan, the Philippines. This form of aquaculture provided about 60 percent of the domestic protein needs of these islands in 1967. (Food and Agriculture Organization)

With world food supplies rapidly becoming critical, it is natural that some have thought to increase them by marine and freshwater aquatic farming of fish, shrimp, oysters, and the like. A large number of the potentially farmable organisms are estuarine, however, and it is hard to see how a viable farm could be established in any major estuary before 1985. Calculations of how the world seafood yield could be increased by aquatic farming are irrelevant unless the estuarine habitat can be saved. To be sure, some aquatic farming, such as that of milkfish, is compatible with a moderate degree of organic pollution and may even benefit, to some extent, from the organic nutrients present in domestic sewage. There can be too much of a good thing, however. We doubt that anything could be raised in New York Harbor, and if it could, we hope that no one could be persuaded to eat it!

## CHAPTER SUMMARY

1   The physical properties of water—principally its ability to absorb light, dissolve polar substances, transport insoluble substances, resist temperature change, solidify at a lighter weight, and form a surface film—determine the characteristics of the aquatic habitat.
2   Aquatic communities include the plankton, free-floating organisms largely at the mercy of currents and tides; the nekton, larger organisms that are strong swimmers; the bottom-dwelling benthos; the intertidal organisms, adapted to periodic immersion; and the estuarine community.
3   Plant nutrients tend to accumulate in many bodies of water, giving rise to eutrophication, that is, nutritional enrichment. Eutrophication is usually accompanied by the deposition of silt, which causes the body to become shallower, warmer, and sunnier. Plant growth is encouraged by all these factors. In the end, a terrestrial climax community is established which resembles the surrounding terrestrial community.
4   Aquatic habitats are almost universally threatened. Specific threats include dam construction, channelization of watercourses, wetland drainage, dredge-fill operations, and varied types of pollution.
5   Aquatic farming, a potential food resource for the future, is threatened by the destruction of estuarine aquatic habitats.

## FOR FURTHER READING

Numerous government publications may be consulted on the fields of oceanography and marine biology. A number of periodicals, such as *Oceans,* are devoted to marine science. Additionally, *Natural History Magazine* and *Scientific American* frequently contain pertinent articles. *Scientific American* has devoted an entire issue (September, 1969) to the oceans.

Carson, Rachel: *The Sea around Us,* Signet, New York, 1954.
Coker, R. E.: *Streams, Lakes and Ponds,* Harper, New York, 1954.
_____: *This Great and Wide Sea,* Harper, New York, n.d.
Davis, Charles: *The Marine and Fresh Water Plankton,* Michigan State College Press, East Lansing, 1955.
Detwyler, Thomas: *Man's Impact on Environment,* McGraw-Hill, New York, 1971.
Ehrlich, Paul, John Holdren, and Richard Holm: *Man and the Ecosphere,* Freeman, San Francisco, 1971.
Hardy, Alistair: *The Open Sea,* Collins, London, 1968.
Hitchcock, Stephen W.: "Can We Save Our Salt Marshes?" *National Geographic,* June 1972, p. 728.
Newell, Norman D.: "The Evolution of Reefs," *Scientific American,* June 1972, p. 54.
Pinchot, Gifford B.: "Marine Farming," *Scientific American,* December 1970, p. 15.
Yonge, C. M.: *The Sea Shore,* Collins, London, 1949.

# WATER POLLUTION

In Köln, a town of monks and bones,
And pavements fang'd with murderous stones,
And rags, and hags, and hideous wenches;
I counted two and seventy stenches,
All well defined, and several stinks!
Ye Nymphs that reign o'er sewers and sinks,
The river Rhine, it is well known,
Doth wash your city of Cologne;
But tell me, Nymphs, what power divine
Shall henceforth wash the river Rhine?
— *Samuel Taylor Coleridge*

***Chapter learning objective.*** The student must demonstrate knowledge of the main varieties of water pollution and the ecologic effects of aquatic pollutants.

*ILO 1*  Define the term "pollution," and differentiate among the major varieties of aquatic pollution.

*ILO 2*  Describe the typical sequence of pollutional events which results from dumping saturating quantities of organic wastes into a watercourse.

*ILO 3*  Describe the process of pollutional eutrophication, differentiating between natural and cultural processes.

*ILO 4*  Identify three types of toxic substances besides biocides that damage the aquatic environment, and discuss some of the specific problems caused by each type.

*ILO 5*  Identify the sources, causes, and effects of thermal pollution, summarize the problems involved in its control, and propose possible ways in which it may be rendered advantageous rather than harmful.

*ILO 6*  Describe the effects of silt and other suspended matter on aquatic ecosystems, and discuss the relationship of dredging to this type of pollution.

*ILO 7*  Outline the dimensions of the agricultural and animal waste problems posed by modern husbandry practices, suggesting means for their alleviation.

*ILO 8*  Give the potential effects, reasons for, and means of controlling oil, saline, and mine drainage pollution.

Pollution is a word on everyone's tongue these days, but it is surprisingly difficult to reach agreement on just what it means. One hears, for example, that aquatic pollution renders water unfit for its intended use. But such a criterion raises questions. If uses conflict, which has priority? What is "use," in any case? If an arctic lake has no obvious practical use, could it then be filled with sewage and still be considered unpolluted?

## WHAT IS POLLUTION?

Since our lives are intimately bound up with those of all other organisms, a meaningful concept of pollution must take them into account. Thus, we suggest that pollution may be said to exist when temperatures or concentrations of substances occur which reduce the suitability of a natural habitat for the organisms that normally inhabit it.[1]

Seemingly noxious substances are not always pollutants; they are pollutants only when they are present in an area where they do not "belong," at least in high concentrations capable of doing biological damage. For instance, silt and many salts may exist normally in a wide variety of aquatic environments but do damage only when present in excess.

Pollution may actually *favor* some organisms, but if it damages any of the normal ecosystem components it is still considered to be pollution. Off the coast of California, for example, certain sea urchins have increased their numbers in recent years because they are able to live on raw sewage. But these sea urchins consume the economically important kelp beds; in some areas the sea urchin population explosion has virtually wiped out the kelp.

## KINDS OF WATER POLLUTION

Varieties of water pollution are almost endless, but two major types stand out:

1 Pollution arising from the deliberate use of the environment as a waste disposal system, and
2 Pollution incidental to agriculture, mining, and other human activities, arising from them as an unintended by-product.

Pollution in the first category arises from fairly well-defined small sources, such as municipal sewers, called *point sources.* Though the source may be small, the total amount of pollution it releases may be very large. Since point sources *are* well defined, theoretically it is easy to identify them and therefore to control them.

Pollution in the second category tends to arise from area or diffuse sources and is frequently difficult to control. A classical example is silt,

---

[1] Although it is a very important form of pollution, air pollution is omitted from this discussion since it has already been considered in Chapter 9.

which arises from such sources as agricultural lands, road building, and construction. Another is land runoff containing chemical fertilizers and pesticides.

Fluid substances such as air and water carry pollution with them as they flow, but obviously they eventually deposit their load somewhere. In most cases, the place of ultimate deposition (called a *sink*) is the ocean and its bed. But pollutants can be temporarily delayed in their journey, for example, in a freshwater lake or a marsh, or in certain soils which may serve as a temporary sink for heavy metals.

## The use of the aquatic environment as a disposal system

Life on earth exists as a closed system in which the waste products of one organism serve as necessary raw materials for the use of others. Unfortunately, the wastes of our modern industrial civilization do not fit well into this ancient pattern. They may be so unnatural that they cannot be metabolized readily by any organism, or they may be of a "natural" type but concentrated to such a highly unnatural degree that the capacity of local organisms to absorb and to deal with them is far exceeded.

*Why pollution has become dangerous.* There is nothing new about the disposal of wastes in water. One is, perhaps, entitled to wonder at the relatively sudden fuss about it. All kinds of organisms, from fish to mammals, defecate in the water, and even pollution from human sewage was until recently pretty much an urban problem. What accounts for the change?

Traditionally, dilution was held to be the solution to pollution. Dilution, combined with microbial action, was thought to render sewage harmless. This point of view was not completely unjustified. If not overwhelmed by gross pollution, decomposing organisms quickly reduce wastes to negligible concentrations, returning their component molecules to the common biologic pool of resources for ultimate recycling. No particular harm results.

But as early as the eighteenth century, the Thames estuary, flowing through London, had become a cesspool, and today the total volume of wastes near urban and industrial centers is so great that no reasonable amount of water can possibly dilute them to tolerable levels anywhere near their point of discharge.

Also until quite recently people were enamored with the apparent fact that the waste disposal services of our habitat were free. Flowing water demands no wages. Yet now we understand that pollution *is* costly—to society at large, to the communities downstream or downwind—but unfortunately, almost always to someone other than the polluters!

Not that the cost of eliminating pollution is low. Quite the contrary! But with reasonable restraint of the growth of manufacturing so as to allow control measures to catch up, it should be possible to render wastes almost innocuous before discharge, to discharge them in a place or in such a manner that they do no harm, or to refrain from discharging them, recycling them instead.

***An example: aquatic pollution by organic wastes.*** The organic wastes of municipal and industrial sewage are the type of pollution that most people think of first, and these wastes at present are probably of more biologic and economic significance than any other kind. How do they produce their effects? We shall consider them in some detail, since organic wastes produce a typical pollution ecology.

Organic wastes provide a rich source of nutrients for decay bacteria and fungi. Hence feces, blood from slaughterhouses, oxygen-demanding wastes from paper mills, and peelings from vegetable processing plants stimulate the growth of bacteria whose metabolism rapidly removes oxygen from the water. Such wastes contain large amounts of sediment, chemically combined nitrogen, phosphorus, carbon dioxide, methane, hydrogen sulfide, and smaller amounts of miscellaneous chemicals, heavy-metal ions, and pesticides. Industrial sewage accounts for most of the water pollution of the United States. It usually is far more concentrated than municipal sewage, producing up to twelve times the amount of pollution per gallon of effluent that municipal wastes do.

A polluted environment is a demanding one. Not surprisingly, few organisms can tolerate it. Yet those that are able to exist in it often attain astronomic numbers and very large biomass. Most aquatic organisms, including, unfortunately, those we hold to be most desirable, are sensitive to the effects of pollution and are destroyed by it.

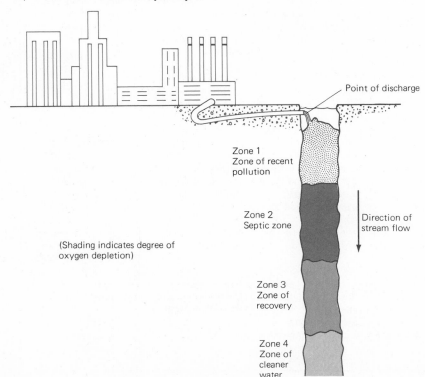

Point of discharge

Zone 1
Zone of recent pollution

Zone 2
Septic zone

Direction of stream flow

(Shading indicates degree of oxygen depletion)

Zone 3
Zone of recovery

Zone 4
Zone of cleaner water

**FIGURE 21-1**

Typical pollution zonation in a stream.

**Table 21-1. Zones of pollution**

| Zone | General description | Water conditions | Typical organisms |
|---|---|---|---|
| 1<br>Zone of recent pollution | Area near point of discharge. May be milky and somewhat evil-smelling. Bacteria have had little time to act upon the sewage, though the sludge at the bottom may approach anaerobic conditions. | High concentrations of particulate matter inhibit photosynthesis, but oxygen content is relatively high. Amount of oxygen that the water will consume in time (*Biologic Oxygen Demand, BOD*) is very high. | Carp, eels, some aquatic insects, crabs, etc. New York City has several raw-sewage outlets where striped bass may be caught. |
| 2<br>Septic zone | Area of high bacterial activity. A noisome course of brown fluid, crusted with rafts of bottom sediments buoyed up by the slowly bursting gases of decomposition. Mosquitoes and flies breed amidst an appalling stench. | Near the zone of recent pollution aerobic bacteria begin to convert sewage into their own body substance, releasing carbon dioxide and inorganic salts. This consumes oxygen, so that farther away from the outlet only anaerobic bacteria are able to work. Oxygen content of water swiftly declines. As bacteria work, organic matter is progressively consumed, reducing BOD. | Bacteria, both aerobic and anaerobic, fungus, some algae, especially blue-green. No higher plants, few animals save those such as mosquito larvae and rat-tailed maggots which breathe atmospheric oxygen (see Figure 21-4). |
| 3<br>Zone of recovery | Bacteria and other specialized organisms begin to taper off, though the area is by no means restored to normality. | Increasing concentration of oxygen; much less turbidity. High content of mineral nutrients released from sewage by previous bacterial action. Low BOD. | Protozoa which consume bacteria. Simplified fish fauna, including carp, bass, etc. Considerable aquatic vegetation. Sludge worms on bottom eat sludge, reducing its BOD by 50% |
| 4<br>Zone of cleaner water | Resembles unpolluted watercourse except for heavier than usual growth of aquatic vegetation. | Clear, high oxygen content, negligible BOD. | Same as in unpolluted habitat, including aquatic insects. Oligotrophic organisms reduced in numbers. |

*Pollutional zonation.* One may divide polluted bodies of water into zones of pollution, each corresponding to a typical resident community of pollution-resistant organisms. The makeup of these populations usually is determined by the type and concentration of nutrient materials present and by the amount of oxygen available. The zones are not marked off from one another in a definite or absolute sense. Rather, they intergrade and may shift their locations in response to such variables as degree of pollution, water temperature, and stream flow. These zones are summarized in Table 21-1 and diagramed in Figures 21-1 through 21-3.

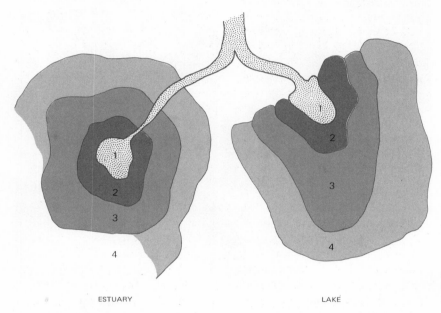

ESTUARY                    LAKE

**FIGURE 21-2**
Pollution zonation in a lake and estuary, simplified.

*Pollutional or cultural eutrophication. Natural eutrophication* proceeds as a result of the slow leaching and seepage of nutrient salts from the surrounding watershed. *Cultural eutrophication,* resulting from pollution, not only occurs far more rapidly than the natural process but differs in several other respects. In natural eutrophication a slow accumulation of silt on the bottom usually also accompanies the nutritional enrichment, so that geologically old lakes are inclined to be shallow and warm. In cultural eutrophication overstimulated plant growth and decay swiftly choke the body of water, leaving a nutrient-rich thick layer of sludge or organic ooze on the bottom.

A nutrient-poor, or *oligotrophic,* lake may contain some of the most highly valued varieties of sport fish, but since it is poor in primary producers, its total productivity is low. Thus from man's point of view a moderate degree of eutrophication may be beneficial, and fishponds are often artificially fertilized for that very reason. It comes as no surprise, then, to learn that the

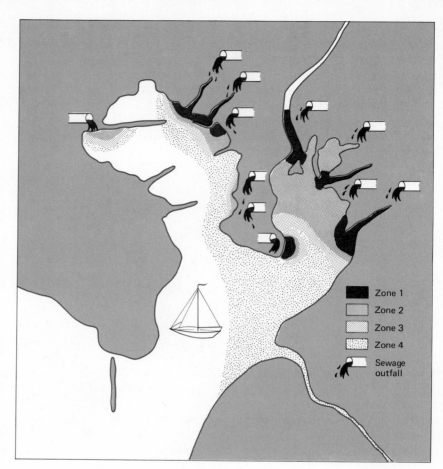

**FIGURE 21-3**
Complex pollution zonation in an estuary, based on an actual case. Note the influence of polluted rivers. The occasional existence of zones 2 to 4 without zone 1 results from the presence of overloaded secondary-treatment plants. Most outfalls, however, deposit either raw or primary-treated sewage in these estuarine waters.

Zone 1
Zone 2
Zone 3
Zone 4
Sewage outfall

years immediately preceding the final pollutional demise of a lake are frequently notable for greatly increased or even record catches of fish. But in the end, unrestrained cultural eutrophication almost always leads to ecologic disaster and the end of the aquatic community's usefulness to man.

The principal reasons for this are twofold. First, green algae tend to be displaced by blue-green in the presence of large quantities of mineral nutrients. Blue-green algae are largely unsuitable for fish food. Second, plants, like animals, must respire, and decomposers feeding upon their dead bodies also consume oxygen when they decay. On a cloudy day or at night, plant respiration far exceeds photosynthesis, and plants compete with fish for oxygen. During the winter, when ice covers a lake or stream, such competition from a jungle of aquatic plants combined with the beginnings of decay of those dead from the last season may deplete the oxygen content of the water virtually to zero. However, massive fish kills resulting from oxygen depletion in eutrophic lakes may occur at any season.

Cultural eutrophication also encourages the growth of floating plants,

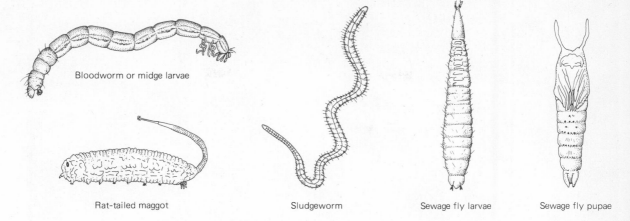

Bloodworm or midge larvae

Rat-tailed maggot

Sludgeworm

Sewage fly larvae

Sewage fly pupae

which cut down on the amount of sunlight able to reach the truly aquatic vegetation. Floating plants, such as water hyacinths found in the South, also respire through their roots, while contributing no oxygen to the water.

*Control of eutrophication.* Eutrophication has proved difficult to control, and all proposed methods must be classed as experimental at best. Sharply reducing pollution at its source is probably the only really adequate measure. Short of that, other suggested treatments include the destruction of aquatic plants with herbicides, the mechanical removal of plants and their transportation out of the aquatic community, the introduction of viruses that attack algae, the encouragement of bacteria that compete with plants for nitrate and phosphate, the mechanical dredging of bottom sludge, drainage, and the introduction of aquatic animals such as fish, insects, or mammals capable of consuming excess plants.

The simple destruction of plants, either by viruses or by herbicides, leaves them in place and will ultimately cause them to rot, reducing oxygen content and releasing the nutrients in their bodies to produce further eutrophication. Mechanical harvesting avoids these objections but is likely to stir up excessive silt, as would dredging. Bacteria and the introduction of herbivorous animals is experimental and sometimes impractical. Sea cows were introduced into Florida waters to consume water hyacinths and other vegetation, but the large, clumsy animals were injured by motorboats and maliciously shot by some local residents.

Drainage is being tried in some areas. In this approach, a eutrophic lake is drained and its plants and sludge are exposed to air oxidation. Perhaps crops might profitably be raised in this very rich soil for a number of years. If further pollution could be avoided, the lake could then be reflooded and restocked with desirable fish and vegetation. Such an approach would be most practical with small, shallow lakes in specially favored geographic locations. Since it involves the temporary destruction of the lake, it will probably be used only in otherwise hopeless cases.

*(a)*

*(b)*

*(c)*

**FIGURE 21-5**

Cultural eutrophication at Clearwater, Fla. (*a*) Aerial view of tremendous growth of sea lettuce (*Ulva latuca*) encouraged by effluent of nearby secondary-treatment plant. (SELBYPIC) (*b*) Piles of dead lettuce seen from the shore. (*c*) Portrait of the alga itself. (Roy Lewis, III)

The shallowest of the Great Lakes, Lake Erie, has been the first to experience serious cultural eutrophication. At one time a very important fishery existed in Lake Erie. Although many less valuable "rough" fishes persist there, and even thrive, the loss of commercially valuable species has almost rendered extinct the once thriving Lake Erie fishery. There is no prospect of remedying the eutrophicated state of Lake Erie in the near future. The other Great Lakes, though less susceptible because larger and deeper, are unfortunately well on their way to a similar state.

Cultural eutrophication can be reversed, at least in some cases. Lake Washington, near Seattle, once experienced serious cultural eutrophication, which cleared up in large part when sewage was diverted to Puget Sound instead.

What mineral substances produce eutrophication? That depends upon which are in shortest supply in relation to the metabolic needs of aquatic vegetation. As you are doubtless aware, phosphorus being a component both of ATP and of DNA, is essential to all life. Since it cannot be made by photosynthesis, phosphorus must be supplied to plants in their water or soil. Nitrate, too, is an essential plant nutrient. Except for blue-green algae, no aquatic plant can fix atmospheric nitrogen. If natural waters are locally low in nitrate, nitrate may be the limiting factor; if they are low in phosphate, then

(a)

(b)

(c)

**FIGURE 21-6**
Pollutional eutrophication encourages the growth of floating aquatic vegetation which contributes little or nothing to the oxygen content of the water, but which does respire oxygen through its roots. Floating vegetation also blocks the light for photosynthesis by submerged plants, and can interfere materially with navigation and with irrigation. (a) Duckweed covering the surface of a pond. (b) Water hyacinths drifting down the Hillsborough River, Fla. (c) Spraying aquatic vegetation with herbicide. A poor method of control.

phosphate is limiting. Thus a superabundance of phosphate will produce eutrophication only in localities where phosphate is scarce. With enough phosphate, more than the local vegetation can use under current circumstances, some other factor will become limiting. That seems to be why blue-green algae often become dominant in eutrophicated situations. With all the phosphate that either one can use, both blue-green and green algae will increase, but the blue-greens will swiftly gain the upper hand, since for them, nitrate is not a limiting factor. Just what factors are locally limiting cannot be discovered in any reference book now available. Since rather little detailed knowledge of local limiting factors is likely to exist in the next 15 years, it seems safest to remove all substances from sewage that are known to produce eutrophication.

An alternative exists, however, and that is to prevent such substances from getting into sewage in the first place. To be sure, each of us contributes a pound of phosphate per year to our surroundings in the form of excrement, and there is no eliminating that from sewage. But each of us also contributes 3 pounds of phosphate per year in the form of detergent formulations, and it has occurred to some that it may be less expensive—for municipalities at least—to prohibit detergents containing phosphate than to install tertiary-sewage-treatment plants to remove the phosphate from the sewage effluent.

When this was first proposed, the detergent manufacturers had just weathered a biodegradability crisis. Previous detergent formulations had foamed distressingly in sewage treatment plants, often putting them out of action. Nauseous chunks of infectious foam blew about towns near rivers that received sewage. When people downstream turned on the faucets, the water foamed like beer. All this reminded folk too forcefully of its passage through someone else's sewage system, or even through someone else. Making new formulations that could be broken down by bacteria had been a very expensive undertaking and had been accompanied by some very undesirable publicity. Now the manufacturers quickly cast about for a phosphate substitute and hit upon several—strongly alkaline carbonates and silicates, for example, and nitriloacetic acid, otherwise known as NTA.

However, these additives generally were adjudged unhealthful and were not widely implemented. The situation continues to be uncertain, but it seems likely that cultural eutrophication is best combatted by sophisticated sewage treatment. This will be many years and a lot of taxes in coming, and we cannot expect to be relieved of our responsibility by a washday miracle.

*Natural pollution treatment.* For many years sewage has been treated (when treated at all) in a fashion that is only a modification of the means used for millenia by natural ecosystems to purify themselves and to recycle the products of decomposition. In artificial sewage treatment the events of miles of septic zone are folded into a few acres of treatment plant. Consequently, if not overburdened, natural aquatic ecosystems can cleanse themselves even from gross pollution in time.

But sufficient time must be allowed for them to do so. A slowly flowing stream in a long tortuous channel is more efficient at waste treatment than a fast-flowing stream in a bed which has been straightened and therefore shortened by channelization. The latter stream is ordinarily far less able to handle pollution. Some streams which were originally not noticeably polluted have exhibited marked deterioration upon channelization (Figure 20-19). Since water flows slowly through swamps and other wetlands, they too function as natural pollution control plants when left undisturbed.

*Toxic substances.* The principal toxic pollutants are chemical industrial wastes, oil or gasoline, acidic (or alkaline) mine drainage, saline waters, and chemical runoff from agricultural lands.

Chemical industrial wastes may be locally very important but are nearly impossible to summarize. Representative toxic wastes include pickling acid and iron salts from steel mills, phenolic compounds from plastic manufacturers, cyanide from photographic processors, spent sulfite liquors from paper mills, and a Pandora's box of wastes from chemical manufacturers. Although the cost of rendering industrial wastes harmless may be very high, specific technologies are available, or are becoming available, to deal with

most of them. These substances must be dealt with, for in the normal course of events all aquatic life will be swept from a watercourse by toxic wastes. Not infrequently, lakes and river beds are permanently damaged by toxicants absorbed in sludge, so that even after the specific nuisance has abated, it proves impossible to return the aquatic habitat to productivity.

***Thermal pollution—is it all bad?*** When a body of water becomes heated to such an extent that it is harmful to the natural inhabitants of that body of water, we speak of the condition as ***thermal pollution.***[2] Increasing water temperature has a number of biologic effects. Oxygen is less soluble in warm water than in cold, so there is less available for the use of aquatic organisms. Despite this, increased temperature causes organic wastes to decompose more rapidly, hastening the local depletion of oxygen. Warmer water also increases the speed of the chemical reactions taking place in the bodies of its inhabitants, heightening their need for the oxygen which is less and less plentiful. In extreme instances of very high water temperatures, the organisms are simply cooked.

Thermal pollution arises from the discharge of water used for cooling power plant condensers and other industrial operations. Because of the universal demand for electric power, thermal pollution arising from this source is probably the most potentially dangerous.

---

[2] Or as *thermal enrichment,* especially by those who produce it!

**FIGURE 21-8**
A conventional coal-fired plant and a nuclear power plant. The cooling canal serves the needs of both. (SELBYPIC)

About two-thirds of the fuel energy, chemical or atomic, that is used to make electric power is released in the form of waste heat. All steam (better called thermoelectric) power plants require cool water to condense the spent steam that is emitted from their turbines. The cooling water absorbs heat from that steam and is exhausted, usually about 20 degrees Fahrenheit warmer, to the surrounding water. Nuclear power plants at present are thermodynamically less efficient than conventional coal- or oil-fired plants, and they exhaust about 50 percent more heat than the former do. At the present rate of increasing demand for electric power, *all* sizable watercourses and coastal waters will be predictably degraded by thermal pollution in the not distant future, unless stringent protective measures are swiftly undertaken.

What might some of the measures be? First, power plants might be constructed at a considerable distance from the shore, on floating or fixed platforms, or even underwater. Surrounded by large quantities of cold ocean water, even large amounts of hot water exhaust from such plants *may* have small effects.

Second, one can render the heated effluent cool before discharge, or even reuse it. Cooling towers exchange heat between waste waters and the atmosphere. Such towers are expensive, however, and something of an eyesore. Nevertheless they are likely to become almost as common a feature of the landscape of the future as factory chimneys. The less expensive "wet" cooling towers drive air through sprayed hot water, cooling it by evaporation. These can be used only with freshwater because of corrosion problems and the possibility of air pollution arising from the salt in salt water. "Dry" towers, which work like an automobile radiator and do not permit the hot water to come into direct contact with the air, are even bigger and more expensive than "wet" towers. Dry towers must be used near the seacoast where only salt water is available for cooling, or where fogging from released water vapor poses a hazard.

The effluent from a cooling tower is by no means completely cool. It may be discharged into a cooling lagoon and held until cold, but usually it is flushed directly into a body of water.

The specific ill effects of thermal pollution and pollution that is related to it comprise a fair-sized catalog. A plume of heated water can serve as a barrier to the migration of such fish as striped bass or salmon. Temperature disruption can interfere with the seasonal breeding of aquatic organisms and, by erratic fluctuation, can prevent physiologic adaptations to seasonal conditions. Fish may be attracted to warm water like a moth to a candle, but this may result in their destruction in cooling canals and other works. The injection of hot water can interfere with the establishment of the thermocline in lakes and rivers, and much aquatic life is destroyed on the screens of power plant water intakes. Not least, the chlorine and toxic substances with which power companies treat their effluent (to discourage slime and algal growth) can damage aquatic life over a large area.

Nevertheless there may be ways of employing thermal pollution as a force for good. Like moderate amounts of eutrophication, moderate thermal pollution (better called *calefaction* where it does no actual harm) of a body of initially cold water may not be an unmitigated evil. Within limits, the pro-

**FIGURE 21-9**
Dwarfed by its cooling tower, the Trojan nuclear power plant rises near Prescott, Ore. (U.S. Atomic Energy Commission, Portland Electric Company, and Jan Fardell)

**FIGURE 21-10**
Cooling lagoons, such as this pleasing example near Chicago, can be esthetically acceptable alternatives to cooling towers in some instances. (Paul Blakemore)

ductivity of Northern waters might be enhanced, or a subtropical fishery established in high latitudes by carefully controlled calefaction. However, even moderate calefaction will certainly change the composition of aquatic communities to some degree, and whether this is desirable is a question that cannot be answered on scientific grounds alone. Moreover, once it has been implemented, such a course implies the responsibility to keep the water warm. If power plants had to be shut down as a result of mechanical failure or other emergency, the warm-water community that had been artificially established would swiftly die of cold stress, at least in the winter. In fact, even fish that normally endure cold temperatures will die of cold shock if suddenly exposed to it. Fish kills of this nature have already been observed.

It is probably better to use the heated effluent in more limited fashions. In Great Britain warm-water fish have been experimentally raised in cooling ponds, and we shall presently discuss how such an approach might be employed in conjunction with municipal sewage treatment. Heated effluent could also be used for warm-water irrigation of land crops, for heating greenhouses, or even for some residential heating.

The attitudes of the American people need to be redirected from their past focus upon increasing and limitless consumption of electric power to a more balanced awareness of the environmental price that must be paid for such consumption.

## Unintentional and incidental pollution

Some forms of pollution are the incidental consequences of other human activities and do not result from the deliberate use of the environment as a waste disposal system. Despite this, accidental or incidental pollution can be just as harmful as deliberate pollution.

**FIGURE 21-11**
(a) Silt escapes from dredge-fill operations. (b) Canvas "diapers" supported on booms are supposed to confine the silt arising from dredging. As this photograph shows, they are far from completely effective. (SELBYPIC)

*(a)*

*(b)*

(a)

(b)

**FIGURE 21-12**

(a) The prevention of soil erosion by careful engineering and prompt planting can help to prevent the entry of silt into aquatic systems. (USDA Soil Conservation Service) (b) This once beautiful lake is now almost completely filled with 20 feet of silt as the result of improper road building and residential construction practices. (*Clearwater Sun*, Fla.)

**Silt.** Silt is a normal component of all waters, as is a certain amount of organic detritus and suspended matter of other sorts. Silts are inorganic particles derived from soils or rock by erosion. Suspended matter of all kinds has three major pollutional effects it reduces photosynthesis because of the turbidity it produces, it tends to bury bottom-dwelling organisms, and it mechanically abrades the gills of fishes.

Organic detritus arises in nature mainly from the decay of plant material and is largely cellulose. Cellulose is also released in staggering quantities by the action of paper mills, and is a major component of municipal sewage. (There it arises from undigested food and, to a surprising extent, from toilet and other paper.)

Silt and siltlike wastes arise from certain industrial processes such as aluminum and fertilizer production, from farm soil erosion, and from erosion in road building and other construction. Industrial silt may be allowed to settle out in basins, and erosion may and should be combatted by proper grading and the planting of ground cover. But there remains one source of silt that is almost impossible to control: dredging.

As we have already discussed, dredging involves the removal of mud, sand, or sediments from the bottom of a body of water. Dredging is a most important source of silt pollution in coastal waters. Many dredging operations involve the separation of fine and coarse particles from the bottom sediments. The coarse ones may be retained to make artificial land or may be deposited as waste (spoil). The fine sediments are dispersed into the surrounding water. No truly effective way of containing them is known. They are likely to persist in sluggish estuarine waters for months. Even if they finally come to rest, every disturbance is likely to stir them up again to do more harm.

Dredging also has certain side effects. It may release local concentrations of phosphorus and nitrogen, which can promote eutrophication and algal blooms. Such release is thought to have contributed to the poisonous red tide outbreaks off the East Coast in 1972.

Environmental scientists are turning increasingly to the point of view that dredging operations should be discouraged and should be undertaken only if there is no reasonable alternative.

**Agricultural wastes.** It is estimated that in the United States alone the wastes produced by domestic animals amount to about 2 billion tons annually, which is greater than that produced by a human population of 2 billion. Formerly, these wastes were allowed to decompose in manure piles and then returned to the soil by spreading in the winter or spring. This is still done to some extent, and the practice should be encouraged, for if animal wastes were *evenly* distributed over *all* agricultural land they would do little pollutional harm, and would reduce the need to use chemical fertilizers.

At the present time about 23 percent of the 115 million cattle raised in the United States are fattened or raised in feedlots. These beef factories, holding from 1,000 to as many as 50,000 animals, produce mountainous wastes which cannot be transported economically for any great distance. There is too much of it to spread it all on nearby land. Generally, it just runs into the nearest creek, although an occasional enlightened operator may provide lagoons or cesspits which may reduce its biologic demand somewhat.

The BOD (see Table 21-1) of feedlot wastes certainly needs to be reduced, for it is five to fifteen times that of municipal sewage. Such concentrated animal wastes should be treated like human wastes and should be recycled after treatment. Unfortunately, this is seldom done. The need for corrective legislation is acute.

Agriculture contributes to water pollution in other ways. In addition to the silt and mud produced by poor soil conservation practices, biocides are washed from fields into lakes and streams, and both phosphate and nitrate fertilizer contribute substantially to eutrophication.

**Oil pollution.** Oil, gasoline, and other liquid hydrocarbons are serious pollutants. Though some oil pollution is deliberate, resulting chiefly from the need of service stations to dispose of gasoline spills and used crankcase oil, most of it is probably accidental. Leaking oil pipelines, wellheads, and tankers are typical instances of accidental oil pollution. Less innocent is that arising from the practice of flushing bilges and tanks with seawater and pumping the oily mixture overboard. Refineries and other industries may also make some contributions to oil pollution.

The more spectacular oil pollution disasters have involved tankers carrying crude or residual oil which have run aground or collided with other vessels, rupturing and spilling the oil; or leaks from offshore oil wells. Oil forms complex gels and emulsions in water and may sink to the bottom, smothering bottom-dwelling organisms. It may also remain afloat atop the water. Some of the chemical constituents of oil are toxic, and some of these may behave like persistent pesticides in aquatic ecosystems.

Oil calms water, so that a patch of oil on the surface produces a smooth

**FIGURE 21-13**
Perhaps the greatest sufferers in
an oil spill disaster are aquatic
birds. (James Roach)

spot that is attractive to aquatic birds. When the birds light on such a spot, their feathers become soaked with oil so that they cannot fly. In addition, they may ingest quantities of the toxic oil. Those not killed outright often die lingering deaths from infection, or they succumb to predators. Although many attempts have been made to cleanse such birds, methods heretofore employed have proved impractical.

Oil-soaked beaches are a poor tourist attraction, and although various specialized microorganisms degrade spilled oil in time, attempts are often made to hasten its removal artificially, usually by spraying with detergents. Unfortunately, past experiences have not been encouraging, for the detergents were often more toxic to aquatic organisms than the oil. Perhaps less harmful formulations may eventually be developed.

A number of other control measures have been tried. Oil can be sunk by treatment with heavy powders, such as chalk or treated sand, but this merely removes it to the bottom of the sea, where it smothers bottom-dwelling organisms. Sometimes attempts are made to confine it with floating booms, but these work well only in quiet harbor waters, and not always there. Specially designed ships are now under development which are purportedly able to skim up spilled oil from the water surface and return it to shore for re-refinement.

But the best attack upon oil pollution is to keep it from occurring in the first place. We are encouraged by some existing state and proposed federal legislation that would hold companies financially responsible for all oil pollution damage resulting from their activities, regardless of degree of negligence and without limitation of liability. If such regulations are vigorously enforced, a marked lessening of the frequency of occurrence of the more acute incidents is likely. Strong but too little enforced legislation aimed against the pollution of harbors and the open sea by tank and bilge cleansing operations has existed for some time. With increased traffic on the

high seas, the development of super-supertankers of staggering capacity, the use of arctic waters for oil transportation (where the action of oil-degrading microbes is minimized by low temperature), the control of oil pollution should be a high-priority item for all nations.

*Mine drainage.* Deposits of coal usually occur in conjunction with the iron sulfide ore known as iron pyrites. Subjected to dampness and the oxygen of the air, the pyrites break down, releasing sulfuric acid. Water polluted by drainage from coal mines or piles of mine wastes is frequently so acidic as to be undrinkable, biologically sterile, and even able to corrode a ship's hull.

Techniques known today could reduce pollution from this source by 70 percent. Among these we may mention sealing off pyrites from the action of the air, flooding the mine with water to exclude oxygen, or treating the mine drainage. In 1969, only 16 percent of acidic mine drainage received any treatment whatever. Mining companies should be made to rectify this situation, and where mines have been abandoned, public funds should be expended.

*Saline pollution.* Those who live in Northern states are familiar with the salt trucks that invade the highways after each snowstorm. The salt spread on roadways to control ice and snow eventually finds its way into adjacent soil and watercourses. Along rights of way it does surprisingly extensive damage to the ecology (and to automobile bodies), but more to our point, it finds its way eventually into creeks and rivers, increasing their salinity.

This may be a problem anywhere but is especially acute in the West, where water is scarce and must be used extensively for irrigation. Many desert soils contain large amounts of salt to begin with, and it is senseless to add to their salinity via the irrigation water. Unfortunately, even the freshest water contains some salt, and this salt is left behind when irrigation water evaporates from desert cropland. The only way to get it out is to flush the soil with an excess of water, which is then drained off, usually into a watercourse which someone else uses for irrigation.

Many oil deposits occur in conjunction with deposits of salt or of brine. Flushing and other operations may bring this salt to the surface in large quantities. Another potential source of awkward quantities of salt is desalinization plants. Designed to produce freshwater from seawater, these plants must operate by removing the salt. But what will they do with the salt?

Saline pollution is not yet as serious a problem as some of the other kinds we have discussed, but it will become a serious problem unless attention is quickly devoted to it.

1 Pollution exists when the environment is degraded in its suitability for any of the organisms that normally would inhabit it.

2 From a practical standpoint, pollution is divided into two main divisions, based upon its source: (a) that which results from waste disposal, and (b) that which arises as an accidental by-product of human activities.

3 Mere dilution of effluent is generally insufficient to render it harmless. If pollutants cannot be discharged at such a rate as to cause no harm, they must be rendered harmless by treatment.

4 Severe aquatic pollution by organic substances produces a characteristic zonation. Eventually, much of the pollution is rendered harmless by natural processes, but only after a portion of the habitat has been badly degraded.

5 Pollutional or cultural eutrophication causes bodies of water to become filled with abnormal quantities of aquatic vegetation, often to become shallower, and in the end, to disintegrate as natural habitats.

6 Thermal pollution usually results from industrial cooling processes. In extreme form, it has highly undesirable consequences. The effects of mild thermal pollution (calefaction) are less clear and perhaps can be turned to good purposes.

7 Silt pollution arises from several sources. It interferes with aquatic photosynthesis, abrades the gills of fishes, and smothers bottom-dwelling organisms.

8 Intensive animal husbandry, such as feedlots, produces concentrated wastes requiring full-scale treatment.

9 Oil pollution damages recreational facilities, releases poisonous substances into the aquatic environment, and is particularly dangerous to aquatic birds.

10 Mine drainage can destroy virtually all aquatic life in a watercourse. Most mine drainage could be controlled through treatment or by reducing oxygen in the mine.

## FOR FURTHER READING

At the present time literally thousands of articles and books are published yearly dealing with water and land pollution. You should consult *Natural History, Audubon, Science, Bioscience, Scientist and Citizen, Environment, Scientific American,* and *Environmental Science and Technology,* to name but a few. Here are a few of the many useful and more interesting books and articles:

*Biology of Water Pollution,* U.S. Department of Interior/FWPCA, 1967.

Cairns, John: "Coping with Heated Waste Water Discharges from Steam-electric Power Plants," *Bioscience,* July 1972, p. 411.

Carter, Luther: "Warm-water Irrigation: An Answer to Thermal Pollution?" *Science,* Aug. I, 1969, p. 478.

Hasler, Arthur D.: "Cultural Eutrophication is Reversible," *Bioscience,* May 1969, p. 425.

Jones, J. R. E.: *Fish and River Pollution,* Butterworth, London, 1964.

Levin, Arthur, Thomas Birch, Robert Hillman, and Gilbert Raines: "Thermal Discharges: Ecological Effects," *Environmental Science and Technology,* March 1972, p. 224.

*The Practice of Water Pollution Biology,* U S. Department of Interior/FWPCA, 1967.

Strobbe, Maurice: *Environmental Science Laboratory Manual,* Mosby, St. Louis, 1972.

# CONTROLLING LAND
# AND WATER POLLUTION

And the waters from the sea will dry up,
And the river will be parched and dry.
And the canals will emit a stench. . . .
The reeds and rushes will rot away. . . .
And the fishermen will lament.
    —*Isaiah 19*

And I saw a new heaven
and a new earth. . . .
And he showed me a pure river of water
of life, clear as crystal. . . .
    —*Revelation 21 and 22*

***Chapter learning objective.*** The student must be able to summarize the main technical, political, and economic problems involved in the control of pollution, and critically to evaluate approaches to their solution.

*ILO 1*   Produce four arguments in favor of committing substantial resources to the control of pollution, and one against.

*ILO 2*   Outline the historic development of water pollution control.

*ILO 3*   Differentiate among and describe the major types of sewage treatment processes, with particular attention to the nature and disposal of end products.

*ILO 4*   Summarize the nature of the solid waste disposal problem, taking affluence, modern technology, and economic factors into account.

*ILO 5*   Give the principal means of solid waste disposal, listing the characteristics, ecologic consequences, advantages and disadvantages of each, and discuss the relevance of recycling to waste disposal.

*ILO 6*   Give five economic, legislative, and social incentives for pollution control, distinguishing between short-term expedients and long-term solutions.

*ILO 7*   Outline the political and economic obstacles to pollution control, and offer concrete suggestions for broad measures to overcome them.

A constructive approach to some of the problems outlined in the previous chapter must be directed to their eventual solution. In this chapter we broaden our discussion of water pollution to include the problem of solid waste disposal, since the sources of and possible solutions to both problems are closely interrelated.

## CONTROL OF WATER POLLUTION

Though it is easy to obtain agreement that pollution should be controlled, pollution control is neither cheap nor easy to obtain. Countless individuals who agree "in theory" with the idea of pollution control, balk at proposed, specific control measures needed to combat it. Therefore it is worth dwelling for a while on why pollution should and must be controlled.

### Why pollution must be controlled

The reasons for controlling pollution may be divided into *hygienic, esthetic, economic,* and *ecologic* categories. Of these, historically the hygienic argument has been most favored.

Even today most of the water we drink, including well water, contains human wastes. It is no wonder that diphtheria, hepatitis, and many other diseases are transmitted chiefly in drinking water. It was a tremendous advance in public health to attack these diseases by purifying water supplies, but such an approach tends to put the cart before the horse. It is vastly preferable not to pollute our drinking water in the *first* place, and failing that, to treat the sewage so that the effluent is harmless before it enters any watercourse.

Unfortunately, there is no reason to believe that common purification methods are any protection against viruses, and even bacteria escape into drinking water when the concentration of wastes exceeds the capacity of the treatment plants to deal with them. Indeed, the chemicals themselves which are added to drinking water to render it bacteriologically safe may become hazardous in the amounts now frequently required. About 30 percent of our drinking water in the United States now contains disinfectants or other chemicals in quantities considered undesirable by the Public Health Service.

In some instances municipalities are able to disregard the hygienic question altogether, and thus they pollute without limit. New York City, for example, draws its drinking water from a far-flung system of reservoirs and other relatively unpolluted sources. Thus 450 *million* gallons of completely untreated sewage are daily discharged by that same city into New York Harbor. Though it is an island completely surrounded by sewage, Manhattan is habitable. Apparently New Yorkers had to be convinced of the need for water pollution control by other arguments.

One such argument is *esthetic*. We enjoy certain surroundings; others, we do not enjoy. It is hard to enjoy a brisk swim in the septic zone, though we have seen it done. Property values on polluted waterfronts are low because people insist, for some perverse reason, on living elsewhere. Tourist revenues suffer from pollution, because people usually like to get *away* from it.

It is esthetic values that help make life worth living. Pollution may smell like money to some, but it does not smell like happiness.

But that brings up the economic argument. It is not at all certain, as we shall see, that pollution *makes* money. Pollution costs money, and a lot of it, when it is uncontrolled. If water is rendered poisonous, corrosive, or unfit for municipal or industrial use, new industry may be repelled from an area, with obvious adverse effects upon the local economy. The loss of fisheries and other aquatic resources may also be immense. We all have a solid economic stake in pollution control.

Ecologically, pollution is a disaster. Entire faunas and floras can be wiped out by pollution, or thrown into disarray. Pollution is an ecologic threat to the whole world of living things, including man himself.

Some do not feel that ecologic action, such as pollution control, is important in comparison with what they take to be more pressing concerns—the elimination of poverty, slum clearance, and the like. As pressing as these matters are, the environment cannot be neglected until they are remedied! By postponing action we only ensure the further deterioration of the environment to the point where the slums are the world, and all are impoverished. A balanced approach will treat the improvement of both social and environmental problems with comparable priority. They are, after all, intimately related aspects of the predicament of modern life.

### The history of water pollution control
Although sewers have been discovered by archaeologists in ancient settlements, most cities waited until the dawn of the industrial revolution to install them. Before then, whatever could flow simply ran through the streets until it reached some stream which more or less carried it away. The digging of ditches in the street helped to channel these wastes somewhat more formally, and the sewer was born! It was a short step to roofing these over, and the stage was set for the advent of Thomas Crapper (1873–1910).

Though sadly neglected in formal histories (if not in the vernacular), Thomas Crapper must be ranked high among the benefactors of humanity, for he invented the modern flush toilet. Let those cavil who will; it would have been a far different world without Mr. Crapper. And he has some tribute. Though not buried in Westminster Abbey, he did install the original facilities there.

But household wastes were not the only traffic borne by the early sewers. Stormwater drainage from the streets also flowed through them. For most of the course of industrial society sewers have served the dual purpose of waste disposal and stormwater drainage. Most of them continue to do so to this day.

Although some communities installed facilities to treat their sewage, none of them attempted to cope with storm drainage, for there was so much of it! However, there was no way to deal with sewage separately, since it flowed through the very same pipes. The only thing to do was to divert the total contents of the sewers from the treatment plant whenever it rained, and to bestow it intact upon whatever body of water was used for its disposal.

Quite recently some communities have awakened to the undesirability of this practice and have gone so far as to provide separate systems for sanitary and storm sewage. Though an undoubted improvement, this does not completely solve the problem. In New York City alone, it is estimated that dogs deposit 50,000 tons of feces and 5 million gallons of urine annually in the streets. This, together with salt, oil, grit, dead leaves, and goodness knows what else, flows through the storm sewers and is released untreated. And even where storm and sanitary sewers are completely separated, rainwater tends to seep into the sanitary system through cracks in the pipes and bad joints. This infiltration of water can increase flow more than twofold, forcing treatment plant managers to divert untreated sewage to the outfall in order to avoid a disaster.

## Sewage treatment

About half the municipal sewage produced in the United States receives no treatment of any kind. That which does may be subjected to primary, secondary, or tertiary treatment before discharge.

*Primary treatment.* In primary treatment, coarse solids are screened out of sewage or are allowed to settle out in large concrete tanks resembling swimming pools. This reduces the BOD of the sewage by about 40 percent. The effluent is then chlorinated, which is supposed to kill disease bacteria but which often does not. It does, however, kill fish. Primary treatment is the *only* treatment received by 30 percent of the treated sewage in the United States.

**FIGURE 22-1**
The primary treatment of sewage involves filtration, settling, of skimming of coarse solids. (*a*) Municipal primary treatment, Tampa, Fla. (*b*) Industrial primary-treatment facility. (Florida Steel)

(a)

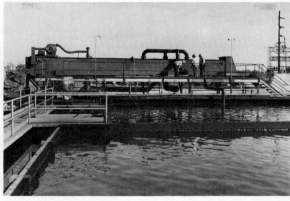

(b)

*Secondary treatment.* Secondary sewage treatment air-oxidizes the effluent issuing from the primary process. By using both primary and secondary treatment in this way, it is theoretically possible to reduce the BOD of sewage by as much as 95 percent, though 65 percent is unfortunately a more usual figure, and even the best of secondary-treatment plants attain their peak efficiency only intermittently.

**FIGURE 22-2**
Secondary sewage treatment by trickling filter. This method is rapidly being displaced by the aerobic activated-sludge method.

There are several types of secondary treatment. One of the oldest, and perhaps still the best known, is the *trickling filter* method. In this method of treatment the effluent from primary treatment is sprayed onto a bed or tank of coarse gravel from an apparatus resembling an overgrown garden sprinkler. The gravel exposes a large surface area to the air for oxidation, and comes to support colonies of protozoa, bacteria, and fly larvae that help to consume the organic wastes in the fluid. The water issuing from this process may be used for irrigation, for many industrial purposes, or, after standing in lagoons, even for recreation and water sports. Generally, however, it is dumped into some convenient body of water. Irrigational use of secondary effluent has proved practical in some localities and has resulted in the al-

**FIGURE 22-3**
Aerobic activated-sludge treatment. Bacterial sludge is mixed with raw sewage and vigorously aerated. The effluent is supposed to be of secondary-treatment quality.

most complete removal of BOD and harmful minerals from the water by the time it seeps into the water table.

The trickling filter, however, is not usually as efficient as the several *aerobic activated-sludge* processes, which in effect combine primary and secondary treatment into one process. In these methods, raw sewage is vigorously aerated by violent mechanical stirring or by the bubbling of compressed air through the fluid. This treatment greatly encourages bacterial growth, which consumes much of the organic content of the sewage. After treatment, the fluid is allowed to settle, and after sludge separation, the effluent is of better-than-average secondary-treatment quality. A portion of the sludge is recycled to the aerating tanks to provide a continuous reinoculation of desirable bacteria. Surplus sludge must, however, be removed from the process.

It has been demonstrated that harmful bacteria are sometimes released from some of these facilities in the form of suspensions in the air (aerosols), a potentially very dangerous form of air pollution. Ironically, the community of microorganisms used in sewage treatment is itself very sensitive to pollution. Toxic substances in the sewage, such as might be produced by industries, can kill the microorganisms, demolishing the effectiveness of the control plant until new cultures can be reestablished.

For these reasons, and also because they promise greater efficiency, chemical methods of digesting sewage have been invented and are being increasingly adopted. The biologic methods of treating sewage may soon be historical curiosities.

***Tertiary treatment.*** Tertiary treatment takes secondary treatment a step further, by treating effluent to remove mineral salts, heavy metals, and toxic substances not removable by ordinary sewage treatment. There are many kinds of tertiary treatment, all of which produce varying results. The confusion of methods has led to the charge that tertiary treatment is an elastic word, stretching to cover whatever the user wishes it to mean. Despite that, there is fairly general agreement that tertiary treatment must remove phosphate and, if possible, chemically combined nitrogen from sewage effluent.

**FIGURE 22-4**
Final treatment of effluent. (*a*) Settling basin, where remaining sludge settles out of suspension before discharge. (*b*) Aeration pond, intended to neutralize any remaining biologic oxygen demand. Note foaming produced by household detergents.

(*a*)

(*b*)

It should therefore have the effect of retarding cultural eutrophication. Particularly efficient forms of tertiary treatment, such as those proposed for use aboard spacecraft, are able to produce water of drinkable quality. This is known as *advanced waste treatment.*

At a time when the amount of available freshwater is decreasing and we stand on the verge of a water crisis in many states, some propose ruinously expensive desalinization plants as a solution. But it is worth bearing in mind that even untreated sewage is something like 98 percent freshwater. It seems silly—to use no stronger word—to do no more with this great potential source of freshwater than to let it run into the ocean. Advanced treatment plants may be one of the best means of providing high-quality water for the year 2000.

*Sludge disposal.* The problem of what to do with the organic sludge that settles out during various stages of sewage treatment is a vexing one. Usually it is partially digested by an anaerobic process which produces methane gas as a by-product. The methane arising from this source could become a fairly important fuel in the future. Even today it is occasionally used for power generation or for heat. Whatever remains of the sludge after such treatment could be dried and sold for fertilizer (the market is very limited), composted, used as landfill, or even incinerated. Alas, all too often it is merely dumped somewhere.

Digested sludge possesses some BOD, and it is by no means free of toxic chemicals and viruses. More important, it mechanically smothers bottom-dwelling organisms if placed in an aquatic habitat. Pumped offshore as semiliquid slurry or transported out to sea in barges, sludge has created extensive biologic deserts near some major cities, especially New York City.

*Septic tanks.* The septic tank is a common alternative to sewer systems, especially in rural areas. A septic tank is a small underground tank, usually made of reinforced concrete, which is used for the partial anaerobic decomposition of individual household wastes. Effluent, equivalent in quality to that of primary treatment, is piped from the septic tank to a cesspool, a drainage field of perforated pipe, or even a nearby creek. We have even seen it piped into a small pond in a suburban yard.

Septic tanks are not a desirable method of sewage disposal. Large numbers of them endanger the purity of the groundwater. It is sound public policy to require builders to provide sewers in all housing projects and to donate them to an appropriate public system.

## SOLID WASTES AND LAND POLLUTION

The garbage explosion is no joke. Each of us produces about 8 pounds of garbage, trash, and other solid waste per day, and the amount is rising steadily. Until quite recently there was little thought of doing anything other than picking it up from one place and putting it down in another. But we are

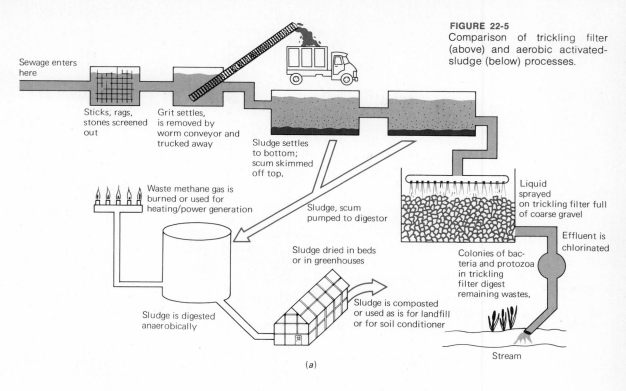

Sewage enters here

Sticks, rags, stones screened out

Grit settles, is removed by worm conveyor and trucked away

Sludge settles to bottom; scum skimmed off top.

Waste methane gas is burned or used for heating/power generation

Sludge, scum pumped to digestor

Sludge dried in beds or in greenhouses

Sludge is digested anaerobically

Sludge is composted or used as is for landfill or for soil conditioner

Liquid sprayed on trickling filter full of coarse gravel

Effluent is chlorinated

Colonies of bacteria and protozoa in trickling filter digest remaining wastes.

Stream

(a)

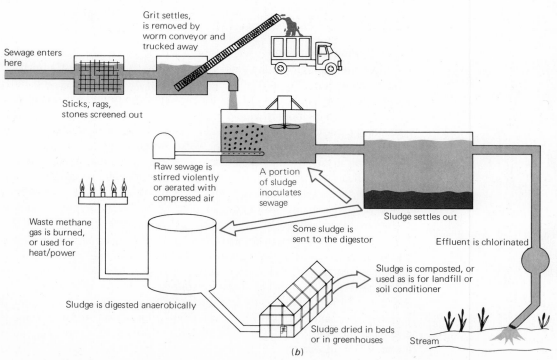

Grit settles, is removed by worm conveyor and trucked away

Sewage enters here

Sticks, rags, stones screened out

Raw sewage is stirred violently or aerated with compressed air

A portion of sludge inoculates sewage

Sludge settles out

Waste methane gas is burned, or used for heat/power

Some sludge is sent to the digestor

Effluent is chlorinated

Sludge is digested anaerobically

Sludge dried in beds or in greenhouses

Sludge is composted, or used as is for landfill or soil conditioner

Stream

(b)

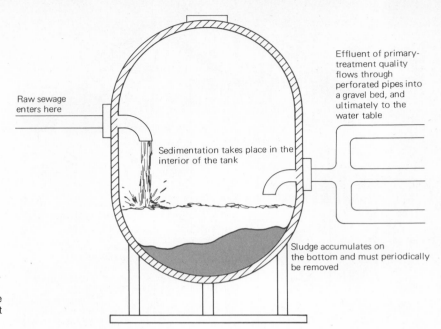

Raw sewage enters here

Sedimentation takes place in the interior of the tank

Effluent of primary-treatment quality flows through perforated pipes into a gravel bed, and ultimately to the water table

Sludge accumulates on the bottom and must periodically be removed

**FIGURE 22-6**
How a septic tank works. The effluent is of primary-treatment quality at best.

rapidly running out of rugs under which the debris of civilization can be swept. The price tag of needed solid waste disposal facilities now stands at $4.2 billion, making solid waste disposal potentially the third most expensive public service, exceeded only by highway and educational spending.

### Why little has been done

It would be enlightening (albeit messy) to sort the contents of one's garbage can into categories. Some of the material would be food scraps and other putrefiable matter, but most of it would turn out to be paper, and a substantial though smaller portion would consist of nearly indestructible plastic, metal, and glass containers. Household waste is a complex mixture.

The very complexity of that mixture works to defeat efforts to dispose of it. It is difficult to design an incinerator that works with equal efficiency on a pound of fish heads and the newspaper in which they are wrapped.

Another reason why we lag in solid waste disposal is that it is a relatively recent problem. In medieval and colonial times the principal household waste other than excrement was organic garbage, and not much of that. Disposable containers were unknown. A bottle or a pot would ordinarily be discarded only if it were broken and unsalvageable. Families consumed very little fiber, using what they did consume almost entirely for clothing. Bones were gnawed clean or boiled down for soup, or both. The little that found its way out into the yard or the midden was rapidly consumed by the traditional garbage disposal of primitive communities—pigs, dogs, and chickens. In turn, such animals were often slaughtered and eaten—a good example of the recycling of wastes.

**FIGURE 22-7**
A perforated pipe usually conducts septic tank effluent to a gravel bed, through which it leaches into the earth.

Today's American consumer discards large quantities of perfectly edible food scraps and an even larger amount of fiber and inorganic trash which is composed mainly of disposable containers such as paper bags, milk cartons, nonreturnable glass and plastic bottles, steel and aluminum cans, and much more, little of which is recycled. Historically speaking, this change has taken place virtually overnight and has caught us unawares.

## Solid waste treatment and disposal

Before solid waste can be disposed of, it must first be collected and transported. The familiar garbage truck is one method, though a none too efficient one. Another involves the use of kitchen sink garbage disposal units, which work only upon the softer wastes and which contribute to sewage disposal problems. Better methods of collection must be devised—perhaps conveyor belts, automated trucks, special sewers, waste precompressors, or other measures. However, it is the final fate of solid wastes that constitutes the most pressing problem connected with their disposal.

*The dump.* One of my fondest boyhood memories is of shooting rats at the town dump. Wary game and sly, these rodents provided many hours of outdoor pleasure to my friends and me. Muskrats, too, can be found in many town dumps, and the sale of their fur provided a livelihood for at least one Staten Island truant, who trapped them in the shadows of the New York City skyscrapers not so long ago. Alas, the simple virtues of the dump are accompanied by some serious vices that spell its doom.

The typical dump is a heap of miscellaneous wastes located outside of town or in a poor neighborhood. Yet the land thus employed is sometimes of considerable value, or would be if the dump were not there. Gulches, canyons, marshes, and other ecologically valuable or even unique areas are often used as dump sites.

Decaying organic matter may pollute both ground- and surface water, and in the dump itself such decaying matter supports a thriving community of scavengers. Flies breed by the pound. One suburban community that set fire to its dump to control, as they hoped, winged cockroaches, was inundated with hordes of the frantic insects buzzing away from their former home. Both sea gulls and rats develop such populations in the vicinity of dumps as to constitute a menace to wildlife and even to public health.[1]

*Sanitary landfill.* A step up from the dump, but very like it, the sanitary landfill method covers wastes with a layer of earth each day. Using this method, abandoned strip mines can be filled and eventually reclaimed, and artificial

---

[1] Not long ago in Great Britain anglers complained of difficulties in catching fish, even where they were abundant. Investigation disclosed the presence of cataracts (opaque lenses) in the eyes of many of the fish, which interfered with the perception of lures and flies. These cataracts were caused by the larvae of parasitic worms, the adults of which live in the intestines of sea gulls. Quite possibly the increased population of gulls was responsible for poor fishing. A simple alteration in the environment may have far-reaching results.

**FIGURE 22-8**
Sanitary landfill. Wastes are covered daily with a layer of earth until the available land is completely filled. Some municipalities have even constructed artificial mountains of waste in this way, which they intend to use for skiing and recreation. (SELBYPIC)

mountains have been constructed for skiing recreation in the Midwestern plains. Unfortunately, sanitary landfill, too, has its drawbacks.

First and most obvious is that not all land *should* be filled. We have already discussed the value of marshes and wetlands. Their destruction may lead to the loss of fin- and shellfish resources, the lowering of the watertable, and reduced populations of migratory birds, to mention but a few results. Yet marshes and even lakes are favorite landfill sites. And it is easy to run out of any kind of land to fill.

Second, sanitary landfill can pollute groundwater. Some localities may be able to evade this by filling strip mines, which tend to pollute groundwater with their acidic drainage anyway. In fact, the anaerobic conditions which often develop at the bottom of landfills could help to counteract the oxidation of pyrites which produces acidic mine drainage.

Third, filled land has a limited number of suitable uses. If used as a building site, settling may cause walls and foundations to crack, and methane gas resulting from anaerobic decomposition in the depths of the fill may seep into buildings and constitute an explosion hazard. All in all, it is probably best to utilize filled land for parks, agriculture, or other nonintensive purposes.

*Incineration.* Incineration is an attempt to burn solid wastes, thus reducing their bulk before ultimate disposal. But incineration, too, has its drawbacks. Incinerators produce both particulate and nitrogen oxide air pollution, some of which is theoretically subject to control measures. Only the larger incinerators can be equipped with efficient control apparatus, so that small apart-

(a)

(b)

ment house and backyard burners are now being discouraged. Even big incinerators have trouble with their pollution control mechanisms, which tend to corrode at unacceptable rates. One common plastic used in household containers, polyvinyl chloride, burns in incinerators, releasing the dangerous and corrosive chemical hydrochloric acid.

Incineration is also expensive and frequently inefficient. The hand labor required to separate the components is far too costly to be considered. Attempts to recover some of the costs of incineration by using the heat to generate steam or electric power have been successful in a few places but have usually been abandoned as impractical.

*Pyrolysis.* A recently developed process, pyrolysis, heats trash in the absence of air, producing charcoal and ash on one hand and an odd mixture of liquid and gaseous products on the other. Since nothing need be exhausted to the out-of-doors, clearly air pollution is not a necessary result of the process. The by-products of pyrolysis have a number of potential uses, including possible use as raw materials for the artificial production of plastic, synthetic rubber, and gasoline. Many experts regard pyrolysis as the most promising method of waste disposal yet envisioned.

*Composting.* Composting oxidizes wastes to produce an inoffensive organic mulch. A nuisance can be converted into an asset by changing garbage to soil conditioner. Usually, garbage is composted by subjecting it to the action of microorganisms, often in special heated chambers. In composting, garbage can even be combined with sewage sludge. Unfortunately, the soil conditioner that is produced is of no great value as fertilizer and is difficult to sell, though it may have a potential use in forestry. Profits are further reduced by the necessity of separating inorganic trash and bones from the wastes to be composted. Since this involves hand labor (at present), it costs a great deal of money.

Composting is probably the most ecologically sound method of municipal waste disposal, but it is likely to require subsidies of one kind or another in order to operate. In the end, compost will probably be discharged over wide areas of agricultural and wild land merely as a way of disposing of it. The capacity of soil to absorb it is very great, so that such practices would probably be harmless.

## POLITICAL AND ECONOMIC CONSIDERATIONS

It is not enough simply to denounce pollution. Pollution is always produced for a reason, and it cannot be stopped until we first understand why it is produced at all.

### Externalities and economics

Pollution is both convenient and profitable, and must be classed, along with many other convenient and profitable operations, as potentially sound business. Generally pollution is not maliciously intentional on the part of polluters; it arises as a by-product of their unconcern with what happens to what they discard. In some neighborhoods and many foreign countries, garbage is not taken out to the curb in containers but is thrown out the window. Now out of sight and smell, the garbage is no longer an inconvenience to those who produced it. Others, true, must take evasive action and may even, if they wish to keep their neighborhood decent, have to gather it and dispose of it, but that is no concern to the one who discarded it. Any costs, direct or indirect, produced by laissez-faire garbage disposal must be assumed by others. They are external costs, or in economic parlance, *externalities.*

Of course, many responsible citizens voluntarily assume their obligations as citizens. But those who will not must be compelled to do so. Otherwise, they infringe on the rights of others. Will business voluntarily assume responsibility for its externalities? As a whole, probably not.

Considering the public benefits to be secured, the costs of pollution control are not excessive, but they are appreciable. In 1972 the Chase Manhattan Bank estimated that the national costs of moderate pollution control would tend to retard production *increase* by about 1.2 percent by 1976. From the viewpoint of manufacturers, these costs will compel either a rise in prices or a reduction of potential profits. So even though society could afford the control costs, institutions such as the businesses and municipalities that comprise society will try, if they can, to avoid paying them as long as possible. Yet they should pay. "The costs," as Rene Dubos writes, "cannot be avoided. The citizen pays either as consumer, or as taxpayer, or as victim."

Not even business escapes victimization by pollution. The cattle industry and citrus groves of Florida have been badly damaged by air pollution from phosphate fertilizer plants. Shrimp fishermen are having extreme difficulty in trolling Tampa Bay because of beds of rotting vegetation produced by cultural eutrophication. Mink farmers around the Great Lakes, who settled there because of the ready availability of fish to feed their mink, must now import fish or feed other meat at much increased expense, on account of contamination of Great Lakes fish with polychlorinated biphenyl chemicals. Good for business? It depends on your business.

Since pollution tends to affect someone other than the polluter, polluters are unwilling to expend their own funds to control it. Yet an increasing, and increasingly affluent, population demands an ever-increasing supply of goods whose manufacture—and often, whose use—produces pollution. The man in the street pollutes in person, and by his consumption of goods he also pollutes by proxy.

Monetary values are preeminent in our society, but although these values have tended to stimulate the production of pollution, they are not solely to blame. Many habits, traditions, and cultural values also seem to contribute to a pollution-producing frame of mind. At a recent pollution control seminar, one executive boasted of the thousands of dollars his company was now saving by recovering valuable chemicals that had formerly been flushed into a river. One of his amazed listeners asked why, if it was that profitable to control company pollution, it had not been done years ago. "I guess nobody made us," was his reply.

## Politics

Politics and economics are closely related in all industrial societies. In ours, the many vested interests who profit from pollution are often able to control the political process to protect themselves. But lest we be misunderstood, we hasten to point out that such interests consist not only of business, big and otherwise, but also of municipal and other governments and their citizens, that is, the body politic itself. The enemy is not confined to boardrooms or city councils. The enemy, in Pogo's immortal words, is us.

Industry contributes about five times as much pollution as residential sources, so industry is, perhaps, the place to begin pollution control efforts. Such control is not easily obtained—particularly in one-industry "company towns" when the principal source of local wages and the mainstay of the local economy may also be the greatest polluter. Not only officers but also rank-and-file employees of such plants may oppose pollution control strongly, out of self-interest. If nothing else serves, management may attempt to cow local governments into submission by threatening to move to another locality if antipollution laws are enforced against them. In our files are records of those who have lost their jobs or even had their lives threatened because of their activities in favor of pollution control. The issue can be deadly serious.

Municipalities are often as much to blame as any industry. The principal source of municipal revenue is, after all, taxes. And citizens would like to pay as little of these as possible. Really adequate pollution control plants cost money, which must be raised by increased taxes, raised water bills, or other politically unpopular means. Those downstream must pay to have their water purified, it is true, but they do not vote in our city elections, so let *them* pay. So goes the usual argument, though perhaps you may not read it in the newspaper couched in just those terms.

## THE SOLUTION TO POLLUTION

It is clear that knowing how to control pollution is a very different thing from actually taking measures to accomplish it. Before anything noteworthy can be done, substantial public sentiment must be developed in favor of doing it. Furthermore, since it is not easy to know what precise action should be taken on any particular issue, there must be public consensus on what corrective measures will best solve the problem.

At the outset, let us recognize that most pollution control measures, however commendable, cannot ultimately solve the problem. These measures are necessary and highly desirable, but they are only palliative. There is a limit to the population and amount of industrialization our planet can support. As we approach this limit we shall have pollution whether we will or no. Long-range pollution control must take population control as well as industrial development into consideration (see Chapter 24). But short of this, certain steps can be taken now. We must fix both responsibility and accountability for pollution control, we must encourage recycling, and we must both develop and enforce the necessary legislation to ensure that these things are done.

### Who should control pollution?

Pollution is generated locally and must be controlled locally, but it does not follow that the responsibility for such control should be locally delegated.

It takes no more scholarly source than the morning newspaper to discover that local politicians tend to be all too responsive to pressures exerted by those nearby who profit from pollution. Larger political divisions are by no means free from such pressures, but they respond more slowly to them and it is more difficult for polluters to bring them to bear.

Then, too, municipalities and other local governments are themselves responsible for much pollution. It is not realistic to expect that they will voluntarily make the necessary changes.

Does pollution respect local political boundaries? If it did, there would be less of it. It is precisely because pollution doesn't stay put that we so blithely discharge it to the world. For that very reason, authorities of broad regional jurisdiction are in the best position to assess the effects of pollution on the nation as a whole.

Last, but by no means least, if despite all these handicaps purely local controls are enacted, they may tend to put local industries at a competitive disadvantage. Manufacturers must either raise their prices or lower their dividends in compensation for the cost of pollution control measures. The former will reduce sales; the latter will drive capital away. These economic problems can be overcome fairly only if regulations are enforced equitably over the nation as a whole.[2]

For all these reasons, state and federal governments should be charged with the responsibility of enforcing effective pollution control standards. In the end, pollution will also have to be regulated through international treaties.[3]

## Regional approaches

Pollution problems tend to develop regionally and are best attacked by a cooperative effort of industries and municipalities in the affected region. It is cheaper to treat waste problems wholesale than retail, and the resources necessary for huge waste disposal facilities are available only to large organizations.

For example, water pollution is best attacked throughout the area drained by a river, that is, its *watershed*. This approach was pioneered in Germany. The Ruhr Valley, one of the most heavily industrialized areas of the world, is still able to use some of its bodies of water for recreation, even though water resources available for industrial use in Germany are extremely small by American standards. Briefly, all industries and municipalities are required to belong to *Genossenschaften,* associations for the regulation and common treatment of sewage effluent. Charges are assessed against each member in proportion to the pollutive "index" of its wastes. All wastes are treated in commonly owned facilities before release to the Ruhr or its tributaries. While the Ruhr and the Emscher Rivers are certainly polluted, they are less so than they would be without the Genossenschaften. Somewhat similar approaches are being developed in the United States today.

One reason why industry resists assuming pollution control responsibilities has already been implied—it is not usually profitable. Industrialists ask why they should invest capital in equipment that often neither makes profits nor reduces costs. But if industries were charged reasonable fees for their pollution, control apparatus would suddenly become profitable.

One caution is in order, however. If pollution charges were set too low, it would be more profitable to pay them than to control pollution; they would constitute, in effect, a license to pollute. Effective fees must be uncomfortably high. Only then will industry be forced to internalize the external costs of its waste disposal.

We certainly do not suggest that industry should be permitted to pass all its environmental costs on to the consumer or taxpayer. A portion of these

---

[2] In the state of Florida, shippers are totally responsible for the effects of all marine oil spills. This unlimited liability has raised the insurance charges of ships making Florida ports.

To avoid the increased insurance costs, many shippers have recently taken to doing business only in states where the laws are more lenient, very much to the economic disadvantage of the state of Florida.

[3] Unfortunately, even the federal government is not above reproach when pollution appears to be in its own best interest. Many military and other federal installations have been notorious sources of pollution.

costs should be deducted from their profits. However, state or federal aid could legitimately be provided to municipalities for use in the construction and operation of pollution control plants.

### Recycling

Many pollutants may be thought of as resources out of place. Solid wastes can be recycled and reused, but recycling can also be applied to sewage. After all, minerals and organic matter originating in Midwestern prairies end up flowing into New York Harbor, where they are only a dangerous nuisance. They could be reclaimed, and so could solid wastes.

Because of the chemical combination of oxygen with waste materials, the addition of dirt, paint, and similar changes that take place in materials before they are discarded, the total bulk of manufactured goods actually increases during use. The job of disposing of them is, in the end, bigger than that of manufacturing them in the first place.

It does not seem that this is the case because wastes are disposed of so casually. If we were to take them apart as carefully as we put them together, we would soon see how gargantuan the task is. Fortunately, you may think, we need not do so. Or do we?

*Solid waste recycling.* During the course of technologic history, metals and other resources were first extracted from their sources by the crudest means. As civilization developed, it became possible to refine less concentrated ores, including those that would have been too poor to work with at all in the early days. There are several examples of modern silver refining based upon the waste rock and tailings discarded by the ancient Romans because they already had extracted as much of the silver from them as was practical with their methods. Fortunately for us, they left the wastes conveniently stockpiled for our use. It has often been observed that if our present civilization were to fall, a technologic culture of the same kind could probably never arise again, because there are now no more deposits of ore having high enough concentrations of metals to permit their utilization with crude methods.

It is characteristic of our civilization that we manufacture goods from resources that have been extracted from the earth with great difficulty, and then disperse them haphazardly in disposal, so that their recovery becomes infeasible even though they may clutter the world. Consider the aluminum can. For each ton of aluminum cans that is discarded, 4 tons of bauxite were mined and 17,000 kilowatt-hours of electricity were consumed. Geologists and prospectors discovered the ore. Miners removed it from the earth. Ships or railroads hauled it to a processing plant. There it was subjected to electrical treatment, flotation, concentration, and other processes until the pure metal was obtained. Pigs, or ingots, of pure aluminum then were shipped to other factories, where they were rolled into sheets. Finally, cans were stamped out of them, and the cans filled with beer or soft drink. Here is the finished product—but no, not quite. To finish it completely, it is discarded: neither we nor our descendants will ever be able to use that aluminum again!

(a)

(b)

(c)

**FIGURE 22-10**
(a) There are about 20 million hulks of stripped and abandoned automobiles strewn throughout the nation, representing more than $1 billion worth of reusable metal. (b) If recycled, this "junker" will probably undergo a scrap-baling process, during which it will be compressed into a compact bundle such as the one being held by the magnetic crane. (c) An alternative and in some ways better method of processing auto hulks is by shredding. This mountain of fragments is the remains of 30,000 old automobiles which have been processed for remelting by steel mills and foundries. (Institute of Scrap Iron and Steel)

The really upsetting thing about all this is that it is deliberate. The container is designed to be wasted, for that sells more cans. The public, for its part, is willing to pay a premium price for goods encased in disposable containers because it is somewhat inconvenient to return deposit bottles and the like. Also, there is evidence that a convenience packaging stimulates demand.

Yet the social benefits of reuse and conservation of resources that are presently discarded are very great. Although already an $8 billion industry, the recycling of wastes has actually *declined* in recent years. Iron and steel are the most extensively recycled wastes, with 45 percent of all steel originating from previously manufactured goods. But the waste with the greatest recycling potential is probably paper. Paper constitutes the largest single component of municipal waste, and tonnage-wise, more paper is recycled even now than any other waste. Yet only 21 percent of all paper is actually

**FIGURE 22-11**
Experimental disposable drinking cups made of caprolactone plastic, which have been buried 0 (left), 2, 4, 6, and 12 months (right). Though worthwhile, degradable containers such as this solve only half the total waste disposal problem—they do not conserve virgin resources. (Dr. R. A. Clendinning and Union Carbide Company)

reused. A figure of 50 percent would easily be possible, and that amount would be equivalent to more than two-and-a-half times the annual production of the four leading wood-producing states! This is to say nothing of the reduction in the solid waste problem which would result from extensive recycling.

One hears a great deal about degradable containers as a solution to the solid waste disposal problem. Yet though degradable containers may be an answer to highway litter and may reduce the bulk of dumps and landfills, they deal with only half the total environmental problem. After all, a degradable container is merely one that will be decomposed in time by the action of microorganisms or oxidation. It is not reusable, and therefore its disintegration only serves to add that much more to the demand for the use of virgin resources. Paper products are fairly degradable at present, but unless they are recycled, their degradability does nothing to help conserve our forest resources. In the end, the recycling of wastes is the only approach that will serve.

*Recycling sewage.* It is well known that Oriental societies have fertilized rice paddies and fish ponds with human and animal waste for centuries. There can be no doubt that this practice is not only malodorous but a menace to public health. Yet it must be admitted that the practice has sustained agricultural land in the Orient for a long time.

Though modern sewage treatment practices would eliminate the odor, parasites, and infectious bacteria that constitute the chief drawbacks to that ancient practice, a better alternative is to design methods of food production explicitly to take advantage of sewage wastes. This also should make sewage treatment easier. It is estimated that by 1990 the oxidation of sewage wastes in the United States will require oxygen equivalent to 10 percent of the daily photosynthetic output of all our green plants. But if green plants are introduced at the point of treatment, they will automatically generate the needed oxygen by their metabolism, while at the same time consuming the troublesome mineral sewage constituents that are responsible for eutrophication. Green algae fill the bill nicely.

The algae produced in the process could then be fed to livestock or, even

FIGURE 22-12
Aquaculture has long been practiced in Japan and in Spain.
These workers are placing pearl
oysters in wire cages suspended
from rafts. Techniques are now
under development to raise
shrimp and finfish in aquaculture
plantations. (Consulate General of
Japan, New York)

better, used as fish food in aquaculture. The final effluent resulting from such
treatment should be of tertiary-treatment quality.

## Economics of recycling

Special means must be employed to encourage these and other potential
means of waste recycling. The recycling of most wastes is usually avoided
because the monetary or energy cost of transporting, re-refining, and remanufacturing them is usually greater than the immediate cost of just dumping
them somewhere. However, as scarce resources become scarcer, antipollution legislation becomes harder to evade, and the economic system is
redesigned for recycling, we will see more of it. If the total cost picture were
to be surveyed over the course of several future generations, recycling
would undoubtably emerge as by far the cheapest and only practical alternative.

At least three courses of action suggest themselves: (1) goods must be
designed durably for continuous and long-term use and for eventual remanufacturing, (2) goods which cannot be reused should be stockpiled, and (3)
we must provide economic and legal incentives to encourage reuse.

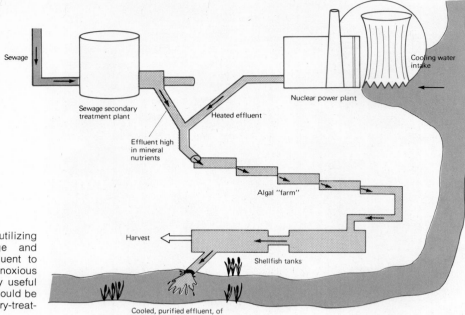

Sewage

Sewage secondary
treatment plant

Effluent high
in mineral
nutrients

Heated effluent

Nuclear power plant

Cooling water
intake

Algal "farm"

Harvest

Shellfish tanks

Cooled, purified effluent, of
tertiary treatment quality

**FIGURE 22-13**
Shellfish aquaculture, utilizing
both municipal sewage and
heated power plant effluent to
produce food. Since the noxious
wastes are consumed by useful
organisms, the effluent should be
approximately of tertiary-treatment quality.

***Remanufacture and reuse.*** Since energy and money must be expended to
transform any object from its original form to some other, goods should be
designed to *last.* Surprisingly, perhaps, many businessmen regard this as
undesirable. Frequent style changes are supposed to motivate increased
public consumption. Whether the style change constitutes a genuine
improvement is relatively unimportant. Long before an automobile is worn
out, it is often unstylish. The situation is even worse in the clothing industry.

Still, goods that last too long are rarely a problem to industry. It is often
possible to build them flimsily enough so that they must frequently be
replaced even in the absence of style changes. Frivolous design changes
and deliberately shoddy construction are sometimes called *planned obsolescence.* Planned obsolescence is incompatible with ecologically sound
business practices. So are goods that are intended to be disposable after
one use. Automobiles, clothing, and containers should be designed for the
longest possible use and reuse.

***Stockpiling.*** Stockpiling is the deliberate accumulation of waste goods in a
limited number of known locations. It is a more worthwhile practice than it
might at first appear. The Romans left us piles of wastes where we could
conveniently process them. The least we could do for our descendants is to
separate our wastes and leave them together in identified and carefully
recorded locations. We could even use them for strip mine landfill or for artificial fishing reefs until they are needed for other purposes. As it now is, we
scatter waste resources to the four corners of the earth, making it impractical
and energetically infeasible for anyone to recover them.

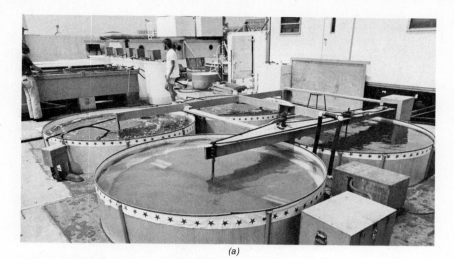

(a)

(b)

(c)

**FIGURE 22-14**
Experimental sewage aquaculture at Woods Hole, Mass. (a) Algae are grown in large tanks of sewage effluent. (b) A tank of mussels raised on the algae produced in the above tanks. (Dr. John Huguenin) (c) Shrimp raised by similar techniques. (Dr. Won Tack Yang and *Environmental Science and Technology*)

**FIGURE 22-15**
The twenty-first century will require unprecedented quantities of freshwater for domestic consumption, for industry, and for agriculture, much of which will be dependent upon irrigation. This photograph of the Gila Project, Ariz., shows a portion of the 115,000 acres of irrigable land in one of the driest parts of the country. (Bureau of Reclamation, USDI)

***Economic and legal incentives.*** Economic and legal incentives are needed before manufacturers will find it worth their while to remanufacture goods, especially complex ones, which have reached the end of their useful life in their original form. An automobile's purchase price could include a deposit refundable only upon presentation at the factory for recycling. Nondeposit beverage containers could be made illegal (as they now are in Oregon). Discriminatory freight rates and taxation policies could be amended to encourage recycling rather than exploitation of fresh resources. In 1972, according to the Scrap Iron Institute, it cost approximately 2½ times as much to ship a ton of scrap iron by rail as it did to ship a ton of newly manufactured iron. Often, because of high shipping rates and excessively low prices of virgin materials, communities that establish recycling centers find it impossible even to give their scrap away. One cannot solve this problem by accumulating more scrap—that just increases supply. At the present time government policy does not allow scrap materials to compete effectively with virgin materials. Clearly, policy should favor recycling, not discourage it.

### Pollution control legislation
Implicit in the foregoing discussion has been the assumption that governmental policies and legal means should be brought to bear upon pollution.

Since the legal picture is changing rapidly, we hesitate to commit much specific information to writing at this point. There are, however, a few broad principles to keep in mind.

First, the environmental impact of major businesses and governmental undertakings should be taken into account *before* they are permitted to proceed. Our current national policy for the environment demands just this. As yet honored more in the breach than in the observance, in time this policy may have a very beneficial effect.

Second, all damage to the environment must be strictly controlled. It is easy to say that benefits of a particular action (e.g., building an airport) must outweigh its environmental costs, but it is not easy to assess just what those costs will be in the long run or how they should be compared with the benefits. A policy of caution must be adopted, and only those projects be permitted whose environmental costs will be *truly* minimal. In other words, the burden of the proof of harmlessness must rest upon the polluter. With continued population and industrial growth, we cannot be satisfied with half measures. Eventually it will become necessary to restrict all pollutional discharges to close to zero, for moderate improvement will not be enough in the face of increased effluent flow. This is, in fact, the 1985 goal of the 1971 Water Pollution Control Act. We must resist the inevitable pressures to weaken this fine legislation.

Third, penalties for violating environmental laws must be realistically severe. Executives of companies convicted of antitrust violations are often

**FIGURE 22-16**

Limited demand, discriminatory freight rates and taxation, and the high cost of hand labor in preparation and separation tend to make the recycling of wastes relatively unprofitable. This pulper, somewhat resembling a gigantic version of a kitchen blender, reduces wastes to a flowable slurry, from which compacted metal containers, glass, and paper fibers can be reclaimed for recycling, thus reducing the costs of separation. (Black Clawson) jailed. If blatant pollution were made a felony for which the officers of guilty corporations and municipalities were held personally accountable, there would be much less of it than if only corporate fines were assessed which often can easily be paid out of company profits.

Fourth, legislation is effective only to the extent that it is enforced. The Federal "Refuse Act" contained in the River and Harbor Act of March 3, 1899, forbids discharge of refuse, except in liquid form, from streets and sewers into navigable waters without a permit from the U.S. Army Corps of Engineers. The act permits citizen suits against violators if United States attorneys will not act. As you could probably guess, this act has scarcely been enforced in the many years of its existence, and federal attorneys still attempt to avoid involvement in suits based upon this 1899 statute.

Only public determination to demand enforcement will put the necessary motive power behind even the finest legislation.

**CHAPTER SUMMARY**

1 Pollution must be controlled for hygienic, esthetic, economic, and ecologic reasons.

2 Sewage disposal arose as a municipal sanitation measure, was greatly expanded partly as a result of the invention of the flush toilet, and only recently has come to encompass the treatment of the wastes carried in the sewers.

3 Sewage treatment processes are classed as primary, secondary, and tertiary. The first two usually involve the biologic treatment of organic wastes. Tertiary treatment disposes of mineral nutrients. Septic tanks are not a satisfactory form of sewage disposal in populated areas.

4 Planned obsolescence and disposable packaging have made solid waste disposal a monumental problem. Together with organic garbage, solid wastes may be disposed of by dumping, sanitary landfill, incineration, pyrolysis, composting, and other methods.

5 One solution both to solid waste disposal problems and to aquatic pollution is the recycling of wastes and their reuse in production.

6 In the long run, pollution can be controlled only as a result of a searching reexamination of economic and population growth trends and public policies.

7 Polluting practices force someone else to pay the costs of waste disposal. Such external costs are known as externalities.

8 For several reasons local political units are not well suited to undertake most of the responsibility for pollution control. Regional, national, and international units have a better chance of success.

9 Waste recycling is the most economically feasible form of pollution control, when considered in toto over the course of several generations. It simultaneously conserves raw material resources and reduces waste disposal problems.

10 Pollution control legislation must take environmental impact into account, must be quite painfully strict, must bear realistically severe penalties for infraction, and must be vigorously enforced.

Dickson, Edward: "Taking It Apart," *Environment,* July-August 1972, p. 36.

Goldman, M.: *Controlling Pollution, the Economics of a Cleaner America,* Prentice-Hall, Englewood Cliffs, N.J., 1967.

Hannon, Bruce: "Bottles, Cans, Energy," *Environment,* March 1972, p. 11.

Reyburn, Wallace: *Flushed with Pride, The Story of Thomas Crapper,* Prentice-Hall, Englewood Cliffs, N.J., 1971.

Ryther, J. H., W. M. Dunstan, K. R. Tenore, and J. E. Huguenin: "Controlled Eutrophication—Increasing Food Production from the Sea by Recycling Human Wastes," *Bioscience,* March 1972, p. 145.

Starbird, Ethel A.: "A River Restored: Oregon's Willamette," *National Geographic,* June 1972, p. 816.

Warren, Charles E.: *Biology and Water Pollution Control,* Saunders, Philadelphia, 1971.

*Waste Management and Control,* National Academy of Sciences, National Research Council, Washington, 1966.

Yee, William C.: "Thermal Aquaculture: Engineering and Economics," *Environmental Science and Technology,* March 1972, p. 232.

Zwick, David, and Marcy Renstock: *Water Wasteland,* Grossman (and Bantam), New York, 1971 and 1972.

# 23

## OUR BIOLOGIC RESOURCES

*Chapter learning objective.* The student must be able to give justification for the conservation of species and living communities and to discuss the means of accomplishing this, with emphasis on urban and regional planning, the preservation and/or management of wild species and their natural habitats, and the causes and prevention of the decline of endangered species.

*ILO 1* Give three justifications for the preservation of biologic resources.

*ILO 2* List three distinguishing characteristics of agricultural communities, and relate these characteristics to the prevalence of pests and wildlife in these communities.

*ILO 3* Give three arguments in favor of, and two against, mixed stands of timber rather than monoculture forestry. Summarize recommended logging practices, and give instances of afforestation, reforestation, and silviculture.

*ILO 4* Give at least one military justification for ecologic warfare as it has been practiced in Indochina and four kinds of ecologic damage resulting from it.

*ILO 5* Give the major environmental problems and priorities encountered in a range management, in highway planning and routing, and in strip mining and reclamation.

*ILO 6* Describe the process and causes of suburban succession, and give its probable consequences if allowed to proceed ungoverned.

*ILO 7* Identify the ecologic aspects of urban blight given by the authors, and list their suggested remedies.

*ILO 8* Distinguish between the preservation of wilderness habitats and species, and the conservation of living communities in relation to their use by man, giving examples of each; and, given an example, be able to make the decision as to whether to preserve or conserve a habitat, giving your reasons.

*ILO 9* List three techniques employed in habitat, fish, and game management, giving an example of each. Provide an evaluation of their relative merits in relation to the requirements of game and other species.

*ILO 10* Summarize the process of extinction, listing factors that can contribute to the decline and extinction of endangered species, and giving an example of each.

Whether one is an emperor or a slave, he cannot exist without green plants. Plants are essential both as food and as a source of oxygen. Our needs do not stop there, however, for even bacteria and protozoa, insects, mammals, and the obscure creatures known only to professional biologists are part of the net of life that holds all of us in its midst. As we have emphasized repeatedly: no living thing is independent of other life.

## ECOSYSTEMS AND CONSERVATION

In a way, to speak of "biologic resources" is misleading, for it implies that our biologic resources are only a few among the many resources that we have, all of more or less equal importance. Not so. Our biologic resources are so important as to be all-important. We could survive after a fashion without iron, steel, glass, or electricity, and indeed we did so for most of the lifetime of the human race. But we could not survive without the presence of the plants and animals with which we share the world.

Though species vary in their ecologic importance, the great majority of organisms can be shown to have an essential or highly desirable place in the ecosystem of which they are parts. Even decomposers such as bacteria play their part, and we could not get along without them. Unfortunately, decay bacteria, mildew, maggots, and worms are offensive to our tastes, but for all that, we must have them. It ought to be clear from this that all creatures should be treated with care and respect. Irresponsible tinkering could produce maladjustments and ultimately could spell the end of the world we know.

An attitude of intelligent respect for the ecosphere is directly related to our own self-interest, even though it is based upon an appreciation of other living things. By preserving other living things, we preserve ourselves and, just as important, preserve the kind of world in which it is pleasant to live.

There is another reason why we should see to the welfare of other species. Some of them may one day have a direct use to us that we cannot yet foresee. The blue whale, now virtually extinct, could have been used as a source of excellent protein for the increasingly meat-hungry world. We can never expect to harvest plankton from the ocean as efficiently as this great beast does. The beef ranching of whales might very well turn out to be the most practical way to make use of the biologic productivity of the high seas. But the possibility of so using the blue whale has been lost to us for ever. The musk ox, a large herbivore which inhabits the arctic tundra, was rendered extinct in Siberia during the Middle Ages, and almost became extinct in recent years in the Western Hemisphere. However, this animal recently has been domesticated in Alaska. It yields a fine wool resembling angora, known as *qiviut,* which can be spun into yarns for luxury and high-fashion garments, providing an important source of income for the Eskimos. Native herbivores in Africa, if domesticated, yield more meat per acre of grazing land than the cattle heretofore employed. Who knows what creatures will one day prove to be tremendously useful? We shall never find out if we render them extinct.

**FIGURE 23-1**
(a) Shown here at play, recently domesticated musk oxen. (J. Bedford) (b) A fine wool, suitable for high-fashion garments, from musk oxen. (H. Ringley)

To this one might well reply, "Wouldn't it be enough to keep a few pairs of each likely kind of animal in a zoo? Why reserve vast tracts of land for their use?" There are several reasons why this is not the most desirable course. In the first place, as we shall see, it is desirable to maintain large gene pools in each species, to minimize the danger of inbreeding and to preserve its range of variation. In the second place, the zoo environment is artificial and probably exerts a selective effect upon the genes of the animals kept there. The genes that fit an animal for survival in a zoo may not be the best ones for it, or for our ultimate purposes. Finally, it is not practical to maintain artificially *all* possible species, but there is no predicting which of them will eventually prove to be of use to us at some future time.

Our ecosystem on earth is a *system*. Just as a spaceship's life support system cannot be expected to function without some of its vital parts, or as a collection of disconnected pieces, so the relationships of the earth's life support system must be preserved in a reasonably intact way and must be harmoniously managed.

Unfortunately, all too few ecosystems can be said to be "reasonably intact." In ecosystems that have been subjected to degradation or stress, whether from pollution or another cause, the only species that survive are those which are resistant to the unfavorable conditions. There are relatively few species in such habitats, though, as with sewage-loving organisms, great numbers of individual organisms may be present. The ecosystem they comprise tends to be simple, unstable, and is often given to wild fluctuations. Although some have specializations fitting them for the local conditions, the organisms tend to be generalists, such as cockroaches and rats, rather than specialists such as nightingales and cheetahs.

It is easy to give lip service to the idea that the conservation of our biologic resources is of utmost importance, but in addition, we need to have the will to act upon that assumption, and the knowledge to act effectively. Time after time, well-meant attempts at the conservation of one resource or other have backfired or have done more harm than good through a lack of understanding of living things, their interrelationships, and their requirements. *Rational conservation is applied ecology.* It can accomplish its ends only if it is scientifically approached.

## THE IMPACT OF MAN UPON ECOSYSTEMS

The very presence of man in a natural community will bring about certain changes, but these changes can and must be kept as minimal as possible. Natural communities are not well suited to man and cannot support large human populations. Only the development of agriculture and husbandry has permitted high human population densities to develop. Modern agriculture is probably the greatest single factor which permits the present world population to exist.

**FIGURE 23-2**
Modern population density requires agricultural land use.
(Don Moores)

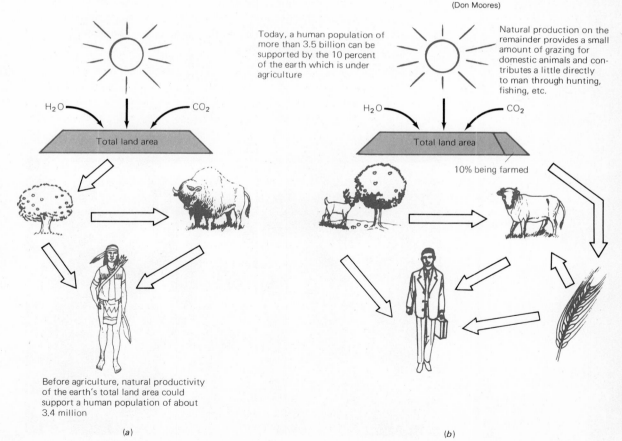

Today, a human population of more than 3.5 billion can be supported by the 10 percent of the earth which is under agriculture

Natural production on the remainder provides a small amount of grazing for domestic animals and contributes a little directly to man through hunting, fishing, etc.

$H_2O$     $CO_2$

Total land area

$H_2O$     $CO_2$

Total land area

10% being farmed

Before agriculture, natural productivity of the earth's total land area could support a human population of about 3.4 million

(a)

(b)

### Agricultural communities and their characteristics

An agricultural community such as a field of corn or of cabbages usually has the following characteristics that set it apart from the original community that it displaced:

1 *It is unstable.*
2 *It is vastly simpler than any natural ecosystem.*
3 *Its chemical cycles are frequently incomplete.*
4 *Interspecific competition is greatly reduced.*

First, it is unstable because it can be maintained only by human intervention. Corn, for example, can reproduce itself only with human aid. Another way of looking at it is that man is an essential part of agricultural communities. This is not true of any natural community.

Second, it is simpler than any natural ecosystem. The plants of an agricultural community may be of only a single species. This *monoculture,* as it is called, greatly simplifies the community but renders it unstable and highly susceptible to insect pests or diseases, which can spread from one host organism to another with little hindrance. In the natural state, host plants (those attacked by pests) are interspersed with others which a particular pest species will not eat. The control of the insects usually requires the use of pesticides, which further unbalances the community. Thus, monocultural agriculture, unfortunately, often demands unnatural pest control methods since monoculture is itself anything but natural.

**FIGURE 23-3**
This vast field of soybeans consists not only of a single species but even of a single variety of that species of plant. (USDA Soil Conservation Service)

Third, the chemical cycles of an agricultural community are often incomplete. That is, the *cultigens* (cultivated organisms, such as maize, of a variety or species which differ considerably from their wild ancestors) are consumed at a spot remote from where they are grown, and their substance is not usually returned to the soil but escapes from the terrestrial ecosystem through some sewage disposal plant or garbage dump. This interruption of nutrient cycles causes the soil to deteriorate unless artifically fertilized—which brings with it problems of its own.

Fourth, competition among species is greatly reduced. Farmers go to great lengths to suppress species considered undesirable (called *weeds*) that compete with cultigens or interfere with their harvest.

It should be obvious by now that an agricultural community is not and cannot be the same as the natural community it has replaced. The demands of agriculture simply do not permit this. Yet some compromise is possible, and with enlightened management, agricultural lands can be made to yield a modest return of wild game, or merely to serve as homes for wild species. However, public policies must be designed explicitly to encourage such management, or it will be practiced much too infrequently.

**FIGURE 23-4**
Even primitive agricultural practices (like the hillside farming shown here) can have substantial environmental impact, sometimes serious impact. (Glenn Westfall)

### Agricultural communities and the requirements of wildlife

Wildlife requires food, water, and "cover" for nesting, escape, and concealment. Even in farmland, the cover requirement for wildlife can be met by allowing weeds and shrubs to grow up around fences or by using multiflora rose bushes instead of fences. European hedgerows have a similar function. Rabbit, quail, and other small game can be encouraged in this way. The preservation of marshes as muskrat habitat can be the most economical use of some poor land through the careful harvest of the fur produced by the rats. Unfortunately, uncultivated areas provide a ready source of weeds, and many farmers will not tolerate them for that reason. Also, hedges and weedy fences tend to break the large areas required by modern farming methods and economy into unmanageably small units. Agricultural subsidies and high prices for produce tend to encourage the use of every last bit of land for farming. Therefore, these marginal habitats are being destroyed almost everywhere they occur. One possible solution to the problem is governmental programs creating "banks" of retired agricultural lands to be managed explicitly for wildlife production, or possibly for timber. The best land is not likely to be banked in this fashion, for the land best suited to agriculture also produces the most wildlife. One cannot expect depleted soil to support lush natural communities any more than it will support productive agriculture. Still, half a loaf is better than none.

### Forestry, afforestation, and reforestation

Hard as it may be for those living in industrialized countries to appreciate, the largest single use of forest products is for fuel (42 percent), for many tropical parts of the world are essentially without fossil fuel resources. Construction (37 percent) and pulp (11 percent) run second and third, but both are of tremendous economic importance. Surely, the preservation of our

**FIGURE 23-5**
In this agricultural landscape, considerable wildlife cover is allowed to persist. (USDA)

forest resources should be a major concern of our civilization. And one important aspect of forest conservation and management is public education as to their value.

One possible answer to the depletion of forests is to manage them as a crop. By analogy with agriculture, such an activity is termed *silviculture,* from the Latin word *silva,* meaning "forest." Some of the things we have said about agricultural communities also apply to silviculture. In some parts of the world, trees of valuable species are managed as a crop and are raised in vast monocultural tracts. These simplified ecosystems have agricultural problems. For example, they support less wildlife than mixed stands of timber, they are far more susceptible to pests than are natural communities, and they are more subject to the depletion of soil minerals. Tree farms, in other words, are not really forests in the usual sense.

The large-scale remote consumption of trees as pulp, lumber, etc., probably will lead eventually to soil depletion. The modern trend in forestry theory, though not yet in practice, seems to be away from these pure stands to a more nearly natural forest community. Mixed stands of timber produce much less soil depletion than monoculture, as the nineteenth-century German foresters who originated monoculture forestry soon discovered.

The biggest cause of loss of raw timber is insect damage, which monoculture encourages. The gypsy moth defoliates vast areas of hardwoods in the Northeast. The pine sawfly damages softwoods. The spruce budworm multiplies prolifically in monoculture spruce forests. Pine blister rust kills sugar pine and other species. Attempts are made to control these pests, especially by spraying biocides from the air. Unfortunately, this spraying

(a)

(b)

FIGURE 23-6

(a) A uniform-aged monoculture forest. (b) Monoculture forests are often more susceptible to the spread of insect pests than are polyculture forests. (*McGraw-Hill Encyclopedia of Science and Technology,* 3d ed.)

has repeatedly produced high mortality rates among the fishes in forest streams, and to some extent has harmed the other animals of the forest community.

But monoculture of forest trees represents convenience in harvesting. In fact, the very diversity of tropical forests leads to the waste of much timber destroyed just to get at the commercially valuable species. However, most commercial species grown by themselves tend to deplete the surface layers of the soil. For that reason, it is well to include the other less commercially valuable species, such as birch, whose roots enrich the earth with nitrogen, and which mine minerals from the deeper layers of forest soils that are inaccessible to most species.

Mixed forests do tend to be unstable, since there is a successional tendency of some species to overtop and suppress others. This disadvantage can be overcome by careful management, and is not always serious, since some mixed woodlands are stable in the natural state.

*Afforestation* is the planting of trees in a habitat in which they do not naturally occur or have not grown in recent history. Communist China and Israel, for instance, have planted vast numbers of trees, both to increase their timber resources and to improve the climate in near-desert areas. Similar plantings, known as *shelterbelts,* have been set up in many countries, including the United States, to produce local windbreaks and to improve the local climate for agriculture through water and soil conservation.

Afforestation is also employed to cover unsightly heaps of slag or other industrial waste, to stabilize soil that is actively eroding, and to cover old strip mines in which soil has been completely removed from the area mined.

*Reforestation* is practiced following forest fires or extensive logging. Its aim is the reestablishment of the original community or one close to it. If successful, it amounts to a great hastening of successional processes. Foresters are attempting to develop techniques of planting and management that will permit the establishment of forest communities to be attained within a lifetime, not, as is natural, in centuries.

(a) (b)

**FIGURE 23-7**
(a) Reforestation to restore timberland. (*McGraw-Hill Encyclopedia of Science and Technology*, 3d ed.) (b) Reforestation to control erosion. (USDA Soil Conservation Service)

***Logging practices.*** Older logging practices destroyed forest communities by removing all trees (clear-cutting). This usually resulted in the loss of the root structure which formerly bound the soil together. Severe erosion resulted. Furthermore, large amounts of brush, branches, and other waste were left about. Such trash is highly inflammable and was the cause of some of the most destructive forest fires in history. Severe fires of this type also can destroy the organic matter in the soil, further rendering it unsuitable for the support of a healthy forest community.

Recently, F. Herbert Bormann and his colleagues undertook studies of clear-cutting and forest regeneration from a different point of view. These workers experimentally clear-cut an entire watershed in a New England forest, and measured the changes in the amount of water running off the cutover land, along with its silt and mineral content. Silt content of the water increased ninefold, indicating a corresponding increase in soil erosion. Mineral concentrations also greatly increased, fixed nitrogen now being present in amounts sufficient to cause eutrophication in the stream drainage of the watershed. As the forest was allowed to regenerate, these trends were reversed. We may conclude that even if cut trees are not removed (as was the case in this study, but which is, of course, not the case in commercial lumbering) clear-cutting damages forested lands. Hopefully, similar studies can be made of rangeland and agricultural land, especially in national forests.

Modern logging practices seek, in a variety of ways, to ensure the regen-

**FIGURE 23-8**
Selective cutting. (a) Forester selects trees to be cut. (b) Immature trees with good growth potential are left standing. (c) Young seedlings grow up to replace those that were cut. (*McGraw-Hill Encyclopedia of Science and Technology*, 3d ed.)

(a) (b) (c)

eration of cut forests. In large part, these practices involve *selective cutting* so that a large part of the climax community is left standing to provide seed and shelter for the new generation of trees. Since it leaves immature trees still standing at all times, this practice also provides a steady, sustained yield of timber, as mature trees are always available for cutting.

Only a few commercial species such as Douglas fir are able to germinate and grow in the open, and require unshaded conditions. With care, such forests may be clear-cut, especially if a few mature trees are left standing to provide seed, or if patches of standing timber are left.

Compromises of various types are possible. Sometimes small portions of forest are clear-cut and closely adjacent areas left undisturbed. If this is done properly, the standing timber will shelter seedlings and aid forest regeneration in the cut areas. Later, the mature trees left standing are completely harvested. This is called the *patch-cutting* system. A somewhat similar technique, the *shelterwood* system, leaves a substantial number of older trees to provide a canopy shading young seedlings. When the seedlings are sufficiently mature, the older trees are cut.

**FIGURE 23-9**
Patch-cutting. (*McGraw-Hill Encyclopedia of Science and Technology,* 3d ed.)

Most standing timber in the United States is owned by small farmers and landholders, who unfortunately do not for the most part manage their woodlots well. In fact, they are often encouraged to convert woodlots into pasture or crop use even if they are not suitable for it. Local taxation often places a higher assessment upon standing timber than on land from which these valuable trees have been removed.

There are vast areas of the world, particularly in the underdeveloped countries, in which rational forestry is unknown. In the more advanced nations state-owned lands or those owned by lumbering firms are managed,

(a)

(b)

**FIGURE 23-10**
Ecologic warfare. Herbicide spraying in Vietnam has devastated these cultivated nipa palms. (a) Before. (b) Several years after. (Dr. A. H. Westing)

to some extent, for sustained yield. The old practices of "cut out and get out" that devastated much of Michigan, Minnesota, and Wisconsin in the nineteenth century are uncommon in the United States today. However, irresponsible exploitation still proceeds elsewhere, especially in Africa and South America. Unfortunately, American companies are often responsible for this practice even to this day in underdeveloped countries.

*Ecologic warfare.* Modern herbicides such as 2,4-D were originally developed during World War II for possible use as biologic weapons to be employed against the crops of enemy nations. After the war, these substances were employed in agriculture for weed control. In the recent Indochina war, the herbicides were put to their original intended use. Herbicides have been used in Vietnam to destroy the crops of those living in communist-controlled areas, thus forcing them to move to government-controlled areas where they might be watched more closely. This course also denied the food they grew to the Viet Cong. Jungles were also sprayed with defoliant to reduce the cover favored by communist forces, whose troop movements required it because of the absolute air superiority enjoyed by the allies. Several study commissions have been sent to Indochina to determine the effects of defoliation on the ecology of that area. The following conclusions seem to stand out:

1 Many climax communities, such as the coastal mangroves so important for the reproduction of fishes and shellfish, simply have been destroyed.
2 Great and probably irreversible damage has been done to croplands and plantations, especially rubber plantations.
3 Subclimax communities have been established in many formerly forested areas—e.g., unbelievably dense stands of bamboo. The former wildlife, though not perhaps directly harmed by the defoliant spray, has had to leave these areas, probably for less suitable habitats.
4 Lateralization of some defoliated areas seems to be proceeding rapidly.
5 Dioxin, an impurity in the 2,4,5-T herbicide that was most widely used in the defoliation program, is a far more potent teratogen even than thalidomide. Though it is not certain, it is likely that this substance has done much damage both to human and animal reproduction in Vietnam.

In response to widespread criticism, the military command discontinued treatment of the Vietnamese landscape with chemical herbicides but continued to destroy the forests by other means. Bulldozers scraped vegetation and topsoil off 800,000 acres of land (equivalent to the area of Rhode Island) by early 1972. In addition, more than 21 million bomb craters, each about 30 feet in diameter and 5 feet deep, have been dug in forest and formerly fertile farmland. Filling these craters, if it proves practicable, will require moving over 3 *billion* cubic yards of earth.

There is nothing new about ecologic warfare. The ancient Romans attempted to sterilize the soil around the city of Carthage by plowing salt into the fields, fortunately without permanent success. Armies have often employed "scorched earth" policies to deny food to enemy military forces and civilians. But it does seem that war against the environment can lead at best to futile victories. Any policy that irreparably damages the areas under contention seems self-defeating, since it destroys the value of the land no matter who wins the war.

Some have attempted to justify such measures by pointing to domestic defoliation, otherwise known as "vegetational-type conversion," in which 2,4,5-T is often used to clear powerline or road rights of way or to kill sagebrush and other species that tend to invade overgrazed land. But the argument cuts both ways. In the words of Senator Gaylord Nelson, "It is hardly believable that after the lessons of Vietnam we would tolerate the same tactic of defoliation to be used in our own backyard."

**FIGURE 23-11**
Cratering in Vietnam. (Dr. A. H. Westing)

## Rangeland

Most of the meat consumed on American tables is beef, lamb, and pork. The cattle and sheep that are the source of the first two are grazing herbivores, and most of them pass from half to all of their pre–pot roast existence on some kind of grassland. Not all of this range is grassland in the ecologic sense. In fact, a majority of it is not—for example, the dairyland that has been made from former Eastern deciduous forest. This would revert to some other vegetational type if allowed to do so. Even so, many of our meat animals are raised on Western grasslands and open forests.

There is a limit to the amount of cattle or sheep that can be supported by an acre of land. That limit will vary according to local conditions of soil fertility, climate, and other factors. Before the limit is reached, rangeland will begin to deteriorate. Just how much deterioration is acceptable is debated, but obviously the time comes sooner or later when it is no longer economical to raise animals on the land.

Range management experts divide grasses and other range plants into three categories: *decreasers, increasers,* and *invaders.* Decreasers are species of climax vegetation that become vastly reduced by grazing. Unfortunately they are also usually the most nutritious species. Increasers are plant species that increase their relative proportions under grazing pressure.

**FIGURE 23-12**
Soil erosion resulting largely from overgrazing: (a) in Illinois, (b) in Lebanon. (USDA Soil Conservation Service)

*(a)*

*(b)*

When grazing becomes too intense even for the increasers, plant regeneration is inhibited, the habitat becomes barren and eroded, or it is invaded by the invaders—weeds of little grazing value that survive because the animals will not eat them. In extreme cases, invaders form stable disturbance climaxes that can be dislodged only with herbicides.

Of all our domestic animals, sheep and goats are the most damaging to rangelands. In moderate numbers they do no harm, but it is hard to persuade sheepherders to exercise restraint in the size of their flocks. Overgrazing by sheep is especially serious because they are able to crop grass closer to the roots than cattle, and their sharp hooves do further damage at the same time. Over the course of centuries, sheep overgrazing has reduced much of the land of Lebanon to a waste. Indeed, throughout much of North Africa and the Mediterranean countries, landscapes resembling those of modern California have been changed to deserts by thousands of years of overgrazing. As the prophet Isaiah foretold, "The land mourns and pines away, Lebanon is shamed and withers; Sharon is like a desert plain, and Bashan and Carmel lose their foliage" (Isaiah 33:9, New American Standard Version).

What Isaiah would say about the probable future of our own grasslands can only be guessed. Though not claiming prophetic gifts, Dr. Alfred Etter has observed that, "It took thousands of years to wreck the grazing lands of North Africa. We have wrecked the Southwest in a hundred." The situation is far more serious than most people realize. Its worst feature may be that much of the destruction is taking place on public land.

A large portion of federally owned land is taken up by our national forests. These are managed for multiple use—grazing, for example, lumbering, recreation, and even mining. Stockmen have often behaved as if they owned this land, and in the past have even tried to fence it in. As things stand today, sheepherders and others lease grazing rights from the government, and in addition must agree to carry out certain responsibilities connected with habitat conservation. Critics of the stockmen contend that the charges assessed against the herders do not compensate the government adequately for the value of the forage they consume, that in many areas erosion and other ecologic damage are resulting from overgrazing, and that stockmen often violate the terms of their agreements with the government.

The time would seem to be ripe to conduct a thorough study of the ecologic impact of traditional grazing practices, and if this study indicates needed changes in livestock density or other practices, to make them, regardless of special interests. Public lands belong to us all.

### Strip or open pit mining

Traditionally coal and other ores were mined by underground tunneling. In more recent years a less expensive and, we are told, safer method has become widespread known as *strip mining*. In strip mining, deep pits are dug—deep enough to reach the coal or other ore from the surface by simply stripping away all the surface soil to do so.

With the prospects for nuclear and thermonuclear power uncertain, coal seems to be the fuel of the future—at least the near future. To help meet the

(a)

(b)

**FIGURE 23-13**

(a) Open pit mining for copper.
(b) Strip mining for coal. (Bureau of
Reclamation, USDI) (c) Surface mining for phosphate. (SELBYPIC)

(c)

much-publicized energy crisis, coal must be mined efficiently and inexpensively. Thus strip mining, more euphemistically termed *open pit mining,* has become widespread in recent years.

As of 1966, *1.5 million* acres of American wildlife habitat had been destroyed by strip and open pit mining, damage to 12,898 miles of streams had been observed, and great destruction of agricultural lands and scenic values had occurred.

A large part of the blame for this situation must rest with state and federal legislatures. Mining companies can purchase public land containing valuable minerals for *$5 an acre,* in accordance with the Federal Mining Law of 1872. A bargain by anyone's standards! It is estimated that a Texas-sized chunk of federal land has been claimed, and while not all of this has yet been exploited, very substantial environmental damage has been done in the majority of our national forests.

In the late nineteenth and early twentieth centuries many Appalachian settlers signed away mineral rights to their lands for from 15 to 25 cents an acre. The deeds did not specify any particular method of mining, but then in those days the gigantic earth-moving machinery that makes strip mining possible was unknown. Today, using these same deeds, mining companies can, in a number of states, remove farmland, knock down homes, and uproot corpses in graveyards without being compelled to pay one cent in compensation. To be sure, when it is all over, the owners can have their land back. Much good that it does them. Clearly, some laws need changing.

Not that strip mining need be this way. For an estimated cost of 10 cents

**FIGURE 23-14**
Reclaimed strip mine site. (Bureau of Reclamation, USDI)

a ton of mined coal, stripped land can be partially reclaimed for agriculture, forestry, or even in some instances for recreation. In Germany coal is often strip-mined, but strict laws compel the careful replacement and regrading of topsoil so that very little environmental damage is produced; so it can be done. But without toothy legislation it will not be done, or will be done only in token amounts. Someone had better do something. We are losing at least 4,650 acres of land per week, less than one-third of which is being reclaimed.

### Highways

It is difficult to estimate the total land area occupied by highway rights of way in the United States. Certainly it is of a state-like order of magnitude, with 200,000 acres being added every year. As long as we continue to add population and dwelling places to our landscapes we shall, of course, need roads, so by themselves these statistics tell us little. But we are entitled to ask two questions: (1) Is all this highway construction necessary, and (2) is it being carried out in the least harmful fashion possible?

(a)

(b)

**FIGURE 23-15**
Two roads.

The answer to the first question is tied up in the general problem of urbanization and regional planning. The answer to the second is a barely qualified "no"! With a few happy exceptions highways are planned to suit not the environment but the convenience of highway engineers. Streambeds and public land are actually prefered highway routes, since the costs of acquisition are low or even nonexistent, and no complications arise from a multitude of private ownerships and the consequent necessary condemnation procedures. In 1965, the then National Park Service director, George B. Hertzog, Jr., put it this way: "Somehow they seem to be able to find a blue pencil that hits the green spot." Not even the constitution of the State of New York was able to prevent highway men from routing the New York Thruway through the Adirondack State Park.

Planners are beginning to realize that highways, by improving access, actually generate traffic in the long run. If this is true we must begin to take a very critical attitude toward new highway construction and place the burden of proof of need upon their proponents as a matter of general policy. When they are constructed, moreover, provision should be made for the preservation or even enhancement of scenic beauty along their routes, such as has been done along New York's Taconic Parkway. Considerable wildlife can be preserved by retaining trees and shrubbery in median strips, and every attempt should be made to preserve habitat traversed by the highway, especially in the case of wetlands.


## THE DEFENSE OF THE LAND

Environmental problems often surface first in cities. Air and water pollution were first perceived as problems in urban areas, and for the most part efforts at their control are even today exercised primarily in cities. There is no doubt that noise, overcrowding, crime, and every imaginable pollution can make the modern city a thoroughly unpleasant place in which to live or work. By any theory of origins, man must be held to be a creature of green spaces and open habitat. Who can blame people and businesses for wishing to leave the inner city for the sunshine, field, and forest of the suburb? The trouble is that there is nowhere to build houses and erect buildings in the suburb except on the former fields and in the previous forest.


### Suburban succession

The typical suburb develops by a sort of successional process. First, a formerly rural area is opened to suburban exploitation by improved transportation facilities for commuter use. A few fairly affluent families settle there, followed swiftly by real estate speculators who proceed to subdivide whatever estates or farmlands they can find. With the rise of cheaper neighborhoods, the influx of young families begins.[1] The new population requires all types of community services which the former village or county government is unable to provide, so taxes are increased. Almost always, the cost of these services outweighs the increased taxes that "improved" land brings into community coffers, so taxes rise even on remaining undeveloped land. Unless special measures are employed to prevent it, any remaining farmers swiftly sell out, because of the combination of sky-high land taxes, difficulty in growing crops on account of air pollution and vandalism, complaints about smelly livestock, and, above all, the very high prices now offered for their land. Thus, the cattle, the woods, or the orange groves that originally

---

[1] "In the 1970's and 1980's, the baby boom generation will marry, have children, and set up house in the suburbs, creating a tremendous demand for the conversion of rural land to urban use." — The Report of the Commission on Population Growth and America's Future.

(a)

(b)

(c)

(d)

**FIGURE 23-16**
Suburban succession. (a) The values that originally attract city dwellers to the suburbs tend to destroy the very values they seek. (b) and (c) Eventually, the suburbs may become almost completely urbanized, with further population growth and industrial development (d).

made the area so attractive swiftly disappear *along with all they could produce,* so that a view from the air often reveals little but a sea of split-level roofs and so much asphalt roadway that rainwater no longer can find its way underground to recharge the water table. The only vestige of the origins of the subdivisions is pathetically preserved in some of the quainter street names: "Fox Lane," "Elm Street," or "Cowpath." In time, the neighborhood may decay, as the houses, often poorly built, deteriorate. Apartments go up where land values remain high, any remaining "waste" land becomes covered with "light" manufacturing fleeing city taxes, and what amounts to another urban area is formed. People flee from it and build more suburbs elsewhere. The growing exurban edges of cities approach one another and coalesce. Soon there is nowhere else to go—a megalopolis now stretches drearily for hundreds of miles, relieved occasionally by some pathetic "greenbelt" or other futility of the regional planners.

Of course we do not mean to imply that regional planning is always futile, but by itself, no amount of planning can deal with the overwhelming problems posed by the basic expansive and exploitative characteristics of

**FIGURE 23-17**
A pattern of growth typical of shore areas is the establishment of large hotels and condominiums, which in the end block off public access to the natural resources that should belong to us all. This is a typical Florida beach scene. (SELBYPIC)

our society. How can planning restore the environment so long as we feel free to overturn or ignore the plans? As long as our population grows without restraint, as long as we can afford to gratify our whims without regard to environmental cost, as long as we place affluence ahead of rational planning, just that long will the decline of our habitat continue. As it goes on, entire ecologic communities are obliterated, the best farms are covered with parking lots, pollution spreads throughout the land, and our national heritage is lost forever.

> When Daniel Boone goes by, at night,
> The phantom deer arise
> And all lost, wild America
> Is burning in their eyes.
> —*Vachel Lindsay*

### The defense of the city
Why do we build the great ant heaps we call cities, and proceed to make them uninhabitable? Can we not produce an urban environment that is pleasant and attractive, where people will *want* to live? The answer is decidedly "yes," but the effort required will be very large and will require sustained determination.

*Why live in the city?* The city offers many advantages available nowhere else. Libraries, operas, museums, zoos, symphony orchestras, and most universities are there. The cultural life of our society is formed and given expression in the great urban centers. Creative people and all kinds of business enterprises concentrate in the city, in large part, just to be near one another and near the great port and transportation hubs that the city provides, and out of which most of them grew. Manufacturing concerns, too, are most efficient when centralized and mutually accessible. The stimulation of coming in contact with all kinds of people and enterprises is irreplaceable. If our cities are permitted to degenerate and decay, we will lose a major seedbed of creativity in our society. Yet it should be obvious that cities, as they are currently constituted, are social and ecologic monstrosities which cannot survive indefinitely.

***Some panaceas.*** Two major schemes have been proposed to combat urban pathology. One is urban renewal, the other is urban decentralization. Their critics contend that neither gets at the heart of the problem, both are at best stopgaps, and both tend to divert public attention from genuine remedies.

*Urban renewal.* Urban renewal operates on the assumption that urban problems are housing problems, and that if bad housing is removed, the problems will go away, perhaps to the suburbs. Too often, the poor people who already live in the slum area are evicted, the tenements are razed, and a business or high-priced residential district is erected in the former slum. Meanwhile, the original inhabitants have crowded into or even created some other slum to the detriment of both it and themselves. If we would begin to call urban renewal "urban removal," perhaps people would begin to see it in a truer light. Frequently the principal beneficiaries of such programs are those whose assets are enhanced through the appreciation of their slum properties by virtue of some nearby urban renewal project.

Of course the inner city should not simply be left to rot. Some kind of remodeling is obviously needed. But until the social, environmental, and demographic root causes of urban decay are dealt with, urban renewal is at best a cosmetic and token solution to the problems of the inner city.

*Urban decentralization.* According to the advocates of this plan, the United States is not overpopulated as a whole. More than 50 percent of the American populace resides on less than 2 percent of the land. If population could be spread more evenly by building new cities in underpopulated areas, the

**FIGURE 23-18**
Urban removal. The Pruitt-Igoe housing project in St. Louis bites the dust. Rising maintenance costs and a burgeoning crime rate forced destruction of this attempt at urban renewal. (Paul Ockrassa)

*(a)*

*(b)*

*(c)*

problem would be solved. Those who dissent claim that this step by itself would not help the already existing problems of the cities; at best, it might keep them from worsening as rapidly as they now do. It would spread urban pollution to areas now free of it, would establish large populations in areas economically and ecologically unsuitable for them, and would take valuable agricultural land out of production. The critics of urban decentralization believe that it is more rational and constructive to make our present cities decent places in which to live.

## What is to be done?

If a stop could be put to the noise pollution, air pollution, and water pollution of cities, the improvement would be so dramatic that the exodus to the suburbs might actually be reversed and an urban middle-class reestablished. As it is, in the 1960s declining central cities lost even more population than rural areas did. But since these rural immigrants must migrate to the urban fringes and to the suburbs, the net effect is vast growth of cities and suburbs. The Commission on Population Growth and the American Future predicts that by the end of the century, fully one-sixth of the American land surface will be urbanized. This process cannot be allowed to continue. The problems of the city can be shown to be of cardinal importance on ecologic grounds alone. Substantial national resources must soon be committed to them. At base, many of our ecologic problems are urban problems.

**FIGURE 23-19**
Cluster development in Reston, Va. By close grouping of residences, considerable natural habitat is preserved in relatively undisturbed state.

## THE CONSERVATION OF NATURAL COMMUNITIES

Recently an African visitor to the United States was asked what impressed him the most. His reply was striking. He had seen more lions confined in zoos in America than he had ever seen free in his own country, to which the lion is native. Must we look forward to the day when the only representatives of wild species will be confined in zoos? Or can we conserve and preserve natural habitats with all their plants and animals intact?

We must distinguish between preservation and conservation. A habitat is *preserved* if it is maintained in its original, natural state with a minimum of change brought about by human use. *Conservation* goes a step farther than preservation. It is practiced to retain the maximum semblance of a natural community in habitats subject to more or less extensive human use. Wilderness is preserved; agricultural land and its wildlife are conserved and managed.

### Parks, government lands, and recreation

Many nations possess tracts of government-owned lands. Sometimes they are set apart as national parks and preserved from exploitation. Occasionally, they are deliberately maintained as wilderness with very limited visitor access and facilities. The national park systems of the United States and Canada approximate this situation. *Theoretically,* they are intended to be preserved forever in an unaltered state.[2] Other lands are merely public and are made available for multiple use—recreation, mining, grazing, and lumbering. In the latter areas, the habitat receives only moderate protection, if any. Most public domain in the Western states falls into this category, as do the "national parks" of India. Many multiple uses simply are not compatible. The use of land for lumbering is not easily reconciled with recreation, or at least, to date, it rarely has been. The use of trailmobiles or the multiplication of campgrounds and facilities as motel substitutes is not compatible with the preservation of wilderness. Grizzly bears do not mix well with tourists, or native herbivores with Hindu cattle. Characteristically, wildlife is subordinated to *all* other considerations on such lands and is frequently persecuted in various ways, sometimes with great vigor.

Most Americans seem to view national parks and similar areas simply as recreation sites; they do not seem to value them for their own sakes. The healing, rest, and relaxation that only wilderness can provide seem to be valued less by the public than water sports, motorboat racing, or noisy and destructive trailbiking. We value paved highways more than pack trails; pontoon planes more than canoe portages.

There is, of course, no valid objection to the recreational use of nature. It is an avidly felt need of urban man which must and should be satisfied. We

---

[2] National Parks were established by the National Park Service Act of 1916 "to conserve the scenery and the natural and historic objects and the wildlife therein and to provide for the enjoyment of the same in such manner and by such means as *will leave them unimpaired for the enjoyment of future generations*" (emphasis supplied).

cannot expect people to retain ecologically healthy perspectives without contact with nature. However, the problem amounts to another version of the suburban theme—the values that draw people to the wilderness are destroyed by their presence. Even dedicated hikers and climbers inadvertently destroy alpine habitats simply by walking over them, and so much human waste is deposited on heavily visited mountain tops that the installation of chemical toilets is seriously contemplated. Recreational use of natural habitats is still *use* and is bound to produce ecologic changes in them. We fully recognize the dilemma this presents. One of the prime motivations for conservation is recreation, so perhaps we should accept the consequences as a necessary evil, better than no conservation at all. However, it is hard to contemplate the destruction this would entail with equanimity. As Shelley wrote:

> Soft sunshine and the sound
> Of old forests echoing round
> And the light and smell divine
> Of all the flowers that breathe and shine:
> We may live so happy there,
> That the spirits of the air,
> Envying us, may even entice
> To our healing paradise
> The polluting multitude.
> —*The Euganean Hills*

## Wise use and management

Perhaps, though, the widespread "development" of national parks to accommodate the horde of visitors that our affluent and expanding population produces is not really necessary. If our everyday lives were in general healthier, perhaps the values that people clumsily seek in "nature" would be accessible to them at home. Furthermore, artificial areas explicitly for recreational use could be designed in the near vicinity of cities, tastefully asphalted and carefully landscaped, concrete-lined, with motel facilities and lunch stands. These areas could be sacrificed as natural habitats so that

**FIGURE 23-20**
Recreational use of land in three different settings. (a) Vest-pocket park in midtown New York; (b) the Sheep Meadow in Central Park; (c) backpacking in the Wind River Range in Wyoming. (OMIKRON)

(a)

(b)

(c)

wild or wilderness areas could be preserved. Such areas could be left deliberately undeveloped and inaccessible, with perhaps, in addition, some limiting visitor quota system. The alternative is to turn our parklands into tent slums and trailer camps. Most more-or-less wild lands are going to be subject to multiple use, of course. However, it is our view that as long as true wilderness areas are preserved inviolate or restored to that condition if necessary, there is no reason why other public and private lands cannot be managed for maximum return. Only let us not define "maximum return" too narrowly. Not infrequently, the economic benefits gained from permitting the habitat to be relatively undisturbed are the greatest even on private lands. After all, consider the revenue brought into an area by hunters, tourists, and fishermen. In the vicinity of certain African national parks, this is actually the principal local source of income. And can a full life be defined in strictly economic terms? Prices are not the same things as values. A reasonable approximation of a natural environment is a prime ingredient of an enjoyable life.

FIGURE 23-21
African wildebeest and hartebeest. In some African countries parks and related enterprises are the mainstay of the local economy. (Pat Gill)

*Preservation and conservation.* For some habitats, the best use is preservation as wilderness. This is especially true if they are unique in the composition of their biota, or if they are of outstanding historic interest. One example of such a habitat is the Galapagos Islands, the investigation of which by Darwin provided a cornerstone for his theory of evolution. Such places should be preserved, even locked up, indefinitely. Not even a remote oceanic island is safe from human exploitation. All pressures for the use of unique ecosystems should be steadfastly resisted.

As our population grows there will be increasing pressures to exploit our national parks and similar areas. As timberland, for example, becomes scarcer, it will become an irresistibly tempting target to all manner of entrepreneurs ready to produce a high-sounding justification for their exploitation. The only way to prevent this is to resolve firmly that national parks and designated wilderness areas will be kept *inviolate,* and to pass on that determination to future generations. A recent survey of the history of New

**FIGURE 23-22**
This alligator, having left his suburban pond, made his way through several backyards to an unknown intended destination. Lassoed by police and returned to the habitat that has been rendered unsuitable for him, he has an uncertain future. No legal protection will preserve species in the face of the destruction of their habitat. (*Clearwater Sun*, Fla.)

York City's Central Park showed that had all the worthy and public-spirited proposals for the use of its land been allowed, the park would not have been large enough to accommodate them. New Yorkers owe the Central Park that exists today to narrow-minded but far-sighted persons who took seriously their responsibility to hold this priceless resource in trust for us, their posterity.

We suggest, then, preservation for wilderness areas, conservation of other lands, and management of lands subject to frequent or substantial use of any kind.

*Habitat management.* Whenever man's activities impinge substantially on a habitat, it changes. Therefore, it is usually necessary to take some kind of counteraction to offset the bad effects of the changes we deliberately or unwittingly produce. For instance, flood control operations have protected coast redwood groves from the periodic flooding they may require. Perhaps we should occasionally flood them deliberately. Also, when we preserve certain forests from fire, we permit the accumulation of brush which, when it *does* burn, burns so fiercely as to destroy the habitat. Under natural conditions, it might only be moderately damaged. Fire prevention may also sometimes change climax vegetation. Ponderosa pine forests develop properly only when competing white fir seedlings are suppressed by burning. Some trees persist mainly because their seedlings are more resistant to fire than those of their competitors. Bristlecone pine actually requires fire to open its seed cones and allow it to reproduce. Regular, controlled burning need not be destructive, but it must be planned carefully and executed on the basis of exact ecologic knowledge.

The problem of controlled burning illustrates the fact that sound management may be contrary to tradition and must be done on the basis of the best professional training.

*Fish and game management.* In colonial America hunting regulations and fees were intended only to restrict harvest, but in the nineteenth and twentieth centuries fees from hunting and fishing licenses became a source of revenue for fish and game management. Of course, fishing licenses also provide a means of regulating exploitation, so that harvest does not exceed natural increase. Hunting is obviously permissible only as a means of removing the excess animal population that the carrying capacity of the habitat is insufficient to support. The determination of what is "excess population" should not be hastily made, nor should it be defined by commercial or narrow interests, but by employing strictly ecologic criteria.

There are four broad techniques of fish and game management. They include the regulation of harvest (including animal "control"), the preservation or manipulation of the habitat, artificial propagation, and the control of pollution (see Chapter 21).

*Preservation and manipulation of the habitat.* If a habitat is preserved, all members of the community will also be preserved. However, if a permanent

(a)

(b)

**FIGURE 23-23**

Habitat improvement. (a) A pleasant instance of cooperation among government agencies, this Minnesota waterfowl wetland was created by flooding 25 acres of lowland. (USDA Soil Conservation Service) (b) Sea bass gather around a newly formed artificial reef. (National Marine Fisheries Service)

increase in the population of some one animal is desired, the habitat must be manipulated in some way to increase its carrying capacity with respect to that particular organism. Such tinkering usually disturbs the former balance of the ecosystem, and some of the effects of that disturbance may be undesirable. For example, one of the best winter foods for quail is the seed of the common ragweed. If a farmer wished to increase the production of quail, he could encourage the growth of ragweed on his fallow land. But pity the poor allergy sufferer nearby!

One of the most successful habitat improvement programs is administered by the United States Fish and Wildlife Service. This program involves the restoration of drained Midwestern marshes and wetlands for the use of migratory waterfowl. So much swampland has been drained, some of it under federal programs, that needed migration stops and breeding grounds are vanishing. Funds derived from the sale of federal duck stamps are used to purchase land for this purpose. Without this program, migrating waterfowl might be a thing of the past today.

*Artificial propagation.* A number of fish and birds are reared commercially or by government agencies for release into natural habitats. The theory behind this is that a higher level of human predation, that is, harvest, can be maintained if the reproduction of the organism is also maintained at an abnormally high level. Also, in the case of some migratory fish (such as salmon or trout), the usual breeding grounds may be rendered inaccessible by high dams, or destroyed by pollution or stream channelization. In such cases, hatcheries may be necessary for the survival of the species.

However, cost analysis of propagation projects frequently reveals that the expenses are far higher than the revenues from licenses that are supposed to support them. Birds or fish are sometimes even released into habitats unable to support them. In extreme instances, it has cost some states several times the price of a hunting license to raise the average hunters "bag" of quail.

It is important to consider artificial propagation in relation to carrying capacity of the environment, that is, its inherent ability to support organisms.

**FIGURE 23-24**
When headwaters are rendered unsuitable by pollution or other damage, or when high dams make it impossible for migratory fish to reach their breeding grounds, artificial propagation may be the only way to preserve populations. Here the eggs of the Atlantic salmon are removed to be artificially fertilized by male salmon sperm. (*Bureau of Sport Fisheries and Wildlife*)

Carrying capacity depends upon the law of the minimum; in other words, a population of organisms must have, at all times, the minimum fulfillment of *each and all* of their requirements in order to maintain themselves. It follows from this that if more game than the habitat can permanently support is released into it, the game simply die off unless almost immediately harvested. This immediate harvest resembles, economically, a purchase in a store. Thus, sometimes animals are deliberately released just a few days before the season begins in numbers far greater than can possibly survive. The ensuing unsportsmanlike spectacle can be almost as ludicrous as shooting animals in a zoo. In one Long Island "trout" stream, fish were released the day before the season opens, for one day was as long as they can live in that polluted water.

In general, artificial propagation is uneconomical and should be avoided in favor of habitat restoration. It can be of use, however, in preserving endangered species such as the whooping crane or the migratory fish we have mentioned, or in restocking game in an area of restored or newly protected habitat from which it had been eliminated.

*Zoos.* It is rarely appreciated that zoos can be a potent force for both the destruction of and the conservation of endangered species. Until quite recently, competition among zoos for specimens of rare wildlife contributed greatly to the endangerment of some species, perhaps the most notable example being the orangutan. But it is also possible for zoos to undertake breeding programs which, if properly administered, can obviate the taking of wild specimens and can provide breeding stock for the eventual restoration of species to their original habitat. This has recently been accomplished with the nene, the rare terrestrial goose of Hawaii, which, though extinct in Hawaii, was preserved and propagated in Great Britain until it could be reintroduced to its native islands.

Zoos have their drawbacks. They exist principally for human entertainment and edification, and some of the worst of them are little more than concentration camps for wild animals. Fortunately, this trend seems to have been reversed of late, so that in some facilities, such as the Busch Gardens in Tampa, Florida, the animals run more or less free, while the visitors view them from monorails or railroad trains. But small roadside "zoos," particularly those run as tourist traps, should be strictly regulated.

*Control of harvest.* Harvesting sanctions are of three types: *prohibitory, regulatory,* and *bounty.* Prohibitory statutes offer endangered or valued species complete legal protection. Regulatory measures attempt to limit the harvest in such a way that reproduction equals losses. Bounties are paid to encourage the reduction or extermination or even extinction of species held to be undesirable.

It is hard to protect migratory species such as whales, many birds, or bighorn sheep, because they may pass beyond international boundaries to countries that may have different hunting regulations, or none, or none that they enforce. Although international treaties are a partial solution to this problem, a complete solution depends upon public education and changes in public attitudes, difficult to achieve in primitive or peasant societies—though not only in them.

Bounties are almost valueless as control measures. In the first place, they are seldom directed against genuinely undesirable species, and in the second place, bounties almost never effectively control a really dangerous pest. Bounties have been offered for years in New Zealand for the introduced deer that have demolished forests there, but there is no evidence that this prolific species has been bothered in the slightest by it. Another pest, the American gray squirrel, has not been reduced in Great Britain even by the device of offering free ammunition to any one who will shoot it.

In the third place, it is easy to cheat on bounties. European muskrat bounty hunters have been known deliberately to preserve colonies of muskrats or even to raise them by hand, so that their source of bounty money would not run out. (The muskrat is an introduced pest in Europe.)

However, of all the objections to bounties, the most serious one is that

**FIGURE 23-25**
In time, some species, especially predatory species, may be found only in zoos. This maliciously shot and permanently crippled bald eagle has found a home at Busch Gardens Zoo, Tampa, Fla.

they sometimes reduce genuinely desirable species (usually large predators persecuted out of ignorance and prejudice) to levels risking extinction. This is a likely result, since predators are usually few in number, for reasons we have discussed. Such has been the case over much of its range with the bald eagle, national emblem of the United States, the timber wolf, the puma, and many other vital and irreplaceable predators.

## ENDANGERED SPECIES

Several organizations maintain lists of endangered species, species of animals or plants in imminent danger of extinction. Every year, with depressing regularity, the lists lengthen. There is nothing new about extinction, of course, as any visit to the paleontology halls of a museum will demonstrate.[3] However, the vast majority of recent extinctions can be shown to result from man's activities, and therefore should be susceptible to man's control. Most or all of these extinctions are undesirable, and a stop could and should be put to them. Extinctions are now taking place at a rate that seems unprecedented in the history of life. Action is urgently necessary to halt this irretrievable loss.

**FIGURE 23-26**
Harpooning a blue whale, Queen Charlotte Strait, British Columbia. High profits to be made from whale products have produced the near-demise of this largest of all animals. Indeed, blue whales may now be virtually extinct.
(Bureau of Commercial Fisheries)

### How species are endangered

The species most vulnerable to extinction appear to be those which are large, predatory, and migratory, such as the polar bear. However, almost every imaginable combination of characteristics occurs among the endangered species, for extinction also pursues animals that require large tracts of wilderness or solitude, that live in very specialized or restricted habitats, that compete with man in any respect, or that yield an economically valuable product. Recently and ominously, environmental pollution has been involved in some extinctions.

*Habitat destruction.* Although it is difficult to be sure, the destruction of habitat appears to account, at least in part, for the bulk of extinctions and endangerments of species. The famous case of the now-extinct passenger pigeon probably resulted from the destruction of the beech forests in which it nested, as much as from the market hunting that is usually blamed. Another extinct species, the ivory-billed woodpecker, required large virgin tracts of cypress forest. Today there is little of that habitat left. Man's modification of habitat does favor some species, on the other hand. The American robin and many other species of songbird, quail, and even red fox have greatly benefited from man's activities. Whether this is good or bad is a matter of opinion. The ecosystem may possibly be harmed by an excess of robins or foxes.

---

[3] Some extinctions of prehistoric *megafauna* (large animals), such as the wooly elephants known as mammoths, may have been directly attributable even then to mankind. Many paleontologists believe that they were exterminated by paleolithic hunters.

***Overexploitation.*** Habitat destruction is not the only danger to which species may be exposed. During the nineteenth century a number of egrets and herons were hunted almost to extinction by the fashion industry; they required only protection from shooting to enjoy rapid recovery.

Excessive hunting or trapping which endangers a species may arise from a variety of causes. In the nineteenth century, game was shot by market hunters to supply food to urban areas. Many wild herds of American bison or buffalo were exterminated in order to starve out the Plains Indians.[4] Commercial users, particularly the fashion industry, may endanger species by their demand for various furs, hides, plumes, or other products such as whale oil. Among the animals most recently threatened by commercial exploitation is the American alligator, a vital component of the ecology of the Everglades and similar swamps. Protecting such species depends upon the prohibition of hunting or trapping the endangered animal plus tight legal restrictions or prohibitions upon the importation, marketing, or other commercial movement of the products of the animal or the animal itself (for use in laboratories, zoos, or the pet trade). Mere legal "protection" is seldom sufficient by itself, for poachers will risk any penalty if the commercial rewards are high enough, as the recent near-demise of the alligator vividly demonstrated.

---

[4] The buffalo, it seems, is not safe yet. Today there are three herds of free-ranging bison, the largest of which is in Yellowstone Park. These animals, like many species of wildlife, harbor the disease brucellosis, a threat to nearby cattle herds. Nearby ranchers wish infected buffalo to be destroyed, but the process would destroy also the character of the buffalo herds and would profoundly disturb park ecology.

**FIGURE 23-27**
Bison, the last of the free-living herds of buffalo, at Yellowstone National Park. The ecology and social behavior of these animals can be studied only in this natural setting. (National Park Service)

*Predator persecution.* Predatory animals have been brought to the verge of extinction everywhere by bounty systems or by general persecution resulting from prejudice and ignorance of the predator's ecologic role. Though a surprising number of states, especially in the West, retain bounty payments on predators, this system is falling into disrepute and is becoming less widespread.

Those few farmers and stockmen suffering losses from predators could be compensated from compulsory predator insurance policies. Most native predators should enjoy complete legal protection. No ecosystem is fully stable without predators.[5] In any case, the numbers of most large wild game are determined not by predation but by other factors. In most Western states every coyote has or recently has had a price on its head, but the services of coyotes as mousers are beyond price. Almost everywhere that coyote "control" (which usually is a euphemism for extermination has been successful, rodent and rabbit control problems have greatly increased. It is sometimes suggested that bounty money or other funds now paid to obtain dead coyotes would be better spent on predator insurance, to pay for whatever livestock these animals might kill. Perhaps we should consider an occasional dead calf as a sort of tax exacted by the coyotes in return for their excellent services in the ecosystem. Despite widespread propaganda to the contrary, such losses are likely to be quite bearable.

*Environmental pollution.* A number of predatory birds have been brought to the verge of extinction by an unexpected and bizarre form of environmental pollution. As previously discussed, chlorinated hydrocarbon biocides (and perhaps the PCB class of industrial chemicals) tend to accumulate in the bodies of the topmost members of a food chain. Although this does occur in predatory mammals, especially if they are aquatic, it is especially marked in predatory birds. Chlorinated hydrocarbons do not seem to kill adult birds directly, in most instances, but they do disturb their reproduction by producing impossibly thin eggshells, as well as severe disturbances in nesting behavior. The peregrine falcon, the bald eagle, and the brown pelican have been disastrously affected by chlorinated hydrocarbon biocides. The brown pelican, the state bird of Louisiana, no longer nests in Louisiana and seems to have drastically curtailed and probably ceased reproduction throughout its range. Even though some live birds now persist in Florida and a few other places, they will become extinct if no more of them are hatched.

Water pollution and air pollution endanger certain species. Coastal or es-

---

[5] In addition to regulating the population density of their prey, many predators, such as wolves, take mainly old and sick animals, thus acting to cull herds and keep them healthy. In areas such as the Northeast where deer predators are largely absent, deer have become a pest which only human predation (hunting) can keep in check. The prohibition of deer hunting in New York and Pennsylvania would produce an ecologic disaster. However, many prey species seem to be limited in their population density by factors other than predation, so that predators take only those individuals which have been forced out of favorable territory and would probably die even without predation.

tuarine fish are in some danger from pollution, and desert species are in many cases in acute peril. Since water of all sorts is so scarce in these arid environments, it takes only a little pollution or other environmental degradation to endanger an entire stream system. Ponderosa pine in the vicinity of Los Angeles has been affected by air pollution, and though the species is by no means endangered by this, we fear that it is a straw in the wind, indicating a possible ominous future trend. Foresters are already talking about attempting to breed varieties resistant to photochemical smog. If air pollution continues to worsen, we may expect further damage to wild and cultivated plants and to animals.

*Exotic species.* Many times exotic or foreign species introduced in an area do no particular harm, as far as one can tell. For example, the Oriental ringnecked pheasant was introduced to England, probably by the Romans in ancient times, and to North America in the nineteenth century.[6] It seems to fit well into the ecology, though it may compete with other desirable gallinaceous (chicken-like) birds such as quail and partridge. The cattle egret of Africa made its way to Florida and other Southern states by unknown means in recent years, and apparently does no harm. However, by far the majority of such introductions have been disasters for native faunas, as in the case of the mongoose.

When the West Indies were first settled by Europeans, they were also settled by rats. The rats were such a threat to plantation agriculture on those islands that the Indian mongoose, a weasel-like predator, was introduced as an attempted biologic control of the rats. The mongooses did eat many rats, and probably forced the rodents to take up a more tree-dwelling habit than had previously been their custom. Unfortunately, the mongoose did not materially reduce the number of rats but did take to eating chickens and native wildlife. Many birds and mammals have been rendered extinct as a probable result of the introduction of both rats and mongooses to a previously isolated habitat. The mongoose is not universally reviled, but its introduction to most countries, including the United States, is now illegal.

Often alien creatures are introduced without the natural enemies or other limiting factors which keep them in bounds at home. The absence of parasol ants and manatees in Florida allows the exotic water hyacinth to flourish there much more than in its native South America. Exotic immigrants also simply may be better adapted to their new environment than the native animals. In any case, if the newcomers have a competitive advantage, they produce extinctions. IMPORTATIONS OF ALIEN PLANTS AND ANIMALS SHOULD BE STRICTLY PROHIBITED EVERYWHERE unless preceded by the most thorough ecologic study of the probable consequences.

---

[6] Ironically, the ring-necked pheasant has suffered recently from the destruction of its habitat. Suburban succession in the lowlands of New York State has endangered it in some areas. Recently the New York State Department of Conservation has tried to introduce Korean and Japanese pheasants, which prefer the hillside habitats less likely to be disturbed by suburban subdividers.

## The extinction process

It is usually not necessary to kill every last representative of a species to render it extinct. Seemingly, all that is necessary is to reduce it to a certain *critical population density*. Below that minimum density, chance happenings may destroy the population, reproduction of the organisms becomes improbable because it is difficult for them to find one another or, in some cases, because courtship is communal and will not take place without the mutual stimulation of large numbers of breeding animals. What is more, a small population has a restricted gene pool, so that inbreeding becomes a danger. Typically, an endangered population declines gradually over a long period of time and then drops all at once to zero. In this way, species may be almost gone before their peril is realized and before anything can be done. A notable example is the heath hen of Martha's Vineyard, which began to be protected when there were still a fair number of individuals left but which became extinct as a result of chance happenings. The same thing may yet happen to the rigidly protected whooping crane or California condor.

There is no evidence that life exists anywhere in the universe except upon our planet. If that is the case, then every extinction is an event of cosmic significance, for nowhere in the galaxies will the skies again be darkened by vast flocks of passenger pigeons. What would we give to see a dinosaur? A mammoth? There is nothing we could have done about their demise, and perhaps there is no place in the world for them today. But we *are* in a position to do something about the mountain lion, which formerly inhabited the Catskills, or the coyote, which now inhabits the West. And we can save a place for them.

## The future

As man increases in numbers and extends his activities throughout the biosphere, even into previously untouched areas such as the deep seas, it is inevitable that one wild species after another will perish, if for no other reason than our need for living space and for the resources that these animals and plants require. In the long run, then, the only solution to this as to most environmental problems will involve both the limitation of human population and an end to the wasteful consumption of resources associated with expanding economies. Meanwhile, we must do everything we can to preserve existing habitats. In some cases this will mean an explicit rejection of the idea of multiple use. In others it will require vigorous artificial management, or even the ultimate management—the preservation of endangered species in zoos until they can be reintroduced into their restored habitat. Enforceable international treaties also seem essential.[7]

---

[7] The lack of such enforcement is now producing the extinction of the blue whale, principally by Japan and the USSR, who refuse to be bound by any sane limitation of their harvest. These nations kill 85 percent of all whales taken yearly. There is no good reason for this slaughter, since satisfactory substitutes are available for all whale products. Even the control measures specified by international agreement are not sufficient to preserve the whales from extinction, but Japanese and Russian ships have, until recently, refused to permit international observers aboard to check their observance of even its inadequate provisions, and now threaten to eject them at whim.

"Bang!"

**FIGURE 23-28**
(William O'Brian, The *New Yorker Magazine,* 1972)

### Spiritual extinction?

Most of all, we must come to realize that the welfare of humanity requires a life-preserving attitude. It may be that our ecosphere could exist somehow even after substantial parts of its living fabric are destroyed, but our descendants would be deprived of so much of the natural world that surrounds us that perhaps they would seem, to us, to be not completely human or, at any rate, not fully alive. René Dubos has expressed the fear that man may adapt to environmental deterioration in much the same way that some modern slum dwellers have adapted to the frightful conditions in which they live. If they can conceive of nothing better for themselves, they are, in a certain sense, happy. Our grandchildren may never miss the sight of the stars or of the lilies of the field. Still, how much genuine kinship can we have with such people? Physically, they may be living, but emotionally and spiritually, a great dimension of what we know as humanity will be absent. We can spare them by measures we take now in our generation. They will not be able to spare themselves.

**CHAPTER SUMMARY**

1  All life is interdependent, and the total welfare of the earth depends on the welfare of all its life. Individual species must be preserved, but it is urgent to preserve them as parts of entire, intact ecosystems.

2  Man changes ecosystems to benefit himself. Generally, he replaces diverse natural communities with those consisting of one or two specially bred species. These artificial communities are unstable, simple, and usually lacking in some important ecologic mineral cycles. Some wildlife can coexist with agricultural communities.

3  To a degree, forests can be managed as crops are and can be artificially established in some areas where they do not naturally occur or from which they have been removed. The widespread military defoliation of forests is likely to lead to severe consequences. Certain forests, such as national forests, should be managed as natural ecosystems.

4  The spread of suburbs and city to the destruction of natural and agricultural communities is undesirable. It can be combatted by limiting population growth, increasing the virtues of urban life, and enforcing the strict and permanent zoning of land for use.

5  Habitats can be preserved or managed, as can fish and game. Management techniques include manipulation of habitat, artificial propagation, and control of exploitation.

6  Species may become endangered or extinct by a variety of processes—habitat destruction, overexploitation, persecution, environmental pollution, or the introduction of exotic species.

**FOR FURTHER READING**

The pertinent literature is enormous. We are able to list only a few outstanding books and no magazine or journal articles at all. We heartily endorse the following magazines: *Natural History, Environment, Audubon, The Bulletin of the Sierra Club, Bioscience, and the New York State Conservationist.* In addition, we particularly recommend the book publications of the Sierra Club.

Dahinden, Justus: *Urban Structures for the Future,* Praeger, New York, 1972.

Dasmann, Raymond F.: *Environmental Conservation,* Wiley, New York, 1968.

Ehrenfield, David W.: *Biological Conservation,* Holt, New York, 1970.

Errington, Paul L: *Of Predation and Life,* Iowa State College Press, Ames, 1967.

Gabrielson, Ira: *Wildlife Conservation,* 2d ed., Macmillan, New York, 1959.

Laycock, George: *The Alien Animals,* Natural History Press, New York, 1966.

————: *The Diligent Destroyers,* Ballantine, New York, 1970.

McHarg, Ian: *Design with Nature,* Natural History Press, New York, 1969.

Olsen, Jack: *Slaughter the Animals, Poison the Earth,* Simon & Schuster, New York, 1971.

Osborn, Fairfield: *Our Plundered Planet,* Little, Brown, Boston, 1948.

Stoddard, Charles H.: *Essentials of Forestry Practice,* 2d ed., Ronald, New York, 1968.

# 24

# ROOTS OF OUR ECOLOGIC CRISIS

I'm glad the sky is painted blue
And earth is painted green
With such a lot of nice fresh air
All sandwiched in between.
— *Nursery Rhyme*

***Chapter learning objective.*** The student must be able to outline the general dimensions of our present ecologic crisis and to summarize critically both the suggested causes and the suggested cures.

*ILO 1*  Describe the growth curve of the human population, and compare it to the growth curves often observed among populations of other organisms.

*ILO 2*  List several possible limiting factors for the human population.

*ILO 3*  Give one major criticism of family planning as a present answer to population growth, and discuss the alternatives from an ethical point of view.

*ILO 4*  Describe and contrast the throughput and closed-system models of economics. Relate both to population growth.

*ILO 5*  Propose a means of preventing the short-term profit of individuals and institutions from population growth and environmental degradation in general.

*ILO 6*  Summarize and criticize an argument based on economic productivity that tends to favor population growth.

*ILO 7*  Summarize the arguments employed to demonstrate that pollutive technology is primarily responsible for the current ecologic crisis. Also summarize the chief criticisms of these arguments.

*ILO 8*  Give reasons for halting and reversing current environmental trends, and propose measures to accomplish this.

As noted in the first chapter, there is now a widespread sense of ecologic crisis. We have spent a large portion of this book in discussing the nature and proportions of this crisis. If the pollution of our environment and the depletion of our biologic resources continue to worsen, it is hard to see how our civilization can be preserved, or if it can be preserved, whether it will in the future be worth preserving. But there is a steadily growing awareness that nothing constructive can be done about these matters until their root causes can be discovered.

On the face of it the causes seem simple enough. The cause of water pollution is inadequate sewage treatment, the cause of most wildlife extinction is habitat destruction, the cause of a good part of the solid waste problem is nondegradable packaging, and so on. But attacking the problems piecemeal at that level does not seem to be sufficient to alleviate them. Somehow they continue to get worse despite all efforts to educate or even to legislate. Apparently we must ask why the city does not treat its sewage, why habitats are relentlessly destroyed, and why people insist on packaging shampoo in polyvinyl chloride containers. Once these root causes are understood, perhaps we may be able to do something about our ecologic crisis.

## SOME POSSIBLE CAUSES OF OUR ECOLOGIC CRISIS

Environmental deterioration is often judged to result from three factors—need, speed, and greed. Need has increased as population has expanded. More people need more food, clothing, houses, schools, roads, and all the other things society holds to be basic to a high standard of living. When consumers are affluent, as in our own society, the realm of need expands subtly into the domain of desire. Great numbers of people can afford to buy luxury items—color TV sets, automobiles, stereos, electric appliances, expensive toys, high-fashion clothes, and other items which require energy to produce. Industry complies by producing ever-greater numbers and varieties of luxury products.

Science and technology have made it possible for goods to be made as fast as the demand appears. This speed in production has in turn created waste and a taste for the new. Obviously, there is less need to care for objects received if they are easily replaced at low cost. Furthermore, with replacements so accessible, affluent people have adapted to the idea of built-in obsolescence. We are thus caught up in the socioeconomic cycle: more people, more need, more work, more wealth, more need, more work, more wealth, more need (real or imagined). And somewhere along the line the "need" becomes more aptly termed "greed."

In numerous cases, the technology that has developed has proceeded with little regard to ecologic considerations. As a result, industrial processes have tended to become increasingly injurious to the environment. And lying beneath the process are the personal and cultural attitudes that promote the entire cycle.

It is the task of this chapter to examine these causes. Perhaps study may suggest ways to deal with them.

## HOW DO WE KNOW WE ARE OVERPOPULATED?

Before proceeding further, we suggest you consult Figure 4-13. It took the human race from the beginning until the nineteenth century to attain a population of 1 billion. But by 1930 we had added a second billion. By 1975 the human population of this earth should stand at *4 billion* people, an increase of *2 billion* in just 45 years. A large fraction of all the people who have ever lived are alive today.

As suburban sprawl stretches out, merging one city into another, as farms are sold and the cost of food skyrockets, as resources dwindle and we experience fuel and other shortages, as water pollution and air pollution produce increasing blight, many individuals no longer need to be convinced that overpopulation is a reality.

### How population grows

Populations of many organisms follow a characteristic pattern of growth when permitted to expand in a favorable medium. Imagine inoculating a flask of broth with a few bacteria, for instance, or placing a pair of fertile mice on an abundantly supplied island free of predators or competitors. The population would increase much as is shown in Figure 24-1. Initially the organisms would multiply rather slowly in the *lag phase,* but they would increase stupendously in the ensuing *logarithmic phase,* in which even the *rate* of

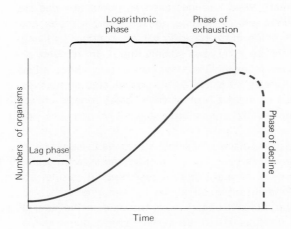

**FIGURE 24-1**
Typical population growth curve.

increase increases. Sooner or later some limiting factor would become effective—perhaps the exhaustion of food or some limited resource, the accumulation of waste products in the environment, or the physiologic effects of stress resulting from overcrowding. At that point, which is known as the *phase of exhaustion* the population would typically decline to some much lower level, where it would remain in a state of equilibrium with the environment. Sometimes, however, the phase of exhaustion results in the *extinction* of a population, for reasons only partially understood, one of which may be the extinction of species used as food.

A comparison of Figure 24-1 with Figure 4-13 is most instructive. The human population of the world was clearly in a lag phase prior to 1850. It is now in the midst of a logarithmic expansion, increasing by 190,000 persons per day. If we do nothing about it, eventually our species may be expected to enter a phase of exhaustion, and almost certainly of decline.

### When is a nation overpopulated?

It seems likely, then, that the world is already overpopulated. Population size results from the dynamic balance which is struck between the opposing forces of reproduction and mortality. Advances in food production and medicine have greatly reduced mortality in most countries, but in underdeveloped countries reproduction has continued at the same high rates. Thus we may think India is overpopulated, or perhaps China. But Western Europe or the United States? Many believe they have a long way to go. But do they?

In the long run the ideal population for a nation, continent, or planet is one that can be sustained indefinitely. Higher population densities than this generally result from temporary expansion of the carrying capacity of the habitat by technologic means. These expansions thus far have always depended on the consumption of nonrenewable resources and the passage of some of the costs to the ecosystem as an externality. Since nonrenewable resources will dwindle and finally cease, and since the ability of the ecosystem to absorb pollution is limited, expansion cannot continue indefinitely. Populations dependent upon present-day mechanical technology eventually will be drastically reduced. Since most resources will by that time have been consumed, the resulting population is likely to be both materially poorer and fewer in number than otherwise would have been the case.

Can the present population of India, the United States, Western Europe, Japan, or China be maintained without technology? Obviously not! Yet the technology employed in all these regions tends in the very directions that we have outlined above. The long-term objectives of population control must involve not only the slowing and stopping of further growth (an almost impossibly difficult task in itself) but also the reduction of population densities to a fraction of their current values. This goal is sometimes known as *zero population growth* (ZPG). ZPG requires even more than the reduction of the American birth rate from the present value of 2.1 to 2.0 children per couple. Even at an average family size of 2.1, calculations show that it would take about two generations for our population growth to level off—at almost 300 million in the United States alone in the year 2037. This is a consequence of the *age distribution* of our populace. The number of women of peak reproductive age (20 to 29) will increase 40 percent by 1980.

Already parts of Western Europe are as densely populated as India, and even now our standard of living depends heavily upon the importation of food and other resources from other countries, often impoverished ones. This situation is precarious *at present,* and in the near future, prosperity even of the industrialized countries of the world is by no means assured.

### Some possible limiting factors

Let us examine the phase of population exhaustion more closely. Essentially this phase occurs because the resources available to a population are always limited. In principle this would be true even if we all lived in space-ships and had access to the resources of the universe. According to Dr. Einstein, even the universe is finite. As it is, we have practical access only to the resources of one planet. Travel to others appears impractical, requiring the expenditure of more material resources than can possibly be returned to us from them. Consider how much money, human ingenuity, and natural raw materials have been expended on each trip to the moon. The return from such journeys, though extremely precious, is largely intangible. In terms of materials, it is limited to a few hundred pounds of rocks. Since all but a very few of us must expect to live upon the face of the earth in the future, we would do well to examine some of the limiting factors that may eventually produce the phase of exhaustion in human population growth.

Perhaps you are skeptical that there is any necessary limit on the expansion of the human race. A little reflection should convince you otherwise. There is no limit to the number of persons we potentially may produce, which means that unrestricted reproduction eventually will lead to an infinite human biomass. Since the earth cannot possibly support an unlimited tonnage of human flesh, let us return to the question of what forces are likely to intervene to prevent the unending expansion of humanity.

*Food.* Writing towards the end of the eighteenth century, the Rev. Thomas Malthus identified several such limiting factors. Although Malthus also recognized disease, warfare, and vice, he felt that the chief limiting factor was food. Malthus agreed that population increase could augment the supply of food, for the more people that work in agriculture, the more food will be produced, at least up to a point. Given the techniques of his time, in other words, increases in food production resulted mainly from having more pairs of hands to do the work. The trouble was that those hands were nourished by stomachs, and, as available agricultural resources became fully exploited, the demands of these stomachs tended to be greater than the production of the hands. As he viewed the matter, population tends to grow geometrically, but food supply tends to increase only at a much slower (arithmetic) rate. Starvation is the ultimate and inevitable result, and starvation, naturally, tends to limit population.

Although Malthus was living in a time of massive technologic and cultural change, he did not foresee that improvements in agricultural technology would make it possible for food supply to keep pace with population growth in the more technologically sophisticated nations. Improvements in agricultural methods which began shortly after Malthus's time have continued to this day, manifesting themselves dramatically in the last few years in the widely heralded "green revolution" (see Chapter 4). Despite this, even today the majority of the human race suffers from chronic malnutrition and from 10 to 20 million persons are estimated to die annually from the effects of starvation. If the world supply of food were to be equally distributed among

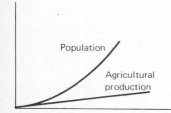

**FIGURE 24-2**
Malthus's concept of population dynamics and agricultural production.

**FIGURE 24-3**
The development of advanced agricultural techniques has postponed the fulfillment of Malthus's prophecy. But such methods require the use of large quantities of increasingly scarce energy. (Soil Conservation Service, USDA)

humanity *today,* all of us would be malnourished. The conclusion is obvious: Malthus was not wrong, but essentially right. Sooner or later human population growth must overtake food supply if population growth is not artificially limited, or if other factors do not intervene first. The question is, when? There is no obvious or clear-cut answer to this. At what stage of famine do we acknowledge that the prophecy of Malthus is upon us? In some parts of the world it seems to be here already.

**Disease.** Other limiting factors are possible. For instance, when a population approaches high densities, it is easier for disease organisms to travel from one individual to another. A theorem of mathematic epidemiology holds that no matter how good the methods of protection against a disease, for every given level of prevention there is a critical population density at which that disease will become epidemic.

**Shelter.** In most parts of the world human beings require shelter. Housing is already in short supply in most countries, and this trend may be expected to continue. In preindustrial Europe housing was deliberately kept in short supply to limit the population of poorer people. In countries such as India, even today a considerable proportion of the urban populace must sleep and perform all bodily functions on the street for want of a better alternative.

**FIGURE 24-4**
Variations in the quality of housing: off the main street in Djakarta. (Arthur B. Gordon)

*Pollution.* It is possible that pollution will also play a large part in limiting population growth. This question will be discussed more fully later in this chapter.

*Depletion of resources.* The general depletion of natural resources may do the job indirectly. If Malthus's day of doom has been postponed by agricultural technology, then the demise of that technology will certainly bring it upon us immediately. Housing, sewage treatment, medical treatment, and other necessities and amenities of modern life are also technologic. Such technologies depend upon a supply of raw material and natural resources.

Economists conventionally divide natural resources into two main categories: *renewable* and *nonrenewable.* A nonrenewable resources is limited in amount; once used, it will not regenerate itself. Obvious examples are coal, oil, and mineral ores. Though new supplies of these goods may be discovered, or more efficient ways of extracting them may be developed, sooner or later they will be used up or will be so depleted that the cost of their extraction will become prohibitive.

Renewable resources are also threatened by population growth. A forest represents a renewable resource, but if the demand for forest products to be used in housing exceeds the forest's regeneration, or if that forest land must be utilized for suburban housing, the renewable nature of the resource is theoretical, not actual. Perhaps it will renew itself when we are once again living in holes; perhaps not even then. Water is also a renewable resource. If you have read Chapters 20 through 22, little more needs to be said about that.

*Energy.* The ready availability of energy is an essential prerequisite both for mechanical technology and for agricultural production. The importance of energy to the production of manufactured goods is obvious, but it is equally important to the production of food. For instance, it has been shown that in general when increased yields of food are derived from agricultural land, approximately as many calories of energy must be invested in that land in terms of fuel burned in tractors, electrical energy consumed in fertilizer production, etc., as are represented in the increased harvests this energy investment makes possible. When our supplies of fossil fuels are used up, what then?

Table 24-1 summarizes some of the principal means of power generation that are now in use or are likely, in our opinion, to come into use by the early years of the twenty-first century. Perhaps thermonuclear power holds the most promise, but all of them probably can be implemented to some extent. However, though it may be possible to maintain our industrial economy through the development of new sources of power, all these sources have more or less serious environmental drawbacks. Even with the most efficient possible form of power generation an upper limit of thermal pollution will be reached. Sufficient release of energy by any means will ultimately raise the temperature of the earth's atmosphere, or at least make local areas uncomfortable or uninhabitable. To some extent this already occurs in central

(a)

(b)

1. Turbine
2a & b. Thermal storage tanks
3. Boiler
4. Oil reserve
5. Fresh water
6. Seawater
7. Desalinating plant
8. Cooling tower

(c)

**FIGURE 24-5**

Some possible sources of power for the future. (*a*) Cutaway view of a nuclear power plant embodying a breeder reactor. (U.S. Atomic Energy Commission) (*b*) Geothermal power plant of the Pacific Gas and Electric Company at the Geysers, Calif. (Pacific Power and Electric Company) (*c*) Artist's conception of a future solar power plant. Heat absorbed from sunlight in the banks of receptors to either side of the complex is used to raise steam and to generate electricity in the central facilities. The complex would also be used for the desalinization of seawater. (Dr. Aden B. Meinel, *Physics Today*)

cities, which may be considerably warmer than outlying areas, partially as a result of their energy consumption.

Like the advances in agricultural technology that Malthus did not foresee, changes in techniques of power generation and energy conservation may give us reprieves of a sort, but they cannot in themselves afford us a permanent solution.

**War.** Although it is perhaps not a limiting factor in quite the same sense as some we have considered, war was recognized by Malthus as a means of reducing population, and, without doubt, it has some effectiveness.

Although in primitive cultures, the enhancement of the death rate may be one of the principal effects of warfare, population reduction is probably an incidental effect of war, rather than its primary cultural function. In modern Western nations war does not seem to be an effective instrument of popula-

**Table 24-1. Energy options for the future***

| Name | Description | Sources | Advantages | Drawbacks |
|---|---|---|---|---|
| Fossil fuel | Coal, oil, or natural gas is burned. The heat produced boils water, producing steam which drives a turbine and generator. Presently in widespread use. | Fossil fuel beds | The technology is currently well developed and can fairly easily be improved by such techniques as coal gasification and magneto-hydrodynamic (MHD) power generation. Likely to provide three-fourths of all electricity even in the twenty-first century. | Considerable thermal pollution and almost intractable air pollution. All fossil fuels will run out in the next few hundred years at the latest, with coal lasting the longest, but coal mining is very damaging to the environment. |
| Nuclear power | A nuclear chain reaction is established in some fissionable material, usually a uranium isotope or plutonium. The heat that is produced boils water or some other fluid, as in fossil fuel techniques. Used to a relatively minor extent in industrialized countries. | Certain radioactive ores. Some uranium occurs in ordinary granite and might be extracted practically if breeder† reactors are perfected. | Essentially no air pollution and in *improved* models, perhaps slightly less thermal pollution than present power generation produces. *If* breeder reactors prove practical, fuel supply limitations are of only academic interest. | Radioactive wastes present an extremely vexing safety and disposal problem. The plutonium produced by breeder reactors, one of the most poisonous substances known, can be used to manufacture nuclear weapons. It will probably be difficult to decontaminate old power plant sites for other uses. It is by no means clear that breeder reactors are practical. Yet other types will run out of fuel early in the twenty-first century. |
| Thermonuclear (fusion) power | All the several versions depend upon the fusion of atoms of light elements such as deuterium, tritium, and lithium to produce heat that would boil water or some other fluid. It may prove possible to eliminate the intermediate fluid and extract power directly from the fusion reaction. As yet, this method of power generation is hypothetical. | Seawater and lithium-bearing rocks | In one version, the fuel supply (deuterium) is essentially limitless. Air pollution is absent, radioactive waste is essentially absent, and thermal pollution is probably minimal. | No one knows whether fusion power can be made to work. If that hurdle can be cleared, the only remaining limiting factor is the possible thermal pollution of the planetary atmosphere. |

* We have omitted some minor techniques of power generation which, while they may be locally employed, are unlikely to have major impact, such as windmills, tidal generators, and hydroelectric power.
† Breeder reactors make more fissionable plutonium fuel than they consume, using uranium as a raw material.

**Table 24-1.** *(Continued)*

| Name | Description | Sources | Advantages | Drawbacks |
|---|---|---|---|---|
| Geothermal power | Reserves of subterranean heat are drilled into. Water or some other fluid usually would be forced into the deposit of hot rock, which would boil it. The vapor would drive a turbine and generator. Not in widespread use at present. | Radioactive decay in the depths of the earth | The energy is free. No air pollution. | Suitable sites may be uncommon, and drilling and site preparation costs may be high. If deep-lying salt water is used as a heat source, saline pollution may be a problem. Thermal pollution probably would be severe. |
| Solar power | Sunlight is focused upon a vessel containing a molten salt or similar heat-exchange medium. The medium then is circulated to a boiler, where steam is raised to drive a turbine and generator. Direct conversion of sunlight to electricity is a possibility. Not employed on large scale at present. | Thermonuclear reactions in the sun | The energy is free and very abundant. No air pollution. Even today, solar house heating might be cheaper than electric heating over most of the United States. | Very large areas of land would be required to absorb enough sunlight to be practical. However, the land would probably be desert. Solar power is impractical in climates that are not sunny. There is a possibility of altering the local heat balance and climate, and of substantial thermal water pollution. As yet, the idea is untested. |

tion control, as evidenced by the crowded conditions now prevailing in most of the nations which participated in World Wars I and II. Future wars, however, may be far more efficient.

With the development of thermonuclear, chemical, and biologic weapons of war, our destructive abilities have been multiplied manyfold. Shortages in food, energy sources, and other supplies may directly motivate armed conflict on a vaster scale than we have seen heretofore, and may well result in the death of a large portion of the human race.

We have given a few possible factors that may limit human population growth. No one can tell which will become acute first, or whether something quite unforeseen may turn out to be the crucial limiting factor. We *can* predict with certainty that eventually it will be stopped, for the earth is finite. There are, we would emphasize, just *two* ways of limiting population growth: the reduction of birth rate and the acceleration of death rate.

## WHAT TO DO ABOUT IT

In the past there was a great deal of debate over the central problem of overpopulation and whether or not it exists, but today there seems to be a growing consensus that this is one of the most pressing issues of our time. How it should be dealt with is far less clear. Every possible course of action, including the course of inaction, is subject to severe objections of a practical or moral nature. Before considering specific suggestions, it would be well to examine some of the reasons for the population problem. In large part, overpopulation has resulted from idealistic and humanitarian programs—the reduction of disease and associated mortality rates, especially in underdeveloped nations. Infant mortality in particular has been drastically reduced. Consequently, the large families once necessary in underdeveloped nations for a variety of reasons are now growing up instead of dying in infancy, as was formerly the case. Though infant mortality has been reduced, the cultural pattern which results in a perceived "need" for a large family has *not* changed. In Mexico 80 percent of the professional, political, and religious leaders recently polled thought that the ideal family should number five or *more* children!

### Family planning

Family planning was the first formal birth control program to be instituted. In terms of its objectives it probably has also been the most successful. Historically, family planning has been relatively unconcerned with the population problem in the Malthusian sense. From the time of Margaret Sanger, founder of the movement, until today, the greatest area of concern in family planning has been the betterment of the individual and his family by avoiding excessive childbirth that might threaten health or economic security. Although some of the educational efforts of this program have been directed toward convincing people of the desirability of limiting their families, for the most part the program has concentrated upon making information and techniques of birth control available. Throughout its history family planning has emphasized the positive, voluntary aspects of birth control.

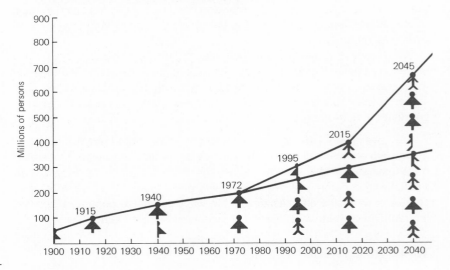

**FIGURE 24-6**
United States population projections, two- and three-child families. Upper line represents families with an average number of three children; lower line, families with an average of two.

Each figure (male and female) represents one hundred million persons

Since family planning reduces unwanted pregnancies and births, it does have the incidental effect of slowing population growth somewhat. From the standpoint of overpopulation, its principal drawback is that it *is* voluntary. Almost everywhere, people desire more than two children. If these desires are realized, there will still be an unacceptable continuance of population growth—slower, perhaps, then we now experience, but ultimately subject to the same phases of exhaustion and decline. These considerations have led to the suggestion that we now must move beyond family planning to programs that will emphasize population control as well as individual goals of health and economic security.

### Population control

The burden placed upon all nations by accelerating population growth is almost incalculable. This burden is particularly severe in underdeveloped countries—precisely where most of the world's population growth is taking place. An annual growth rate of only a very few percent can double populations in 20 or 30 years. Such doubling times are characteristic of the underdeveloped nations of the world, and for that matter, of the world as a whole. Rapid doubling times would be bad enough in a developed nation, for it would have to double *everything* (as Ehrlich has pointed out) in those 20 years if the same standard of living was to be maintained. This would involve doubling housing, schools, highways, sewage disposal plants, electrical generating capacity, and almost everything else you might think of. Even in America, whose growth rate is only about 0.8 to 1 percent, a general lowering of living standards by the year 2000 is likely. There are those who perceive such a decline already in education, diet, medical care, police protection, and other areas. But imagine trying to double such facilities in a nation such as India, the Philippines, or Haiti. Yet most nations aspire not only to maintain their present standards of living but to improve them.

In addition to likely decreases in standard of living, increased pollution and other environmental deterioration, resulting both from increased demands for space and food and inefficient pollution control efforts are predictable. Inevitably, unlimited or continuing population growth will cancel gains in pollution control. (Review the case of air pollution in the Los Angeles area in Chapter 9 for an example of this.)

Changes in the general economic pattern of life may also be expected from rapid population growth. The age distribution of a rapidly growing population differs from that of a reasonably stable population in a country with good public health measures. Figure 24-7 shows the age distribution typical of a rapidly growing population, together with the age distribution of an advanced nation with a relatively stable population. Notice that in the rapidly growing population about half the people are in economically unproductive age ranges (too young or too old to be gainfully employed.) These individuals must be supported by the remainder of the population. There are fewer such dependents in the stable population, hence less of an economic burden to the productive workers. The services and industries required by the two populations also differ. In the rapidly growing population, the

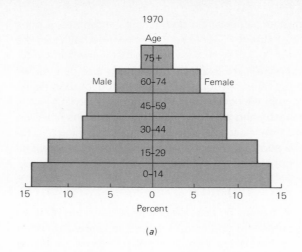

1970

Age

75+

Male 60-74 Female

45-59

30-44

15-29

0-14

15    10    5    0    5    10    15

Percent

(a)

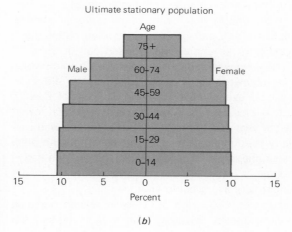

Ultimate stationary population

Age

75+

Male 60-74 Female

45-59

30-44

15-29

0-14

15    10    5    0    5    10    15

Percent

(b)

**FIGURE 24-7**
(a) Age distribution of a rapidly growing population. (b) Age distribution of a relatively stable population. (U.S. Bureau of the Census, *Current Population Reports,* ser. P-25, no. 480, "Illustrative Population Projections for the United States: The Demographic Effects of Alternate Paths to Zero Growth," 1972)

emphasis must be upon infant and child care; in the stable population, more resources can be invested in the needs of mature and elderly persons, a certain proportion of whom will be contributing to the advancement of the society. It is clear from this that in a country with a doubling rate of 20 years, some commodities such as education must be *more* than doubled to maintain present standards of adequacy. A rapidly growing society will also understandably have increased taxes and other expenses as a result of such increased needs.

These considerations have not escaped the notice of some underdeveloped countries. The oldest and probably most determined population control program in the world has been in India. In addition to the employment of the usual family planning techniques, the Indian government has offered a number of material incentives to those undergoing vasectomies, IUD insertions, etc. Other programs exist in other countries, notably Taiwan and, until recently, Japan. None of these programs, however, has kept pace with the rate of population growth, and the majority of the nations most troubled by overpopulation have done little or nothing in the way of population control. Singapore is an exception. In the fall of 1972, the Health Ministry of Singapore announced that discriminatory measures would be taken against new additions to large families. "For the Singapore of the 1970's," it said, "the third child is a luxury. The fourth and fifth are anti-social acts."

A suggested remedy for overpopulation is some form of coercion, and a number of coercive schemes have been advanced for use not only in underdeveloped countries but in developed ones as well. After all, the United States may well possess a population equal to that of present-day India by the year 2100. Such remedies include sales taxes upon baby supplies, taxation on the number of children above two, mandatory sterilization of parents with large families, mass administration of fertility depressants (perhaps in drinking water), the issuance of government licenses for reproduction, and many more.

### Reservations and objections

The natural reaction to these and similar proposals is one of aversion. One is moved, however, to reflect upon the human misery that would be inflicted upon future generations in an overpopulated world. Failure to act is an ethical decision, too. Is there some way to solve this problem without violating the consciences or the persons of our fellow human beings?

The most serious ethical objection may be that the only methods which appear to be effective involve the totalitarian coercion of people. This is a corollary of the fact that people want more children than society can support. The only alternative to making them have fewer children whether they want to or not, is to persuade them to want fewer children. It is to be hoped that as education concerning the impact of family size is disseminated, people will limit their families voluntarily. Even after several years of education, only a small percentage of the United States population appreciates the gravity of the problem, but as the impact of books such as Ehrlich's becomes widespread, and as environmental education penetrates deeply into the American educational system, this percentage might be expected to increase. Techniques of mass education are having some impact in both underdeveloped and developed countries.

## AFFLUENCE AND ECONOMICS

Posters in the New York subways a few years ago extolled the number of babies born every minute in the United States. "Your future is great," they said, "in a growing America." Such slogans have been rarer of late, but some of the thinking still persists. Is population growth economically desirable, or even a necessity for a healthy economy?

## Two views of economics

In order to attack the question of economics we must examine two analytical concepts which represent attempts to view economic problems from an ecologic standpoint. They are intended to apply with similar force to both capitalist and collectivist systems.

***The throughput theory.*** As the throughput theory views it, raw materials and natural resources are extracted from their place of origin, processed and manufactured, distributed, consumed, and finally discarded and dispersed. Since people are employed at every stage of the process, and since they are paid for their efforts, a portion of the cash value of the manufactured goods is siphoned off for various markets at every stage of production, distribution, and waste disposal. Thus, stockholders, assembly line workers, designers, engineers, truck drivers, executives, and many others have an interest in the passage of the goods they handle through the manufacturing and distribution pipeline to ultimate use.

The larger the total economic value of the goods which travel through the manufacturing-distributing pipeline, the larger the total value of the cash which can be extracted by various middlemen. The rate at which goods pass through the manufacturing-distributing pipeline is called ***throughput***.

Throughput can be increased in a number of ways. One obvious way is by increasing demand. If more people buy goods, demand is increased. Those people must, of course, want or need the goods, they must have money or credit to buy goods, and they must exist. Advertising can generate increased desire to buy, increased throughput can increase general prosperity and the ability to pay, and, to come to the crux of it, increased population growth helps to increase throughput, and, it is commonly held, general prosperity.

The principal problem with the throughput theory is that those who employ it do not necessarily use it in conceptually complete form. For example, complications are introduced by the shrinking supply of most raw materials

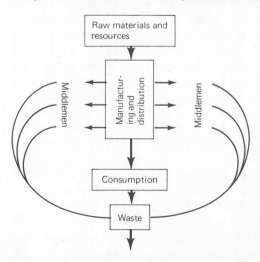

**FIGURE 24-8**
Throughput economics.

and resources. In the short run this may lead to increased transportation costs from remote localities not previously exploited, and increased costs of processing lower grade sources (such as metallic ores). In the long run it may lead to their total depletion. Then what? If the world is a pie, it can be cut into a lot of slices, but the more slices there are, the smaller each person's slice becomes. Furthermore, the throughput theory tends to ignore the costs of waste disposal (though it can be modified to take these into account), and the potential value of waste as a resource. Finally, since we are all middlemen in *some* throughput process or other, the man who is helping to produce one set of goods is simultaneously consuming others. Since all manufacturing involves some waste, the end result is the wastage and dispersal of more materials than one produces. If both raw materials and the ability of the environment to absorb wastes were unlimited, that would not constitute a problem. In actual fact, neither is unlimited. In the long run, then, population growth certainly will be uneconomical to society *as a whole*.

**The closed-system theory.** The earth *is* a closed system, and a finite one, whether we like it or not. The closed-system model is designed to take this fact into account. A throughput economy can operate only for a limited time, and then only in areas where all resources are large in relation to the demand placed upon them. Since we have now filled the earth with our own numbers, it seems clear that we must permanently dispense with the over-simplified throughput model of economics.

An economy based upon the closed-system theory would contain extracting, manufacturing, and distributing activities and industries much as in the throughput model. However the closed-system theory does not contain any category for consumption, since goods really are not consumed but *used,* and eventually disposed of when they are no longer useful. Instead of consumption, the closed-system model substitutes a category of primary

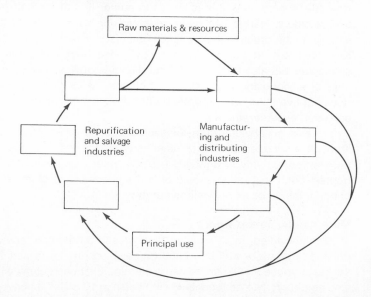

FIGURE 24-9
A closed-system economy.

use. Once goods have been used they are not dispersed but are placed in recycling channels which convert them into forms that can be used in manufacturing new goods. Even waste from manufacturing processes is recycled wherever practical. Since reclamation processes are not and cannot be completely efficient, there will still be some consumption of natural resources. However, such consumption will be *minimal—if,* and only if, primary use does *not* increase. If use increases, recycling will not greatly alleviate the depletion of raw materials, and pollution by waste products will increase despite it. Thus the potential of even a closed economic system will be realized only if population is stabilized; it is *incompatible* with population growth. Yet the closed system is the only economic system that our civilization ultimately can employ, since it is the only one that recognizes and counteracts resource depletion and the voiding of wastes to the environment.

## The commons

A few paragraphs ago we made the point that population growth is uneconomical for society as a whole. It is worthwhile to ask ourselves if population growth benefits some portions of society even if it is at the expense of the remainder. The answer would appear to be "yes." Among those profiting we must list virtually the entire generation living during a time of population expansion in an industrialized nation at a time when resources are still essentially undepleted. In addition, those who are in a position to push some of their costs of doing business off on others will tend to benefit not only from population growth but also from other ecologically unsound practices, such as the discharge of waste without treatment (at least where recycling is not yet economically profitable), the destruction of forest and wildlife resources for their own profit, and so on. In his now classic paper, "The Tragedy of the Commons," Garrett Hardin presented the thesis that where any resource is held in common, individuals will tend to exploit it to the maximum extent possible, even though this is known to render it unfit for their own use and that of others. Hardin has been criticized for his position, to be sure, but usually on the grounds that he is too optimistic! His critics say that common resources are degraded not only by those who must suffer the consequences but also by those who need not. An obvious example is an executive of an industry that vigorously pollutes the air of an urban ghetto. He does not live there, but he does profit from his unfair exploitation of a free, common resource—the air.

There is no obvious cure for such practices other than legal restraint. Through equitable taxation or other measures it may be possible to stop people from unfairly or unwisely exploiting our resources, at least to some degree. Such legislation should take the needs of our children and grandchildren into account, as well as our own.

## More hands, lighter work?

As Malthus noted, more workers do increase production, at least up to a point. If I can dig a well in a hundred days working by myself, it is reasonable to suppose that it can be completed in 50 if I can get my wife to help. If nine neighbors can be recruited, will I finish in 10 days? If I have 99 helpers, will I have it tomorrow? Obviously, there is some practical limit to the number of people I can fit inside the well as it is being dug, and before that, some point at which they interfere with one another's digging to such an extent that it is not profitable to add more. Thus we must ask whether there is gainful employment available for the people we add to our world, and at what point it becomes impractical or ceases to be profitable to add more. As Malthus might have put it, they tend to eat more than they can produce. Finally, there will come a time when all resources will be fully committed and there will be none to spare for newcomers!

Of course, individual productivity can often be increased by technologic means, and much of the affluence of industrial nations results from this very situation. However, in most of the world such technology simply is not widely available; even in the developed nations there are limits to its application.

## Local costs of population growth

Obviously, advantages accrue to nations and areas of the world which are moderately populated, that is, which are not underpopulated. Military power depends to some extent upon available manpower, although not so much now as in the time of Genghis Khan. Moreover, one cannot have modern agriculture or industry without a labor force. However, there must, it would seem, be some optimum level of population that would ensure sufficient manpower for the needs of society without producing excessive undesirable effects. In time, this problem may become susceptible to computer analysis. Meanwhile, several studies have established that the costs of population growth in large communities considerably exceed whatever benefits may accrue.[1]

## Economic systems

Some have attributed environmental degradation to the economic systems employed by the societies in which they occur. As we have seen, a case can be made for the general thesis that pollution can be alleviated by certain basic economic changes. However, no economic system in the world today has made them. Soviet and other communist regimes are equally plagued with pollution and a deteriorating environment and seem even less able to combat environmental degradation than nations employing what we call "free enterprise." This is clear evidence that merely blaming the profit motive misses the point. There is no reason why a closed-system economy could not be developed in conjunction with either public or private ownership of the means of production. Without such an economy, neither system can be expected to operate successfully for much longer. As Marshall Goldman has written, "There seems little to choose between private greed and public greed."

---

[1] The following statistics are reported by Richard Lamm: Education of children in cities of less than 50,000 costs $12 per capita, but $85 per capita in cities of 1,000,000 or more. There are 12.8 robberies per 100,000 population in cities of less than 10,000, but 117.6 robberies per 100,000 in cities of over 250,000. At the same time, the cost of police protection doubles. Public health measures of all kinds are twelve times more expensive per capita in cities of over a million than in those of less than 50,000. Similarly, welfare costs are 88-fold increased.

## THE ROLE OF TECHNOLOGY

Most, if not virtually all, environmental scientists would agree in principle with what has been called "Ehrlich's equation," stated by Barry Commoner in the following form:

Population size × per capita consumption × environmental impact
(e.g., pollution)
per unit of production

= level of pollution

If the equation is correct, then it follows that pollution can be increased by increasing any of the three factors, or reduced by reducing any of the three, providing that the other two do not greatly change. The difference in opinion among environmentalists arises concerning the relative weight to be assigned to each.

### Commoner's arguments

We have already presented the basic position of those whose primary concern is overpopulation. Some believe that more emphasis should be placed upon means of reducing per capita consumption—sumptuary laws, luxury taxes, restriction of advertising, and similar measures. 1971 saw the publication of Barry Commoner's important book, *The Closing Circle,* extracts from which also appeared in *The New Yorker.* In it, Commoner minimized the roles of population and affluence and emphasized that of pollutive forms of technology in producing environmental degradation. His basic thesis may be summarized as follows:

1 Most forms of pollution have increased more rapidly than population.
2 Most forms of pollution have increased more rapidly than general affluence. In fact, adjusting for inflation and certain other factors, one may make the case that no average per capita increase in standard of living has taken place in recent years.
3 Therefore, most pollution results neither from population nor from affluence but, by elimination, from technologic changes which increase the environmental impact of production.

Commoner then advances the concept that many traditional products and technologies were less ecologically destructive than their modern counterparts. For example, degradable cotton and wool have been largely replaced by far less degradable synthetic fibers, and soaps by synthetic detergents. Not only do these finished products cause more ecologic harm than their older equivalents, but the processes of their manufacture also tend to be more dangerous to ecosystems. However, since they require less investment or overhead cost per unit of production, these newer products are more profitable to manufacturers. Through advertising, and simply by denying the public other options, these products have been marketed to the virtual exclusion of their predecessors. Furthermore, an attractive investment for a capitalist is also an attractive investment for a communist planning com-

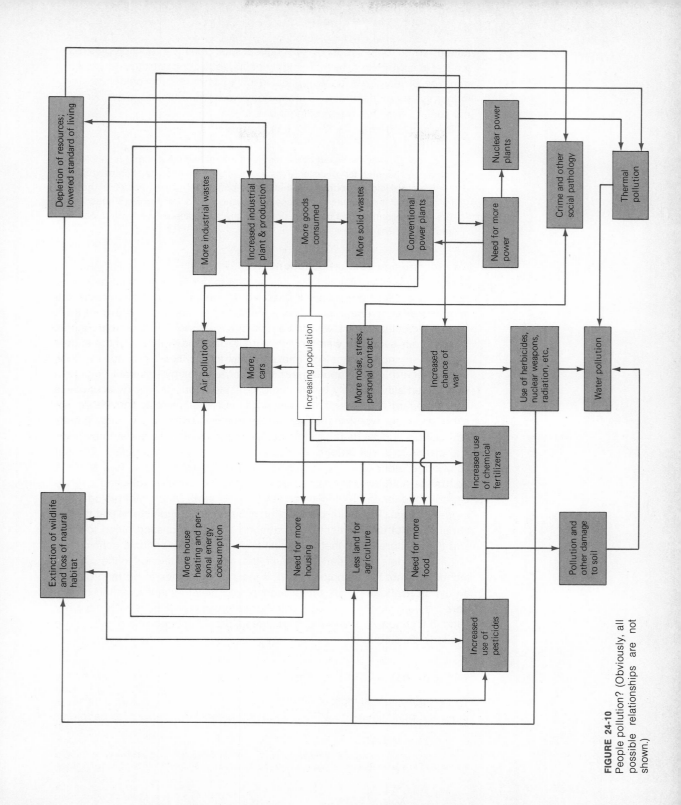

**FIGURE 24-10**
People pollution? (Obviously, all possible relationships are not shown.)

mittee, so in both kinds of economies less pollutive technologies have tended to be replaced by more highly polluting successors. It is not merely a question of increased consumption resulting from general affluence. Commoner seems to feel that if its technology were well planned, even our affluent society need not be an effluent one.

But, as Dr. Bertram G. Murray has expressed it,

> Surely, Commoner would agree that an ecologically-responsible technological society is headed for disaster if its population continues to increase. Surely, Ehrlich and Holdren understand that a zero-population-growth society is headed for disaster if it continues to increase its consumption of non-renewable resources. The ecological health of mankind demands the simultaneous attention to both aspects of the environmental crisis.

### Can technology combat pollution?

Technology is obviously not the sole answer to our environmental problems, but without doubt it can be of vast aid. No technology can increase our supply of raw materials, but in some cases it can permit us to extract them more efficiently from lower-grade sources. Technology probably cannot increase our supply of water, but it can alleviate the pollution of the supply we have. In short, technology can do many things that it has not yet been called upon to do. By and large, the use of technology has until now been determined almost solely by economic considerations, that is, by economics that disregard external cost. Technology must be redirected to the reconstruction of our world industrial and agricultural plants along ecologically sound lines. Almost certainly, this transformation must be mandated by political means, and will require an educated and committed populace. Factual knowledge alone will not suffice, particularly if limited to a few. Many a deliberate polluter today is ecologically quite knowledgeable and entirely capable of debating biologists on their own ground. Others hire biologists to argue for them. The entire body politic must come to understand that the external costs of environmental degradation are being passed along to them and their children, and that they must take appropriate political action.

Let us understand clearly, however, that the blame belongs to the *users* of technology, not to technology itself. If technology comes to be regarded as an evil in itself, and if all further technologic development is halted or hindered, the environmental crisis will be made worse, not better. On the other hand, if technology is allowed to pursue its present course, the result will be similarly disastrous.

### THE UPSHOT

The value system of most of us assumes that human welfare is an ultimate or at least a penultimate good.[2] We doubt if it is realistic to appeal to people on

---

[2] Some see the root causes of the ecologic crisis in our value systems themselves. In the past, the Judeo-Christian religious heritage of Western culture was often held to have been an impedi-

other grounds, and so, throughout this book we have tried to show that ecologically irresponsible behavior is also socially irresponsible, and that it ultimately diminishes the dimensions and quality of human life. Eventually, we stand to lose not only the quality of our lives and a satisfactory standard of living, but life itself.

## What we stand to lose

If present environmental trends continue, we stand to lose the following, among others:

1 Clean air and water
2 An adequate diet
3 A large part of our present standards of health
4 An uncrowded way of life and, concomitant with this, adequate housing
5 Good education and schooling
6 The freedom to escape to pleasant surroundings, not to mention wilderness
7 The irreplaceable genetic and ecologic resources of species that become extinct
8 Privacy
9 Our present standard of living — beginning with the luxuries, then what we now consider to be the necessities.

## Changes we must make

In order to gain or maintain desirable lives for ourselves and our descendants, it seems logical to assume that we must refrain from having too many of them. It is also possible that we will have to forego the privilege of becoming rich at the expense of the ecosystem. This may well limit some of the prospects for individual advancement that now exist. On the other hand, it should also result in greater economic security. Economic readjustments resulting from pollution control may necessitate a temporary sacrifice of a portion of our affluence. But if our desires become redirected toward ecologically sounder ends, we may find that our changed styles actually will provide increased satisfaction.

Local and temporary difficulties may be acute, but with adequate planning and enabling legislation, it should be possible to alleviate them. The price we pay for a clean, decent world is far less than the price that will be exacted from us if we do not. But as we have noted in Chapter 23, we may not realize the price we are paying.

---

ment to "progress," but now that progress is appearing to be more of a mixed blessing than was formerly apparent, there has been a recent disposition to reverse that position and to hold the Judeo-Christian world view responsible not only for progress but for ecologically irresponsible exploitation of the environment.

According to Lynn White and others, such documents as the book of Genesis (for example, Genesis 1:28) encourage an exploitative attitude toward nature and have helped to create a general cultural attitude encouraging such exploitation.

White's critics reply that many societies (such as ancient Rome, ancient China, and modern Japan) exploited and exploit their environment at least as ruthlessly as do we. Moreover they contend that close examination of biblical documents does not support the conclusions White draws from them. In other words, they hold that the exploitative atmosphere of Western culture has developed in spite of its Judeo-Christian heritage, not because of it.

***Human adaptability.*** The human organism is extremely adaptable. If his water is polluted, he may purify it. If his air is polluted, he may filter it. If food and mineral resources run short, he may find substitutes. If reserves of fossil fuels are exhausted, he may develop controlled thermonuclear fusion power. In the end, though, he will be living a life of unprecedented restriction and squalor. As Kenneth E. F. Watt has written, "Unfortunately, *Homo sapiens* is a remarkably adaptable species, so much so that we may well have adapted to our ultimate doom before we are aware that it is upon us." We do face such a danger. If the degradation of our lives is sufficiently gradual, we may simply get used to it, make adjustments that deal only with some of the symptoms and not with the disease, and slide imperceptibly into the apocalypse.

***Treating the symptoms and the disease.*** Perhaps realism would predict just that, since it is all too similar to what has happened up till now. We must avoid token solutions that are too little and too late. But how?

Any and all measures are bound to fail in the absence of personal, moral commitment. In the absence of widespread consensus, no law is effective, no treaty is honored, no constitution is more than a piece of paper. Each of us contributes to the ecologic crisis, either directly (as by our waste products) or by proxy (as by utilizing electric power.) Changes in personal life styles are going to be necessary, as well as legal restraint of public and private degraders of the environment. When one considers the likely political and economic opposition, it becomes clear that nothing but the strongest public sentiments will prevail. And yet, sadly, many indications point to public apathy rather than to widespread public commitment.

Not long ago, the Chrysler Corporation sent 15,000 automobile antipollution kits to its dealers. The cost was about $20 apiece, low enough so that most Chrysler owners ought to have been able to afford having them installed. Approximately 50 have been sold. At this writing, 13,000 have been sent back by the dealers to the company. Who should bear the blame? The Chrysler Corporation?

It is easy to castigate a company that pollutes in another state or does not employ one's husband. It is harder to support measures that cost us money—new sewage treatment plants that raise the water or sewer rates to pay for their financing, for example. The armchair ecologist is likely to get cold feet at this point.

**FIGURE 24-11**
Meaningful reforms can be accomplished only by vigorous action, for even trivial changes may be powerfully opposed by vested interests.

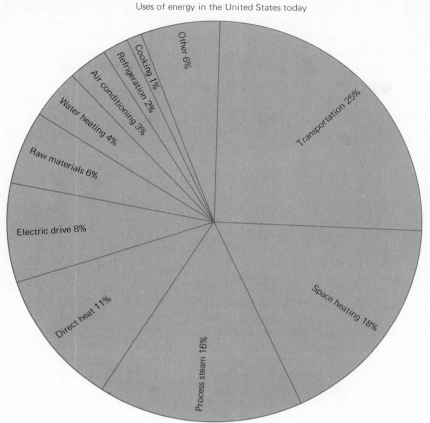

Uses of energy in the United States today

Other 6%

Cooking 1%

Refrigeration 2%

Air conditioning 3%

Water heating 4%

Raw materials 6%

Electric drive 8%

Direct heat 11%

Process steam 16%

Transportation 25%

Space heating 18%

**FIGURE 24-12**
Uses of energy in the United States. How might we work to conserve our energy resources? What changes in our economy or life styles might we make?

## THE LAST WORD

Knowing the true dimensions of our environmental predicament tempts one to shout, and loudly. No other reaction to the facts seems appropriate. Though the facts do speak for themselves, facts do not motivate to action. For that, commitment is necessary. Commitment does not rise from any purely intellectual root, but from the basic spiritual and emotional forces that underlie our lives. Faith, hope, love, awe, compassion, and exaltation are generated not by computer programs but by human hearts.

The defense of our world of life will require unprecedented breadth of vision, and a concern embracing not only the billions of today but those of tomorrow as well. It is irrational to suggest that this defense can or will be accomplished without emotion. How many worthwhile things *can* be accomplished without emotion? And how many can be accomplished without reason? The whole man is both a rational and emotional being. Neither trait exists in a vacuum.

Though the ecosphere is composed of a multitude of organisms, for better or for worse man is the steward of the ecosphere. Its disease is of human creation; its healing is a human problem. To understand the nature of the problem or to plan a solution is a rational process. But only idealism,

*"May I kiss you goodbye?"*

**FIGURE 24-13**
(William P Hoest, Saturday Review, Inc., 1970)

decency, humanity, and faith will motivate us in what well may be the last days of the late, great planet earth.

There is no sense pretending that the prospects are bright. They are not. It is likely that irreparable damage will be done before the environmental crisis is treated as the national and international emergency that it is. Yet many opportunities are ours. It is still a privilege to be alive in this, one of the great and crucial ages of world history. There has never been a less appropriate time to drop out or to cop out.

**CHAPTER SUMMARY**

1 The human population is now in a stage of logarithmic growth. Eventually some limiting factor may be expected to cause its decline, unless it is first stopped by artificial means.

2 Among possible limiting factors for human population growth, we may list shortages of food, shelter, resources, energy, and living space, as well as disease, pollution, and war.

3 A nation may be considered overpopulated if its total population is too great to be sustained permanently by its resources. By this criterion, virtually all areas of the modern world must be judged to be overpopulated, *at the present time.*

4 Population growth may be humanely limited only by reducing the number of births. As presently practiced, family planning is not an adequate instrument, for people tend to desire larger families than would be required to halt or reverse population growth. On the other hand, coercive measures are open to objections. Perhaps educational campaigns can be mounted to convince the public that a low birth rate is desirable.

5 Population growth is economically beneficial only if one employs the throughput model of economics. The throughput model disregards external costs, the depletion of natural resources, and the welfare of future generations. The closed-system model emphasizes recycling of used materials and is the only theoretical basis upon which a technologic civilization can be permanently erected. It is, practicable, however, only under conditions of population stability. In the long run population growth and civilization are economically incompatible.

6 Increased population has the potential of increased productivity, but in time a point of diminishing returns will be reached which, together with diminished resources, will depress rather than enhance affluence. Population growth produces increased costs for goods and services, especially public services. This offsets economic gains temporarily accruing from that growth.

7 It has been argued, notably by Barry Commoner, that pollution has been encouraged greatly by the attractiveness of pollutive technologies as investments. Commoner's critics dispute the weight that he attaches to technologic causes of the environmental crisis. Though pollutive technologies must be restrained or redesigned, technology also may be employed to alleviate pollution. No conceivable technology can patch up the situation indefinitely in the absence of population control.

8 If current environmental trends are not reversed, the human race stands to lose first the amenities of life and finally technologic civilization or perhaps life itself. We are likely to adapt to some of the stages in the deepening crisis, but such adaptation is not really a desirable response. Perhaps education and personal commitment afford some hope for the alleviation of our ecologic pathology.

A number of important papers are listed here individually; most of them also occur in the collections of papers that are listed. Nowadays there are almost enough collections of reprinted papers to constitute a considerable waste-disposal problem, however, and we have made no attempt to cite them all. Works dealing with the legal, economic, and practical aspects of environmental degradation are starred (*). For additional references, consult the bibliographies of the other environmental chapters.

Borgstrom, Georg: *Too Many,* Macmillan, New York, 1969.

Brooks, Harvey, and Raymond Bowers: "The Assessment of Technology," *Scientific American,* February 1970, p. 13.*

Burby, John: *The Great American Motion Sickness,* Little, Brown, Boston, 1971.*

Commission on Population Growth and the American Future: *Population and the American Future,* New American Library, New York, 1972.

Commoner, Barry: *The Closing Circle,* Knopf, New York, 1971.*

————: *Science and Survival,* Viking, New York, 1967.

Crocker, Thomas D., and A. J. Rogers: *Environmental Economics,* Bryden, Hinsdale, 1971.*

Derrick, Christopher: *The Delicate Creation,* Devin-Adair, Greenwich, Conn., 1972.

Dolan, Edwin: *Tanstaafl, The Economic Strategy for Environmental Crisis,* Holt, New York, 1971.*

Ehrlich, Paul R. (ed.): *The Population Bomb,* Ballantine, New York, 1971.

———— and Anne Ehrlich: *Population, Resources and Environment,* 2d ed., Freeman, San Francisco, 1972.

Elder, Frederick: *Crisis in Eden,* Abingdon, Nashville, 1970.

Goldman, Marshall: *The Spoils of Progress: Environmental Pollution in the Soviet Union,* Massachusetts Institute of Technology, Cambridge, Mass., 1972.*

Hardin, Garrett: "The Tragedy of the Commons," *Science,* December 1968, p. 1243.

Hinrichs, Noel (ed.): *Population, Environment and People,* McGraw-Hill, New York, 1971.

Holdren, John, and P. Herrera: *Energy,* Sierra Club Books, New York, 1972.*

Kesteven, G. L.: "A Policy for Conservationists," *Science,* May 24, 1968, p. 857.

Klotz, John W.: *Ecology Crisis: God's Creation and Man's Pollution,* Concordia, St. Louis, 1971.

Langer, William L.: "Population Growth 1750–1850," *Scientific American,* February 1972, p. 93.

McCaull, Julian: "The Politics of Technology," *Environment,* March, 1972, p. 3.*

Moncrief, Lewis: "The Cultural Basis for Our Environmental Crisis," *Science,* Oct. 30, 1970, p. 508.

Murphy, Earl: *Man and His Environment:* Law, Harper, New York, 1971.*

Osborn, Fairfield: *Our Plundered Planet,* Pyramid, New York, 1968.

Saltonstall, Richard: *Your Environment and What You Can Do about It,* Walker, New York, 1971.*

Sax, Joseph L.: *Defending the Environment, a Strategy for Citizen Action,* Knopf, New York, 1971.*

Solow, Robert: "The Economist's Approach to Pollution and Its Control," *Science,* Aug. 6, 1971, p. 498.*

Stein, Richard G.: "A Matter of Design," *Environment,* October 1972, p. 17.*

Weaver, Kenneth F.: "The Search for Tomorrow's Power," *National Geographic,* November 1972, p. 650.

White, Lynn: "The Historical Roots of Our Ecological Crisis," *Science,* Mar. 10, 1967, p. 1203.

Wright, Richard T.: "Responsibility for the Ecological Crisis," *Bioscience,* Aug. 1, 1970, p. 851.

# APPENDIX A  BASIC PRINCIPLES OF CHEMISTRY

*Unit learning objective.* The student must be able to apply the basic chemical facts and concepts discussed in this appendix to cells and to living systems. For example, he should be able to summarize the chemical composition of living things, list and describe the main classes of organic chemical compounds of biologic importance, and analyze situations involving osmotic principles.

*ILO 1*  Diagram the basic structure of the atom in accordance with the conventions presented in this unit, showing the position of the main subatomic particles: electrons, protons, and neutrons.

*ILO 2*  Given their chemical symbols, identify the 11 biologically significant elements given in Table A-1, and summarize the main functions of each in living organisms.

*ILO 3*  Interpret simple chemical formulas, including structural formulas.

*ILO 4*  Give the physical basis for the ability of atoms to combine in the formation of molecules.

*ILO 5*  Characterize ionic and covalent bonds, distinguish between them, and give examples of each.

*ILO 6*  Interpret and analyze a simple chemical equation, and describe in words the chemical reaction which it denotes.

*ILO 7*  Define oxidation and reduction in terms of electron relationships, and given examples, identify both.

*ILO 8*  Define pH in terms of hydrogen ion concentration, and be able to identify any given pH as acid, alkaline, or neutral.

*ILO 9*  Distinguish between inorganic and organic compounds, list the four main classes of organic compounds found in living systems, and identify each from a given example.

*ILO 10*  Given examples, distinguish the various components of solutions and colloids (i.e., solvent, solute); and distinguish between solutions and colloids;

*ILO 11*  *a.* Given concentrations and membrane characteristics, predict the direction of diffusion of solutes and solvents across differentially permeable membranes.

   *b.* Solve simple problems involving osmosis and active transport in cells; for example, describe the behavior of the contractile vacuole of protozoa in solutions of various osmotic strengths, or predict swelling or shrinking of cells under various osmotic conditions.

*ILO 12*  Define the basic metric units as given in Table A-2.

*ILO 13*  Define each chemical term (such as atom, molecule, electron, hypertonic) introduced in this unit.

Life depends upon the precise interaction of chemical substances. Reduced to its simplest components, a living system is an aggregation of atoms and molecules. Thus, a basic knowledge of chemistry is essential to an understanding of metabolic processes. For those who have never enjoyed a formal chemistry course, this introduction will provide the basic definitions and concepts necessary to an understanding of biology. Those who have studied chemistry may find it a useful review.

## THE ATOM

All matter, living and nonliving alike, is composed of chemical *elements,* substances which cannot be broken down into simpler substances. There are 103 different types of elements. Of these, 92 occur naturally in the universe. The others have been produced in the laboratory. About 98 percent of cytoplasm by weight is composed of only six elements—carbon, hydrogen, oxygen, nitrogen, calcium, and phosphorus. About 14 other elements are consistently present in cytoplasm, but in small quantities.

Instead of writing out the name of each element, chemists use a system of abbreviations called *chemical symbols.* For example, the chemical symbol for carbon is simply C, and for hydrogen, H. Table A-1 lists the chemical symbols for several biologically important elements.

The smallest unit of a chemical element that retains the characteristic chemical properties of that element is an *atom.* Atoms are the basic units of matter which participate in chemical reactions.

**Table A-1. Some important elements**

| Name | Chemical symbol | Biologic importance |
|------|-----------------|---------------------|
| Carbon | C | Present in all organic compounds; can form four bonds with other atoms; forms backbone of organic molecules |
| Hydrogen | H | Can form more chemical combinations than any other element; present in all organic compounds |
| Oxygen | O | Required for cellular respiration; present in many organic compounds |
| Nitrogen | N | Present in all proteins and nucleic acids |
| Calcium | Ca | Structural component of teeth and bones; important in processes of muscle contraction, nerve transmission, and blood clotting |
| Phosphorus | P | Found in proteins and nucleic acids; component of bones and teeth |
| Sodium | Na | Ions needed in blood and other body fluids |
| Chlorine | Cl | Ions needed in blood and other body fluids |
| Iron | Fe | Component of hemoglobin molecule and various enzymes |
| Magnesium | Mg | Component of chlorophyll molecule; needed in blood and other body tissues |
| Sulfur | S | A component of most proteins |

An atom itself is composed of smaller components called *subatomic particles*. Although several subatomic particles are known, only the *proton, neutron,* and *electron* will be discussed here. Protons and neutrons are located in the central portion of the atom, called the *nucleus* (Figure A-1). A proton is a positively charged particle, and each type of atom has a characteristic number of protons in its nucleus. This number is known as its *atomic number.* For example, an atom of hydrogen always has one proton in its nucleus, so its atomic number is 1. An oxygen atom, atomic number 8, always has eight protons in its nucleus.

A neutron is an electrically neutral particle. Ordinarily the number of neutrons is constant for any type of atom. Occasionally the number of neutrons

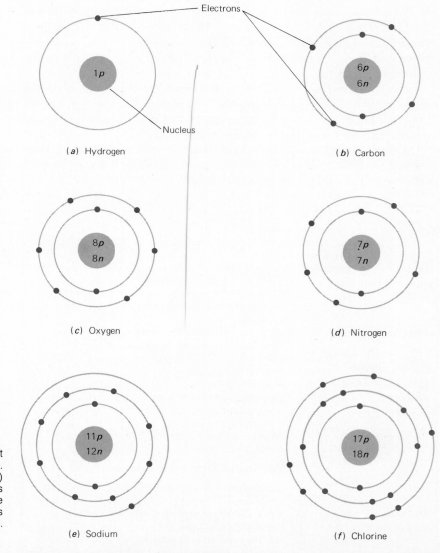

**FIGURE A-1**
Structure of some important atoms. (*a*) Hydrogen. (*b*) Carbon. (*c*) Oxygen. (*d*) Nitrogen. (*e*) Sodium. (*f*) Chlorine. (Electrons are indicated by dots on the circles. Each circle represents an electron shell or energy level. *p*, proton; *n*, neutron.)

(*a*) Hydrogen

(*b*) Carbon

(*c*) Oxygen

(*d*) Nitrogen

(*e*) Sodium

(*f*) Chlorine

varies among the atoms of an element. Forms of an element with different numbers of neutrons are called *isotopes.* Some isotopes are stable; others, called *radioisotopes,* break down, releasing stable products. The latter are the basis for radioactive dating of fossils and artifacts, and are useful in biochemical and physiologic research.

Electrons are negatively charged particles which revolve in definite orbits, or *shells,* better thought of as *energy levels,* about the nucleus. Most of the atom's volume is occupied by space, through which the electrons revolve. Normally, the number of electrons orbiting about the nucleus is equal to the number of protons within it, so that the atom is electrically neutral. A hydrogen atom consists of one proton and one electron (but has no neutrons). A carbon atom consists of six protons and six electrons, as well as six neutrons.

Electrons are held in their specific shells by their attraction for the positive protons in the nucleus. Each shell remains at a fixed distance from the nucleus. Most atoms have more than one electron shell, and each shell holds a characteristic maximum number of electrons. For the first shell the maximum is two, for the second it is eight, for the third 18, and so on. An atom of oxygen which has eight electrons would have two in its first shell, and six in the second shell.

The number of electrons revolving about its nucleus determines the types of chemical associations that an atom is able to make. An atom is most stable when each of its orbits contains the maximum number of electrons. Eight is a stable number for the outermost shell of most atoms except for hydrogen and helium (which require only two electrons). A few gases, including helium, argon, and neon (Figure A-2), naturally possess a stable number of electrons in their outer shell. Because of this stability these gases do not normally participate in chemical reactions and are known as *inert gases.* Only atoms whose outermost shell is not stable are capable of chemical combination. Atoms tend to give, take, or share electrons with other atoms until they reach a stabilized state.

## MOLECULES AND THEIR BONDS

A *molecule* consists of two or more atoms chemically combined with one another. The atoms in a molecule may be of similar type (as in molecular oxygen, which consists of two atoms of oxygen), or they may be composed of different elements. Water, for example, consists of two atoms of hydrogen chemically combined with one atom of oxygen. A molecule which consists of two or more different types of atoms in definite proportion is called a *chemical compound.*

### Chemical formulas

A shorthand method for describing the composition of a molecule is the *chemical formula.* In writing a chemical formula, chemical symbols are used to indicate the *types* of atoms in the molecule, and subscript numbers

(*a*) Helium

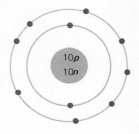

(*b*) Neon

**FIGURE A-2**
Atomic structure of two inert gases. These atoms are stable because their outer electron shells are filled. They do not enter into chemical reactions. (*a*) Helium. (*b*) Neon.

are used to describe the *number* of each type of atom present. The chemical formula for molecular oxygen, $O_2$, reveals at a glance that this molecule is composed of two atoms of oxygen. The chemical formula for water, $H_2O$, tells us that each molecule of water consists of two atoms of hydrogen and one atom of oxygen.

Another type of formula, the **structural formula,** discloses not only the types and number of atoms in a molecule, but also their *arrangement.* In any specific type of molecule the atoms are always arranged in the same way. From the chemical formula for water, $H_2O$, a novice could only guess whether the atoms were arranged H—H—O or H—O—H. The structural formula, H—O—H settles the matter, indicating that the two hydrogens are chemically attached to a central oxygen atom. In the case of water this arrangement is the only one chemically possible. But there are substances which consist of the same atoms, yet have different chemical properties as a consequence of alternative arrangements. Such substances are known as structural *isomers.* The sugars glucose and fructose are examples (see Chapter 4).

### Valence

The number of electrons in its outer shell determines whether a given atom will accept electrons or donate them. If an atom normally has more than four electrons in its outer shell it behaves as an *electron acceptor,* gaining electrons from other atoms in order to complete its outer shell. If an atom has fewer than four electrons in its outer shell it behaves as an *electron donor,* acting to rid itself of all the electrons in its unstable outer shell. The number of electrons that an atom can give or accept is known as its *valence,* or combining power. Valence is designated by a number from 1 to 7, indicating the number of electrons which can be accepted or donated. The valence number is preceded by a plus or minus sign which indicates whether the atom will give away electrons (thus becoming positively charged) or accept them (becoming negatively charged). For example, sodium has a valence of +1, meaning that it can donate one electron, thereby becoming positively charged. (Having done so, it now has one more proton than electron.) Chlorine's valence of −1 indicates that a chlorine atom can accept one electron, thereby gaining a negative charge. Oxygen has a valence of −2 because it can accept two electrons. Atoms which tend to donate electrons are known as metals, while those which accept electrons are nonmetals. By knowing the valence of each type of atom, chemists are able to predict how atoms will react with one another.

### Chemical bonds

The atoms of a molecule are held together by forces of attraction between them termed *chemical bonds.* A chemical bond consists of a specific amount of *chemical energy.* When drawing structural formulas on paper it is convenient to represent a chemical bond by a line drawn between two atoms. There are two main types of chemical bond, *ionic* and *covalent.* (Although there are other types of bonds, only these will be discussed here.)

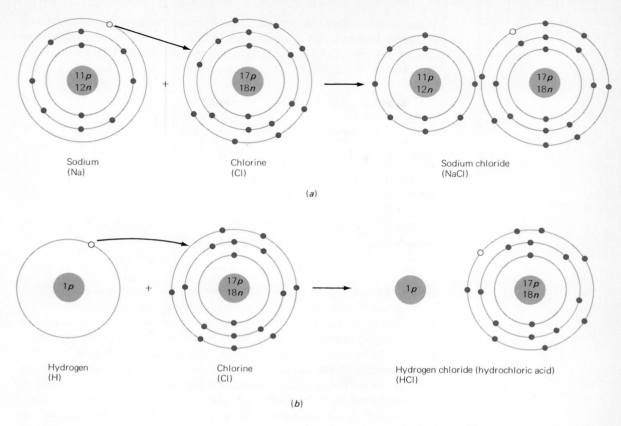

Sodium
(Na)

Chlorine
(Cl)

Sodium chloride
(NaCl)

(a)

Hydrogen
(H)

Chlorine
(Cl)

Hydrogen chloride (hydrochloric acid)
(HCl)

(b)

**Ionic bonds.** An ionic bond (often called an electrovalent bond) is formed when one atom donates an electron to another, each atom thereby becoming charged and attracted to the other by virtue of these electric charges. An atom of sodium forms an ionic bond with an atom of chlorine when the sodium ion donates the single electron in its outer shell to the chlorine, which requires only that one more in order to fill its own outer shell (Figure A-3). Each atom thus achieves stability by this transaction. But the sodium now has one more proton than electrons, while the chlorine has one more electron than protons. Thus, each now has an electric charge, the sodium a positive charge resulting from its excess proton, and the chlorine a negative charge resulting from its excess electron. As a consequence of these charges the two atoms are attracted to each other and stay together, forming an ionic compound. This compound is NaCl, sodium chloride, or table salt. To sum it up, ionic bonds are established by *electron exchange*.

A *chemical reaction* occurred between the sodium and the chlorine atoms. Such a change may be described on paper by means of a *chemical equation*.

$$2Na \ + \ Cl_2 \ \longrightarrow \ 2NaCl$$
sodium    chlorine      sodium chloride

**FIGURE A-3**
Formation of ionic compounds. (*a*) Sodium donates an electron to chlorine, forming the compound sodium chloride. (*b*) Hydrogen donates its electron to chlorine, forming hydrogen chloride (hydrochloric acid).

Three types of ionic compounds are *acids, bases* (alkaline substances), and *salts.* When placed in solution the atoms of an ionic compound tend to dissociate (separate). The free, charged atoms which result are called *ions.* Thus $H^+$ is a hydrogen ion, $Cl^-$ is a chlorine ion.

$$HCl \xrightarrow{\text{in } H_2O} H^+ + Cl^-$$

hydrochloric acid       hydrogen ion   chlorine ion

$$H_2SO_4 \xrightarrow{\text{in } H_2O} 2H^+ + SO_4^{-2}$$

sulfuric acid       hydrogen ions   sulfate ion

$$NaOH \xrightarrow{\text{in } H_2O} Na^+ + OH^-$$

sodium hydroxide       sodium ion   hydroxyl ion
(a base)

$$NaCl \xrightarrow{\text{in } H_2O} Na^+ + Cl^-$$

sodium chloride       sodium ion   chlorine ion
(a salt)

***Covalent bonds.*** In a covalent bond atoms neither gain nor lose electrons; atoms share electrons. An atom of hydrogen which has only one electron orbiting about its nucleus is not stable. But when two atoms of hydrogen chemically combine by sharing their electrons, they become chemically stable (Figure A-4). The two electrons orbit about the nuclei of both hydrogen atoms of the molecule. Oxygen atoms also achieve stability by forming covalent bonds. Each atom of oxygen contains only six electrons in its second shell. By sharing two pairs of electrons ( a double covalent bond), each oxygen atom becomes chemically stable, and molecular oxygen is formed. Similarly an atom of oxygen can combine with two atoms of hydrogen to form water. Each hydrogen shares an electron with the oxygen, so that the oxygen atom achieves the magic number of eight. At the same time each hydrogen shares one of the oxygen's electrons, so that its electron stability is also enhanced. To sum it up, covalent bonds are established by the *sharing of electrons.*

## CHEMICAL REACTIONS

A chemical reaction occurs whenever atoms or molecules interact to form different chemical combinations. Substances which participate in the reaction are termed ***reactants;*** those formed by the reaction are ***products.***

### Chemical equations

The shorthand method for denoting a reaction is termed a *chemical equation.* A reaction such as

$$H_2 + Cl_2 \longrightarrow 2HCl$$

is a *synthetic reaction* because two substances have combined to form a

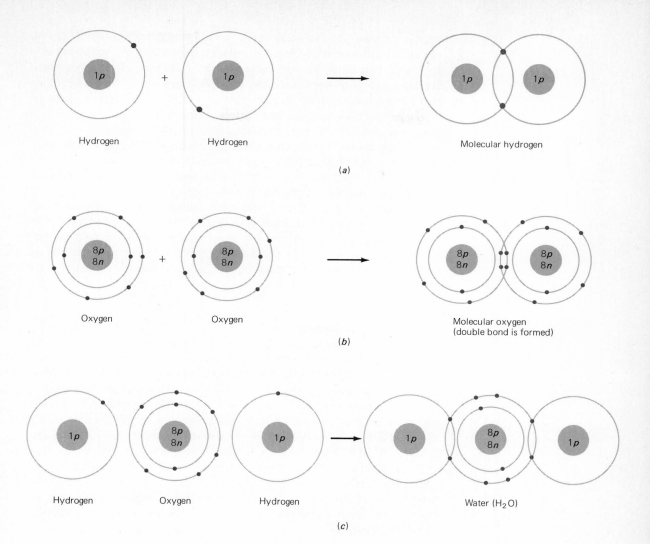

$(a)$

$(b)$

$(c)$

new substance. The reverse reaction would be an example of a **decomposition reaction.** Still another type of chemical reaction involves the exchange of atoms within a molecule, or between molecules, to yield a new substance.

One of the most important chemical equations considered in a basic biology course is the summary equation for photosynthesis.

$$6H_2O + 6CO_2 \xrightarrow[\text{chlorophyll}]{\text{light energy,}} C_6H_{12}O_6 + 6O_2$$

water    carbon            glucose    oxygen
         dioxide

Though learning to balance chemical equations is beyond the scope of this course, it is useful to understand the meaning of equations such as this. The

**FIGURE A-4**
Formation of covalent compounds. (a) When two hydrogen atoms share electrons, molecular hydrogen is formed. (b) When two oxygen atoms share electrons, molecular oxygen is formed. In this case two pairs of electrons are shared, and the compound is said to possess a double bond. (c) When two hydrogen atoms share electrons with one oxygen atom, water is formed.

large numbers preceding some of the molecules indicate the number of molecules needed for the reaction. Thus $6H_2O$ means that 6 molecules of water are needed in order for the reaction to be balanced. Since there are 2 hydrogen atoms in each molecule of water, there would be 12 hydrogen atoms in the 6 water molecules. As there is a total of 12 hydrogen atoms on the left side of the equation, there must also be a total of 12 on the right side of the equation. Note that this is exactly the case. Count the number of each type of atom on each side of the equation, and determine whether the equation is balanced. Now note that above the yield arrow we have written "light energy, chlorophyll." This is the conventional way to indicate that these substances are necessary for the reaction to proceed.

In living systems almost every reaction requires a specific *enzyme,* a protein which regulates the reaction (see Chapter 5). Often a *coenzyme,* a vitamin or substance derived from a vitamin, or a specific mineral is also required for the reaction to proceed. All these substances which do not become a permanent part of the product but are needed for the reaction are indicated above the arrow.

### Oxidation-reduction reactions

Oxidation-reduction reactions are chemical reactions in which one atom gains electrons while another simultaneously loses them. When an atom loses electrons by donating them to another atom it is said to be *oxidized.* The atom that gains the electrons is *reduced.* When sodium, for example, reacts with chlorine, sodium loses an electron, becoming oxidized. The chlorine gains the electron, becoming reduced. Every oxidation is accompanied by a reduction, for something must accept the electrons lost. In living systems oxidation almost always involves the removal of a hydrogen atom from a molecule, while reduction involves the addition of a hydrogen atom. The term oxidation originated with early chemists, who defined it as combination of a substance with oxygen. Though this is indeed a particular example of oxidation, the term is now used more broadly. Cellular respiration depends upon a series of oxidation-reduction reactions (Chapter 8).

### ACIDS, BASES, AND pH

An *acid* is a compound which dissociates in solution to form hydrogen ions. Hydrochloric acid (HCl) and sulfuric acid ($H_2SO_4$) are familiar examples. A compound which dissociates in solution to form hydroxyl ions, $OH^-$, is called a *base* (or alkali). Sodium hydroxide is an example. Acids taste sour; bases are bitter. Whether a solution is acidic or basic is determined by testing with litmus paper or the use of a pH meter. Litmus paper, an acid-base indicator found in every basic chemistry set, turns red when dipped into an acid solution and blue when dipped into a basic solution. A pH meter is an instrument which measures the relative acidity or alkalinity of a solution. Its scale goes from 0 to 14. Seven is neutral. The more acidic a solution is, the lower its pH. Thus a solution with a pH of 1 would be highly acidic, while a solution with a pH of 6 would be only slightly acidic. The higher the number above 7, the greater the alkalinity. The concept of pH is important in biology because living systems contain many acids and bases which must be kept in appropriate balance.

## THE CARBON ATOM

Chemists divide chemical compounds into two groups: organic and inorganic. *Organic compounds* always contain carbon and are generally quite complex, as compared with the simpler *inorganic compounds*. Carbon atoms form very strong covalent bonds with one another, and comprise the *main chain* of atoms in all organic compounds. Organic compounds are generally covalently bonded.

With four electrons in its outer shell a carbon atom can form four bonds with other atoms. For example a carbon atom can combine with four hydrogen atoms

$$\begin{array}{c} \text{H} \\ | \\ \text{H}-\text{C}-\text{H} \\ | \\ \text{H} \end{array}$$

or it can combine with a variety of atoms,

$$\begin{array}{c} \diagdown\ \diagup \\ \text{N} \\ | \qquad | \\ -\text{C}-\text{C}-\text{C}- \\ | \qquad | \qquad | \\ \quad\ \ \text{H} \end{array}$$

Sometimes carbon forms a double bond (two pairs of electrons are shared) either with another carbon atom or with an oxygen or other type of atom

$$\text{H}-\text{C}=\text{C}-\text{H} \qquad \begin{array}{c} | \\ -\text{C}-\text{C}=\text{O} \\ | \quad\ | \\ \quad\ \text{H} \end{array}$$

When doubly bonded to another atom, carbon can combine with only two additional atoms. (The total number of bonds formed is always four.)

Four main classes of organic compounds found in living systems are carbohydrates, lipids, proteins, and nucleic acids. The first three are discussed in Chapter 4, and nucleic acids are discussed in Chapter 14. Each type of compound is composed of specific kinds of molecular subunits, usually linked together in repeating units. Thus, a protein is built from molecular subunits called amino acids, and large carbohydrates consist of repeating units of simple sugars. These subunits, linked together by the chemical removal of water, can form very large molecules, called *macromolecules.* The process of removing water during synthesis is called *dehydration.* During digestion these macromolecules are broken down into their molecular subunits by the chemical addition of water, a process called *hydrolysis.* Figure A-5 explains how some biologically important organic compounds may be identified.

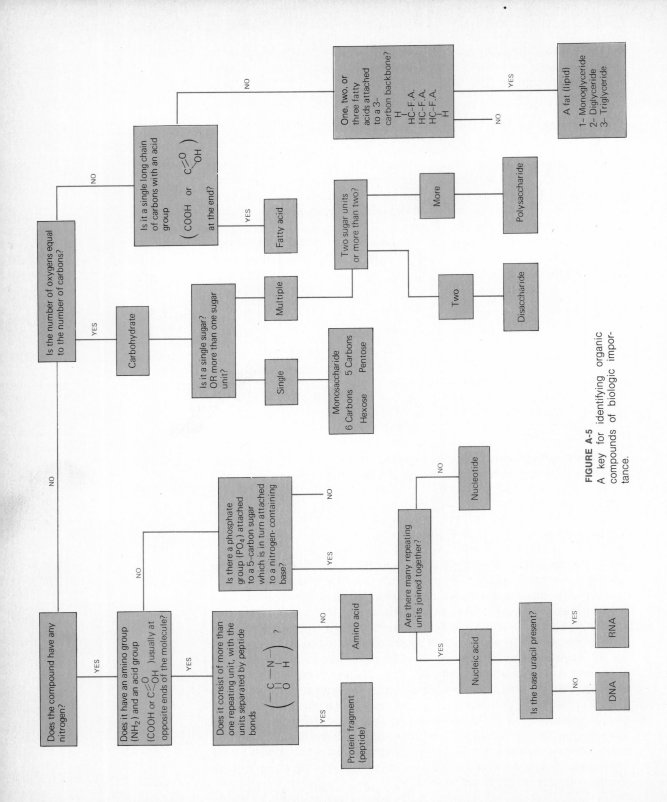

**FIGURE A-5**
A key for identifying organic compounds of biologic importance.

## SOLUTIONS AND COLLOIDS

When you dissolve sugar in water you are making a *solution,* a chemical system in which one substance (or more) is uniformly dissolved in another. In this case water is the *solvent,* the substance doing the dissolving, and sugar is the *solute,* the substance which is being dissolved. Molecules of the solute occupy spaces between the molecules of the solvent. In a true solution the solute remains dissolved, that is, does not settle to the bottom. Furthermore, in a true solution the solute particles are of molecular size.

When dispersed particles are too large for the system to be considered a solution, it is called a *colloid.* There are many different types of colloids. One familiar example is gelatin dessert. Hot gelatin is a solid dispersed in a liquid, that is, the solid gelatin is dispersed in the boiling water. As the gelatin cools there is a reversal of phase, so that the liquid water becomes dispersed in a lattice of solid gelatin. The dispersed substance of a colloid is called its *discontinuous phase,* and the substance in which it is dispersed is the *continuous phase.* Often the discontinuous phase (or dispersed particles) of a colloid consists of molecular aggregates.

## MOVEMENT OF MOLECULES

If a bottle of perfume, or ammonia, were opened in a corner of a large closed room with no air current, a few hours later the odor would be obvious throughout the room. Some of the perfume molecules would have moved out of the bottle and distributed themselves from the corner of the room where they emerged, evenly all over the room. This is *diffusion,* the net movement of molecules from a region of greater concentration to regions of less concentration. Diffusion depends upon the random movement of individual molecules, propelled by collision with the vibrating molecules of the solvent.

Diffusion is an important process in living systems. Oxygen moves from the air sacs in the lungs into the blood by diffusion, and many other substances pass in and out of cells by this simple physical process. On the other hand cells also have mechanisms for interfering with diffusion. A cell contains many chemicals in its cytoplasm which are present in very different concentrations from those which occur outside the cell. If these differences were not maintained, the cell would quickly die. So in some cases the cell must struggle to prevent diffusion from occurring. *Active transport* is an important mechanism by which the cell can keep out materials in its environment which would normally enter by diffusion. Some substances are taken into the cell against a concentration gradient by the process of active transport. Many nutrients are taken into the cell in this way. Substances moving from a region of greater concentration to one of lesser concentration are moving along a concentration gradient, but substances moving in the opposite direction are moving *against* the concentration gradient, a process which requires energy. Thus, active transport is a type of work performed by the cell and requires energy.

A special kind of diffusion is *osmosis,* the passage of water through a semipermeable membrane[1]. Cell membranes permit water to move freely in and out of the cell, but they selectively regulate the passage of solutes. Water always moves from a region where dissolved substances are less concentrated to a region where they are more concentrated. We may say that the solution or colloid with the more concentrated solute exerts an *osmotic pressure* upon the less concentrated solution or colloid on the other side of the membrane.

Many substances are present in greater concentration inside a cell than outside, creating an osmotic pressure inside so that water tends to diffuse into the cell. Quite a bit of pressure may be produced by the entering water, and this pressure is exerted against the cell membrane. In animal cells, excess water is removed by active transport. In plant cells, the presence of the cell wall permits the development of as much as a hundred atmospheres of pressure inside the cell. This pressure helps to support the plant cell. When plants wilt, or when a vegetable such as celery becomes limp, the rigidity has been reduced because of water loss.

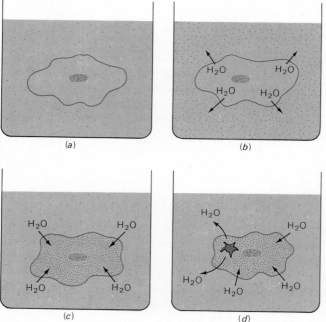

**FIGURE A-6**
Osmosis. (*a*) When a cell is placed into an isotonic solution, the flow of water molecules in and out of the cell is equal. There is no net change. (*b*) When a cell is placed into a hypertonic solution, more water molecules pass out of the cell than enter. The cell shrinks. (*c*) When a cell is placed into a hypotonic solution, water flows into the cell much more rapidly than it leaves. The cell may swell and burst. (*d*) When a saltwater ameba is placed into a hypotonic solution, the ameba forms a contractile vacuole and pumps the excess water out.

An *isotonic solution* has exactly the same concentration of dissolved substances per unit volume as a living cell, say an ameba, placed in it. The net movement of water through the cell membrane will remain constant in

---

[1] This explanation of osmosis is generally accepted at the present time but may not be the most adequate possible model. Our understanding of osmotic mechanisms may be greatly modified in future years.

such a situation. If a salt water ameba is placed in a container of fresh water, the water is said to be *hypotonic,* that is, it contains less dissolved material than the cell. Now the osmotic pressure in the cell causes water to enter. The ameba is able to form a contractile vacuole which bails out the excess water. Cells which cannot perform this trick remove the excess water by active transport, but if the water moves in too quickly, the cell may burst.

If living cells are placed in a *hypertonic* solution (one containing a higher concentration of dissolved substances than is found in the cells) water tends to flow *out* of the cells. In this case the cells may become dehydrated, shrink, and die.

## THE METRIC SYSTEM

The metric system is the decimal system of measurement used by scientists throughout the world. In many countries the metric system of weights and measures is used in commerce and everyday living as well as in the laboratory. For some time now the United States has been considering changing to this system and most likely will gradually make this change. Even now in the United States the commonly used English units are officially defined in terms of metric measurement. A few relationships useful to the beginning biology student are given in Table A-2.

**Table A-2. The metric system**

| | |
|---|---|
| *Standard metric units* | |
|    Standard unit of mass | gram |
|    Standard unit of length | meter |
|    Standard unit of volume | liter |
| *Some common prefixes:* | |
|    kilo      1,000 | e.g., a kilogram is 1,000 grams |
|    centi    0.01 | e.g., a centimeter is 0.01 meter |
|    milli     0.001 | e.g., a milliliter is 0.001 liter |
|    micro   one-millionth | e.g., a microgram is one-millionth of a gram |
| *English equivalents* | |
|    1 pound = 453.6 grams | 1 quart = 0.946 liter |
|    2.2 pounds = 1 kilogram | 1 liter = 1.06 quarts |
|    1 inch = 2.54 centimeters | 1 meter = 1.094 yards |

# APPENDIX B DIVERSITY OF LIVING SYSTEMS

Although it looks like a cookie monster from Sesame Street, this fellow is a spider.

***Unit learning objective.*** The student must be able to describe the general characteristics of each of the five kingdoms given in this unit, and should be able to give examples of organisms which belong to each kingdom and phylum.

*ILO 1*   Give the basic characteristics and ecologic roles of bacteria and blue-green algae, and justify their placement in the kingdom Monera.

*ILO 2*   Give the characteristics and ecologic roles of protozoa and eukaryotic algae.

*ILO 3*   Characterize the fungi; give their ecologic significance and specific importance in human affairs.

*ILO 4*   Give the general characteristics and ecologic role of plants.

*ILO 5*   Compare bryophytes with tracheophytes from the standpoint of complexity and biologic success.

*ILO 6*   Distinguish between ferns and seed plants; between gymnosperms and angiosperms; between monocots and dicots.

*ILO 7*   Describe the level of organization and characteristic features of each phylum of animals.

*ILO 8*   Give specific examples of animals which belong to each phylum.

*ILO 9*   List the three main characteristics of a chordate.

*ILO 10*   List and give examples of each of the seven classes of vertebrates, and give distinguishing characteristics for each class.

"Still believe in two kingdoms, Professor?"

FIGURE B-1

Traditionally living things have been designated either plant or animal, but many organisms do not fit neatly into these categories. Prokaryotes, for example, are very different from eukaryotes—even more so than plants are from animals. And some unicellular forms have both plant and animal characteristics. For these and other reasons many biologists prefer assigning living things to more than two kingdoms. Some use three, some four. Here we shall use a recently proposed system of five kingdoms (Figure B-2). For want of space, only the major groups of organisms in each kingdom are discussed.

**Table B-1. Classification of the house cat and of man\***

| Categories | Classification of cat | Classification of man |
|---|---|---|
| Kingdom | Animalia | Animalia |
| Phylum | Chordata | Chordata |
| Subphylum | Vertebrata | Vertebrata |
| Class | Mammalia | Mammalia |
| Subclass | Eutheria | Eutheria |
| Order | Carnivora | Primates |
| Family | Felidae | Hominidae |
| Genus | *Felis* | *Homo* |
| Species | *domestica* | *sapiens* |

\* The main categories of classification are given here. In accordance with the Binomial System of Nomenclature, each organism is assigned two names, a genus and a species name. Man, for example, is *Homo sapiens;* the common house cat, *Felis domestica.*

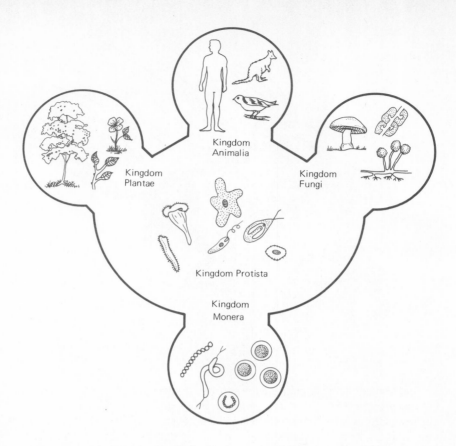

**FIGURE B-2**
The five-kingdom system.

Even the five-kingdom system does not comfortably accommodate the viruses, the only living things which are not cellular. Most viruses consist of a nucleic acid core of RNA or DNA, surrounded by a protein coat. Recently, infective particles, called *viroids,* consisting of naked nucleic acid have been described. The life style of viruses has been described in Chapter 14.

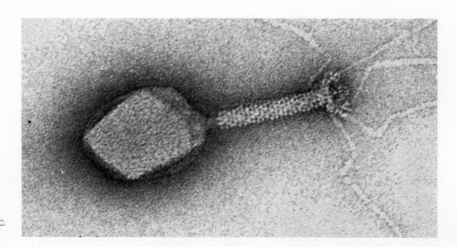

**FIGURE B-3**
Tobacco mosaic virus. (Dr. Lyle C. Dearden)

**Table B-2. Three kingdoms: Monera, Protista, and Fungi**

| Kingdom and phyla | Characteristics | Role in ecosystem | Comments |
|---|---|---|---|
| Monera | Prokaryotes (lack distinct nuclei and other membranous organelles); all single-celled (see Chapter 17) | | |
| Bacteria | Cell walls composed of murein (a substance derived from amino acids and sugars); many secrete a capsule made of a polysaccharide material. In pathogenic bacteria capsule may protect against defenses of host. Cells spherical (cocci), rod-shaped (bacilli), or coiled (spirilla) | Decomposers; some chemosynthetic autotrophs; important in recycling nitrogen and other minerals. Some pathogenic (e.g., cause "TB," "strep" throat, gonorrhea). Some utilized in industrial processes | Sensitive to penicillin because this drug inhibits murein, thus interfering with cell division |

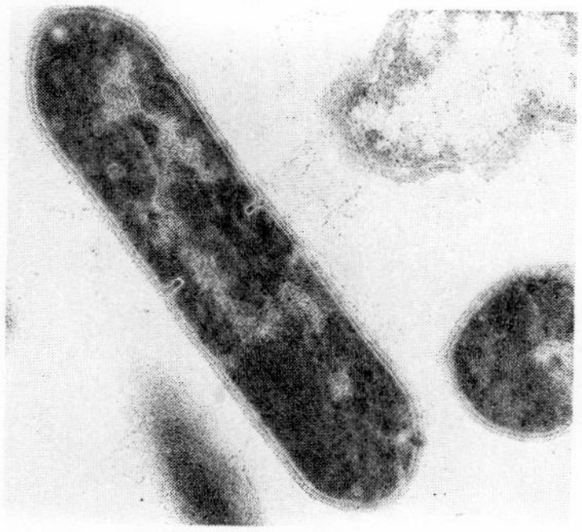

**FIGURE B-4**
Cell division of a bacterium. Note absence of a clearly defined nucleus. (Dr. John Betz)

| | | | |
|---|---|---|---|
| Blue-green algae | Specifically adapted for photosynthesis; chlorophyll and associated enzymes organized into layers in cytoplasm; some can fix nitrogen | Producers; blooms (population explosions) associated with water pollution | |

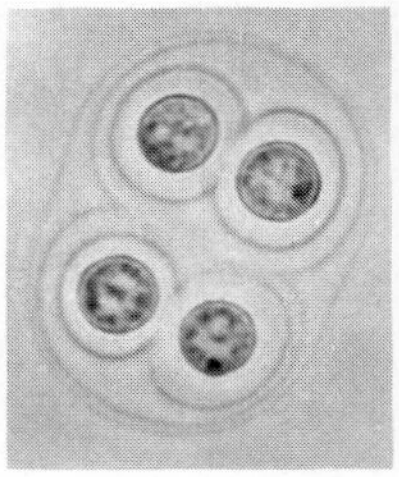

**FIGURE B-5**
Blue-green algae (*Gleocapsa*). (×2,000) (Dr. Roy M. Allen and Dr. Wayne Frair)

| | | | |
|---|---|---|---|
| Protista | Eukaryotes; mainly unicellular or colonial | | |
| Protozoa | Microscopic; unicellular; depend upon diffusion to support many of their metabolic activities | Part of zooplankton; serve as vital links in many food chains. Some pathogenic (e.g., cause malaria, amebic dysentery) | Formerly classified as animals |

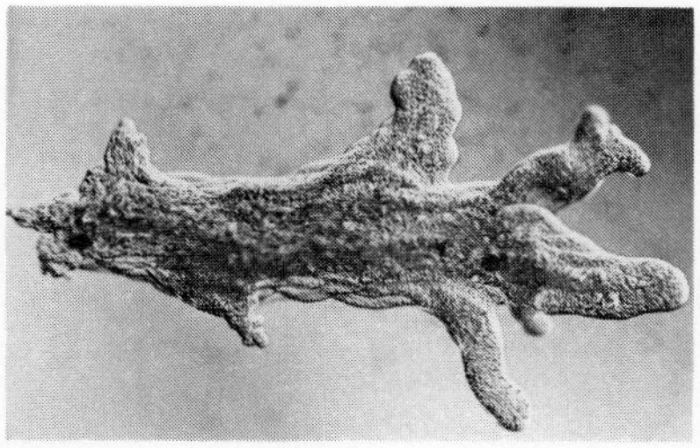

**FIGURE B-6**
*Amoeba proteus,* a protozoon which moves by means of pseudopods, i.e., temporary extensions of its cytoplasm. (Dr. Robert D. Allen and *Science*)

| Kingdom and phyla | Characteristics | Role in ecosystem | Comments |
|---|---|---|---|
| Eukaryotic algae | Carry on photosynthesis; on basis of type of photosynthetic pigment, algae are designated green, brown, or red. Many unicellular, but some large with differentiated bodies [e.g., giant seaweeds (kelps), over 100 ft in length] | Important producers; form trophic base of many food chains | |

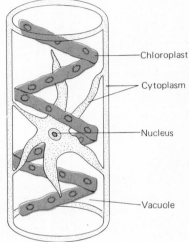

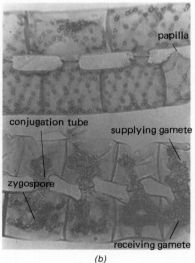

**FIGURE B-7**
*Spirogyra,* a filamentous green alga (i.e., one composed of a string of cells). (*b*) Sexual reproduction is shown here. Cells of two adjacent filaments come into contact. One cell becomes ameba-like and moves into the adjacent filament, fusing with the other cell to form a zygote.

*(a)*

*(b)*

(Labels in figure a: Chloroplast, Cytoplasm, Nucleus, Vacuole)

(Labels in figure b: papilla, conjugation tube, supplying gamete, zygospore, receiving gamete)

| | | | |
|---|---|---|---|
| Fungi | Plantlike but cannot carry on photosynthesis; cell wall usually composed of chitin rather than cellulose | Decomposers | |
| True fungi (molds, yeasts, mildews, mushrooms, rusts, smuts) | Body composed of threadlike hyphae, rather than discrete cells. Hyphae may form tangled masses called *mycelia,* which infiltrate whatever the fungus is eating or inhabiting. (In mushrooms, bulk of body is a nearly invisible network buried in the rotting material it invests.) | Some (mushroom, morel, truffle) used as food. Yeast used in making bread and alcoholic beverages. Some used to make industrial chemicals. Penicillin is product of blue mold, *Penicillium*. Responsible for crop losses, damage to stored goods and building materials, and some diseases (e.g., "ringworm," athlete's foot, and some serious lung diseases) | Possess powerful digestive enzymes for breaking down wood, food, and other materials |

**FIGURE B-8**
*Morchella,* the morel, a gourmet's delight. Spores are borne on the convoluted surface of the cap.
(Carolina Biological Supply Company)

**Table B-2.** (*Continued*)

| Kingdom and phyla | Characteristics | Comments |
|---|---|---|
| Slime molds | Animal characteristics during part of life cycle, plant characteristics during the other part | Cellular slime molds are subject of research by developmental biologists. |

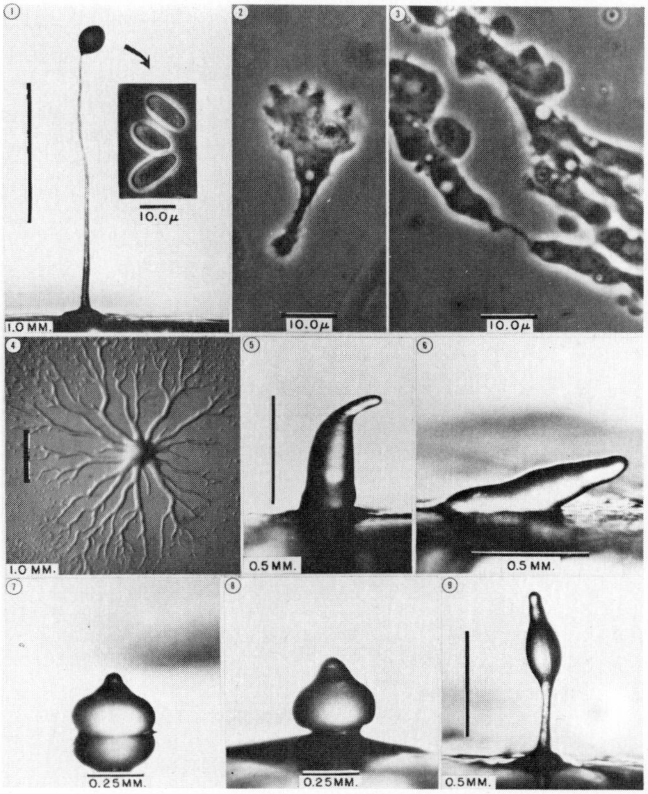

**FIGURE B-9**
Life cycle of the cellular slime mold, *Dictyostelium discoideum*. (1) Mature fruiting body releases spores. (2) Each spore cell opens to liberate a new ameba. Amebas eat, grow, and multiply by binary fission. (3) After the food supply (usually soil bacteria) in their immediate vicinity has been depleted, amebas stream together. (4) Pattern of aggregation. (5) Amebas join to form a multicellular heap. (6) Cells of each heap arrange themselves to form a slug-shaped organism (migrating pseudoplasmodium), which behaves like a multicellular organism. (7) After a period of migration which may involve search for a favorable habitat in which to continue its life cycle, the slug begins to form a fruiting body. (8) Cells which had formed the anterior one-third of the slug differentiate to become stalk cells, while those from the posterior two-thirds form spore cells. (9) Supported by a thin stalk, a mass of spores is raised above the surface. (Dr. J. Gregg, *The Fungi*, Academic, New York, 1966)

# Table B-3. Plant kingdom

| Kingdom and division | Characteristics | Role in ecosystem |
|---|---|---|
| Plant | Multicellular; complex; adapted for photosynthesis; have reproductive tissues or organs; pass through distinct developmental stages; alternation of generations—sporophyte generation produces spores which develop into gametophyte plants which produce gametes | Producers |

**FIGURE B-10**
(a) *Polytrichum*, a common moss. The sporophytes are the elongated threadlike structures capped with spore cases. They are completely parasitic upon the gametophytes beneath them. (Carolina Biological Supply Company) (b) Life cycle of a moss.

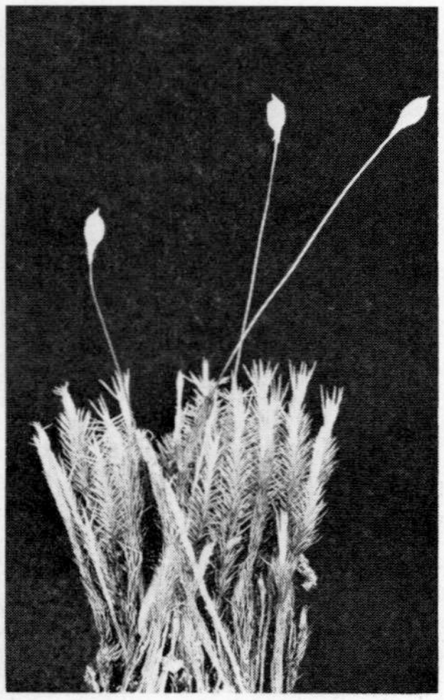

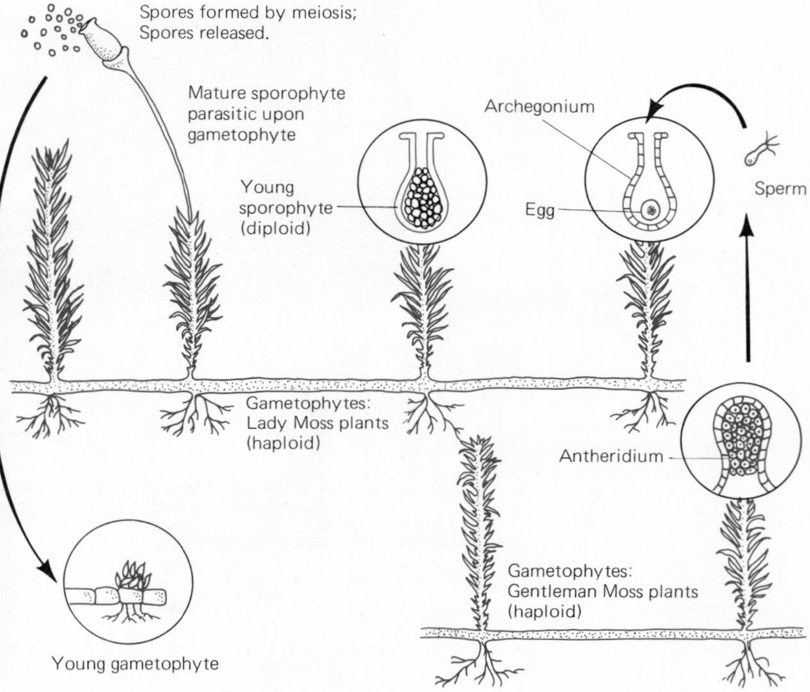

Spores formed by meiosis; Spores released.

Mature sporophyte parasitic upon gametophyte

Archegonium

Young sporophyte (diploid)

Egg

Sperm

Gametophytes: Lady Moss plants (haploid)

Antheridium

Young gametophyte

Gametophytes: Gentleman Moss plants (haploid)

(a)                    (b)

| | | |
|---|---|---|
| Bryophytes (mosses, liverworts, hornworts) | Small because they lack true stems, roots, or leaves, and any sort of circulatory system (and must depend upon diffusion for distribution of materials). Gametophyte is dominant form; smaller sporophyte is attached to gametophyte in mosses and is partially parasitic upon it. | Often pioneer organisms in colonization of new habitats |
| Tracheophytes (vascular plants) | Have circulatory systems which enable them to attain large size; have true roots, stems, and leaves; multicellular embryos; sporophyte dominant form | Adapted for life on land |
| (1) Ferns | Gametophyte develops as a tiny, independent plant. Spores are found in little clusters on underside of some sporophyte leaves. | |

**FIGURE B-11**(a)
Leaves of a fern sporophyte showing clusters of spores.

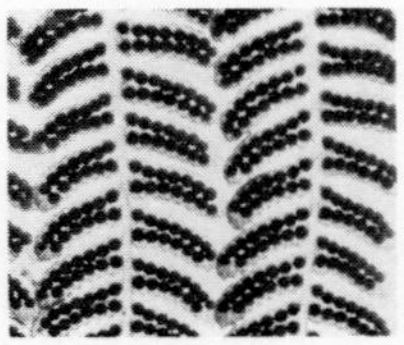

(a)

| Kingdom and division | Characteristics | Role in ecosystem |
|---|---|---|

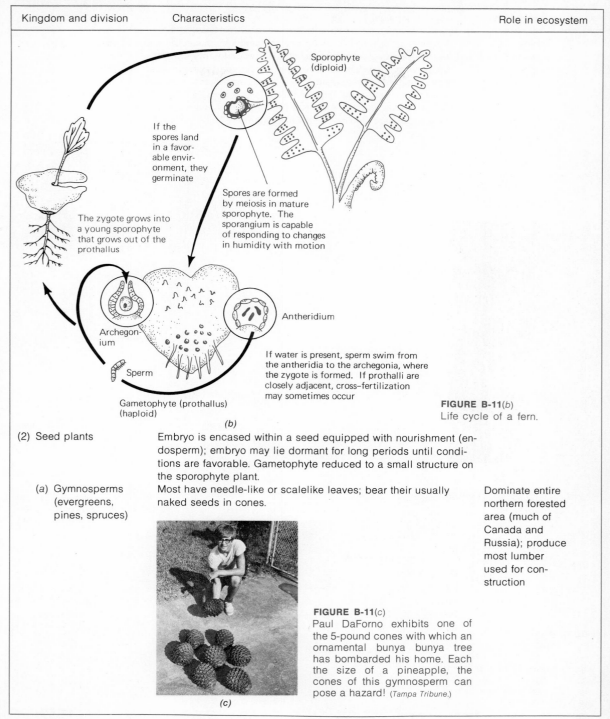

Sporophyte (diploid)

If the spores land in a favorable environment, they germinate

Spores are formed by meiosis in mature sporophyte. The sporangium is capable of responding to changes in humidity with motion

The zygote grows into a young sporophyte that grows out of the prothallus

Archegonium

Antheridium

Sperm

If water is present, sperm swim from the antheridia to the archegonia, where the zygote is formed. If prothalli are closely adjacent, cross–fertilization may sometimes occur

Gametophyte (prothallus) (haploid)

(b)

**FIGURE B-11**(*b*)
Life cycle of a fern.

(2) Seed plants — Embryo is encased within a seed equipped with nourishment (endosperm); embryo may lie dormant for long periods until conditions are favorable. Gametophyte reduced to a small structure on the sporophyte plant.

(a) Gymnosperms (evergreens, pines, spruces) — Most have needle-like or scalelike leaves; bear their usually naked seeds in cones.

Dominate entire northern forested area (much of Canada and Russia); produce most lumber used for construction

**FIGURE B-11**(*c*)
Paul DaForno exhibits one of the 5-pound cones with which an ornamental bunya bunya tree has bombarded his home. Each the size of a pineapple, the cones of this gymnosperm can pose a hazard! (*Tampa Tribune.*)

(c)

**Table B-3. Plant kingdom** (*Continued*)

| Kingdom and division | Characteristics | Role in ecosystem |
|---|---|---|
| (b) Angiosperms (flowering plants) | Possess sexual organs called flowers; seeds often surrounded by fleshy or hard envelope called a fruit (comprised mainly of the ovary wall). Either annual (live only one season) or deciduous (lose their leaves seasonally). Two types: monocots—seed has single small embryonic leaf and large endosperm, leaves have parallel veins; dicots—seed has two large embryonic leaves and little endosperm, leaves have netlike veins | Provide food for man and other consumers |

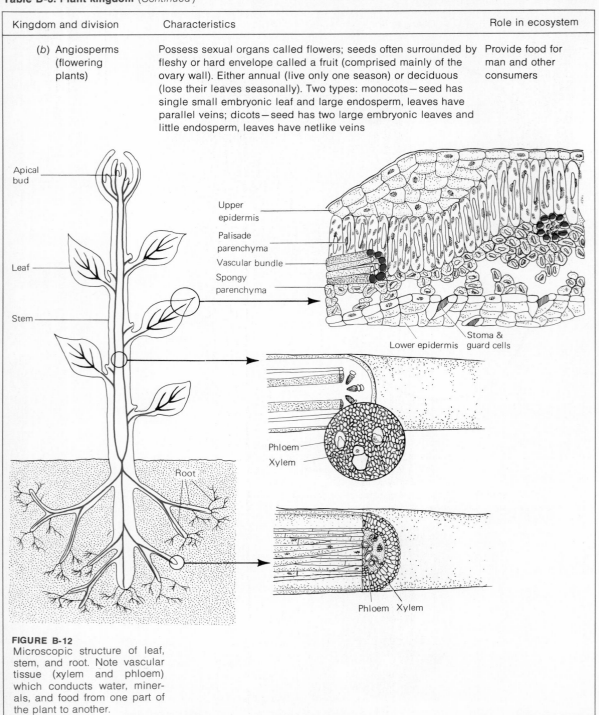

Apical bud

Leaf

Stem

Root

Upper epidermis

Palisade parenchyma

Vascular bundle

Spongy parenchyma

Lower epidermis   Stoma & guard cells

Phloem
Xylem

Phloem  Xylem

**FIGURE B-12**
Microscopic structure of leaf, stem, and root. Note vascular tissue (xylem and phloem) which conducts water, minerals, and food from one part of the plant to another.

| Kingdom and division | Characteristics | Role in ecosystem |
|---|---|---|

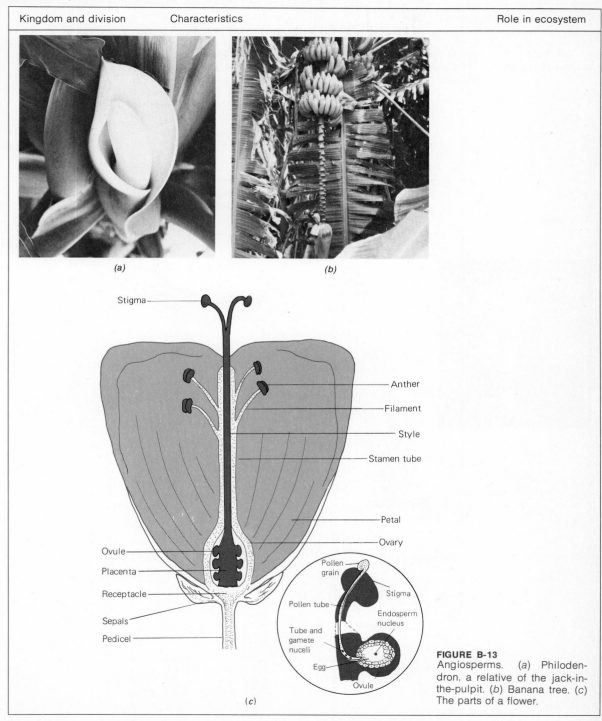

*(a)*        *(b)*

Stigma

Anther

Filament

Style

Stamen tube

Petal

Ovary

Ovule

Placenta

Receptacle

Sepals

Pedicel

Pollen grain

Stigma

Pollen tube

Endosperm nucleus

Tube and gamete nuceli

Egg

Ovule

*(c)*

**FIGURE B-13**
Angiosperms. (*a*) Philodendron, a relative of the jack-in-the-pulpit. (*b*) Banana tree. (*c*) The parts of a flower.

**Table B-4. Animal kingdom**

| Phyla | Level of organization | Symmetry | Digestion | Circulation |
|---|---|---|---|---|
| Porifera (pore bearers): sponges | Multicellular but tissues loosely arranged | Radial or none | Intracellular | Diffusion |
| Cnidaria<br>　Hydra<br>　Jellyfish, coral | Tissue level | Radial | Digestive cavity with only one opening; intra- and extracellular digestion | Diffusion |

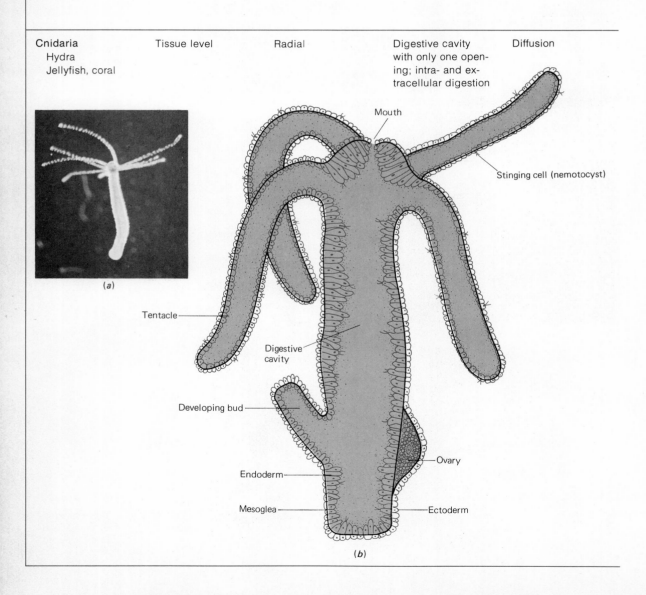

Mouth

Stinging cell (nemotocyst)

(a)

Tentacle

Digestive cavity

Developing bud

Endoderm

Mesoglea

Ovary

Ectoderm

(b)

| Respiration | Excretion | Nervous system | Reproduction | Comments |
|---|---|---|---|---|
| Diffusion | Diffusion | Irritability of cytoplasm | Asexual, by budding; sexual—may be monoecious (both sexes in one organism) or dioecious (separate sexes). Larvas swim by cilia; adults incapable of locomotion | Filter feeders; skeleton of chalk, glass, or spongin (a protein material) |
| Diffusion | Diffusion | Nerve net; no centralization of nerve tissue | Asexual, by budding; sexual—dioecious | Have stinging cells along their tentacles |

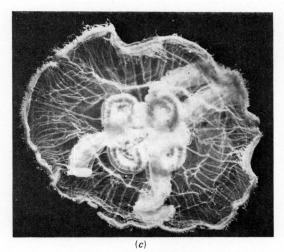

**FIGURE B-14**
A cluster of *Grantia* sponges. These sponges have a chalky (calcereous) skeleton. (Carolina Biological Supply Company.)

(*c*)

(*d*)

**FIGURE B-15**
(*a*) *Hydra.* (Carolina Biological Supply Company) (*b*) Anatomy of hydra. (*c*) *Aurelia,* a common jellyfish. (Carolina Biological Supply Company) (*d*) Coral polyps. (Roy Lewis, III)

**Table B-4. Animal kingdom** (*Continued*)

| Phyla | Level of organization | Symmetry | Digestion | Circulation |
|---|---|---|---|---|
| Platyhelminthes (flatworms) Planarians, flukes, tapeworms | Organ level | Bilateral; rudimentary head | Digestive tract with only one opening | Diffusion |

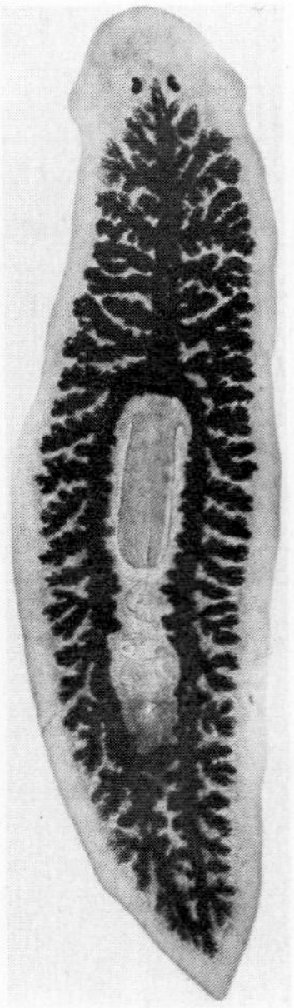

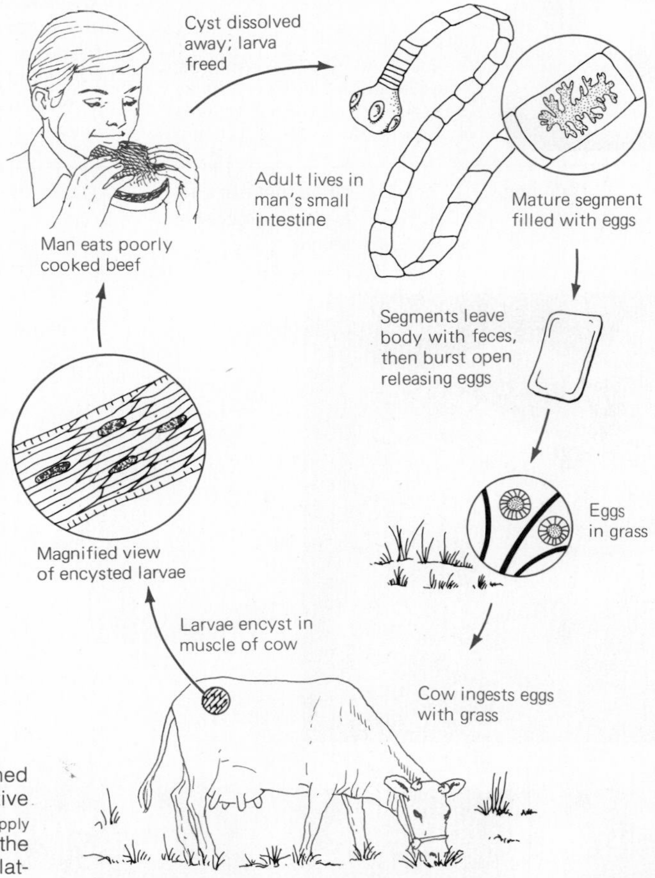

Cyst dissolved away; larva freed

Man eats poorly cooked beef

Adult lives in man's small intestine

Mature segment filled with eggs

Segments leave body with feces, then burst open releasing eggs

Magnified view of encysted larvae

Eggs in grass

Larvae encyst in muscle of cow

Cow ingests eggs with grass

**FIGURE B-16**
(*a*) Planarian flatworm stained to show branched digestive system. (Carolina Biological Supply Company) (*b*) Life cycle of the beef tapeworm, a parasitic flatworm.

(*a*)

(*b*)

| Phyla | Level of organization | Symmetry | Digestion | Circulation |
|---|---|---|---|---|
| Aschelminthes (roundworms) Ascarids, hookworms, nematodes | Organ system | Bilateral | Complete digestive tract (mouth to anus) | Diffusion |

| Respiration | Excretion | Nervous system | Reproduction | Comments |
|---|---|---|---|---|
| Diffusion | Flame cells and ducts | Simple brain; two nerve cords; ladder-type system; simple sense organs | Asexual, by fission; sexual—monoecious, but some cross-fertilize | |
| Diffusion | Excretory canals | Simple brain; dorsal and ventral nerve cords; simple sense organs | Sexual—sexes separate | Have body cavity (space between internal organs and body wall) |

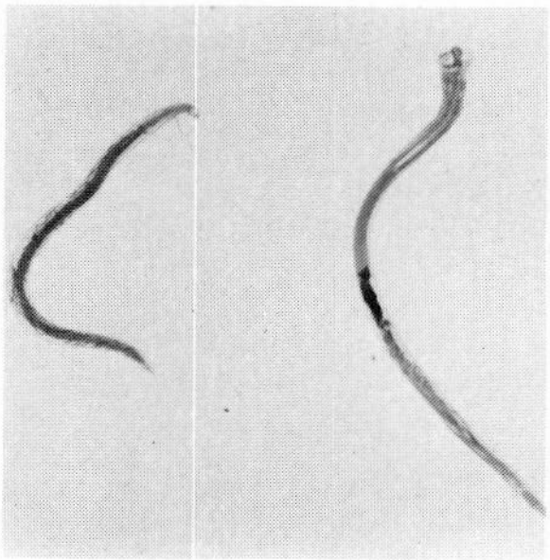

(a)

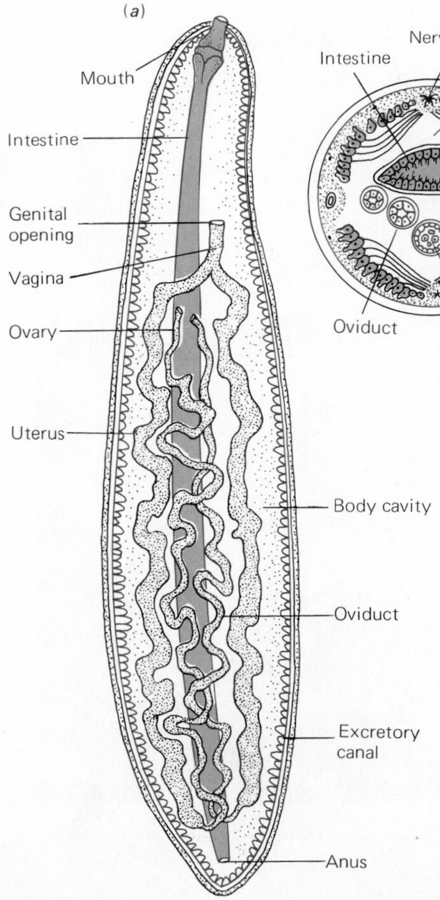

**FIGURE B-17**
Structure of a roundworm, *Ascaris*. (a) Male and female hookworm (*Necator americanus*). (b) Internal anatomy. (c) Cross-section. (Carolina Biological Supply Company) (d) Life cycle of a hookworm.

(b)

Labels for (b): Mouth, Intestine, Genital opening, Vagina, Ovary, Uterus, Body cavity, Oviduct, Excretory canal, Anus

(c)

Labels for (c): Nerve cord, Body cavity, Intestine, Excretory canal, Cuticle, Muscle, Epidermis, Uterus, Oviduct

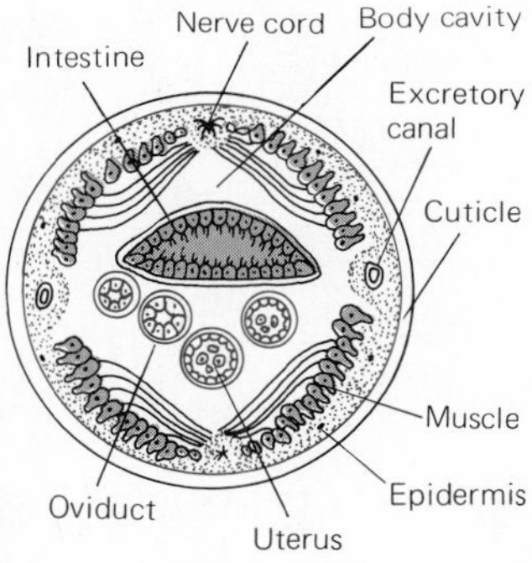

(d)

Labels for (d): Reach throat and are swallowed, Lung, Pass through heart, Larvae make tour through body, Larvae enter circulation, Adults live in small intestine, Eggs leave body with feces, Larvae hatch in soil; feed on bacteria, Infective larvae bore through skin

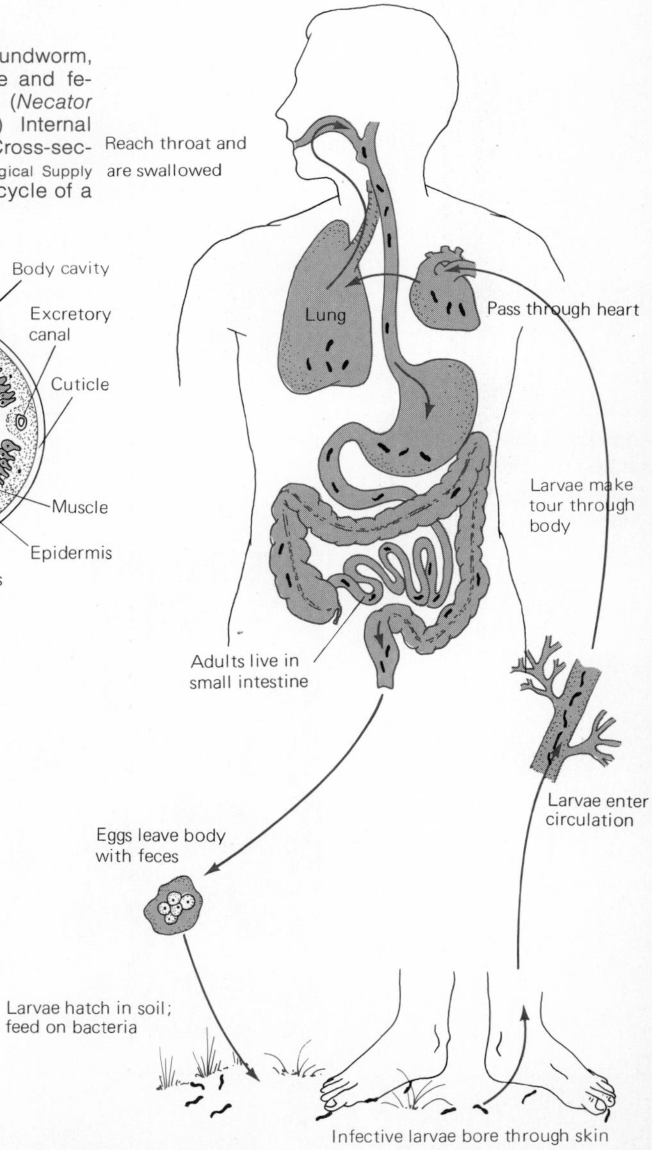

| Phyla | Level of organization | Symmetry | Digestion | Circulation |
|---|---|---|---|---|
| Mollusca<br>Clams, snails, squids | Organ system | Bilateral | Complete digestive tract | Open system |

**FIGURE B-18**
(*a*) Tulip snail. (Carolina Biological Supply Company) (*b*) Internal anatomy of a clam.

(*a*)

| | | | | |
|---|---|---|---|---|
| Annelida (segmented worms)<br>Earthworms, leeches | Organ system | Bilateral | Complete digestive tract | Closed system |

(*a*)

**Table B-4.** (*Continued*)

| Respiration | Excretion | Nervous system | Reproduction | Comments |
|---|---|---|---|---|
| Gills and mantle | Kidneys | Three pairs of ganglia; simple sense organs | Sexual; sexes separate; fertilization in water | Soft-bodied; usually have shell and ventral foot for locomotion; coelom (true body cavity) |

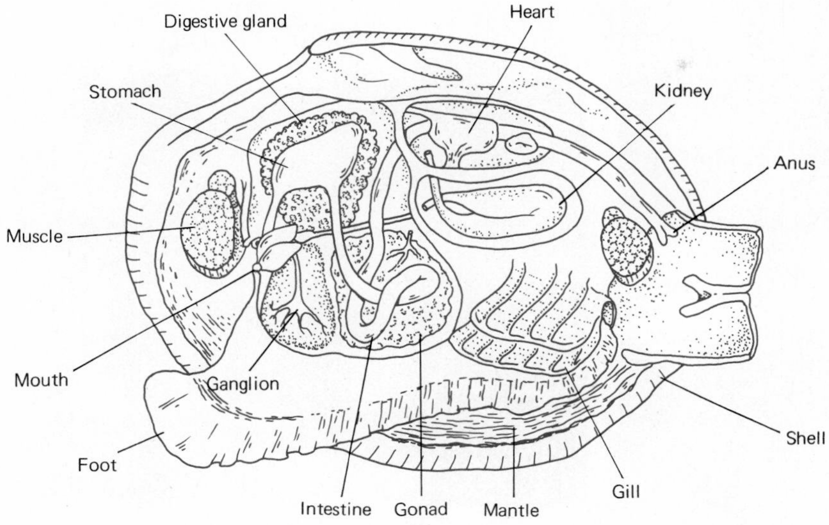

Digestive gland

Heart

Stomach

Kidney

Anus

Muscle

Mouth

Ganglion

Foot

Intestine   Gonad   Mantle   Gill   Shell

(*b*)

| Diffusion through moist skin; oxygen circulated by blood | Pair of excretory organs in each segment | Simple brain; ventral nerve cord; simple sense organs | Sexual; monoecious but cross-fertilize | True coelom; earthworms till soil |
|---|---|---|---|---|

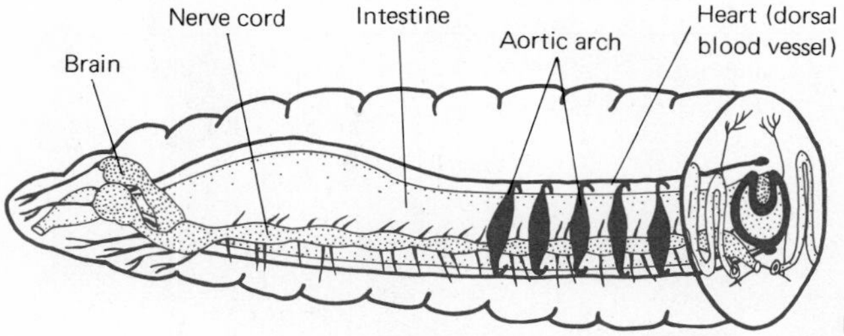

Nerve cord   Intestine   Aortic arch   Heart (dorsal blood vessel)

Brain

(*b*) Internal anatomy

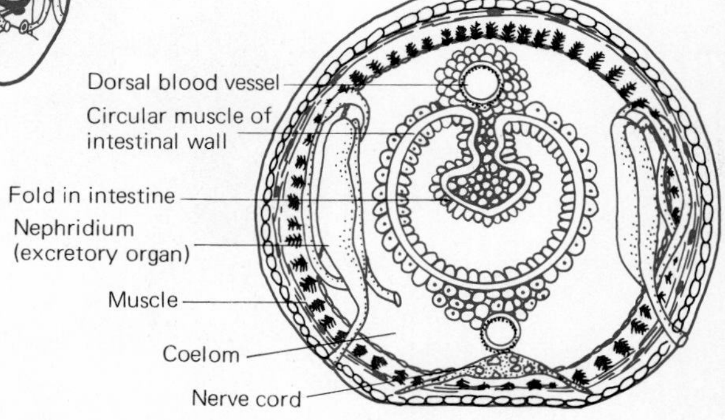

Dorsal blood vessel
Circular muscle of intestinal wall
Fold in intestine
Nephridium (excretory organ)
Muscle
Coelom
Nerve cord

(*c*) Cross section

**FIGURE B-19**
(*a*) Earthworms. (Roy Lewis, III)
(*b*) Internal anatomy of an earthworm. (*c*) Cross section through the body of an earthworm.

**Table B-4. Animal kingdom** (*Continued*)

| Phyla | Level of organization | Symmetry | Digestion | Circulation |
|-------|------------------------|----------|-----------|-------------|
| Arthropoda (joint-footed animals)<br>Crustaceans, insects, spiders | Organ system | Bilateral | Complete digestive tract | Open system |
| Echinodermata (spiny-skinned animals)<br>Sea stars, sea urchins, sand dollars | Organ system | Embryo, bilateral; adult, radial | Complete digestive tract | Open system; reduced |

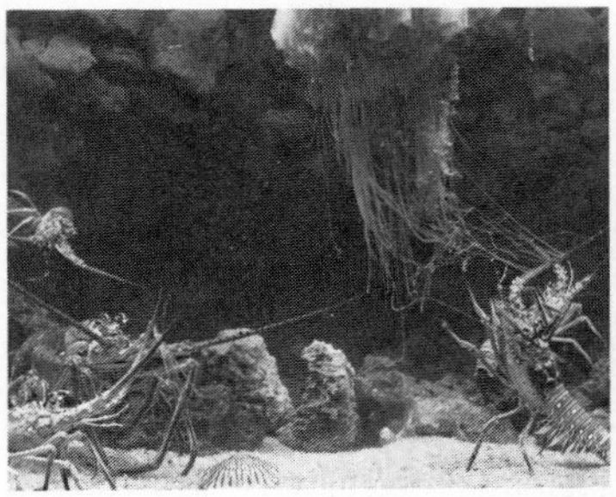

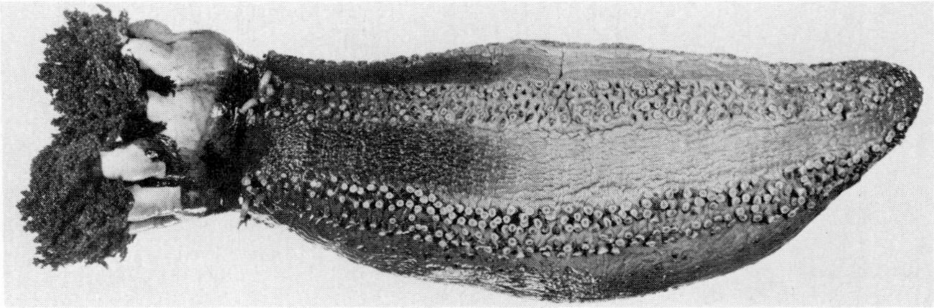

(a)

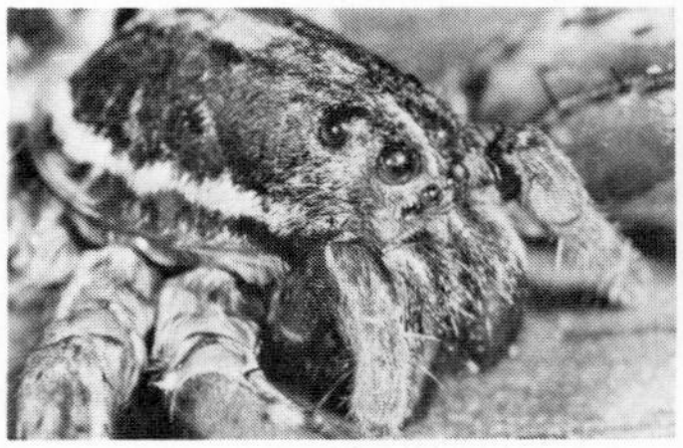

(b)

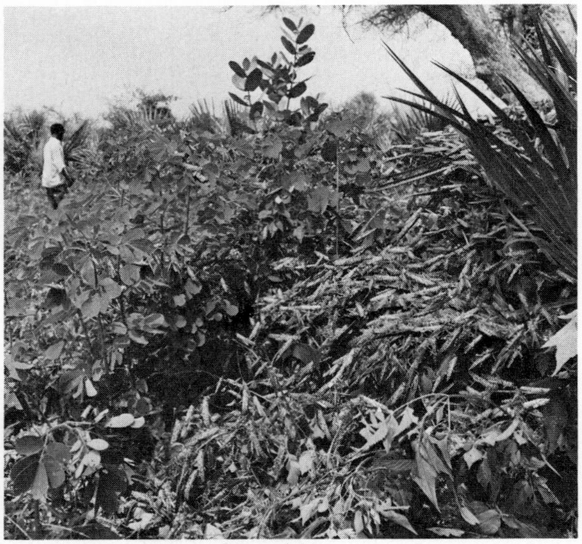

(c)

**FIGURE B-20**
(a) These spiny lobsters (crustaceans) appear to be enmeshed in the tentacles of a jellyfish. (Wometco Miami Seaquarium) (b) Wolf spider. (c) Swarms of migratory locusts which have wreaked havoc on man's crops since before the time of the Pharaohs. (Food and Agriculture Organization) (d) External anatomy of a grasshopper. (e) Internal anatomy of a grasshopper.

(a)

**FIGURE B-21**
(a) Sea cucumber. (Carolina Biological Supply Company) (b) Anatomy of a sea star.

| Respiration | Excretion | Nervous system | Reproduction | Comments |
|---|---|---|---|---|
| Trachea in insects; gills in crustacea; lungs or trachea in spider group | Malpighian tubules in insects | Simple brain; ventral nerve cord; well-developed sense organs | Sexual; sexes separate | Hard exoskeleton; true coelom; most successful group of animals |

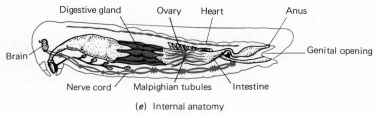

(*d*) External anatomy

(*e*) Internal anatomy

| Respiration | Excretion | Nervous system | Reproduction | Comments |
|---|---|---|---|---|
| Skin gills | Diffusion | Nerve rings; no brain | Sexual; sexes almost always separate | True coelom; water vascular system |

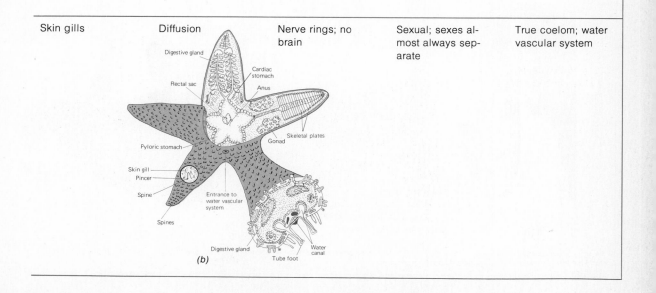

(*b*)

**Table B-4. Animal kingdom** (*Continued*)

| Phyla | Level of organiza-<br>tion | Symmetry | Digestion | Circulation |
|---|---|---|---|---|
| Chordata<br>Fish, man | Organ system | Bilateral | Complete digestive<br>system | Closed system;<br>ventral heart |

**Table B-5. Characteristics of vertebrate classes**

| Class | Skin | Breathing | Heart<br>chambers | Blooded* | Other |
|---|---|---|---|---|---|
| Agnatha (lam-<br>prey) | No scales | Gills | 2 | Cold | No jaws; cartilage, notochord persists |

**FIGURE B-22**
A West Coast sea lamprey holds
onto a rock with its mouth. (Carolina
Biological Supply Company)

* A cold-blooded animal must seek a habitat with a favorable temperature, for the body temperature fluctuates with changes in environmental temperature. A warm-blooded animal is able to regulate its body temperature.

**Table B-4.** (*Continued*)

| Respiration | Excretion | Nervous system | Reproduction | Comments |
|---|---|---|---|---|
| Gills or lungs | Kidneys and other organs | Dorsal nerve cord with brain at anterior end | Sexual; sexes separate | (1) Notochord; (2) hollow, tubular, dorsal nerve cord; (3) pharyngeal grooves. True coelom |

**Table B-5.** (*Continued*)

| Class | Skin | Breathing | Heart chambers | Blooded* | Other |
|---|---|---|---|---|---|
| Cartilage fish (Chondrichthyes) (sharks, rays, skates) | Scales | Gills | 2 | Cold | Skeleton of cartilage |

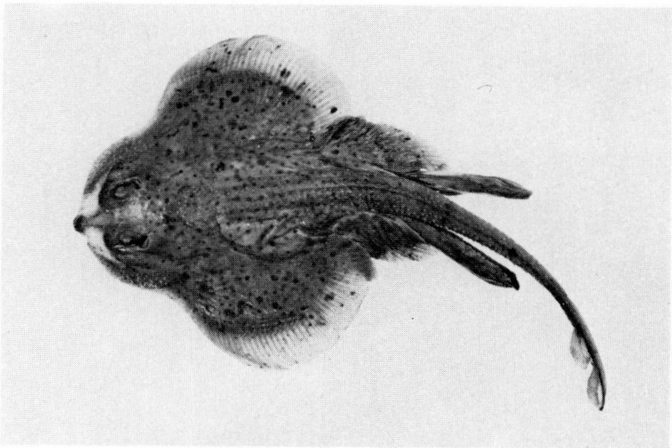

**FIGURE B-23**
The skate, a cartilaginous fish. (Carolina Biological Supply Company.)

| Class | Skin | Breathing | Heart chambers | Blooded* | Other |
|---|---|---|---|---|---|
| Bony fish (Osteichthyes) | Scales | Gills (lungs in some) | 2 | Cold | Skeleton of bone; most species, most individuals |
| Amphibia (frogs, toads) | No scales; smooth, moist | Gills (in larvae); lungs (adult) | 3 | Cold | Intermediate group; must return to water to reproduce |
| Reptilia (snakes, turtles) | Dry, epidermal scales | Lungs | 3.5 (4 in crocodile) | Cold | Embryos have amnion; adapted for life on land |

**FIGURE B-24**
Alligator. (Dr. Sonja Lessne)

| Class | Skin | Breathing | Heart chambers | Blooded* | Other |
|---|---|---|---|---|---|
| Birds (Aves) | Skin with *feathers* | Lungs | 4 | Warm | Forelimbs modified as *wings* |

**FIGURE B-25**
Sharp-tailed grouse. (Frank Martin)

**Table B-5.** (*Continued*)

| Class | Skin | Breathing | Heart chambers | Blooded* | Other |
|-------|------|-----------|----------------|----------|-------|
| Mammalia | Skin with *hair* | Lungs | 4 | Warm | *Mammary glands* 3 main branches: egg-laying (e.g., duck-billed platypus); marsupials (e.g., kangaroo); placental (e.g., cow, man) |

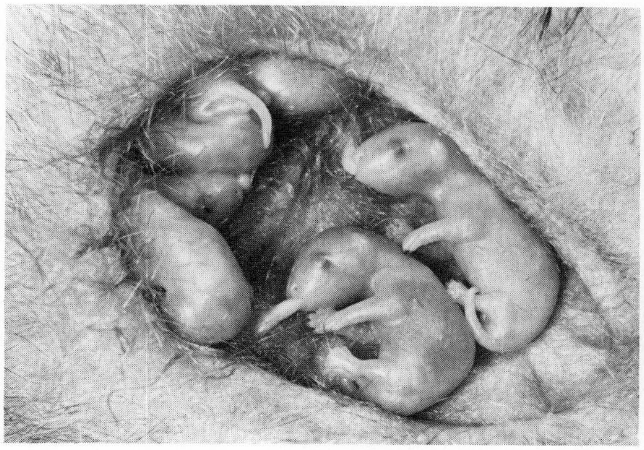

(a)

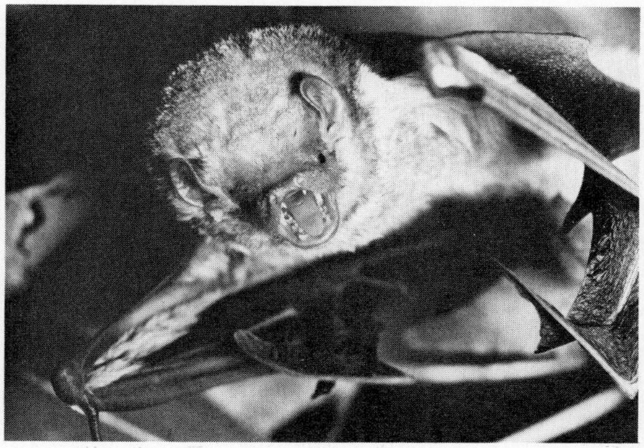

(b)

(c)

**FIGURE B-26**
(*a*) Opossum young in pouch of mother. (Carolina Biological Supply Company) (*b*) Brown bat, a flying mammal. (Carolina Biological Supply Company) (*c*) Chimpanzee.

# INDEX

Page numbers in *italic* indicate main definition.

Abortion, 292–294, 296
Absorption, 101, 102, 106–107
Acetabularia, 333–335
Acetylcholine, 220, 225, 255
Acetyl-coenzyme A, 173–174
Acid, sulfuric: in air pollution, 189
    in water pollution, 528
Acids, 624, *626*
Acromegaly, 248
ACTH, 246, 257–258
Actin, 12, 229–231
Active transport, 86, 107, *629*
Adaptation, *14*–15, 67, 460–461
Addison's disease, 257
Adenine (A), 337–338
Adenoids, 137
Adenosine diphosphate (ADP),
    84–85, 170–173, 178, 229,
    231
Adenosine monophosphate (AMP),
    170, 171
Adenosine triphosphate (ATP), *11*,
    84–86, 170–173, 175, 177–178,
    229, 231
Adipose tissue, 75, 110–111
Adrenal glands, 241, 246,
    254–259
    cortex of, 246, 254, 256–258
    medulla of, 246, 254–259
    and noise, 258–260
    and stress, 255, 257–260
Adrenaline (*see* Epinephrine)
Aeration pond, figure, 535
Aerobic activated sludge process,
    535
Afforestation, 563
Aggregations of animals, 442
Agriculture, 94–97
    and community structure,
        560–561
Air pollutants, 181, 188–201
    sources, 188–193
    (*See also* specific types of
        pollutants)
Air pollution, 6, 181–206
    in Chicago, 194
    and climate, 197
    cost, 183
    emergencies, 181
    in Florida, 194
    and human health, 184–188,
        202–206

Air pollution:
    industrial, 192, 199–200
    in Los Angeles County, 196,
        199
    in New York, 181, 194
    in Tokyo, 181
Air sacs of lungs (*see* Alveoli)
Albinism, *375*
Albumins, 119
Alcohol, 233, 238
    production of, 176–177
Aldosterone, 256–257
Algae:
    blooms of, 469, 635
    blue-green, 469, 516, 518, 635
    eukaryotic, 636
Alimentary canal, 101
Alkaptonuria, *375*
Alleles, *356*
    figure, 357
Allergens, 141
Allergy, 141–142, 318
Allopatric species, 404
Alpha wave patterns, 228
Alternation of generations in plants,
    638–639
Alveoli, 163–165, 183–186
Amebas, 12–13, 101, 147, 209,
    630
    (*See also* Amoeba)
Amino acids, 78–80, 105,
    107–109
    essential, 80, 89, 95
    figure, 628
Ammonia, 467–468, 487
Amniocentesis, 383
Amnion, 306, 315, 328
Amniotic fluid, 306, 315
Amoeba, 635
    excretion in, 147
    figure, 630
    responsiveness in, 209
    (*See also* Amebas)
Amphetamines, 112, 234, 238
Amphibians, 652
    figure, 265
Anaphase, 57, 59
Anemia, 92, *121*
    hemolytic, *121*
    sickle cell, 121, *376*–377, 399
Angina pectoris, 133
Angiosperms, 640–641

Animals:
  evolutionary tree, figure, 409
  kinds, 642–653
  phyla of, 642–651
Annelida, 646–647
Antibiotics, 139, 343
Antibodies, 120, *137*–141, 369
Antidiuretic hormone (ADH), 152,
  244–246
Antigen, *138*–143, 369
Antihistamines, *141*
Ants, 425
Anus, 101–102
Aorta, 131
Appendicitis, 109
Appendix, 102, 103
Aquaculture, 508–509
Aquifer, 506
Arteries, *124*–125
  aorta, 130–131
  carotid, 130–131
  coronary, 131–133
  iliac, 131
  mesenteric, 131
  pulmonary, 130–131
  renal, 130–131
Arthritis, 50, 142
Arthropods, 648–649
Artificial inovulation, 319
Artificial insemination, 319
Artificial propagation:
  and carrying capacity, 581
  of fish, 581
Artificial respiration, 135, 167
Asbestos:
  bodies, 191
  as pollutant, 191
Ascaris, 644–645
Aschelminthes, 644–645
Aspirin:
  effects upon embryo, 325
  mode of action, 260
Aswan dam, 504
Atherosclerosis, *76*, 131–134
Atoms, 9, *619*–621
  nucleus of, 620
  structure of, 620
ATP (*see* Adenosine triphosphate)
Atrium of heart, *125*–131
Automobiles (*see* Motor vehicles)
Autonomic nervous system, *211,*
  212
Autosomes, *381*
Autotrophs, 80–81
  chemosynthetic, 81
  photosynthetic, 80–81

Auxin, 422
Axon, 212–213, 219–220

Backcross, *368*
Bacteria, 635
  figure, 407
  in pollution treatment, 534–535
  role as decomposers, 4, 514
Bacteriophage viruses, 344–345
  figure, 345
Barbituates, 235, 238
Barr bodies, 380
Basal body, 52
Bases, 624, *626*
Bees:
  behavioral genetics of, 425
  division of labor, 431
Behavior, *424*
  development of, 431
  displacement, 428
  genetics and, 425
  redirected, 429
  social, 440–444
Benthos, *495*–498
Bias, experimental, 23
Bile, 102, 105, 109
  pigments of, 157, 378
Biochemistry:
  creationist interpretation, 415
  as evidence for evolution, 412
Biocides, 219–220, 470–476
Biologic clocks, 429–430
Biologic magnification, 470–473
Biologic oxygen demand (BOD),
  *514,* 526, 533
Biomass, *87*–88
  figure, 470
Biome types, 463–464
Biosphere (*see* Ecosphere)
Bird song:
  development of, 433
  in territoriality, 450
Birds, 652
Birth, 314–315
  figure, 316
Birth defects, 323–328
Bladder, urinary, 148–149
Blastocyst, *303*–304
  inner cell mass of, 303
  trophoblast of, 303–306
Blastula, *303*
Blending theory of inheritance, 33,
  *353*
Blood:
  clotting of, 123

Blood:
  deoxygenated, 129–131
  hemoglobin, 120–121
  oxygenated, 130–131
  plasma, 119–120
  platelets, 123
  pressure, 128
    high, 134
  red cells of, 120–121, 129
  serum, 120
  transfusion, 123, 370, 379
  types: ABO, 369–371
    Rh, 377–378
  white cells of, 121–122
Blue-green algae, 469, 516, 518
Bombardier beetle, figure, 410
Borlaug, Norman, 97
Bottleneck effect, 400
Bounties, 583, 586
Boutons, 213, 219
Bowman's capsule, 149, 150
Brain, 221–229
  action systems of, 223–225
  of amphibian, 224
  development of, effect of
      environment upon, 227–228
  functions of, 223–229
  generalized vertebrate, 224
  meninges of, 222
  of shark, 224
  stroke, 134
  structure of, 221–224
Breasts, 276, 317–319
Breathing, 165–167
  expiration, 165
  inspiration, 165
Bronchi, 161, 163–165
Bronchioles, 164–165
Bronchitis:
  acute, *185,* 205
  chronic, *185*–186, 204
Brown tooth enamel, 366
Bryophytes, 638
Budding, 263
Bumpus, Hermon, 405

Caffeine, 152, 234
Calcitonin, 251
Calcium, 69, 123, 229, 231, 250–252
Calefaction (*see* Pollution, thermal)
Calories, *73*
  table, 112
Cancer, 114, 115, 142, 187–188, 205
  lung, 187–188, 205
  of prostate, 270

Capillaries, 124–*125,* 135, 136
  in lung, 165, 168
  lymph, 136–137
Carbohydrates, 10, 73–75, 86, 105,
      107, 247
Carbon, 627
  figure, 467
  table, 619
Carbon cycle, figure, 467
Carbon dioxide, 5–6, 84, 85,
      166–169, 173–178, 467
  as pollutant, 193
Carbon monoxide, 188–189, 192,
      201, 203
Carcinogens, 115, 187, 190, 205,
      206
Cardiovascular disease, 110,
      131–134, 203–204
Carnivores, 81
  (*See also* Predators)
Carrying capacity, 466, *582*
Carson, Rachel, 8
Castration, 273
Catecholamines, 254
Cecum, 102, 103
Cell, 9, *41*
  division of, 54–60
  eukaryotic, 55, 407–408
  general characteristics of, 41–43
  membrane, 41, 46, 55
  origin, 406–408
  plant, 55
  prokaryotic, 55
  size, 43, 55–56
  theory, 43, 56
Cell plate, 59
Cellular differentiation, 301–302,
      307
Cellular respiration, *5,* 10–11,
      85–86, *161,* 169–179
  regulation of, 178
Cellulose, 75
Central nervous system (CNS), *211,*
      221–229
Centriole, 51, 55
Centromere of chromosome,
      57–59
Cerebellum, 221–223
Cerebral vascular accident (CVA),
      134
Cerebrum, 221–225
Cervix, 275, 282
  figure, 274
Channelization of streams,
      505–506, 520
Chapparal, 484

Chemical bonds, 622
  covalent, 624
  ionic, 623–624
Chemical compound, *621*
  inorganic, 627
  organic, 627, 628
Chemical energy, 5, 622
Chemical equations, 623–625
Chemical formula, 621
Chemical reactions, 623
Chemical symbols, 619
Chemotaxis in bacteria, figure, 423
Chlorinated hydrocarbons, 219
  (*See also* DDT)
Chlorophyll, 51, 84–85
Chloroplasts, 51, 408
  and air pollution, 195
Cholesterol, 77, 132–134, 203
Cholinesterase, 220
Chordates, 650–653
Chorionic gonadotropin (CG),
  279–281, 305
Chromatids, *58*–59
Chromatin material, 53
Chromosomes, 53–54, 57, 59
  abnormalities, 59, 379–384
  duplication of, 57–58
  homologous, 267, *356*
  maternal, 267
  paternal, 267
Chyme, 102
Cigarette smoking, 200–206
  and atherosclerosis, 133, 203
  and cancer, 205
  and chronic respiratory disease,
    204
  and death rates, 202–203
  and emphysema, 204–205
  and heart disease, 134, 203–204
  and pregnancy, 205–206, 327
  and ulcers, 206
Cilia, *51*–52, 162–164
Circulation:
  in animals, 642–653
  coronary, 131
  hepatic portal system, 131
  in plants, 119
  pulmonary, 129–131
  systemic, 129–131
Circulatory system, 118–145
  closed versus open, 124
  in embryo, 306, 315
    figure, 305
  and gas transport, 168–169
  lymph, 135–137
Circumcision, 271

Cirrhosis of liver, 109
Cities, 574–576
Citric acid, 173–174
Citric acid cycle, 172–174, 177
Clam, 496, 647
Clean air act, 198–199
Clear-cutting, 564
Cleavage, 302–303
Climate and air pollution, 197
Climax community (*see*
  Communities)
Clitoris, 275, 282
Closed system theory of economics,
  607
Clostridium botulinum, 220
Cluster development, figure, 576
Cnidaria, 642–643
Coacervates, 406
Cocaine, 238
Coenzymes, 71, 103, 173
Coitus, 281–282
Coleoptile, 419
Colloids, *629*
Colon, 102
Color blindness, 364–365
Colostrum, *317*
Commensals, 82
Commoner, Barry, 610–612
Commons, the tragedy of, 608
Communication, animal, 444–450,
  453
Communities, *478*
  aquatic, 489–503
  climax, *480*–481
    disturbance, *484*, 568
    edaphic, 480*n.*
  subclimax, 482
Compensation point, figure, 493
Composting, 541–542
Compounds, chemical: inorganic,
  *9,* 627
  organic, *9,* 627
Conception, 282–283
  contraception, 285–296
Conditioned stimulus, 434
Conditioning, 434–437
  operant, 435–437
  pavlovian, 434
Condom, 289
  table, 291
Conduction of nerve impulses, 214,
  218–222
Connective tissue, 61, *62*
  intercellular substances of, 62
Consanguinity, 384–385
Conservation of energy, law of, 80

Conservation of resources, *559,* 577
Constipation, 109
Consumers, *4,* 67, 81–82, 466
   primary, 86–88
   secondary, 87–88
Contraception, 285–296
Controlled burning, 580
Controls, experimental, 34
Convergent evolution, figure, 413
Cooling lagoons, figure, 523
Cooling towers, 522–523
Coral, 642–643
Coronary circulation, 131–134
   and atherosclerosis, 131–134
Coronary thrombosis, 133
Corpus luteum, 277, 279–281
Cortisol, 256–257
Cortisone, 256
Cough, 162, 184
Courtship behavior, 443–444
   as isolating mechanism, 404
Covalent bonding of molecules, 624
Cowper's glands, 270
   figure, 269
Cranial nerves, 211
Crapper, Thomas, 532
Creatine in muscle contraction, 231
Creationism, 413–416
Cri-du-chat syndrome, 382
Cristae, 50
Crossing-over of chromosomes and
   linked genes, 267, 368
Crowding, 595–597
Crustaceans, 496–497, 648
Culture, 453
Cushing's disease, 257
Cyanide, 105, 175, 201
Cyclamates, 114
Cycle, nitrogen, 467
Cycles:
   ecochemical, 467–469
   toxicant, 470–478
Cyclic AMP, 242–244, 249, 253,
   256
Cytochromes, 173, *175*
Cytokinesis, 59
Cytoplasm, 9, 41
   irritability of, 12, 209

Dams, 503–505
Darwin, Charles, 396
DDT, 113, 219, 401, 470–472
   in milk, 319
Deamination, 107–108, 148
Deciduous woodland, 463

Decomposers, *4,* 6, 67, 81, 87, 88,
   176, 466, 479, 635, 636
Decomposition, 468, 479
Decompression sickness, 166
Decreasers, *567*
Deduction, logical, 21
Defoliation, 566–567
Dehydration reactions, 75, 627
Dendrites, 212, 213
Denitrifying bacteria, 468
Deoxyribonucleic acid (*see* DNA)
Deoxyribose, 74
Depolarization, 219–221
DES, 114
Desert, 464
Detergents:
   biodegradable, 520
   in oil pollution control, 527
   role in eutrophication, 519–520
Development, 14, 301–328
Diabetes insipidus, 152–153
Diabetes mellitus, 110, 246,
   253–254
Diaphragm (contraceptive), 289
   table, 291
Diaphragm (respiratory organ), 161,
   165–166
Diarrhea, 92, 109
Diastole, *126*
Diet:
   average American, 112
   balanced, 78
   reducing, 111–113
   role in atherosclerosis, 133
   vegetarian, 91
Diffusion, 107, 135, 629
Digestion, 101, 105
Dipeptide, 79, 105
Disease, role in population
   limitation, 597
Displacement behavior, 428–429
Distribution, as evidence for
   evolution, 413
Diuretics, 152
DNA, 10, 50, 51, 53–54, 74,
   336–338
   base-pairing in, 337
   and protein synthesis, 338–339
   replication of, 338
   structure of, 336–337
Dogma, central, 344
Dominance, genetic, *355*
Dominance hierarchies, 433, 448
Dopamine, 225, 324
Douching, as contraceptive
   method, 290

Drainage in treatment of
    eutrophication, 517
Dredging, 506–508, 525–526
Drives, 428
Drugs, 232–238
    alcohol, 233, 238
    depressants, 235, 238
    effects upon embryo, 324–327
    hallucinogenic, 236–238
    marijuana, 237, 238
    methaqualone, 238
    opiates, 235, 238
    stimulants, 234, 238
Dumps, 539
Duodenum, 102


Ear, 216–217
Earthworm, 647
Echinoderms, 648–649
Ecologic niche, 405, 466
Ecologic warfare, 566
Ecology, *460*
Economics:
    of natural habitats, 579
    of pollution, 532, 542–543, 549
    systems of, 605–608
Ecosphere, *4,* 615
    figure, 41
    man's influence upon, 6–8,
        181–200, 503–508, 511–527,
        536–543, 557–588, 593–615
Ecosystems, *4,* 466
    degraded, 558
Ectoderm, 306, 308, 322–323
Eel grass, 496
Effection, 214, 229–231
Effectors (*see* Effection)
Egocentric problem, 24
Ejaculation, 282
Ejaculatory ducts, 270
    figure, 269
Electrocardiogram (EKG), 127
Electroencephalogram (EEG), 228
Electron micrograph, *48*
Electron transport system, 173–175
Electrons, 173–175, *620–627*
Elements, chemical, *9,* 68, *619*
Elimination, 101, 147
Embolus, 134
Embryo, 303–312
    environmental influences upon,
        323–328
    tissues of, 306–308
Emphysema, 186, 204–205
Endangered species, 584–590

Endergonic reactions, 170–171,
    178–179
Endocrine glands, 240–260
Endoderm, 306, 308
Endoplasmic reticulum (ER), *47*–48
Endosperm, 640
Energy, 5, 10, 11, *80, 83*–88
    and cellular respiration, 169–179
    chemical, 5, 80, 83
    consumption of, 598–601
        figure, 615
    light, 80, 84–85
    and muscle contraction, 229, 231
    nuclear, 600
    as population limiting factor,
        598–599
    solar, 601
    and synthesis, 86
    and thermodynamics, 80, 87
Engineers, Army Corps of, 505
Environment:
    effect on development, 227, 228,
        322–328
    man's impact upon, 3, 6–8,
        181–200, 503–508, 511–527,
        536–543, 557–588, 593–615
Enzymes, 10, 78, 103–105
    induction of, 347–348
Epididymis, 269
Epigenesis, 301
Epinephrine, 162, 246, 254–255
Episiotomy, 315
Epithelial tissue, 60
Erection of penis, 271, 281–282
Ergosterol, 77
Erythroblastosis fetalis, 378
Erythrocytes, 120–121, 129
ESB (electrical stimulation of the
    brain), 237
Esophagus, 101–103
Estrogen, 279–281
    as factor in atherosclerosis, 133
    and menopause, 281
    and oral contraception, 286–287
    and placenta, 305
Estuaries, 498–502
    productivity of, 496
    and real estate development, 499,
        506–508
    varieties, 498
Ethics of science, 36–37
Ethology, *424*
Eugenics, 387
Eukaryotic cells, *55,* 635–636
    origin of, 406–408
Euthenics, *385*

Eutrophication, 503
  cultural, 515–517
    control of, 517–520
    and mineral enrichment, 518
Evolution, *395*
  Darwinian, 397
  evidence, 411–413
  micro-, 396, 411
  organic, 395
Excretion, 146–159
  in various phyla of animals,
    642–649
Excretory tubule, 149–152
Exercise, physical: energy
    expenditure in, 112
  heart beat during, 127
  and heart disease, 134
Exergonic reactions, 170–171
Exotic species, 587
Experiments, *34*–35
External heart massage, 135
Externalities, economic, 542
Extinction:
  in learning, 434
  of species, 557–558
    by environmental pollution,
      586
    by habitat destruction, 584
    by overexploitation, 585
    phase of, 594
Eye, 215–217
  structure of, 216

Family planning, 602–603
  (*See also* Contraception;
    Population, control)
Famine, 99, 597
Farming, aquatic (*see* Aquaculture)
Fat tissue, 75
  (*See also* Adipose tissue)
Fats, 75–76, 105
  diglycerides, 76
  monoglycerides, 76
  triglycerides, 76, 134
Fatty acids, 76–77
  saturated, 76–77, 133
  unsaturated, 76–77, 133
Feces, 102, 109, 157
Federal mining law of 1872, 570
Feedlots, 526
Fermentation, 176–177
Ferns, 638–639
Fertilization, 283
  external, 264–265
  internal, 265

Fertilizers, effect on environment of,
    95, 526
Fetus:
  circulation in, 305–306, 315
  development of, 312–314
Fibrillation, 128
Fibrin, 123–124
Fibrinogen, 119–120, 123
Filter, trickling, 534
Filtration in kidney, 150
Fish, 494–495, 499, 650–653
Flagella, *52,* 55
Flagellates, 82–83
Flatworms, 644–645
Fluoridation, 70
Fluoride pollution, 191, 192,
    195–196
Fluorine, 70
Follicle, *276,* 279, 280
Follicle Stimulating Hormone (FSH):
  in female, 278–280, 286–287
  in male, 273
Food:
  calories in, table, 112
  from petroleum, 98
  production of, 94–98
  from the sea, 97
  from sewage, 97–98, 550
  as world problem, 67, 89–99,
    596–597
Food additives, 113–115
Food chain(s), 83, 86–88
  and biologic magnification, 470
  trophic levels of, 86–87
Food and Drug Administration,
    113–115, 326
Food pyramid, 87–88
  figure, 470
Food vacuole, 50
Food webs, 87
Forelock, white, figure, 392
Forests, 561–567
  damage to, 562, 564, 566–567
  management of, 564
  products, use of, 561
  silviculture of, 562
  soil depletion, 563
Fossil record:
  creationist interpretation, 416
  as evidence for evolution, 413
Founder effect, 399
Four o'clocks, color inheritance in,
    33, 353–355
Frameshift mutation, 341
Freshwater habitat, 487–497
Freud, Sigmund, 228

Fructose, 74–75
  in seminal fluid, 270
Fuel, 600
  (*See also* Energy)
Fungi, 636–637
  role as decomposers, 4, 636

Galactosemia, 385
Gall bladder, 102–103
Gallstones, 109
Gametes, 264
Gametophyte, 638–639
Gamma globulin, 141, 379
Garrod, Sir Archibald, 375
Gastric glands, 102, 105
Gastrin, 246, 260
Gastrula, *307*
Gastrulation, *306*–308
General Adaptation Syndrome, 258
Genes, *53,* 56, 350, *353*
  alleles of, *356*
  and behavior, 425
  heterozygosity of, *356*
  linkage of, 362–368
  maps of, 368
  operons, 347–349
  recombination of, 403
  role in development, 320–323
  and sexual reproduction,
    265–266
Genesis, book of, 395
Genetic code, 338
Genetic drift, 399
Genetic engineering, 386–*387*
Genetics, *353*
  creationist interpretation, 415
  as evidence for evolution, 412
Genossenschaften for pollution
    control, *545*
Genotype, *356*
Geographic isolation (*see*
    Allopatric species)
Gigantism, 248
Gill structures, 310
Glands:
  endocrine, 240–260
  exocrine, *240*
Glial cells, 213, 228
Globulins, 119–120
Glomerulus, 149–151
Glucagon, 246, 252–253
Glucocorticoids, 256
Glucose, 10, 73, 105, 107, 109, 151
  and insulin, 252–254
Glycerol, 76, 105

Glycogen, 73, 107–109, 178, 231
Glycolysis, 170, 172–174, 176–177
Goiter, 250
Golgi apparatus, 48–49
Gonadotrophic hormones, 273
Gonads, *268*
  (*See also* Ovaries; Testes)
Gonorrhea, 296–297
  and pregnancy, 328
Graft rejection, 142–144
Grana, 51
GRAS list, 114
Grasshopper, 648–649
Grassland, 463
Green revolution, 95–97
Growth, 13–14, 301–303
Growth hormone, 246–248, 257
Guanine (G), 336, 338
Guano, 494
Gymnosperms, 639

Habitats (*see* Biome types)
Habituation, 440
Haemmerling, J. (Hämmerling), 332
Hamster, naked, figure, 390
Hardy-Weinberg law, 398
Health foods, 115–116
Heart, 125–134
  attack, 133–134
  beat, 126–128
  disease, 110, 131–134, 203–204
  external massage of, 135
  structure of, 125, 127
Heartburn, 109
Heat:
  in environment (*see* Pollution,
    thermal)
  production of: body, 178, 231
    and energetics, 80, 231
    and receptors, 215
Height, human, 362
Hemoglobin, 69–70, 79, 120–121,
    169
  and carbon monoxide, 189
  and sickle-cell anemia, 121, *376*
  and thalassemia, 377
Hemophilia, 123, *365*
Hemorrhoids, 129
Hepatic portal vein, 107, 131
Hepatitis, 238
Herbicides, 566–567
Herbivores, 81
Heredity, 33, 352–354
  laws of, 359
  and Mendel, 358–359

Heroin, 235–236, 238
Heterotrophs, 81, 406
Heterozygous genotypes, 356
Highways, 571–572
Histamine, 141
Histocompatibility genes,
    142–143
Home range, 450
Homeostasis, 157
Homozygous genotypes, 356
Honeybees (see Bees)
Hooke, Robert, 43
Hookworm, 645
Hormones, 240–260
    chemical structure of, 242
    mechanisms of action, 242–243
    plant, 242
    regulation of, 243–245
    reproductive, 273, 277–281
    tropic, 248–249
    (See also specific types of
        hormones)
Host mothering, 319–320
Human adaptability, 614
Humboldt current, 493
Humus, 480
Hunting licenses, 580
Hybrid vigor, 403
Hydra, 209–210, 242
    nerve net of, 209–210
Hydrocarbons, unburned, as air
    pollutants, 190, 192
Hydrogen, 85, 172–177
    figure, 620
    table, 619
Hydrolysis, 75, 627
Hymen, 276
    figure, 275
Hypothalamus, 221, 223, 241,
    244–245, 247–250
    releasing factors of, 245,
        247–248
    and reproductive hormones, 273,
        280
Hypotheses, 27–33
    disproof of, 30–33
    unfalsifiable, 29–30
    untestable, 30

Immune response, 137–144, 257
    graft rejection in, 142–144
Immunity, 137–144
    active, 141
    artificial, 140
    passive, 141

Implantation of embryo, 304
Impotence, sexual, 282
Imprinting, 439
Incineration, 540–541
    and air pollution, 192
Increasers, 567
Independent assortment, law of,
    359–361
Induction, 22
    of enzymes, 347–348
Industrial melanism in moths, 15,
    402
Industry:
    and air pollution, 192, 199–200
    and the ecologic crisis, 606–607,
        610
    and water pollution, 522,
        542–543, 545
Ingestion, 101
Inguinal canal, 269
Inguinal hernia, 269
Inheritance (see Heredity)
Insect societies, 448, 452
Insecticides (see Biocides)
Insects, 648–649
    behavior of, 425–426, 428
    (See also Insect societies)
Insight, 26, 438
Instinct, 431
Insulin, 246, 252–255
Intelligence, 362
    environmental influences upon,
        227
Interferon, 140
Interphase, 57–58
Intertidal community, 497–498
Intestine, 101–103, 157
    large, 101–103
    small, 101–103, 105–107
Intrauterine devices (IUDs),
    288–289
Inversion layer, 197–198
Invertebrates, 642–649
Iodine, 70, 250
Ions, 624
Iris, 216
Iron, 69–70
    table, 619
Irrigation and saline pollution,
    528
Irritability of cytoplasm, 12, 209
Isolating mechanisms, 404–405
    postzygotic, 404
    prezygotic, 405
Isomers, 622
Isotopes, 621

Jacob-Monod hypothesis, 347–348
Jellyfish, 642–643
Joint laxity, congenital, figure, 392
Judeo-Christian world view, 613*n.*

Karyotype, *366*
Kelp, 636
Kelsey, Frances, 326
Kidney, 147, 155, 158, 244, 260
    dialysis, 153–155
    disease, 153–155, 252
    failure, 153
    transplant, 155
Kingdoms in classification, 633
Krebs cycle (*see* Citric acid cycle)
Kwashiorkor, 91–92

Labia, 275
Labor in childbirth, 314–315
Lactic acid, 176–177, 255
Lactogenic hormone, 248, 317
Lakes:
    eutrophication of, 502–503,
        515–518
    thermocline in, 493
La Leche league, 319
Lamarck, Jean Baptiste de, 396
Landsteiner's law, 369
Langerhans, Paul, 252
Language, 453
Larynx, 161–163
Laterite, 94, 464, 482
Laws, scientific, 35
Lead pollution, 191, 476
Learning, 226–228, *434*–440
Leaves, 640
Leeches, 646
Legumes, 90, 469
Leukemia, 122–123
Leukocytes, 121–122, 138–139
    agranular, 120, 122
    granular, 120–122
Lichens, 480
Life, *16*
Light energy, 80, 84–85
Limbic system, 224–227
Limiting factors, 466, 491
    of human population growth, 596
Lipase, 104, 105
Lipids, 10, 75–77, 105, 108–110
    in atherosclerosis, 131–133
Liver, 101–103, 105, 107, 109–110,
    129, 147–148
Liverworts, 638

Lizards, 438
    figure, 443, 447
Logging practices, 564
LSD, 229, 236–238, 326
Lungs, 147–148, 157, 161, 163–169
Luteinizing hormone (LH):
    in female, 278–281, 286–287
    in male, 273
Lymph, 136–137
Lymph nodes, 137
Lymph system, 107, 135–142
Lymphocytes, 120, 122, 129, 135,
    137–140, 142–143
Lyon's hypothesis (of sex
    determination), 380
Lysine, 90, 95
Lysosomes, *49*–50, 101, 257

McConnell, James, 228
Macromolecules, origin of,
    405–406
Macrophages, 122
Malaria (*see* Anemia, sickle cell)
Malnutrition, 67, 89
Malthus, Robert Thomas, 596
    figure, 397
Mammals, 653
Mammary glands, 317–318, 653
Management:
    of communities, 578–580
    of game, 580–584
Mangroves, 499–502
Marasmus, 91–92
Marijuana, 237–238, 326
Marsupials in Australia, 413
Martians, 30
Mating, 265, 451
Measles, 89, 140
Medulla (of brain), 221, 223–224
Meiosis, 266–268
    crossing over in, 267, 368
    synapsis in, 267
Melanin, 375
    (*See also* Industrial melanism in
        moths)
Memory, 226–227
Memory cells, 140
Mendel, Gregor, 358–359
    postulates of, *359*
Menopause, *277,* 281
Menstrual cycle, 277–281
Menstruation, *275*
Mental illness, 228–229
Mercury pollution, 476
Mesoderm, 306, 308

Metabolic rate, 249
Metabolism, 10–12
  wastes of, 147–148
Metals, 622
  heavy, poisoning, 476–477
Metaphase, 57, 59–60
Methadone treatment, 236
Methaqualone (see Drugs)
Metric system, *631*
Microorganisms, 469
  (*See also* Bacteria; Fungi; etc.)
Microscope, 43
  electron, 44
  light, 45
Microtubules, 51
Miescher, Friedrich, 335–336
Migration, *430*–431
Milk, human, 317–319
Mimicry, batesonian, 402
Minamata disease, 476
Mine drainage, 528
Minerals, 68–71
  dietary requirement for, 68–71
Minimum, law of, 462–466
Miracle grains, 95–97
  Opaque-2 corn, 95–96
Mites, 473
Mitochondria, 50
  function of, 50
  possible origin of, 407–408
  in sperm cells, 272
  structure of, 50
  in zygote, 321
Mitosis, *55*–60
  in early embryo, 302–303
  and meiosis, 266
  timing of, 60
Mitotic spindle, 58
Molecules, *621*
  covalent bonding of, 624
  and diffusion, 629–631
  ionic bonding of, 623–624
Molluscs, 646–647
Monera, 635
  figure, 634
Monkeys, 453–454
Monoculture, 560
  of forests, 562–563
Monocytes, 120, 122
Monohybrid cross, *358*
Monosaccharides, 74
Moral commitment, 614
Morphogenesis, 301–302, 307
Morula, *303*
  figure, 302
Mosses, 480, 638

Motor vehicles, 188–189, 193
  (*See also* Transportation)
Mouth, 101–102
Mucus, 162–163, 184–185, 204
Multiple alleles, 368–371
Multiple sclerosis, 213
Multiple use, 568, 577
Muscle tissue, 62, 63, 177–178,
    229–231
  cardiac, *62,* 229, 230
  fatigue, 177
  function of, 62, 229–231
  protein of, 229–231
  skeletal, *229*–231
  smooth, *62, 229*–230
  striated, *62,* 229–231
  structure of, 229–231
Mushrooms, 636
Mutation, *14,* 193, 340–342
  evolution and, 398, 415
  in viruses, 141
Mutualistic partners, 82
Mycelium of fungus, 636
Myelin sheath, 212–213
Myocardial infarct, 133–134, 143
Myofibrils, 229
Myofilaments, 229–231
Myosin, 12, 229, 231
Myxedema, 250
Myxomatosis, 484

NAD, 172–177
Nader, Ralph, 113
NADP, 85
Narcotics and pregnancy, 327
  (*See also* Drugs)
National forests, 568
Natural selection, 14, 397, 399
Nekton, *494*–495
Nerve impulse:
  conduction of, 218–222
  direction of, 220
  integration of, 221–222
  speed of, 220
Nerve net, 209–210
Nerve tissue, 62
Nerves, *214*
  cranial, 211
  spinal, 211
Nervous system:
  autonomic, 211–212
  brain, 211
  central, 211
  development of, 308–309,
      312–313

Nervous system:
  parasympathetic, 211–212
  and receptors, 214–217
  spinal cord, 211
  sympathetic, 211–212
Neural plate, 308–309
  figure, 309
Neural tube, 308–309, 322
  figure, 309
Neurohormones, 245
  (*See also* Transmitter
    substances)
Neurons, *62, 210,* 212
  axon of, 212–213
  cell body of, 62, 212–213
  cellular sheath of, 212–213
  connector, 214–215
  dendrites of, 212–213
  fibers of, 62
  motor, 214–215
  myelin sheath of, 212–213
  regeneration of, 213
  sensory, 214–215
Neutrons, 620
Nicotiana (*see* Four o'clocks)
Nicotine, 133, 201, 203, 206
Nitrates, fixed, 467
  (*See also* Eutrophication;
    Nitrogen cycle; Sodium
    nitrate)
Nitrifying bacteria, 468
Nitrites, 468
  (*See also* Sodium nitrite)
Nitrogen:
  in air, 168
  cycle, 467–469
  and decompression sickness,
    166
  excretion of, 147–158
  fixed, 467
  and pollution, 191–194
    (*See also* Eutrophication)
Nitrogen oxides, 191–194
Nitrogenase, 469
Nitrosamines, 115
Noise, 258–260
Norepinephrine, 225, 229, 234,
    236–238, 246, 254–256, 324
Nose, 162, 215, 217
Notochord, 651
  and development of nervous
    system, 308, 322
Nuclear power plants, 193, 522
Nucleic acids, 10, 336–340
Nucleolus, 53–*54*
Nucleoplasm, 45

Nucleotides (*see* Nucleic acids)
Nucleus of cell, 9, 41, 53, 121
  and cell division, 55–59
  envelope of, 53
Nursing an infant, 317–319
Nutrition, *10*–11, 66–99, 179
  basic food groups for good,
    78
  effects of, on embryo, 324

Obesity, 110–113
Oceans:
  and food supply (*see*
    Aquaculture)
  as habitat, 489–500
  and pollution, 183
Olfactory epithelium, 217
Oligosaccharides, 74–75
Omnivores, 81
Open pit mining (*see* Strip mining)
Operons, *347*–348
Opium, 235
Opsonins, 139
Organelles, 41, 43–44, 407
  centrioles, 51
  chloroplasts, 51, 55
  Golgi apparatus, 48–49
  lysosomes, *49*–50
  mitochondria, 50
  nucleolus, 53
  nucleus, 41
  plastids, 51
  ribosomes, *48,* 342–344
Organism, 41, 63
Organ systems, 41, 63–65
Organ transplants, 142–144, 155
Organizers, 322
Organs, 41, *63*
Orgasm, 282
Osborne, W. F., 8
Osmosis, *107, 629*–630
Osmotic pressure, 135, 630–631
Ova, 276–277
  production of, 276–277
    figure, 272
Ovaries, 274
  hormones of, 279–281
Overgrazing, 568
Oviducts, 274
Ovulation, *277*–280
  changes in body at time of,
    277–278
Oxidation, 169–170, 173, 175,
    626
  and reduction, 173, 175, 626

Oxygen, 5, 11, 161, 165, 169–171, 173–179, 206
  atomic structure of, figure, 620
  deprivation in pregnancy, 327
  transport of, 120–121, 135
Oxygen debt, 177
Oxyhemoglobin, 120, 169
Oxytocin, 244, 246, 317–318
Ozone, 193–194

Pacemaker, 126
  artificial, 127
Pacinian corpuscles, 216–217
Pain, 221
Pancreas, 101–102, 105
  islets of, 252–254
Panmixis, 398
Paramecium, 51
Parasites, 82, 92
Parasitoids, 81
Parathyroid glands, 241, 246, 250–252
Parental care, 227
  in chimpanzee, figure, 445
Parkinson's disease, 225
Particulate air pollution, 185, 190–192, 201
Passenger pigeon, 584
Patch cutting, 565
Pauling, Linus, 200
Pavlov, I. P., 226
Peck order (*see* Dominance hierarchies)
Penicillin, 635
  discovery of, 26–27
  and venereal disease, 297, 328
Penicillium, 27
Penis, 270–271, 281–282
  circumcision of, 271
  erection of, 271, 281–282
  figure, 269
  glans of, 271
Pepsin, table, 105
Peptide bond, 79, 342
Pericardial cavity, 125
Pericardium, 125
Peripheral nervous system (PNS), 211
Peristalsis, *62,* 102, 104
Perspiration, 157
Pest control, alternative methods of, table, 474–475
Pesticides (*see* Biocides)
PGAL (phosphoglyceraldehyde), 172–173, 176

Phagocytes, *138,* 184
Pharynx, 102–103, 161–163
Phenotype, *356*
Phenylalanine, 376
Phenylketonuria, 376
Pheromones, 447, 452, 474
Phocomelia, 325–326
Phosphates (*see* Eutrophication)
Phosphorus, 69, 250–251
  (*See also* Eutrophication)
Photosynthesis, *5,* 12, 169
Phytoplankton, *8,* 489–494
Picard, Jacques, 196
"Pill," contraceptive, 286–287
Pines, 639
Pituitary gland, 241, 244–250
  anterior, 244–250
  in dwarfs, 248
  posterior, 152, 244–247
  and reproductive hormones, 273, 278–280
Placenta, 305–306, 315, 327
Planaria, 210, 644
  behavior, 423–424
  responsiveness in, 210
Plankton, 489–494
Plants, 638–641
  alternation of generations in, 638
  cells, 55
  distribution of (*see* Biome types)
  internal transport in, 119
  reproduction, 638–641
  responsiveness in (*see* Tropisms)
Plasma, blood, 119–120, 135
Plasma cells, 138–140
Plastids, 51
Platelets, 120, 123
Platyhelminthes, 644–645
Play, 433
Pleiotropism, *371*
Pleura, *165*
Pleurisy, 165
Pneumatophores, figure, 500
Poisons, 105, 201, 219–220
  botulism toxin, 220
  curare, 220
  cyanide, 105, 175, 201
  DDT, 219
  malathion, 220
  parathion, 220
Politics and pollution, 543–544
Pollen, 356
Pollution, *511*
  air, 6, 186–206
  by biocides, 470–473

Pollution:
  control, 531–536
    ecologic justification for, 532
    economic justification for, 532
    esthetic justification for, 531
    hygienic justification for, 531
    natural, 520
  effects on aquatic communities,
    513–515
  industrial, 513, 520
  legislation, 552–554
  by mine drainage, 528
  by oil, 526–527
  political aspects, 543–544
  regulatory authorities, 544–546
  saline, 528
  by silt, 511, 525
  sink, 512
  sources of, 511
  thermal, 493, 521–524
  water, 6–8, 511–514
  zones, 514–515
Polychlorinated biphenyls, 472
Polygenes, *361*
Polypeptides, 79, 105
Polyploidy in evolution, 412
Polyribosomes, figure, 343
Polysaccharides, 75
Population:
  age distribution in, 595
  control, 99, 294–296, 603–605
  cost of, 603, 609–610
  genetics of (*see* Hardy-Weinberg
    law)
  growth rates in, 93–94, 594–595
  human, 93–94, 594–605
    size of, 93–94, 594–605
  ideal, 595
Porifera, 642–643
Porpoise (dolphin) and
  communication, 447
Power plants, 189, 192
  nuclear, 193, 522
  (*See also* Pollution, thermal)
Predators, 81
  persecution of, 586
  ultimate, 470
Preformation theory, 301
Pregnancy, 312, 314, 317, 320,
  323–328
  and calcium metabolism, 252
  risk of, 285, 287, 291
Preservation of habitats, *577*
Pressure receptors, 215
  figure, 216
Primates, use of, in testing drugs,
  326

Producers, *4,* 67, 466, 635–636
Progesterone, 279–281
  and oral contraceptives, 286–287
  and placenta, 305
Prokaryotic cells, *55,* 635
Prophase:
  of meiosis, 267
  of mitosis, 57–58
Prostaglandins, 241, 243–244, 260,
  293–294
Prostate gland, 270
  figure, 269
Protective coloration, 401–402
Proteins, 9–10, 77–80
  animal, 89–91
  deficiency of, 91–92
  digestion of, 105
  functions of, 78
  plant, 89–91
  poverty, 89–92
  structure of, 78–80
  synthesis, 339–343
Prothrombin, 123
Protista, 635
Protocultural behavior, 453
Protons, *620*
Protozoa, 635
Psammon, *497*
Puberty, 271, 273
Pulmonary arteries, 130
Pulmonary circulation, 129–131
Pulmonary veins, 130
Pulse, *128*
Punnett square, 356
Pupil of eye, 216, 445
Purines, 336
Pus, 335–336
Pylorus, 102
Pyrimidines, 336
Pyrolysis, 541
Pyruvic acid, 172–173, 176–177


Radial symmetry, 642, 648
Radiation, effects on embryo, 328
  (*See also* Radionuclides)
Radioisotopes, *621*
  (*See also* Radionuclides)
Radionuclides, as pollutants, 193,
  477–478
Rain forest, 464
Rangeland, deterioration of,
  567–568
Rationality, law of, 20
Reabsorption in kidney, 150–152,
  244
Rebound effect, 162

Reception of stimuli, 214–217
Recessive genetic traits, *355*
Rectum, 103
Recycling, 546
  economics of, 549–552
  of sewage, 548–549
  of solid wastes, 546–548
Red blood cells (*see* Erythrocytes)
Red tide, 491–492
Redirected behavior, 429
Reduction, oxidation and, 173, 175
Reefs:
  coral, 497
  oceanic, 496
Reflex actions, 214–215
  conditioned, 434–435
Reflex pathways, 214–215
Reforestation, 563
"Refuse act" of 1899, 554
Regeneration:
  of Acetabularia, 332
  of neurons, 213
Reinforcement, 435
Releasers, *426*
  chains of, 427
Remanufacturing and reuse in
    recycling, 550
Renal artery, 130–131, 149–150
Reproduction, 13–14
  in aminals of each phylum,
    642–651
  asexual, 263–264
  in plants, 638–641
  sexual, 13–14, 264–298
    advantages of, 265–266
  in Spirogyra, 636
Reptiles, 652
Resemblances, morphological:
    creationist interpretation, 415
  as evidence for evolution, 411
Resources:
  depletion of, 598
  nonrenewable, 598
  renewable, 598
Respiration:
  anaerobic, 176–178
  artificial, 135, 167
  cellular, *5,* 10–11, 85–86, *161,*
    169–179
  organismic, 11, *161*–196, 178,
    179
Respiratory diseases, 185–188,
    202–206
  asthma, 185
  cancer, 205
  chronic bronchitis, 185–186,
    204–205

Respiratory diseases:
  emphysema, 186, 204–205
  lung cancer, 187–188, 190, 205
Responsiveness, 12–13, *209*–237
  in plants, 210
  in simple animals, 209–210
Resting potential of neuron, 219
Reticular activating system (RAS),
    224–226, 234, 236
Retina, 216
Rh blood type (*see* Blood types)
Rhodopsin, 216
Rhythm method of contraception,
    278, 289–291
Ribonucleic acid (RNA), 10,
    338–343
  chemical structure of, 338
  messenger, 339, 342–343
  ribosomal, 339, 342–343
  transfer, 339, 342–343
Ribose, 74, 338
Ribosomes, *48,* 342–343
Rickets, 72
Ringworm, 636
RNA (*see* Ribonucleic acid)
Road salt, 528
Roots, 638, 640
Roundworms, 644, 645
Rubella, 327–328


Salivary glands, 101–102, 105
Salmon, 581
Sampling error, 23
Sanger, Margaret, 602
Sanitary landfill, 539
Sarcoplasmic reticulum, 229–231
Schizophrenia, 229, 388
Schleiden, M. J., 43
Schwann, T., 43
Scientific method, 19–20, 24–35
Scopes, John Thomas, 414
Scrotum, 268–269
Scurvy, 73
Sea (*see* Oceans)
Sea urchin, early development in,
    figure, 321
Secretion, tubular, 152
Seed plants, 639
Seeds, 639–640
Selection:
  natural, 397
  stabilizing, 405
Selective cutting, 565
Self-regulation, 12–13, 126,
    243–244
Semen, 270, 282

Semicircular canals, 217
Seminal vesicles, 270
  figure, 269
Septic tanks, 536
Serotonin, 225, 235–238
Sewage treatment:
  advanced, 536
  primary, 533
  secondary, 533–535
  tertiary, 535–536
Sewers, 532–533
Sex:
  determination of, 283–285
  education, 286
  planning, 285
Sex-linked inheritance, 363–365
Shelter, as possible limiting factor
      for human population, 597
Shelterbelts, 563
Shelterwood, 565
Shettles, Landrum, 285
Sickle cell anemia (*see* Anemia)
Signs, 445
Sinuses, nasal, 162
Sinusitis, 162
Skin, 156–157
  color, 361–362
  functions of, 156
  structure of, 156
  sweat glands in, 156–157
Sleep, 225
  and EEG patterns, 228
  REM sleep, 225
Slime molds:
  acellular, figure, 422
  cellular, 637
Sludge, sewage, 535–536
Smell, sense of, 215–217
Smog:
  photochemical, 193–196
  SOP, 193
Societies, *440*–452
Society, human, 452–455
Sodium, 256–257
  figure, 620
  table, 619
Sodium nitrate, 115
Sodium nitrite, 115
Sodium pump, 218–219
Soil, lateritic, 94, 464, 482
Soil Conservation Service, 505
Solid wastes, 536–542
  disposal, 539–542
  recycling, 546–548
Solutions, 692
  hypertonic, 631
  hypotonic, 631

Solutions:
  isotonic, 630
  solute of, 629
  solvent of, 629
Spacecraft, 30, 459–460
Spartina, 499
Speciation, 403
Species, *403*
  allopatric, 404
  sympatric, 404
Spemann, Hans, 322
Sperm, 271–273
  numbers ejaculated, 282
  production of, 271
  and sex determination, 283–
    285
  structure of, 272
Spermicidal substances, 288–289
  table, 291
Sphygmomanometer, 128
Spiders, 648
Spinal cord, 222
Spirogyra, 636
Spleen, 129
Sponges, 642–643
Spores, 263–264, 638–639
Sporophyte, 638–639
SST (supersonic transport), 259
Stabilizing selection, 405
Starvation, 67, 91–92
Steady state, *12*–13, 147, 157–158,
    209
Stem of plants, 640
Sterility, 269, 282
Sterilization, 291–292
Steroids, 77
Stewart, William H., 200
Stockpiling in recycling, 550
Stomach, 101–103
Strip mining, 568
Stroke, 134
Substrate, 103–104
Suburbs, 572–574
Succession, ecologic, *478*–484
    in aquatic habitats, 502–503
    microsuccession, 479*n*.
    primary, 480
    secondary, 481
Sucrose, 74–75
Suits, paternity, 371
Sulfur oxides, 189–190, 192–194
Swamps, drainage of, 506
Sweat, 148, 156–157
Sweat glands, 156–158
Symbionts, 81–83, 480
Symbiosis, 81–82
Symbols, 444

Sympathetic nervous system, 211–212
Synapse, 219–220
Synthesis, 10–*12,* 86
Syphilis, 297–298
  and pregnancy, 328
  primary chancre of, 297
Systole, 126

Taiga, 463
Tapeworms, 644–645
  life cycle of, 644
Taste, 217
  buds, 217
Taxes, 424, 572
Tay-Sachs syndrome, 386
Technology:
  and air pollution, 192, 199
  role in environmental crisis, 610–612
Telophase, 57, 59
Temperature, body, 156–158, 178
  at ovulation, 277–278
  table, 430
Teratogens, *324*–327
Termites, 82–83
Territoriality, *450*–451
Testes, 268–270, 273
  interstitial cells of, 268
  seminiferous tubules of, 268
Testosterone, 268, 273
Tetany, 220, 251
Thalamus, 221, 223–224
Thalidomide, 325–326
Theories, scientific, *35*
Thermocline, 493
Thermodynamics, laws of, *80,* 87
Thrombin, 123
Thromboembolism, 287
Thromboplastin, 123
Thrombus, 132–134
Throughput theory of economics, 606–607
Thymine (T), 337–338
Thymus, 140, 260
Thyroid, 241, 246, 249–251
Thyroxine, 12, 249–250
Timber, ownership, 565
  (*See also* Forests)
Tissue fluid, 135–136
Tissues, 41, *60*–62
  (*See also* specific types of tissues)
Tobacco smoke, 201
Tonsils, 137
Tool use, 453–454

Toxicity, acute and chronic, 476*n.*
Trace elements, 69
Trachea, 161–163
Tracheophytes, 638
Transmitter substances, 220, 225, 234, 236–238
Transplantation of organs (*see* Organ transplantation)
Transportation, 188–189, 192, 199
Trial-and-error learning, 437
Tricarboxylic acid cycle (*see* Citric acid cycle)
Trophic levels of foof chains, 87–88
Trophoblast, 303–306
Tropical forests (*see* Rain forest)
Tropisms, *419*–424
Truth, *20*
Tubal ligation, 292
Tundra, 463
Turner's syndrome, 379
Twins, 303–304
  and organ transplants, 142

Ulcers, 109, 206
Umbilical cord, 306, 315
  figure, 305
Unconditioned stimulus, 434
Undernutrition, 89
Uniformitarianism, 415
Unit membrane, 46
United States of America:
  contraception in, 285–286
  diabetes in, 253
  food additives in, 113–115
  heart disease in, 131
  kidney disease in, 153
  malnutrition in, 67, 91
  obesity in, 110
  pollution in, 6–8
  population growth in, figure, 602
  pregnancies in, 285
U.S. Army Corps of Engineers, 505
  figure, 507
U.S. Department of Agriculture, 67
U.S. Fish and Wildlife Service, 581
U.S. Supreme Court (and abortion), 293
Uracil (U), 338
Urban decentralization, 575–576
Urban renewal, 575
Urea, 147–148, 151–152
Uremia, 153
Ureter, 148–149
Urethra, 149, 270
Uric acid, 147–148
Urination, *149*

Urine, 148–153
  composition of, 152
Uterus, 244, 246, 274–275
  implantation of embryo in, 304
  and placenta, 305–306
  structure of, 274–275


Vaccines, 140–141
  and venereal disease, 298
Vacuoles:
  food, 50
  water, 52–*53,* 55
    figure, 630
Vagina, 275, 282
Vagus nerve, 126
Valence, chemical, 622
Validity of syllogisms, 21
Value systems, 579
Valves:
  of heart, 125–126
  of veins, 128
Vas deferens, 269–270
Vasectomy, *291*–292
Veins, 124–*125*
  coronary, 131
  hepatic, 130–131
  hepatic portal, 107, 130–131
  iliac, 130
  jugular, 130–131
  postcaval, 130–131
  precaval, 130–131
  renal, 130–131
  subclavian, 130–131
  varicose, 129
Venereal disease, 296–298, 328
Ventricles of heart, *125*–131
Vertebrate societies, 452
Vestigial structures:
  creationist interpretation, 415
  as evidence for evolution, 411
Vietnam, 566–567
Villi:
  intestinal, 102, 106
  placental, 306
    figure, 305
Viroids, *634*
Virus:
  classification of, 634
  effects upon embryo, 327–328
  infections, 122
  structure and function, 344–347
Vision:
  blindness of, 93
  color, 216
    genetics of, 364–365
Vitamin A, 71–73, 93, 216

Vitamin B, 71–73, 102, 121, 173
Vitamin C, 71–72
  and air pollution, 190
Vitamin D, 71–73, 77, 325
Vitamin E, 71–72
Vitamin K, 71–72, 102, 123
Vitamins, 71–73
  effects of deficiency, 72–73
  fat-soluble, 71
  functions, 72–73
  minimum daily requirements, 72–73
  sources, 72–73
  water-soluble, 71
Vomiting, 109
Vulva, 275–276
  labia of, 275


Wallace, A. R., 397
Wasps, 448–449
Wastes, nitrogenous, 147–148
  and uremia, 153
Water, 68
  as basic nutrient, 68
  density of, 488
  formation in cellular respiration, 173, 175
  light transmission of, 489
  oxygen content, 487
  as oxygen source, 85
  pollution of, 6–8
    (*See also* Pollution)
  properties of, 68, 487–489
  as solvent, 487
  surface tension of, 488
Watson, John D., 227
Watt, K. E. F., 94
Wilderness, 579
Wildlife:
  management, 580–584
  requirements of, 561


Xylem, 640


Yeast and fermentation, 176–177
Yucca moth, 461


Zero population growth, 595
Zooplankton (*see* Plankton)
Zoos, 558, 583
Zugunruhe, *430*
Zygote, 264, 302–303, 320–321